Cardiomyocytes in Health and Disease

Chandrasekharan C. Kartha

Cardiomyocytes in Health and Disease

 Springer

Chandrasekharan C. Kartha
Amrita Institute of Medical Sciences
Cochin, Kerala, India

ISBN 978-3-030-85538-3   ISBN 978-3-030-85536-9   (eBook)
https://doi.org/10.1007/978-3-030-85536-9

This Springer imprint is published by the registered company Springer Nature Switzerland AG
The registered company address is: Gewerbestrasse 11, 6330 Cham, Switzerland

To
Professor Sankaran M. Valiathan and Professor Naranjan S. Dhalla
For inspiration and guidance during my career

M. S. Valiathan   N. S. Dhalla

Preface

Cardiomyocytes are the most important cell for the pumping function of the heart. They have several features distinct from other muscle cells. During the last two decades, significant progress has been made in our understanding of the cellular and molecular basis of cardiomyocyte functions and their relation to the normal action of the heart. Response of these cells to diverse forms of physiological stress and injury has been delineated as well. There is also more clarity on the alterations in the biological mechanisms in cardiomyocytes (CMs) that lead to pathological states and the changes in the cells secondary to diseases conditions. This text contains concise yet comprehensive reviews of the developments in biology and pathobiology of the cardiomyocyte. There are 20 chapters in this book, which is divided into five parts. The initial part has 7 articles. Chapter 1 contains a description of the structural components of cardiomyocyte and the role of these components in the contractile mechanism of the cell.

Heart is one of the first organs formed in the developing foetus. Cardiomyocytes are the first cells to terminally differentiate. They are present in the primordial heart, which starts beating around 22 days after fertilization. Several developmental signalling pathways, cardiac-specific transcription factors and transcriptional regulators control the formation of the four-chambered mammalian heart. These processes are discussed in Chap. 2. The ultimate structure of the cardiomyocyte is the result of several mechanisms which operate during the entire prenatal period as well as in the early phase after birth.

Cell cycle activity is an intrinsic component of differentiation of CMs and heart development. Several studies have investigated the expression of known mammalian cell cycle regulatory genes during cardiac development. Chapter 3 contains a survey of the currently known regulatory mechanisms of cell cycle in the cardiomyocyte. Microenvironment in the heart and signals arising from it as well as paracrine signals from cardiac fibroblasts and endocardium modulate the developmental program of the heart and direct the transcription mechanisms required for the proliferation and terminal differentiation of cells of cardiomyocyte lineage. The role of cardiac fibroblasts and endocardium in promotion of the growth and proliferation of embryonic CMs during development of the heart and the interacting mechanisms are reviewed in Chaps. 4 and 5.

Cardiomyocytes of the mammalian adult heart differ significantly from CMs in the foetal heart, both in structure and biochemical properties. Adult CMs are terminally differentiated and the rate of DNA synthesis is extremely low. The cells exit the cell cycle resulting in a postnatal decline in cell proliferation. Chapter 6 surveys the hitherto known mechanisms for the cell cycle exit and permanent postnatal growth arrest in mammalian CMs.

Development, differentiation and postnatal growth of CMs are accompanied by substantial changes in its energy metabolism. These changes influence cardiomyocyte proliferation during the early phases of heart development and also the terminal differentiation of CMs during later stages. Indeed, energy metabolism drives the postnatal development of CMs. Energy metabolism in CMs is discussed in Chap. 7.

Chapters in part II have descriptions of the effects of various physiological stimuli and genetic defects on cardiomyocyte structure and function and consequently on heart structure and function. How CMs remodel in response to mechanical stress, both during cardiac development as well as in disease conditions is narrated in Chap. 8. Mechanical stretch activates several intracellular signalling networks, modulates gene expressions and induces secretion or synthesis of molecules with autocrine and paracrine effects, all of which lead to structural and functional remodelling in CMs.

Several hormones and their receptors have a critical role in the homeostasis of CMs. They also modulate pathophysiological alterations in the cells and are also therapeutic targets in heart diseases. Chapter 9 illustrates the influence of insulin, thyroxine, growth hormone, oestrogen and aldosterone on growth, metabolism and function of CMs.

The consequences of genetic defects in CMs are congenital heart defects or inherited cardiomyopathies. Linkage analysis, candidate gene screening and next generation sequencing technologies have led to the identification of hundreds of mutations in genes encoding various subcellular components of the cardiomyocyte, in patients with congenital heart defects and inherited cardiomyopathies. These are chronicled in Chap. 10.

Chapter 11 discusses the studies that indicate the role of endogenous cardiac progenitor cells and their niche in cellular homeostasis in the heart and the response of these cells to ischemic injury to the myocardium.

Part III has four chapters. Three of them pertain to cardiomyocyte senescence, different forms of cell death and the mechanisms of cell death in ischemic injury to the heart. Senescent associated defects in metabolism of CMs and pathways for acute or progressive cell death in myocardial infarction as well as ischemia/reperfusion are the themes in these chapters.

The cause for heart failure (HF) is the death or dysfunction of a significant number of CMs, which can result from several cardiovascular diseases. Given that the pathophysiology of HF involves death or dysfunction of CMs, understanding their mechanisms is imperative for identifying strategies for the treatment of HF and reducing related complications. The distinct processes that characterize cardiomyocyte remodelling in failing hearts and the molecular mechanisms that lead to the progressive loss of CMs are specified in Chap. 15.

The first chapter of part IV deals with known endogenous mechanisms for regeneration and proliferation of CMs. These include the regulators of cell cycle, signal transduction pathways, epigenetic mechanisms and microRNAs. Endogenous mechanisms for renewal of adult CMs are insufficient for repair of extensive injury to the myocardium, as adult CMs have limited capacity for proliferation. Several recent studies suggest that latent regenerative potential in the adult heart can be exploited; in the injured heart, regeneration can be induced by modulating the endogenous molecular signals that stimulate proliferation of CMs. These are discussed in Chap. 17.

Chapter 18 has the cell transplantation-based strategies attempted thus far for repairing extensively damaged heart. Many sources of cells have been investigated for their potential use in regenerating the injured myocardium. They include induced pluripotent stem cells, cardiac fibroblasts, endogenous cardiac progenitor cells, cardiosphere-derived cells, side population cells, bone marrow cells, embryonic stem cells and mesenchymal stem cells.

The last chapter in this part narrates the methods for genetic and pharmacologic reprogramming of fibroblasts into cardiomyocytes. Several studies have demonstrated the feasibility of transdifferentiating fibroblasts into induced cardiomyocytes (iCMs) or induced cardiac progenitor cells (iCPCs) in vitro as well as transdifferentiating fibroblasts into iCMs in vivo. Cardiac reprogramming is currently recognized as a promising option for regenerative therapy in heart diseases.

The concluding part of the book is a summary of the progress that has been made in the translation of the discoveries in cardiomyocyte biology and pathobiology into developing diagnostic and prognostic biomarkers and treatment avenues for heart diseases. Several biochemical, molecular and genetic biomarkers of heart diseases are in clinical use or are under evaluation. Novel targets and strategies which enhance clinical efficacy in patients with heart diseases are also currently available. The exciting treatment approaches on the horizon include the use of molecules such as micro RNAs, transcription factors and growth factors, cell therapy, gene therapy, gene editing, epigenetic therapy, modulation of cardiomyocyte death, metabolic therapies and senolytics.

I hope that this compilation based on both original study reports and recent masterly reviews could benefit cardiologists, cardiovascular surgeons and cardiovascular scientists who wish to update their awareness about CMs. Cell biologists, biochemists and molecular biologists who desire to set foot in cardiovascular research may also find the book informative and useful.

I wish to dedicate this book to Prof. M. S. Valiathan and Prof. N. S. Dhalla. As a young pathologist, cardiovascular pathology was not my first choice for a career. I was initiated into cardiovascular pathology and later encouraged to pursue Molecular Cardiology by Prof. Valiathan, an eminent cardiovascular surgeon. The idea for this book was of Prof. Dhalla, a renowned cardiovascular scientist and editor of the popular book series *Advances in Biochemistry of Health and Disease*. He initially nudged me to engage in this project and later monitored the progress regularly. Both of them have been my mentors and benefactors for several decades now. They have been my inspiration for persistent academic pursuits.

There are 41 illustrations in this volume. Nine of the figures for part I were done by Diya K. Prasad and Maidhili Narayanan who are graduate students at St. Xavier's College, Mumbai and B. J. Amogh, a young physician. Surya Ramachandran, my former colleague and scientist at Rajiv Gandhi Center for Biotechnology did 9 of the figures for parts II and III. I am much obliged to them for providing appealing artwork. I acknowledge the painstaking effort of my former colleague A. Vinitha of Rajiv Gandhi Center for Biotechnology, in compiling and formatting all the references. Shammy S. and Vikas Kumar prepared the tables.

My wife Mira Mohanty, a pathologist prepared 23 diagrams for this book. She has also been my pillar of strength during moments of anxiety and despair during the last one year of my endeavour of composing this text.

I also acknowledge the guidance and support I had from Ambrose Berkumans and Gonzalo Cordova at *Springer Nature.*

Thiruvananthapuram, India   Chandrasekharan C. Kartha

Contents

Part V   Translational Aspects of Cardiomyocyte Biology

About the Author

Chandrasekharan C. Kartha is a pathologist engaged in the investigation of cellular and molecular mechanisms of pathogenesis of human cardiovascular diseases, aimed at discovery of biomarkers and drug targets.

Kartha had his medical education at Government Medical College, Thiruvananthapuram and All India Institute of Medical Sciences at New Delhi. He is presently an Honorary Professor at Amrita School of Medicine, Amrita Vishwa Vidyapeetham at Cochin, Kerala.

He has earlier served as Senior Professor and Head of the Division of Cellular and Molecular Cardiology and Dean of Academic Affairs at Sree Chitra Tirunal Institute for Medical Sciences & Technology, as Professor of Eminence at Rajiv Gandhi Centre for Biotechnology, and as Senior Adviser for Society for Continuing Medical Education & Research at KIMSHEALTH, Thiruvananthapuram.

Kartha is an elected Fellow of Royal College of Physicians (London), International Academy of Cardiovascular Sciences (Canada), National Academy of Medical Sciences (India), Indian Academy of Sciences, National Academy of Sciences (India), and Indian College of Pathology. He has received Makoto Nagano Award for distinguished achievements in cardiovascular education and Lifetime Achievement Award in Cardiovascular Science, Medicine and Surgery from International Academy of Cardiovascular Sciences.

List of Figures

Part I
Normal Cardiomyocyte and its Growth

Chapter 1
Structure and Function of Cardiomyocyte

Abstract Cardiomyocyte is the fundamental contractile cell of the heart. Sarcomere, the fundamental contractile unit of cardiomyocytes is composed of thick and thin interdigitating filaments of myosin and actin, tropomyosin, titin and the troponin complex. Interactions among these proteins, initiated by a rise in extracellular Ca^{2+} result in the hydrolysis of adenosine tri phosphate and alterations in physical–chemical dynamics leading to the development of tension within the cardiomyocyte. Cyclic increases and decreases in intracellular Ca^{2+} initiated by depolarization of the sarcolemma and Ca^{2+} release and re-uptake by the sarcoplasmic reticulum coordinate the contraction–relaxation cycle of cardiomyocytes. Several proteins in the sarcolemma, sarcoplasmic reticulum and mitochondria regulate cytosolic levels of Ca^{2+}.

Keywords Cardiomyocyte · Sarcomere · Myofilaments · Sarcolemma · Sarcoplasmic reticulum · Mitochondria · Action potential · Excitation–contraction coupling

Introduction

The human heart is a muscular organ and consists of several types of cells such as cardiomyocytes (CMs), endothelial cells, fibroblasts and smooth muscle cells. The fundamental contractile cell of the heart is the cardiomyocyte. Specialized CMs form the cardiac conduction system, a collection of nodes and cells which initiate and co-ordinate the rhythmic beating of the heart. A contractile cardiomyocyte in the adult human heart is cylindrical in shape and about 100 μm long and 10–25 μm in diameter. Structural organization of a cardiomyocyte can be seen in Fig. 1.1. Heart diseases are associated with alterations in structure and function of CMs [1].

© The Author(s), under exclusive license to Springer Nature Switzerland AG 2021
C. C. Kartha, *Cardiomyocytes in Health and Disease*,
https://doi.org/10.1007/978-3-030-85536-9_1

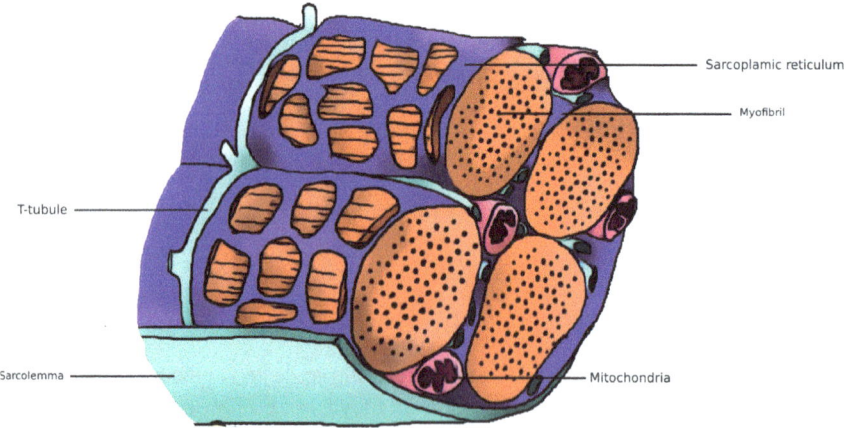

Fig. 1.1 A schematic diagram of a cross section of a cardiomyocyte

Structural Components of Cardiomyocytes

Contractile Apparatus

Sarcomere is the fundamental contractile unit of cardiomyocytes. It has a resting length of 1.8–2.4 μm. The sarcomere is composed of thick and thin interdigitating filaments of myosin and actin, tropomyosin, titin and the troponin complex, component proteins of the contractile apparatus (Fig. 1.2) [2]. Interactions among these proteins, initiated by a rise in extracellular Ca^{2+} result in the hydrolysis of adenosine tri phosphate (ATP) and alterations in physical–chemical dynamics leading to the development of tension within the cardiomyocyte. Myosin, a thick filament, has a

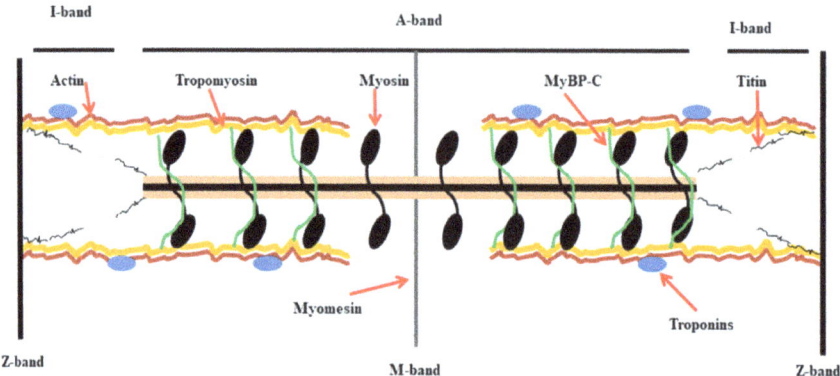

Fig. 1.2 Structural organization of cardiomyocyte depicting organization of myofilaments. MyBP—Myosin binding protein

filamentous tail and a globular head. The sites for actin binding and for catalyzing ATPase activity are located in the head region. Actin, a thin filament has two forms, G and F. F-actin is the backbone and G-actin is the stabilizing protein. Each monomer of G-actin has two binding sites for myosin. The myosin globular head and the G-actin monomer interact in the presence of ATP, resulting in the formation of a crossbridge and the sarcomere shortens. Tropomyosin on either side of actin engages the actin-myosin cleft, averts Ca^{2+} binding and thus impacts formation of the actin-myosin crossbridge [3]. The troponin complex, which is part of the thin filament, has three proteins: troponin T, I, and C. Troponin T binds the troponin complex to tropomyosin, anchoring the complex to the thin filament. Troponin I add to the structural integrity of the sarcomere as well as regulates the extent of crossbridge formation. Actin-myosin interaction and formation of crossbridge is initiated by the conformational change of the complex caused by Ca^{2+} binding to troponin C.

Basement Membrane

Each cardiomyocyte has a membrane, which separates it from the surrounding fibrillar collagen matrix. Type IV collagen, the glycoproteins laminin and fibronectin, and proteoglycans are the constituents of the membrane [4]. The basement membrane regulates the transport of macromolecules between the cardiomyocyte and the extra-cellular matrix (ECM). Integrins and receptor transmembrane proteins together bind the cardiomyocyte to the basement membrane and the ECM. The membrane is also adhered to the ECM through anchoring fibres. The collagen-integrin-cytoskeletal attachment is important for the transduction of cardiomyocyte shortening into contraction of cardiac ventricle [5].

Cardiomyocyte Cytoskeleton

At the interface between the ECM and the contractile system is the cytoskeleton of the myocyte [6–8]. Several cytoskeletal proteins such as α-actinin, talin and desmin are assembled at the site where integrins enter the cytosol. When they are phosphorylated, the structural geometry and function of the cardiomyocyte change. Titin, a determi-nant of the viscoelastic properties of the cardiomyocyte restricts overstretch of the myofilaments [9, 10]. Tubulins are the other cytoskeletal proteins in the myofibrillar assembly. They are important for transduction of mechanical signals to the nuclear membrane [11].

Sarcolemma

The basement membrane along with a distinct plasma membrane forms a specialized structure, the sarcolemma. The sarcolemma composed of a lipid bilayer controls the diffusion of molecules across it. The hydrophobic core of the lipid bilayer causes the sarcolemma to be impermeable to charged molecules. Specialized intercalated disks in the sarcolemma provides strong mechanical linkages between CMs. They are paths of low resistance and aid rapid transmission of the action potential between CMs [3]. The membrane proteins of the sarcolemma include receptors, ion pumps, and ion channels.

Ion Pumps and Ion Channels in Sarcolemma

The pumps and channels of the sarcolemma are best understood in the context of their function during the different phases of an action potential. An action potential in a ventricular cardiomyocyte is shown in Fig. 1.3. During phase 4 of the action potential, the sarcolemma is permeable only to K^+. During this phase, the K^+ equilibrium determines the resting membrane potential in the cardiomyocyte. K^+ diffusion into the cell is allowed by the inward K^+ rectifier. For the entry of two K^+ ions, three Na^+ ions are extruded by the Na^+/K^+ ATPase, generating a net outward current. The Na^+/Ca^{2+} exchanger and the sarcolemmal Ca^{2+} ATPase regulate Ca^{2+} efflux from the cell. The balance between Ca^{2+} efflux and influx is maintained by removal of cytosolic Ca^{2+}. The resting potential is thus maintained. When the membrane potential is at a pre-set threshold voltage, the Na^+ channels are rapidly activated (<1 ms), but remain activated for only a short duration of 2–10 ms. The activation of the fast Na^+ channel permits influx of Na^+ to the cell along both electrical and

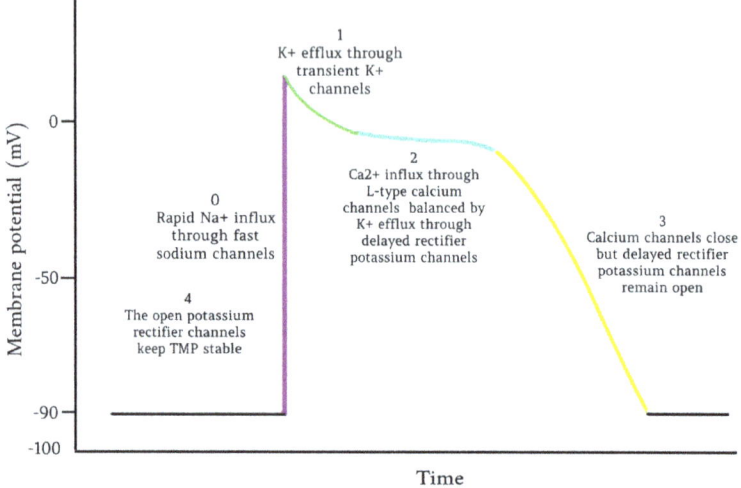

Fig. 1.3 Schematic diagram of action potential showing 4 phases and ionic transients

chemical concentration gradients. Influx of Na^+ initiates the ionic processes for the rest of the phases of the action potential. Rapid inactivation of the Na^+ channels and slower activation of two outward currents are the sources for early repolarization. The Cl^- concentration gradient and increased membrane permeability to Cl^- cause the cellular entry of Cl ions. A transient efflux of K^+ through specific channels also occurs along the K^+ electrochemical gradient. These three events during phase 1 of the action potential contribute to a repolarization of the membrane potential for a short duration. The influx of Ca^{2+} through the L type Ca^{2+} channels and a counter outward K^+ current flow through the 'anomalous' K^+ rectifier determine the action potential plateau or phase 2 [3, 12, 13]. These two channels which are activated during the upstroke of the action potential synchronously reach peak current during the plateau phase. Delayed rectifier K^+ channels are activated towards the end of the plateau phase. They cause K^+ ions to flow along the concentration gradient which results in repolarization or phase 3. Other inward currents such as of Na^+ and Ca^{2+} are inactivated. The membrane potential is restored to the resting state by the delayed rectifying K^+ current [14].

Receptors in Sarcolemma

The β-adrenergic, muscarinic α, and endothelin receptors have been identified in the cardiomyocyte [15–17]. The β-adrenergic receptor system is mostly in the inactivated state [18, 19]. Binding of endogenous catecholamines to the β-adrenergic receptor activates a guanine nucleotide–dependent coupling protein, leading to stimulation of adenylate cyclase and increased production of cyclic adenine monophosphate (cAMP). cAMP stimulates the active catalytic subunits of protein kinase A and in turn phosphorylates sarcolemmal L-type Ca^{2+} channel, the sarcoplasmic reticulum regulatory protein, phospholamban, and troponin I within the cardiomyocyte [20–22]. These phosphorylation sites are relevant to the excitation–contraction coupling process.

Sarcoplasmic Reticulum

Sarcoplasmic reticulum (SR) is a network of intracellular membranes, which regulates Ca^{2+} homeostasis in the cytoplasm. Thus, SR function is important for excitation–contraction coupling [20, 23, 24]. SR has regions juxtaposed to invaginations of the sarcolemma into the cytoplasm of CMs (transverse tubules or T-tubules). These contacts result in the proximity of the L-type Ca^{2+} channel with the Ca^{2+} discharge system and thus aid excitation–contraction coupling [24]. SR has three important components: the sarcoplasmic reticulum Ca^{2+} ATPase (SERCA-2), phospholamban, the regulatory protein of SERCA-2 and the Ca^{2+} release channel. SERCA-2, an ATP-dependent Ca^{2+} pump is a fundamental determinant of Ca^{2+} accumulation within the myocyte [3]. During the excitation–contraction coupling process, Ca^{2+} in the cytosol can be altered more than 100- times by SERCA-2 and the Na^+/Ca^{2+} exchanger [2]. The phosphorylated state of phospholamban, which is colocalized with SERCA-2 is a critical determinant of Ca^{2+} uptake in the SR and thus the rate and extent of Ca^{2+} removal from the cytosol [24, 25]. The calcium release channel, also known as the ryanodine receptor channel is located at the interface between the SR and the T tubular system of the sarcolemma [26]. This channel is sensitive to changes in cytosolic Ca^{2+} and regulates Ca^{2+} release from SR stores. An influx of Ca^{2+} through the L type Ca^{2+} channel can induce a large release of Ca^{2+} from the calcium release channel into the cytosol, and thus activate the contractile machinery [3].

Mitochondria

Mitochondria occupy nearly 40% of the cardiomyocyte volume. They meet the large energy demands of cardiomyocyte [3]. Cytosolic phosphocreatine which is higher in concentration compared to adenosine diphosphate, is a high energy reserve. It also functions as a rapid shuttling system of high-energy phosphate between mitochondria and cytosol [3]. The high-energy phosphate of ATP is transferred to phosphocreatine, which diffuses through the cytosol to be reconverted to ATP for use as cell energy, for example, in excitation–contraction coupling [4]. Mitochondria can also bind and take up large amounts of cytosolic Ca^{2+}, as well as contribute to buffering of cytosolic Ca^{2+}. Thus, they protect the cell from the effects of Ca^{2+} overload [3, 27].

Excitation–Contraction Coupling

Contraction of cardiomyocyte is initiated by an action potential. The process of transduction of an action potential to contraction of the cardiomyocyte is known as excitation–contraction (EC) coupling. Salient components of the excitation–contraction coupling mechanism are represented in Fig. 1.4. An action potential results in depolarization at the T-tubular system and activates voltage-sensitive L-type Ca^{2+} channels and Ca^{2+} conductance in the sarcolemma [12, 13]. A small influx of Ca^{2+} triggers the Ca^{2+} release channel, resulting in release of large amounts of Ca^{2+} into the cytosol [12, 20, 26, 28–32]. Cytosolic Ca^{2+} increase from nanomolar (100 nmol/L) to micromolar (10 μmol/L) levels [2]. The release of Ca^{2+} from the SR initiates a series of interactions among the contractile proteins of the sarcomere.

The Sliding Filament Theory

The sliding filament theory explains the interactions among the contractile proteins in the sarcomere [2, 3, 22, 23]. At resting conditions, concentrations of cytosolic Ca^{2+} are low and troponin I is in the phosphorylated state; affinity of cytosolic Ca^{2+} for troponin C is less and troponin I—actin interaction is stable. Actin-myosin

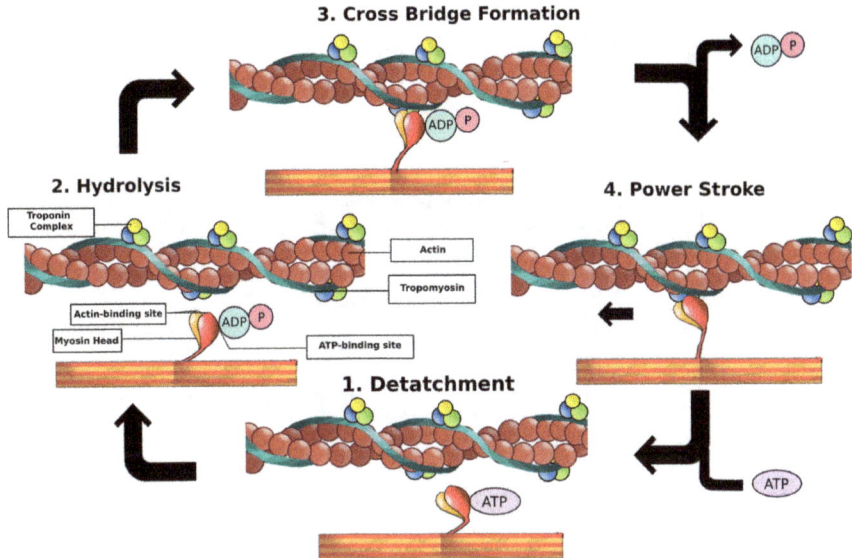

Fig. 1.4 Salient components of the excitation–contraction coupling mechanism in cardiomyocyte. LTCC- L-type calcium channel. NCX-Sodium-calcium exchanger. NK ATP- Sodium–potassium ATPase. PLB—Phospholamban. RyR-Ryanodine receptor. SERCA—Sarcoplasmic reticulum calcium ATPase

interaction is prevented by the troponin-tropomyosin complex, which is shifted to the outer grooves of the filament. An increase in cytosolic Ca^{2+} permits binding of Ca^{2+} to troponin C. The result is a shift in the affinity of troponin I, from actin to troponin C. A conformational shift of the troponin-tropomyosin complex from the actin-myosin binding site leads to crossbridge formation. The hinge regions in the crossbridge allows the myosin head to swing towards the actin filament. Consequence of the ensuing change in conformation of the myosin head is hydrolysis of ATP. Conformational change in the crossbridge generates a force to move the actin filament relative to myosin filament. Release of the crossbridge is brought about by binding of new ATP to the myosin head. Each crossbridge cycle can move the filaments approximately 10 nm with an average velocity of 0.98 μm/s. Until energy dependent removal of Ca^{2+} from the cytosol is completed or ATP stores are exhausted, the cycle of crossbridge formation would continue. Figure 1.5 depicts the mechanism of formation and release of the cross bridge.

Active relaxation depends on the function of SERCA-2, and is also high energy–dependent. Two mol of Ca^{2+} is transported back into the SR for each mol of ATP that is hydrolyzed. While SERCA-2 and phospholamban are the important players in the active relaxation of the cardiomyocyte, to a less extent, Na^+/Ca^{2+} exchanger, the sarcolemmal Ca^{2+} ATPase, and Ca^{2+} binding proteins such as calmodulin and calsequestrin in the cytoplasm also influence the removal of Ca^{2+} from the cytosol [22, 30, 32, 33].

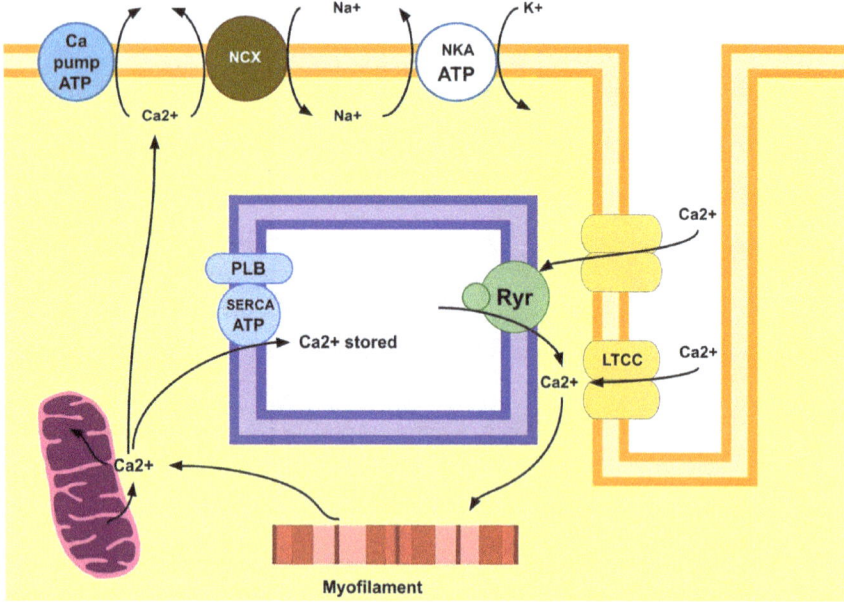

Fig. 1.5 Schematic diagram of mechanism of cardiomyocyte contraction

Conclusion

The fundamental contractile unit of cardiomyocyte is the sarcomere, composed of thick and thin interdigitating filaments of myosin and actin, tropomyosin, titin and the troponin complex. Interactions among these proteins result in the hydrolysis of ATP and alterations in physical–chemical dynamics leading to the development of tension within the cardiomyocyte. Contraction of cardiomyocyte is initiated by an action potential. Excitation–contraction coupling refers to the mechanism by which an action potential leads to contraction of the myocyte. Ca^{2+} is the fundamental ion for inducing the excitation–contraction coupling complex. Several proteins in the sarcolemma, sarcoplasmic reticulum and mitochondria regulate cytosolic levels of Ca^{2+}. The sliding filament theory explains the interactions among the contractile proteins in the sarcomere, during contraction of cardiomyocyte.

References

1. Woodcock EA, Matkovich SJ. Cardiomyocytes structure, function and associated pathologies. Int J Biochem Cell Biol. 2005;37(9):1746–51.
2. Pappano AJ, Gil Wier. Cardiovascular physiology. Mosby physiology series. Elsevier; 2018.
3. Katz AM. Physiology of the heart. Lippincott Williams & Wilkins; 2010.

4. Walker CA, Spinale FG. The structure and function of the cardiac myocyte: a review of fundamental concepts. J Thorac Cardiovasc Surg. 1999;118(2):375–82.
5. Hsueh WA, Law RE, Do YS. Integrins, adhesion, and cardiac remodeling. Hypertension. 1998;31(1):176–80.
6. Borg TK, Terracio L. Interaction of the extracellular matrix with cardiac myocytes during development and disease. In: Robinson TF, Kinne RKH, editors. Cardiac myocyte-connective tissue interactions in health and disease. Basel: Karger; 1990. p. 113–29.
7. Factor S. Role of the extracellular matrix in dilated cardiomyopathy. Heart Failure. 1993;9:260.
8. Robinson TF, Cohen-Gould L, Factor SM. Skeletal framework of mammalian heart muscle. Arrangement of inter-and pericellular connective tissue structures. Laboratory Investigation. 1983; 49(4):482–498.
9. Horowits R, Kempner ES, Bisher ME, Podolsky RJ. A physiological role for titin and nebulin in skeletal muscle. Nature. 1986;323(6084):160–4.
10. Morano I, Hädicke K, Grom S, Koch A, Schwinger RH, Böhm M, et al. Titin, myosin light chains and C-protein in the developing and failing human heart. J Mol Cell Cardiol. 1994;26(3):361–8.
11. Eble D, Spinale F. Effects of chronic supraventricular tachycardia on contractile and non-contractile mRNA expression: relation to changes in myocyte structure and function. Am J Physiol. 1995;268:H2426–39.
12. Balke CW, Shorofsky SR. Alterations in calcium handling in cardiac hypertrophy and heart failure. Cardiovasc Res. 1998;37(2):290–9.
13. Mukherjee R, Spinale FG. L-type calcium channel abundance and function with cardiac hypertrophy and failure: a review. J Mol Cell Cardiol. 1998;30(10):1899–916.
14. Nuss H, Johns D, Kääb S, Tomaselli G, Kass D, Lawrence J, et al. Reversal of potassium channel deficiency in cells from failing hearts by adenoviral gene transfer: a prototype for gene therapy for disorders of cardiac excitability and contractility. Gene Ther. 1996;3(10):900–12.
15. Brodde O-E, Broede A, Daul A, Kunde K, Michel M. Receptor systems in the non-failing human heart. Basic Res Cardiol. 1992; 187(Suppl 1):1–14.
16. Schmitz W, Eschenhagen T, Mende U, Moller F, Scholz H. The role of alpharadrenergic and muscarinic receptors in cardiac function. European Heart Journal. 1991; 12 (suppl_F):83–87.
17. Thomas PB, Liu E, Webb ML, Mukherjee R, Hebbar L, Spinale FG. Exogenous effects and endogenous production of endothelin in cardiac myocytes: potential significance in heart failure. Am J Physiol Heart Circ Physiol. 1996;271(6):H2629–37.
18. Johnson M. The β-adrenoceptor. Am J Respir Crit Care Med. 1998; 158 (supplement_2):S146-S53.
19. Onaran H, Costa T, Rodbard D. Beta gamma subunits of guanine nucleotide-binding proteins and regulation of spontaneous receptor activity: thermodynamic model for the interaction between receptors and guanine nucleotide-binding protein subunits. Mol Pharmacol. 1993;43(2):245–56.
20. Williams A. The functions of two species of calcium channel in cardiac muscle excitation-contraction coupling. Eur Heart J. 1997; 18(suppl_A):27–35.
21. Alberts B, Johnson A, Lewis J, Raff M, Roberts K, Walter P. Molecular biology of the cell. New York: Garland Publishing; 2008.
22. Katz AM, Lorell BH. Regulation of cardiac contraction and relaxation. Circulation. 2000; 102: iv-69-iv-74.
23. Kadambi VJ, Kranias EG. Phospholamban: a protein coming of age. Biochem Biophys Res Commun. 1997;239(1):1–5.
24. Aubier M, Viires N. Calcium ATPase and respiratory muscle function. Eur Respir J. 1998;11(3):758–66.
25. Tada M, Toyofuku T. SR Ca^{2+}-ATPase/phospholamban incardiomyocyte function. J Cardiac Fail. 1996;2:S77–85.
26. Franzini-Armstrong C, Protasi F. Ryanodine receptors of striated muscles: a complex channel capable of multiple interactions. Physiol Rev. 1997;77(3):699–729.

27. Crompton M. The role of Ca^{2+} in the function and dysfunction of heart mitochondria. In: Langer GA, editor. Calcium and the heart. New York: Raven Press; 1990.

28. Hasenfuss G, Meyer M, Schillinger W, Preuss M, Pieske B, Just H. Calcium handling proteins in the failing human heart. Basic Res Cardiol. 1997;92(1):87–93.

29. Bers D. Excitation-contraction coupling and cardiac contractile force. Norwell: Kluwer Academic Publishers; 1991.

30. Bers DM, Bassani JW, Bassani RA. Na-Ca exchange and Ca fluxes during contraction and relaxation in mammalian ventricular muscle[a]. Ann N Y Acad Sci. 1996;779(1):430–42.

31. Pollack GH, Ishiwata S, Sugi H. Molecular and cellular aspects of muscle contraction. Adv. Exp Med Biol. 2003; 538: 647–654 and 687–688.

32. Bers DM, Perez-Reyes E. Ca channels in cardiac myocytes: structure and function in Ca influx and intracellular Ca release. Cardiovasc Res. 1999;42:339–60.

33. Blaustein MP, Lederer WJ. Sodium/calcium exchange: its physiological implications. Physiol Rev. 1999;79:763–854.

Chapter 2
Development of the Cardiomyocyte

Abstract Heart is one of the first organs formed in the developing foetus. Heart development begins after embryo gastrulation when the mesoderm is formed. The earliest cardiac progenitors arise in the anterior region of the mesoderm and from the pharyngeal mesoderm. Cells derived from these first and second heart fields contribute to definitive sections of the adult heart. Cardiomyocytes are present in the primordial heart, which starts beating around 22 days after fertilization. Several developmental signalling pathways, cardiac-specific transcription factors and transcriptional regulators regulate the formation of the four-chambered mammalian heart. Microenvironment in the heart and signals arising from it as well as paracrine signals from the epicardium and endocardium modulate the developmental program of the heart and direct the transcription mechanisms required for the proliferation and terminal differentiation of cells of cardiomyocyte lineage. The ultimate structure of the cardiomyocyte is thus the result of several mechanisms which operate during the entire prenatal period as well as in the early phase after birth.

Keywords Cardiac development · Embryonic heart · Cardiomyocyte proliferation · Cardiomyocyte specification · Cardiac transcription factors · Developmental signalling

Introduction

Given its vital function in supplying oxygen and nutrients to the various tissues of the organism, heart is one of the first organs formed in the developing foetus [1]. The primordial heart in humans begins to beat at around 22 days after fertilization and cardiomyocytes are present in the heart during that period. Cardiomyocytes (CMs) ensure proper contractile function of heart.

© The Author(s), under exclusive license to Springer Nature Switzerland AG 2021   13
C. C. Kartha, *Cardiomyocytes in Health and Disease*,
https://doi.org/10.1007/978-3-030-85536-9_2

The Embryonic Heart Development

The process of heart development begins after embryo gastrulation when the mesoderm is formed between the ectoderm and the endoderm. The earliest cardiac progenitors arise in the anterior region of the mesoderm as two group of cells on either side of the mid line [1]. These cells migrate, extend across the midline and form the cardiac crescent [2, 3]. Cardiac crescent is the first anatomically distinct heart structure and the location of the progenitor cells of the first heart field (FHF) [4, 5]. FHF is the initiator of heart formation and is exclusively committed to a cardiomyogenic cell fate [6, 7]. Progenitor cells of the FHF express markers such as NK2 homeobox 5 (Nkx2.5), Gata 4/5/6, Mef2b/c, Hand1/2, Tbx5/20, and myocardin. These cells mainly contribute to the formation of the left ventricle and atria (Table 2.1) [8, 9].

Another group of progenitor cells, originating from the pharyngeal mesoderm form the second heart field (SHF) [10]. The cells of the SHF contributes primarily to the right ventricle, outflow tract, and atria (Table 2.1) [8, 11, 12]. Progenitor cells of the SHF express Islet1, Nkx2.5, and Flk1 [13].

FHF cells in contrast to SHF progenitors, express the ion channel hyperpolarization-activated cyclic nucleotide-gated channel 4 (HCN4) and are committed very early on towards a cardiomyogenic cell lineage [7].

While the mesoderm-derived FHF and SHF are the predominant sources of cardiomyocytes, lineage-tracing experiments and clonal analysis have suggested that Isl1+ SHF progenitors can contribute to CMs of the right ventricle and atria [13, 14].

The initial structure of the embryonic heart is tubular, with an inner layer of noncontractile endothelial-like cells (endocardium) and an outer layer of contractile muscle cells (myocardium). The embryonic tubular heart is made of cells from the

Table 2.1 Cardiac precursor cells and signaling molecules which contribute to the development of the heart

Progenitor cell population	Origin	Contributes to	Important signaling molecules and pathways
First heart Field	Anterior lateral mesoderm	Entire left ventricle, adds to atria and inflow tract	Nkx2.5, GATA4, Hand1, TBX5, BMP, Wnt
Second heart Field	Pharyngeal and splanchnic mesoderm	Entire right ventricle, outflow tract	Islet1, Nkx2.5, GATA4, Hand2, Mef2c, Pitx2c, BMP, Wnt, Retinoic acid, Hedgehog (Hh), Foxh1
Cardiac neural crest cell	Dorsal neural crest	Aorticopulmonary septum, great arteries, semilunar valves, parasympathetic nerves	Snail2, Sox9, FoxD3, Wnt-β catenin, BMP, FGF, Retinoic acid
Proepicardium	Coelomic mesothelium	Epicardium and coronary vessels	TBX18, TCF21, GATA4, BMP

cardiac crescent [15, 16]. The organ sans valves starts functioning as a pump with the initiation of the heart beat in its inflow region. [17, 18].

Looping and differential growth transforms the tubular heart into a four-chambered structure with the ventricles lower to the atria [15]. The formation of the multichambered heart commences with the migration of neural crest cells, which principally contribute to the conotruncus and great vessels [15]. Cardiac muscle cells form the atria, ventricles, the conduction system and septa. Endocardium and valves are derived from endothelial cells. Smooth muscle cells aid formation of the veins and arteries. The outer epithelial layer (epicardium) of the embryonic heart is assembled from the proepicardial organ (PEO). A cluster of mesothelial cells near the venous end of the embryonic heart tube create the PEO [19, 20]. Epicardial cells can undergo an epithelial-to-mesenchymal transition, migrate into the myocardium and contribute to a small proportion of CMs [21–24].

Cardiac neural crest cells-CNCCs) which originate from the dorsal neural tube and migrate to the heart form the parasympathetic innervation of the heart. They also contribute to the formation of the valves as well as to the septation and modelling of the outflow tract [25–27].

Niche Interactions in the Developing Heart

Microenvironment in the heart has a key role in the regulation of cardiac development. The signals arising from the microenvironment direct the transcription mechanisms for differentiation and proliferation of cells during various stages of heart development, beginning from the early stage of specification of progenitors to the final formation of heart chambers [28]. Both soluble paracrine signals and physical contacts through integrins and cadherins mediate the niche interactions. Cytoplasmic receptors transduce these signals to the nucleus, to regulate transcription [14].

Endoderm derived bone morphogenic protein (BMP) induces the specification of cardiac progenitors; this effect is inhibited by neural Wnt signals [28]. While transcription factors Tbx5 and ISL1 are specific respectively for cardiac progenitors of the first and second heart fields, Gata4 and Nkx2.5 are important for all types of cardiac progenitors [14, 28]. Ontogenesis of the heart is coordinated by signalling among several types of cardiac progenitors. These cells include progenitors of cardiac fibroblasts, epicardial cells, pericytes, and resident immune cells [29].

Extra cellular matrix (ECM)/β1 integrin signalling from embryonic cardiac fibroblasts stimulates the proliferation of CMs. Differentiation of CMs and trabecular growth within the ventricles are regulated by Neuregulin1 (NRG-1) released from the endocardium, NOTCH1, VEGFR-2, and FGF signalling [30]. Paracrine signals which include endothelin-1, aid differentiation of contractile cardiomyocytes into conduction cells [31].

Hemodynamics of the early circulation and contractility of the myocardium also control heart development [32, 33]. Various cell sensors that respond to flow, pressure, stretching and rhythmicity transduce the mechanical signals [34]. Epigenetic mechanisms such as chromatin remodelling, histone acetylation and methylation, and DNA methylation aid differential gene expression requisite for heart development [35–40].

Cardiomyocyte Specification

Several developmental signalling pathways, cardiac-specific transcription factors and transcriptional regulators regulate the formation of the four-chambered mammalian heart (Fig. 2.1). Evolution of the heart fields and their specification into various cell lineages are modulated by multiple interactions between growth factors and transcriptional regulators [8, 41]. Several transcription factors are expressed during initiation of myocardial differentiation at the cardiac crescent. These factors include the zinc finger factors Gata4 and Gata5, the homeodomain factor Nkx2.5, T-box factors such as Tbx5, the MADS-box factor Mef2c, the basic helix-loop-helix factors and neural crest derivatives expressed 1 (Hand1) and Hand2. Mutations in Gata4, Nkx2.5 and in Tbx1, Tbx5 and Tbx20 are associated with congenital heart anomalies in humans [42–44].

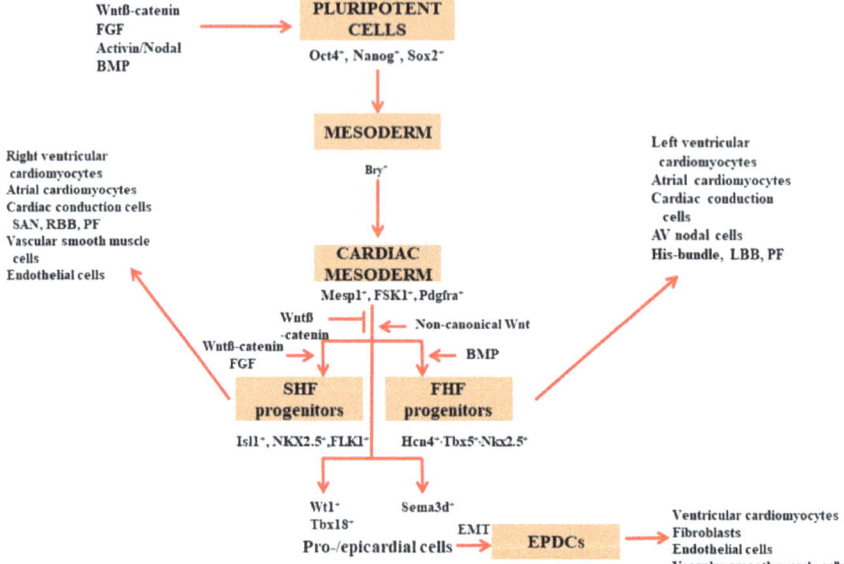

Fig. 2.1 Cell lineages, signalling pathways and transcriptional regulators, which contribute to the development of various mature cell types in the heart

Extracellular signalling molecules such as wingless integrated (Wnt), fibroblast growth factor (FGF) and transforming growth factor-beta (TGFβ) superfamily ligands, which include Wnt3a, bone morphogenetic protein 4 (BMP4), Nodal and activin A primarily regulate the induction of the mesoderm, further specification of cardiac mesoderm and formation of cardiac progenitor cells [45]. Gradient spatiotemporal expression of the ligands and resulting activating or inhibitory cues for the cells, aid patterning the developing embryo. Signalling factors and downstream transcriptional events in concert contribute to the expression of cardiac-specific factors in a defined group of cells in the mesoderm and thus specification of the cardiac field [45].

Formation of primitive streak and pre-cardiac mesoderm is dependent on active Wnt/β-catenin signalling [46–49]. Inhibition of Wnt/β-catenin signalling is necessary for subsequent specification into cardiac progenitor cells. This is achieved via frizzled-related protein 2 (FRZB2)] and Dickkopf 1 (DKK1), expressed in the adjacent endoderm [50, 51]. Active Wnt signalling is vital for the specification and expansion of SHF progenitors as well [13, 52–58]. FGF, BMP, Hedgehog (Hh), Notch and non-canonical Wnt signalling are also participants during several stages of specification, expansion and differentiation of FHF and SHF progenitors [3, 45, 54, 59–66].

Paracrine signals from the epicardium and endocardium orchestrate the proliferation and terminal differentiation of the newly formed cells of cardiomyocyte lineage. Insulin growth factor (IGF) secreted from the epicardium and canonical Wnt signalling determine the number of cardiomyocytes and in mice, the size of the embryonic heart [45, 60, 67]. Both Wnt and IGF pathways are linked to the Hippo signalling pathway via YAP. WNt/β-catenin and the Hippo/YAP pathways are recognized to be important for development and growth. The transcriptional co-activator Yes-associated protein (YAP) is a key mediator of cardiac growth and size [68–70]. Proliferation of CMs during heart development is also regulated by FGFs and BMPs as well as Notch signalling from the endocardium [71–73].

Extracellular signalling pathways and a network of transcription factors also regulate the specification of the cardiomyogenic lineage. The T-box transcription factor brachyury (Bry), expressed in the mesoderm is a direct target of canonical Wnt signaling [74]. Bry positive cells give rise to subpopulations which are positive for vascular endothelial growth factor receptor 2 (VEGFR2) and kinase insert domain receptor (KDR)] and these cells develop into not only hematopoietic and vascular progenitors but also cardiac progenitor cells [75–78]. Downregulation of Bry is accompanied by the upregulation of the T-box transcription factor eomesodermin (Tbr2) and the long noncoding RNA Braveheart (Bvht), which induce the expression of the transcription factor Mesp1 along the primitive streak [40, 79]. Bvht causes epigenetic changes that aid cardiac lineage specification.

Mesp1 is known to act upstream of GATA binding protein 4 (Gata4), Isl1, myocyte enhancer factor 2c (Mef2c) and Nkx2.5, which are cardiac-specific transcription factors necessary for cardiac development [80–82].

Several novel and important transcription factors and related molecular signalling pathways that are spatiotemporally active during the early stages of heart development have been discovered [83].

Chromatin remodelling is also important for initiation of the program for cardiac development. Cardiac development program is initiated by Gata 4 by triggering the expression of Nkx 2.5. The chromatin remodelling component BRG1-associated factor 60c (Baf60c—matrix associated, actin dependent regulator of chromatin, subfamily d, member 3) also triggers the expression of Nkx 2.5. Tbx5 is required for further differentiation. To induce beating in ectopic myocardial tissue, if Baf60c is available, expression of Gata4 and Tbx5 are sufficient [84]. Gata4 and Nkx2.5 also interact with the MADS-box transcription factor serum response factor (Srf). Srf along with myocardin (Myocd), has a key role in creating and preserving the contractile mechanisms of CMs [45, 85–87].

Intracellular Complexity and Increased Efficiency

There is significant evolution of the intracellular structure of the cardiomyocyte during development. The ultimate structure of the cardiomyocyte is the result of several mechanisms which operate during the entire prenatal period as well as in the early phase after birth. Growth during the early post-natal period is achieved by hyperplasia of CMs. Later growth is by hypertrophy of the cardiomyocyte and is achieved within a few days [88]. A polygonal cardiomyocyte is converted to an elongated shape [88, 89]. A significant increase in intracellular structures such as the myofilaments, sarcoplasmic reticulum and mitochondria as well as spatial rearrangements of these organelles result in a complex yet highly organized cellular architecture, which suit the functions of the adult cardiomyocyte [90, 91]. Development of structures involved in excitation–contraction coupling imparts a Ca^{2+} handling system and an efficient mechanism for amplification of the calcium signal. Expanded mitochondria in the adult heart organize into a reticular network creating energy micro-domains [92].

Post-natal Development

Shortly after birth, most CMs exit the cell cycle permanently; they undergo terminal differentiation. Subsequently, they have acytokinetic cell division which results in multinucleation in the cells [93–97]. Further growth of the heart in humans is thought to be mostly by hypertrophy of CMs. A proliferation burst at postnatal day 15 has however been observed in mice. This is initiated by a surge in thyroid hormone (T3) levels. T3 activates the IGF1/IGF1R/AKT pathway and can increase the number of cardiomyocytes by ~40% [98]. In mice also, later cardiac growth is achieved through hypertrophy of CMs.

Conclusion

Cardiomyocyte differentiation and proliferation during development of the heart is regulated by well-recognized transcription factors such as Gata4, Nkx2-5, and Tbx5, signalling pathways including WNT and BMP, and epigenetic networks that are modulated by histone modifications and or DNA methylation as well as chromatin remodelling. These networks orchestrate precise gene expression patterns and differentiation of cardiovascular progenitors in an appropriate spatial and temporal manner for the normal cardiac development. Extracellular signals and paracrine signalling from the epicardium and endocardium are also involved in the modulation of the of cardiomyogenic lineage specification.

References

1. Buckingham M, Meilhac S, Zaffran S. Building the mammalian heart from two sources of myocardial cells. Nat Rev Genet. 2005;6(11):826–35.
2. Brade T, Pane LS, Moretti A, Chien KR, Laugwitz K-L. Embryonic heart progenitors and cardiogenesis. Cold Spring Harbor Perspect Med. 2013; 3(10):a013847.
3. Vincent SD, Buckingham ME. How to make a heart: the origin and regulation of cardiac progenitor cells. Curr Top Dev Biol. 2010;90:1–41.
4. Latif S, Masino A, Garry DJ. Transcriptional pathways direct cardiac development and regeneration. Trends Cardiovasc Med. 2006;16(7):234–40.
5. Wu SM, Chien KR, Mummery C. Origins and fates of cardiovascular progenitor cells. Cell. 2008;132(4):537–43.
6. Liang X, Wang G, Lin L, Lowe J, Zhang Q, Bu L, et al. HCN4 dynamically marks the first heart field and conduction system precursors. Circ Res. 2013;113(4):399–407.
7. Später D, Abramczuk MK, Buac K, Zangi L, Stachel MW, Clarke J, et al. A HCN4+ cardiomyogenic progenitor derived from the first heart field and human pluripotent stem cells. Nat Cell Biol. 2013;15(9):1098–106.
8. Später D, Hansson EM, Zangi L, Chien KR. How to make a cardiomyocyte. Development. 2014;141(23):4418–31.
9. Garry DJ, Olson EN. A common progenitor at the heart of development. Cell. 2006;127(6):1101–4.
10. Rochais F, Mesbah K, Kelly RG. Signaling pathways controlling second heart field development. Circ Res. 2009;104(8):933–42.
11. Masino AM, Gallardo TD, Wilcox CA, Olson EN, Williams RS, Garry DJ. Transcriptional regulation of cardiac progenitor cell populations. Circ Res. 2004;95(4):389–97.
12. Tzahor E, Evans SM. Pharyngeal mesoderm development during embryogenesis: implications for both heart and head myogenesis. Cardiovasc Res. 2011;91(2):196–202.
13. Bu L, Jiang X, Martin-Puig S, Caron L, Zhu S, Shao Y, et al. Human ISL1 heart progenitors generate diverse multipotent cardiovascular cell lineages. Nature. 2009;460(7251):113–7.
14. Moretti A, Caron L, Nakano A, Lam JT, Bernshausen A, Chen Y, et al. Multipotent embryonic isl1+ progenitor cells lead to cardiac, smooth muscle, and endothelial cell diversification. Cell. 2006;127(6):1151–65.
15. Abu-Issa R, Kirby ML. Heart field: from mesoderm to heart tube. Annu Rev Cell Dev Biol. 2007;23:45–68.
16. Christoffels V, Jensen B. Cardiac Morphogenesis: Specification of the Four-Chambered Heart. Cold Spring Harb Perspect Biol. 2020. https://doi.org/10.1101/cshperspect.a037143.

17. Bressan M, Liu G, Mikawa T. Early mesodermal cues assign avian cardiac pacemaker fate potential in a tertiary heart field. Science. 2013;340(6133):744–8.
18. Forouhar AS, Liebling M, Hickerson A, Nasiraei-Moghaddam A, Tsai H-J, Hove JR, et al. The embryonic vertebrate heart tube is a dynamic suction pump. Science. 2006;312(5774):751–3.
19. Männer J, Perez-Pomares J, Macias D, Munoz-Chapuli R. The origin, formation and developmental significance of the epicardium: a review. Cells Tissues Organs. 2001;169(2):89–103.
20. Schlueter J, Brand T. Epicardial progenitor cells in cardiac development and regeneration. J Cardiovasc Transl Res. 2012;5(5):641–53.
21. Cai C-L, Martin JC, Sun Y, Cui L, Wang L, Ouyang K, et al. A myocardial lineage derives from Tbx18 epicardial cells. Nature. 2008;454(7200):104–8.
22. Christoffels VM, Grieskamp T, Norden J, Mommersteeg MT, Rudat C, Kispert A. Tbx18 and the fate of epicardial progenitors. Nature. 2009;458(7240):E8–9.
23. Katz TC, Singh MK, Degenhardt K, Rivera-Feliciano J, Johnson RL, Epstein JA, et al. Distinct compartments of the proepicardial organ give rise to coronary vascular endothelial cells. Dev Cell. 2012;22(3):639–50.
24. Zhou B, Ma Q, Rajagopal S, Wu SM, Domian I, Rivera-Feliciano J, et al. Epicardial progenitors contribute to the cardiomyocyte lineage in the developing heart. Nature. 2008;454(7200):109–13.
25. Hildreth V, Webb S, Bradshaw L, Brown NA, Anderson RH, Henderson DJ. Cells migrating from the neural crest contribute to the innervation of the venous pole of the heart. J Anat. 2008;212(1):1–11.
26. Hutson MR, Kirby ML, editors. Model systems for the study of heart development and disease: Cardiac neural crest and conotruncal malformations. Semin Cell Dev Biol. 2007; 18(1):101–110.
27. Waldo KL, Hutson MR, Stadt HA, Zdanowicz M, Zdanowicz J, Kirby ML. Cardiac neural crest is necessary for normal addition of the myocardium to the arterial pole from the secondary heart field. Dev Biol. 2005;281(1):66–77.
28. Tirosh-Finkel L, Zeisel A, Brodt-Ivenshitz M, Shamai A, Yao Z, Seger R, et al. BMP-mediated inhibition of FGF signaling promotes cardiomyocyte differentiation of anterior heart field progenitors. Development. 2010;137(18):2989–3000.
29. Tian Y, Morrisey EE. Importance of myocyte-nonmyocyte interactions in cardiac development and disease. Circ Res. 2012;110(7):1023–34.
30. Gassmann M, Casagranda F, Orioli D, Simon H, Lai C, Klein R, et al. Aberrant neural and cardiac development in mice lacking the ErbB4 neuregulin receptor. Nature. 1995;378(6555):390–4.
31. Gourdie RG, Wei Y, Kim D, Klatt SC, Mikawa T. Endothelin-induced conversion of embryonic heart muscle cells into impulse-conducting Purkinje fibers. Proc Natl Acad Sci USA. 1998;95(12):6815–8.
32. Bartman T, Hove J. Mechanics and function in heart morphogenesis. Dev dyn Off Publ Am Assoc Anat. 2005;233(2):373–81.
33. Hove JR, Köster RW, Forouhar AS, Acevedo-Bolton G, Fraser SE, Gharib M. Intracardiac fluid forces are an essential epigenetic factor for embryonic cardiogenesis. Nature. 2003;421(6919):172–7.
34. Ando J, Yamamoto K. Vascular mechanobiology. Circ J. 2009;73(11):1983–92.
35. Meehan RR, Dunican DS, Ruzov A, Pennings S. Epigenetic silencing in embryogenesis. Exp Cell Res. 2005;309(2):241–9.
36. Gilsbach R, Preissl S, Grüning BA, Schnick T, Burger L, Benes V, et al. Dynamic DNA methylation orchestrates cardiomyocyte development, maturation and disease. Nat Commun. 2014;5(1):1–13.
37. Kathiriya IS, Nora EP, Bruneau BG. Investigating the transcriptional control of cardiovascular development. Circ Res. 2015;116(4):700–14.
38. Schlesinger J, Schueler M, Grunert M, Fischer JJ, Zhang Q, Krueger T, et al. The cardiac transcription network modulated by Gata4, Mef2a, Nkx2. 5, Srf, histone modifications, and microRNAs. PLoS Genet. 2011; 7(2):e1001313.

39. Takeuchi JK, Lou X, Alexander JM, Sugizaki H, Delgado-Olguín P, Holloway AK, et al. Chromatin remodelling complex dosage modulates transcription factor function in heart development. Nat Commun. 2011;2(1):1–11.
40. Klattenhoff CA, Scheuermann JC, Surface LE, Bradley RK, Fields PA, Steinhauser ML, et al. Braveheart, a long noncoding RNA required for cardiovascular lineage commitment. Cell. 2013;152(3):570–83.
41. Junion G, Spivakov M, Girardot C, Braun M, Gustafson EH, Birney E, et al. A transcription factor collective defines cardiac cell fate and reflects lineage history. Cell. 2012;148(3):473–86.
42. Bruneau BG. The developmental genetics of congenital heart disease. Nature. 2008;451(7181):943–8.
43. McCulley DJ, Black BL. Transcription factor pathways and congenital heart disease. Curr Top Dev Biol. 2012;100:253–77.
44. Srivastava D. Genetic regulation of cardiogenesis and congenital heart disease. Annu Rev Pathol Mech Dis. 2006;1:199–213.
45. Noseda M, Peterkin T, Simões FC, Patient R, Schneider MD. Cardiopoietic factors: extracellular signals for cardiac lineage commitment. Circ Res. 2011;108(1):129–52.
46. Barrow JR, Howell WD, Rule M, Hayashi S, Thomas KR, Capecchi MR, et al. Wnt3 signaling in the epiblast is required for proper orientation of the anteroposterior axis. Dev Biol. 2007;312(1):312–20.
47. Haegel H, Larue L, Ohsugi M, Fedorov L, Herrenknecht K, Kemler R. Lack of beta-catenin affects mouse development at gastrulation. Development. 1995;121(11):3529–37.
48. Liu P, Wakamiya M, Shea MJ, Albrecht U, Behringer RR, Bradley A. Requirement for Wnt3 in vertebrate axis formation. Nat Genet. 1999;22(4):361–5.
49. Rivera-Pérez JA, Magnuson T. Primitive streak formation in mice is preceded by localized activation of Brachyury and Wnt3. Dev Biol. 2005;288(2):363–71.
50. Schneider VA, Mercola M. Wnt antagonism initiates cardiogenesis in Xenopus laevis. Genes Dev. 2001;15(3):304–15.
51. Foley AC, Mercola M. Heart induction by Wnt antagonists depends on the homeodomain transcription factor Hex. Genes Dev. 2005;19(3):387–96.
52. Ai D, Fu X, Wang J, Lu M-F, Chen L, Baldini A, et al. Canonical Wnt signaling functions in second heart field to promote right ventricular growth. Proc Natl Acad Sci USA. 2007;104(22):9319–24.
53. Cohen ED, Wang Z, Lepore JJ, Lu MM, Taketo MM, Epstein DJ, et al. Wnt/β-catenin signaling promotes expansion of Isl-1–positive cardiac progenitor cells through regulation of FGF signaling. J Clin Investig. 2007;117(7):1794–804.
54. Klaus A, Saga Y, Taketo MM, Tzahor E, Birchmeier W. Distinct roles of Wnt/β-catenin and Bmp signaling during early cardiogenesis. Proc Natl Acad Sci USA. 2007;104(47):18531–6.
55. Kwon C, Arnold J, Hsiao EC, Taketo MM, Conklin BR, Srivastava D. Canonical Wnt signaling is a positive regulator of mammalian cardiac progenitors. Proc Natl Acad Sci USA. 2007;104(26):10894–9.
56. Kwon C, Qian L, Cheng P, Nigam V, Arnold J, Srivastava D. A regulatory pathway involving Notch1/β-catenin/Isl1 determines cardiac progenitor cell fate. Nat Cell Biol. 2009;11(8):951–7.
57. Lin L, Cui L, Zhou W, Dufort D, Zhang X, Cai C-L, et al. β-catenin directly regulates Islet1 expression in cardiovascular progenitors and is required for multiple aspects of cardiogenesis. Proc Natl Acad Sci USA. 2007;104(22):9313–8.
58. Qyang Y, Martin-Puig S, Chiravuri M, Chen S, Xu H, Bu L, et al. The renewal and differentiation of Isl1+ cardiovascular progenitors are controlled by a Wnt/β-catenin pathway. Cell Sstem Cell. 2007;1(2):165–79.
59. Cohen ED, Tian Y, Morrisey EE. Wnt signaling: an essential regulator of cardiovascular differentiation, morphogenesis and progenitor self-renewal. Development. 2008;135(5):789–98.
60. Kerkela R, Kockeritz L, MacAulay K, Zhou J, Doble BW, Beahm C, et al. Deletion of GSK-3β in mice leads to hypertrophic cardiomyopathy secondary to cardiomyoblast hyperproliferation. J Clin Investig. 2008;118(11):3609–18.

61. High FA, Jain R, Stoller JZ, Antonucci NB, Lu MM, Loomes KM, et al. Murine Jagged1/Notch signaling in the second heart field orchestrates Fgf8 expression and tissue-tissue interactions during outflow tract development. J Clin Investig. 2009;119(7):1986–96.
62. Hoffmann AD, Peterson MA, Friedland-Little JM, Anderson SA, Moskowitz IP. Sonic hedgehog is required in pulmonary endoderm for atrial septation. Development. 2009;136(10):1761–70.
63. Nagy II, Railo A, Rapila R, Hast T, Sormunen R, Tavi P, et al. Wnt-11 signalling controls ventricular myocardium development by patterning N-cadherin and β-catenin expression. Cardiovasc Res. 2010;85(1):100–9.
64. Park EJ, Watanabe Y, Smyth G, Miyagawa-Tomita S, Meyers E, Klingensmith J, et al. An FGF autocrine loop initiated in second heart field mesoderm regulates morphogenesis at the arterial pole of the heart. Development. 2008;135(21):3599–610.
65. Zhou W, Lin L, Majumdar A, Li X, Zhang X, Liu W, et al. Modulation of morphogenesis by noncanonical Wnt signaling requires ATF/CREB family–mediated transcriptional activation of TGFβ2. Nat Genet. 2007;39(10):1225–34.
66. Zhang J, Lin Y, Zhang Y, Lan Y, Lin C, Moon AM, et al. Frs2α-deficiency in cardiac progenitors disrupts a subset of FGF signals required for outflow tract morphogenesis. Development. 2008;135(21):3611–22.
67. Buikema JW, Mady AS, Mittal NV, Atmanli A, Caron L, Doevendans PA, et al. Wnt/β-catenin signaling directs the regional expansion of first and second heart field-derived ventricular cardiomyocytes. Development. 2013;140(20):4165–76.
68. Heallen T, Morikawa Y, Leach J, Tao G, Willerson JT, Johnson RL, et al. Hippo signaling impedes adult heart regeneration. Development. 2013;140(23):4683–90.
69. von Gise A, Lin Z, Schlegelmilch K, Honor LB, Pan GM, Buck JN, et al. YAP1, the nuclear target of Hippo signaling, stimulates heart growth through cardiomyocyte proliferation but not hypertrophy. Proc Natl Acad Sci USA. 2012;109(7):2394–9.
70. Xin M, Kim Y, Sutherland LB, Qi X, McAnally J, Schwartz RJ, et al. Regulation of insulin-like growth factor signaling by Yap governs cardiomyocyte proliferation and embryonic heart size. Sci Signal. 2011; 4(196):ra70-ra.
71. Collesi C, Zentilin L, Sinagra G, Giacca M. Notch1 signaling stimulates proliferation of immature cardiomyocytes. J Cell Biol. 2008;183(1):117–28.
72. Hotta Y, Sasaki S, Konishi M, Kinoshita H, Kuwahara K, Nakao K, et al. Fgf16 is required for cardiomyocyte proliferation in the mouse embryonic heart. Dev Dyn. 2008;237(10):2947–54.
73. Qi X, Yang G, Yang L, Lan Y, Weng T, Wang J, et al. Essential role of Smad4 in maintaining cardiomyocyte proliferation during murine embryonic heart development. Dev Biol. 2007;311(1):136–46.
74. Yamaguchi TP, Takada S, Yoshikawa Y, Wu N, McMahon AP. T (Brachyury) is a direct target of Wnt3a during paraxial mesoderm specification. Genes Dev. 1999;13(24):3185–90.
75. Huber TL, Kouskoff V, Fehling HJ, Palis J, Keller G. Haemangioblast commitment is initiated in the primitive streak of the mouse embryo. Nature. 2004;432(7017):625–30.
76. Kattman SJ, Huber TL, Keller GM. Multipotent flk-1+ cardiovascular progenitor cells give rise to the cardiomyocyte, endothelial, and vascular smooth muscle lineages. Dev Cell. 2006;11(5):723–32.
77. Kouskoff V, Lacaud G, Schwantz S, Fehling HJ, Keller G. Sequential development of hematopoietic and cardiac mesoderm during embryonic stem cell differentiation. Proc Natl Acad Sci USA. 2005;102(37):13170–5.
78. Yang L, Soonpaa MH, Adler ED, Roepke TK, Kattman SJ, Kennedy M, et al. Human cardiovascular progenitor cells develop from a KDR+ embryonic-stem-cell-derived population. Nature. 2008;453(7194):524–8.
79. Costello I, Pimeisl I-M, Dräger S, Bikoff EK, Robertson EJ, Arnold SJ. The T-box transcription factor Eomesodermin acts upstream of Mesp1 to specify cardiac mesoderm during mouse gastrulation. Nat Cell Biol. 2011;13(9):1084–91.
80. Bondue A, Lapouge G, Paulissen C, Semeraro C, Iacovino M, Kyba M, et al. Mesp1 acts as a master regulator of multipotent cardiovascular progenitor specification. Cell Stem Cell. 2008;3(1):69–84.

81. Kitajima S, Takagi A, Inoue T, Saga Y. MesP1 and MesP2 are essential for the development of cardiac mesoderm. Development. 2000;127(15):3215–26.
82. Saga Y, Miyagawa-Tomita S, Takagi A, Kitajima S, Miyazaki J, Inoue T. MesP1 is expressed in the heart precursor cells and required for the formation of a single heart tube. Development. 1999;126(15):3437–47.
83. Liu Y, Lu P, Wang Y, Morrow BE, Zhou B, Zheng D. Spatiotemporal gene coexpression and regulation in mouse cardiomyocytes of early cardiac morphogenesis. J Am Heart Assoc. 2019; 8(15):e012941.
84. Takeuchi JK, Bruneau BG. Directed transdifferentiation of mouse mesoderm to heart tissue by defined factors. Nature. 2009;459(7247):708–11.
85. Balza RO, Misra RP. Role of the serum response factor in regulating contractile apparatus gene expression and sarcomeric integrity in cardiomyocytes. J Biol Chem. 2006;281(10):6498–510.
86. Niu Z, Li A, Zhang SX, Schwartz RJ. Serum response factor micromanaging cardiogenesis. Curr Opin Cell Biol. 2007;19(6):618–27.
87. Evans SM, Yelon D, Conlon FL, Kirby ML. Myocardial lineage development. Circ Res. 2010;107(12):1428–44.
88. Leu M, Ehler E, Perriard J-C. Characterisation of postnatal growth of the murine heart. Anat Embryol. 2001;204(3):217–24.
89. Hirschy A, Schatzmann F, Ehler E, Perriard J-C. Establishment of cardiac cytoarchitecture in the developing mouse heart. Dev Biol. 2006;289(2):430–41.
90. Piquereau J, Novotova M, Fortin D, Garnier A, Ventura-Clapier R, Veksler V, et al. Postnatal development of mouse heart: formation of energetic microdomains. J Physiol. 2010;588(13):2443–54.
91. Porter GA Jr, Hom JR, Hoffman DL, Quintanilla RA, de Mesy Bentley KL, Sheu S-S. Bioenergetics, mitochondria, and cardiac myocyte differentiation. Prog Pediatr Cardiol. 2011;31(2):75–81.
92. Mühlfeld C, Urru M, Rümelin R, Mirzaie M, Schöndube F, Richter J, et al. Myocardial ischemia tolerance in the newborn rat involving opioid receptors and mitochondrial K^+ channels. Anat Rec A Discov Mol Cell Evol Biol. 2006;288(3):297–303.
93. Schmid G, Pfitzer P. Mitoses and binucleated cells in perinatal human hearts. Virchows Archiv B. 1985;48(1):59.
94. Brodsky VY, Chernyaev A, Vasilyeva I. Variability of the cardiomyocyte ploidy in normal human hearts. Virchows Archiv B. 1992;61(1):289.
95. Li F, Wang X, Capasso JM, Gerdes AM. Rapid transition of cardiac myocytes from hyperplasia to hypertrophy during postnatal development. J Mol Cell Cardiol. 1996;28(8):1737–46.
96. Olivetti G, Cigola E, Maestri R, Corradi D, Lagrasta C, Gambert SR, et al. Aging, cardiac hypertrophy and ischemic cardiomyopathy do not affect the proportion of mononucleated and multinucleated myocytes in the human heart. J Mol Cell Cardiol. 1996;28(7):1463–77.
97. Bergmann O, Zdunek S, Alkass K, Druid H, Bernard S, Frisén J. Identification of cardiomyocyte nuclei and assessment of ploidy for the analysis of cell turnover. Exp Cell Res. 2011;317(2):188–94.
98. Naqvi N, Li M, Calvert JW, Tejada T, Lambert JP, Wu J, et al. A proliferative burst during preadolescence establishes the final cardiomyocyte number. Cell. 2014;157(4):795–807.

Chapter 3
Cell Cycle Regulation in Cardiomyocytes

Abstract Cell cycle activity is an intrinsic component of cardiomyocyte differentiation and heart development. Several studies have investigated the expression of known regulatory genes of the mammalian cell cycle during cardiac development. In embryonic hearts which have a high cardiomyocyte cell cycle activity, there is increased expression of positive cell cycle regulators such as cyclins, the cyclin-dependent kinases and protooncogenes. In the adult heart, in which cell cycle activity is largely absent in cardiomyocytes, the expressions of positive cell cycle regulators are downregulated. Expressions of negative cell cycle regulatory genes such as Cdk inhibitors are increased in adult hearts. This article contains a review of the currently known regulatory mechanisms of cell cycle in the cardiomyocyte.

Keywords Cardiomyocyte cell cycle · Cell cycle regulation · Cyclins · Pocket proteins · Transcription factors · Signalling pathways · Growth factors · Cytokines · Cytokinesis

Introduction

Synthesis of deoxy ribonucleic acid (DNA) in cardiomyocytes (CMs) occurs in two phases, first during early embryonic development and the second in the early neonatal period [1]. CMs proliferate rapidly during foetal life. Their proliferation ceases soon after birth. Prior to their exit from the cell cycle, CMs in the perinatal period however have an added cycle of DNA synthesis and nuclear mitosis sans cell division (acytokinetic mitosis) resulting in binucleated cardiomyocytes in most species [2, 3]. The number of binucleated CMs increases during the postnatal period and in humans they may form 25–75% of all CMs by the third week [4, 5]. The number of CMs which re-enter the cell cycle in the normal adult heart is very low and evidence for DNA synthesis is seen only in about 0.005% of CMs in the ventricle [6–8]. In response to growth signals adult CMs only increase in cell size or hypertrophy. In the injured human heart, the reported mitotic index is 0.015–0.08% [9, 10].

At different stages of life, CMs have distinct forms of cell growth [11]. In the foetal heart, CMs proliferate rapidly. CM proliferation ceases during the perinatal period; they permanently exit the cell cycle. CMS of the perinatal period undergo acytokinetic mitosis. Adult CMs do not re-enter the cell cycle. In response to growth signals, they increase in cell size or hypertrophy. Cell cycle exit in the postnatal heart is associated with downregulation of the positive regulators of the cell cycle and upregulation of the retinoblastoma protein (pRb), which has a key role in the negative control of the cell cycle and cyclin-dependent kinase inhibitors (CdkI) p21 and p27 [12–15].

Cell Cycle Regulators

Cell cycle progression is dependent on regulated transduction of mitogenic signals to cyclically expressed proteins known as cyclins and active targets, the cyclin-dependent protein kinases (Cdks) (Fig. 3.1). A series of checkpoints between different phases ensure smooth progression of the cell cycle. These checkpoints govern various

Fig. 3.1 Regulatory factors involved in cell cycle exit in cardiomyocytes. Cdk—Cyclin dependent kinase. Cyc—Cyclin. DP—Dimerization partner. E2F—Transcription factor involved in cell cycle regulation. Ink—Inhibitor of cyclin dependent kinases. Rb—Retinoblastoma protein

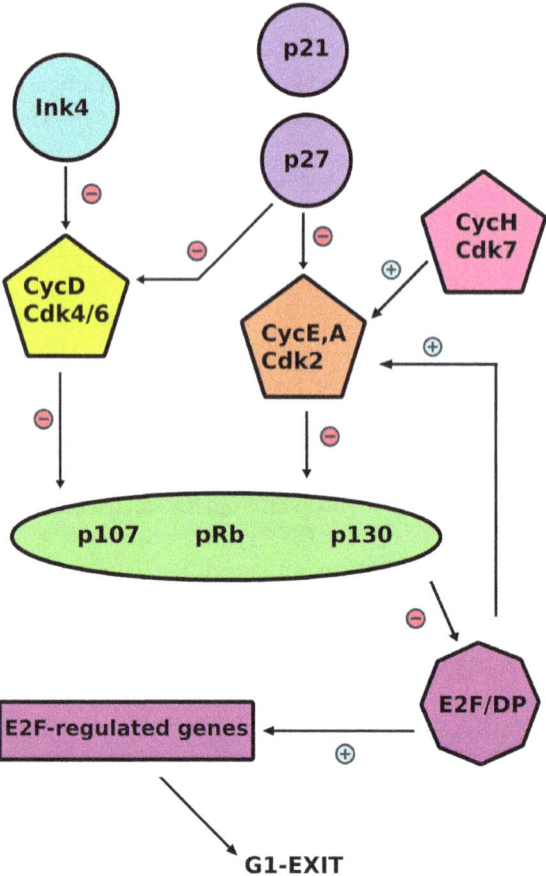

cyclin-Cdk complexes necessary for unique cell cycle events. CAK (cyclin H/Cdk7) and Cdk inhibitors (CKI) regulate their activities. Among the key regulators in the major cell cycle checkpoint in late G1 phase of the cell cycle are Cdk4 and Cdk6. Cdk4 and Cdk6 complex with cyclin D1, D2, or D3 [16] and preferentially phosphorylate members of the Rb family (Rb, p107, and p130) resulting in the release of E2F, required for the transcription of genes that mediate G1 exit and DNA synthesis (Fig. 3.1) [17]. Cyclin E which is mainly expressed at the G1-S transition phase, accelerates the phosphorylation of the Rb proteins by forming complexes with its catalytic partner Cdk2. In S phase, Cyclin A and Cdk2 complexes play an important role. In the G2/M phase, cyclin B and Cdc2 are required.

Cyclin/Cdk/CdkI

The embryonic heart express cyclins such as D1, D2, D3, A, B1, and E in high levels. These are involved in G1, S, G2, and M-phase [18–22]. Proliferating cell nuclear antigen (PCNA), the cyclin-dependent kinases (Cdks) Cdc2, Cdk2, Cdk4, and Cdk6 and their associated kinase activities are also highly expressed in the embryonic heart [18, 19, 21, 22]. These genes are also required for DNA replication. Cyclin D expression seems critical for the development of the embryonic ventricle [23–25]. A combined loss of Cdk2 and Cdk4 causes hypophosphorylation of Rb, leading to repression of E2F target genes, such as Cdc2 and cyclin A2 and can result in cardiac abnormalities. A combined loss of Cdk4 and Cdk6 does not however affect cell cycle initiation and progression [26]. Evidences so far indicate a specific role for Cyc D/Cdk4 complexes in normal cardiac development.

The exit of CMs from the cell cycle during the postnatal period is associated with alteration in the expression pattern of several molecules involved in the regulation of the cell cycle. In the CMs of the postnatal heart, in comparison with CMs of the embryonic heart, there is significant down regulation of protein expression of cyclins D1, D2, D3, A, B1, and E and their associated kinases. In adult CMs, the protein levels of cyclin A, B, D1, E, and Cdc2 are undetectable [18, 19, 21, 22]. The downregulation in expression of cyclins and Cdks during normal development of CMs is linked with a reciprocal upregulation of CdkIs. One family of Cdk inhibitors, the INK4 family which comprises p15, p16, p18, and p19 is specific for Cdk4/6. Another, the Cip/Kip family comprising p21, p27, and p57 inhibits Cdk4/6 as well as Cdk2 and Cdc2 [16]. INK4 proteins binds to cyclin D-dependent kinase and prevents cyclin D interaction. The Cip/Kip family negatively regulate activity of cyclin E and cyclin A-dependent kinases [27]. It is also an inhibitor of cyclin D-dependent kinases. Expression of p16 and p18 of the INK4 family is seen in the embryonic heart. Their levels are low in young adult hearts [12, 28, 29]. p16 positive myocytes increases with age [30, 31]. In contrast, expression of Cip/Kip family members is high in CMs during the perinatal period and are higher in adult myocardium [32, 33].

Pocket Proteins (Rb, P130, P107)

The primary target of G1 Cdks is the product of the retinoblastoma susceptibility gene (Rb). Rb, p107 and p130, members of the family of pocket proteins are expressed in the developing myocardium [34]. Rb is negligible in foetal myocardium, upregulated during the neonatal stage, and is the major pocket protein expressed in adult, terminally differentiated CMs [35, 36]. Expression pattern of p107 is highest in the embryonic heart and lowest in adult heart. p130 expression peaks in the neonatal period and in adult heart, its expression is at low levels. These proteins inhibit cell cycle progression by controlling E2F-responsive genes [37]. Hypophosphorylated Rb proteins bind to E2F complexes and recruits transcriptional repressors such as histone deacetylases (HDACs) or Jumonji (a member of Jmj-type chromatin modifying enzyme and transcriptional regulators) [38]. Phosphorylated Rb releases E2F complexes, resulting in activation of transcription and the expression of genes required for DNA synthesis and cell cycle regulators such as cyclin E, cyclin A, the mitotic kinase Cdc2 (p34/Cdk1), as well as E2F-1.

Rb proteins has a key role in the regulation of cell cycle exit and possibly differentiation of heart muscle [39–41]. While there are several evidences for the critical role of Rb family members in the cell cycle exit of the normal cardiomyocyte, it is unclear whether they are also responsible for non-proliferation of adult CMs in response to growth stimuli.

Transcription Factors

E2Fs

E2F family of transcription factors has eight members, E2F-1 through -8 [42]. There are activator E2Fs (E2F-1, E2F-2, and E2F-3) and repressor E2Fs (E2F-4, E2F-5, E2F-6, E2F-7, and E2F-8). Activator E2Fs activate transcription or push quiescent cells into the cell cycle. At the cell cycle phases, E2F proteins are present at E2F-regulated promoters. Repressor E2Fs also activate transcription when they are over-expressed. E2Fs form obligate heterodimers with DP-1 and -2, members of a second family of transcription factors, to bind DNA and activate transcription.

E2F-1, -2, and -3 are structurally similar and are expressed in proliferating cells [43]. They associate with Rb [44]. E2F-4 and E2F-5 are both expressed in quiescent and differentiating cells [45]. E2F-4 form complexes with Rb, p107, and p130 [46]. E2F-5 binds with p130 [47]. E2F-6, -7, and -8 also function as transcriptional repressors; their physiological role is unclear. E2F-1, -2, and -3 are considered to drive quiescent cells through G1 into S phase [48–50]. E2F-1 also induces apoptosis [51]. E2F-4 or E2F-5 promotes differentiation [52]. They are also associated with the pocket protein-dependent downregulation of a number of genes involved in cell cycling including E2F-1 [53], cdc2 [54], and β-myb [55]. Thus, E2F family members and pocket proteins are part of a complex network which regulates proliferation and differentiation (Fig. 3.1).

E2F-1, E2F-3, and DP family members are downregulated from foetal to adult stages of development [36]. E2F-4 and E2F-5 are upregulated in heart during development [36]. A combined loss of E2F-1, E2F-2 and E2F-prevent entry of cells into S phase and cell proliferation [56]. S-phase entry can be induced by overexpressing E2F-1, E2F-2, E2F-3, and E2F-4 in neonatal CMs. Increased expression of E2F-1 and E2F-3 induces apoptosis in addition to progression in the cell cycle [57]. While deletion of E2F-1, -4, or -5 do not result in any cardiac defect [58, 59], deletion of E2F-3 results in either embryonic lethality or early death of the surviving embryos [60].

Myc

Myc is thought to act as 'third messengers' for ligand-dependent signals and thus regulate growth in several tissues [61]. The Myc family consists of c-Myc, N-Myc, and L-Myc. They are members of the basic helix-loop-helix-leucine zipper (βHLHZ) family of proteins. They activate transcription forming a complex with a protein, Max. It is generally accepted that for both proliferation and cellular growth during normal development, Myc is essential [62, 63]. Be that as it may, there is also a view that Myc is only involved in cellular division [64]. The mechanism of its action is not completely known.

In the heart, Myc is critical for both cardiomyocyte division and hypertrophy. Myc is expressed in embryonic CMs. Myc-deficient mice have heart defects and they die prematurely [65]. Myc-null embryos are retarded in development and have enlarged heart and pericardial effusions. In these embryos, neither the size nor the number of CMs were studied. Whether the heart defects are primarily because of Myc deficiency in CMs is unsettled. In transgenic mice with overexpression of Myc in the foetal heart, ventricles are enlarged and cardiomyocyte hyperplasia has been observed [66].

Hypoxia Inducible Factor-1

Hypoxia inducible factor-1 (HIF-1) is a transcription factor complex activated in response to hypoxia and or specific signalling pathways. HIF-1 activates or represses genes containing hypoxic response elements. HIF-1 stability is regulated by van Hippel-Lindau (VHL), a tumour suppressor protein [67, 68]. Deletion of HIF-1 results in lethality in embryos which develop obstruction between outflow tract and ventricle secondary to hyperplasia of CMs [69]. Several studies indicate that stabilization of HIF-1 inhibits cell proliferation [11]. The mechanistic basis for growth arrest induced by HIF-1 is not understood. HIF-1 has been shown to induce cell cycle arrest by removing Myc from its binding with the p21 promoter and thus derepressing p21 [70]. Neither HIF-1-transcriptional activity nor its DNA binding is necessary for cell cycle arrest. There is a suggestion that HIF-1 may also be an endogenous antagonist of Myc.

Signalling Pathways

Tuberous Sclerosis Complex 1 and 2

The genes causing tuberous sclerosis complex (TSC) have been identified and encode for hamartin [71] and tuberin [72]. Fifty percent of patients with TSC have primary tumours of the myocardium [73]. The predisposition for cardiac tumours suggests that the TSC gene products may regulate cell cycle in CMs. CMs isolated from TSC2EK/EK embryos continue to proliferate and synthesize DNA even after as many as eight passages while CMs from heterozygous or wild-type embryos exit the cell cycle. TSC2EK/EK cardiomyocytes are highly differentiated and have a phenotype similar to normal embryonic or neonatal CMs from rats. TSC2 gene product hence is thought to be important for cell cycle exit and terminal differentiation of normal CMs. When the growth inhibitory activity of the endogenous TSC2 in the heart is blocked by overexpressing a mutant TSC2 in transgenic mice, CMs exit cell cycle and cardiac development is normal.

P38 Mitogen-Activated Protein Kinase

In the mammalian heart, mitogen-activated protein kinase (MAPK) signalling pathways modulate the growth of CMs in response to diverse developmental signals [74]. The MAPK signalling pathways activate extracellular signal regulated kinases (ERK), c-Jun NH2-terminal kinases (JNK), or p38 MAPKs. All four isoforms of p38 are expressed in the heart. p38α and p38γ are the predominant family members in CMs [75, 76]. The major upstream activators of p38 MAPKs are MAPKKs. Substrates of p38 MAPKs include other protein kinases and several transcription factors (TFs). These TFs are MEF2, MAPKAPK2 and MAPKAPK -3, ATF-2, ELK-1, Chop, TEF-1, C/EBP and Max [77, 78]. p38 induces differentiation and also control proliferation of many cell types [79–81]. The regulatory action of p38 MAPKs on cell cycle progression depends on the cell type and the stimulus [80, 82–85].

There is some evidence that p38 may be a key negative regulator of the cell cycle in mammalian CMs [85–87]. Genetic activation of p38 in vivo reduced proliferation of foetal CMs, whereas targeted disruption of p38 and growth factor stimulation promoted cell cycle re-entry in cultured adult cardiomyocytes [86]. There is also a report that in transgenic mice, cardiac-specific expression of a dominant negative mutant p38 generates a hypertrophic rather than proliferative response in adult hearts [88]. Further studies are necessary to clarify the role of p38 in the proliferation and terminal differentiation of CMs.

Effects of Growth Factors and Cytokines on Cardiomyocyte Proliferation

Growth Factors

Several growth factors seem to have a role in cardiomyocyte proliferation.

Insulin Growth Factor (IGF)

Insulin-like growth factor I (IGF-I) is known to have insulin-like short-term metabolic effects and growth factor-like long-term effects on cell proliferation and differentiation of many cell types. The mitogenic activity of IGF-I is mediated primarily through binding with the IGF-I receptor [89]. Overexpression of IGF-I in the hearts of transgenic mice results in an increase in heart weight by 50% and an increase in the number of cells in the heart [90]. A similar phenotype was also seen in mice lacking the IGF-II receptor [91]. The signalling components downstream of the IGF-I receptor such as the IGF-phosphatidylinositol 3-kinase (PI3K)-Akt-p70S6K signalling pathway has been shown to regulate cardiac hypertrophy, viability, and homeostasis [92]. The role of Akt, the downstream effector, is less clear [93–96]. Overexpressing a dominant-negative Akt during development does not affect either cardiomyocyte size or number. This observation indicate that Akt may not play a role in the proliferation of normal CMs [97]. IGF-I may be useful to enhance myocardial repair after injury.

Fibroblast Growth Factor (FGF)

FGF is a known regulator of the growth and differentiation of CMs. When FGF signalling is blocked in the embryonic heart, proliferation of CMs is inhibited [98]. FGF-2 stimulates proliferation of embryonic and neonatal CMs [99], as well as cardiac stem cells [100]. High-affinity receptors (FGFR) on the cell surface mediate the effects of FGF [101]. FGFR-1 is the dominant isoform in CMs at all developmental stages [102]. Lavine et al. observed that epicardial and endocardial FGF signalling through FGFR-1 and FGFR-2 is necessary for proliferation and differentiation of CMs in vivo [103]. The critical role of FGF signalling in the developing myocardium and the functional requirement for both FGFR1 and FGFR2 are evident from the studies which revealed severe hypoplasia and thinning of the ventricular wall in embryos with deficiency of both FGFR-1 and FGFR-2 in the myocardium [103].

Cytokines

There is much evidence to indicate that cytokines regulate proliferation of CMs. These cytokines are interleukin (IL) -6, IL-11, leukemia inhibitory factor (LIF), cardiotropin-1, ciliary neurotrophic factor (CNTF), and oncostatin M. Mice deficient for IL-6, LIF, CNTF and IL-11 or their receptors are seen to have a large number of developmental defects [104–108]. Specific association of any deficiency with a distinct cardiac defect is however not seen in them.

These cytokines signal through receptor complexes containing the gp130 subunit. Germline deletion of gp130 results in death of embryos. The embryos have hypoplastic ventricles, suggesting that gp130-dependent signals are critical for proliferative growth of the heart [109]. Deleting gp130 in the perinatal period results in thin walls of the heart [110]. Mice with cardiac specific gp130-deficiency may also have normally appearing hearts; they have however increased apoptosis in CMs and develop dilated cardiomyopathy under hemodynamic stress [111]. This finding suggests that gp130- mediated signalling is possibly not requisite for normal cardiac development, but is necessary for adaptive hypertrophic growth in the adult heart.

Regulation of Cytokinesis

Cytokinesis results in partitioning and separation of cytoplasm between daughter cells. It is the final step of cell division to complete mitosis [112]. Cytokinesis is dependent on the contraction of the actomyosin cytoskeleton. Actin is considered to be the scaffold onto which the other components of the cytokinesis machinery assemble. Contractile ring formation is regulated not only by actin and myosin, but by small GTPases such as RhoA and its effectors ROCK I and ROCK II, citron kinase, GTPase Cdc42, Rac, and septins [113].

In dividing neonatal myocytes, actin gets disassembled during the early stages of mitosis. During anaphase, actin concentrates at the equator of the spindle. Finally, in telophase, actin forms a circumferential band. Cytoplasmic myosin is evenly distributed in the cytoplasm as small spots. In anaphase, myosin concentrates in association with the cortical membrane in the equator region and later forms a ring like structure. Myosin stays associated with adjacent membranes at the cleavage furrow until telophase [2].

Polo-like kinase, is a protein involved in spindle formation and chromosome segregation during mitosis. Anillin is another known regulator of the cleavage furrow formation. Defective localization of anillin in the midbody region could lead to failure of cytoplasmic division and binucleation [114].

Septins, a family of cytoskeletal GTPases are important for cytokinesis. They have a stage specific expression during heart development [115]. Expression of small Rho GTPases such as RhoA, Cdc42, Rac1, ROCK-I, ROCK-II, and p-cofilin, which are coupled to the formation of actomyosin ring are regulated in CMs during development. In embryonic hearts in which cytokinesis occurs, these proteins are expressed at high levels.

Conclusion

Several studies have investigated the expression of known mammalian cell cycle regulatory genes during cardiac development. Synthesis of deoxy ribonucleic acid (DNA) in cardiomyocytes occurs in two phases, initially during early embryonic development and later in the early neonatal period. CMs have three developmentally determined forms of cell cycle control and growth, namely, proliferation, binucleation, and hypertrophy. In embryonic hearts which have a high cardiomyocyte cell cycle activity, there is increased expression of cyclins, cyclin-dependent kinases and their associated activities. Transcription factors such as E2Fs, Myc and HIF-1 as well as signalling pathways involving TSC2 gene products and pMAPKs regulate cell cycle activity in the heart. Several cytokines and growth factors such as IGF-1 and FGF also seem to have a role in cardiomyocyte proliferation. In the adult heart, in which cell cycle activity is largely absent in CMs, the expressions of positive cell cycle regulators are downregulated and the expressions of negative cell cycle regulatory genes such as Cdk inhibitors are increased.

References

1. Soonpaa MH, Kim KK, Pajak L, Franklin M, Field LJ. Cardiomyocyte DNA synthesis and binucleation during murine development. American Journal of Physiology-Heart and Circulatory Physiology. 1996;271(5):H2183–9.
2. Li F, Wang X, Bunger PC, Gerdes AM. Formation of binucleated cardiac myocytes in rat heart: I. Role of actin–myosin contractile ring. Journal of Molecular and Cellular Cardiology. 1997; 29(6):1541–51.
3. Li F, Wang X, Gerdes AM. Formation of binucleated cardiac myocytes in rat heart: II. Cytoskeletal organisation. Journal of Molecular and Cellular Cardiology. 1997; 29(6):1553–65.
4. Olivetti G, Cigola E, Maestri R, Corradi D, Lagrasta C, Gambert SR, et al. Aging, cardiac hypertrophy and ischemic cardiomyopathy do not affect the proportion of mononucleated and multinucleated myocytes in the human heart. J Mol Cell Cardiol. 1996;28(7):1463–77.

5. Schmid G, Pfitzer P. Mitoses and binucleated cells in perinatal human hearts. Virchows Archiv B. 1985;48(1):59–67.
6. Rumyantsev P, Borisov A. DNA synthesis in myocytes from different myocardial compartments of young rats in norm, after experimental infarction and in vitro. Biomed Biochim Acta. 1987;46(8–9):S610–5.
7. Soonpaa M, Field LJ. Assessment of cardiomyocyte DNA synthesis in normal and injured adult mouse hearts. American Journal of Physiology-Heart and Circulatory Physiology. 1997;272(1):H220–6.
8. Soonpaa MH, Field LJ. Survey of studies examining mammalian cardiomyocyte DNA synthesis. Circ Res. 1998;83(1):15–26.
9. Beltrami AP, Urbanek K, Kajstura J, Yan S-M, Finato N, Bussani R, et al. Evidence that human cardiac myocytes divide after myocardial infarction. N Engl J Med. 2001;344(23):1750–7.
10. Kajstura J, Leri A, Finato N, Di Loreto C, Beltrami CA, Anversa P. Myocyte proliferation in end-stage cardiac failure in humans. Proceedings of the National Academy of Sciences USA. 1998;95(15):8801–5.
11. Ahuja P, Sdek P, MacLellan WR. Cardiac myocyte cell cycle control in development, disease, and regeneration. Physiol Rev. 2007;87(2):521–44.
12. Koh KN, Kang MJ, Frith-Terhune A, Park SK, Kim I, Lee CO, et al. Persistent and heterogeneous expression of the cyclin-dependent kinase inhibitor, p27KIP1, in rat hearts during development. J Mol Cell Cardiol. 1998;30(3):463–74.
13. Poolman RA, Gilchrist R, Brooks G. Cell cycle profiles and expressions of p21CIP1 AND P27KIP1 during myocyte development. Int J Cardiol. 1998;67(2):133–42.
14. Walsh K, Perlman H. Cell cycle exit upon myogenic differentiation. Curr Opin Genet Dev. 1997;7(5):597–602.
15. Wang J, Nadalginard B. Regulation of cyclins and p34CDC2 expression during terminal differentiation of C2C12 myocytes. Biochem Biophys Res Commun. 1995;206(1):82–8.
16. Sherr CJ. G1 phase progression: cycling on cue. Cell. 1994;79(4):551–5.
17. Nevins JR. E2F: a link between the Rb tumor suppressor protein and viral oncoproteins. Science. 1992;258(5081):424–9.
18. Brooks G, Poolman RA, McGill CJ, Li J-M. Expression and activities of cyclins and cyclin-dependent kinases in developing rat ventricular myocytes. J Mol Cell Cardiol. 1997;29(8):2261–71.
19. Flink IL, Oana S, Maitra N, Bahl JJ, Morkin EJ. Changes in E2F complexes containing retinoblastoma protein family members and increased cyclin-dependent kinase inhibitor activities during terminal differentiation of cardiomyocytes. J Mol Cell Cardiol. 1998;30(3):563–78.
20. Kang MJ, Kim J-S, Chae S-W, Koh KN, Koh GY. Cyelins and cyclin dependent kinases during cardiac development. Mol Cells. 1997;7(3):360–6.
21. Kang MJ, Koh GY. Differential and dramatic changes of cyclin-dependent kinase activities in cardiomyocytes during the neonatal period. J Mol Cell Cardiol. 1997;29(7):1767–77.
22. Yoshizumi M, Lee W-S, Hsieh C-M, Tsai J-C, Li J, Perrella MA, et al. Disappearance of cyclin A correlates with permanent withdrawal of cardiomyocytes from the cell cycle in human and rat hearts. J Clin Investig. 1995;95(5):2275–80.
23. Fantl V, Stamp G, Andrews A, Rosewell I, Dickson C. Mice lacking cyclin D1 are small and show defects in eye and mammary gland development. Genes Dev. 1995;9(19):2364–72.
24. Kozar K, Ciemerych MA, Rebel VI, Shigematsu H, Zagozdzon A, Sicinska E, et al. Mouse development and cell proliferation in the absence of D-cyclins. Cells. 2004;118(4):477–91.
25. Pasumarthi KB, Nakajima H, Nakajima HO, Soonpaa MH, Field LJ. Targeted expression of cyclin D2 results in cardiomyocyte DNA synthesis and infarct regression in transgenic mice. Circ Res. 2005;96(1):110–8.
26. Malumbres M, Sotillo Ro, Santamaría D, Galán J, Cerezo A, Ortega S, et al. Mammalian cells cycle without the D-type cyclin-dependent kinases Cdk4 and Cdk6. Cells 2004; 118(4):493–504.

27. Sherr CJ, Roberts JM. Inhibitors of mammalian G1 cyclin-dependent kinases. Genes Dev. 1995;9(10):1149–63.
28. Akli Sd, Zhan S, Abdellatif M, Schneider MD. E1A can provoke G1 exit that is refractory to p21 and independent of activating cdk2. Circulation Research. 1999; 85(4):319–28.
29. Burton P, Raff M, Kerr P, Yacoub M, Barton P. An intrinsic timer that controls cell-cycle withdrawal in cultured cardiac myocytes. Dev Biol. 1999;216(2):659–70.
30. Kajstura J, Pertoldi B, Leri A, Beltrami C-A, Deptala A, Darzynkiewicz Z, et al. Telomere shortening is an in vivo marker of myocyte replication and aging. Am J Pathol. 2000;156(3):813–9.
31. Krishnamurthy J, Torrice C, Ramsey MR, Kovalev GI, Al-Regaiey K, Su L, et al. Ink4a/Arf expression is a biomarker of aging. J Clin Investig. 2004;114(9):1299–307.
32. Burton P, Yacoub M, Barton PJ. Cyclin-dependent kinase inhibitor expression in human heart failure. A comparison with fetal development. European Heart Journal. 1999; 20(8):604–11.
33. Li J, Brooks G. Downregulation of cyclin-dependent kinase inhibitors p21 and p27 in pressure-overload hypertrophy. American Journal of Physiology-Heart and Circulatory Physiology. 1997;273(3):H1358–67.
34. Classon M, Salama S, Gorka C, Mulloy R, Braun P, Harlow EJ. Combinatorial roles for pRB, p107, and p130 in E2F-mediated cell cycle control. Proceedings of the National Academy of Sciences USA. 2000;97(20):10820–5.
35. Jiang Z, Zacksenhaus E, Gallie BL, Phillips RA. The retinoblastoma gene family is differentially expressed during embryogenesis. Oncogene. 1997;14(15):1789–97.
36. MacLellan W, Garcia A, Oh H, Frenkel P, Jordan M, Roos K, et al. Overlapping roles of pocket proteins in the myocardium are unmasked by germ line deletion of p130 plus heart-specific deletion of Rb. Mol Cell Biol. 2005;25(6):2486–97.
37. Cobrinik D. Pocket proteins and cell cycle control. Oncogene. 2005;24(17):2796–809.
38. Jung J, Kim T-g, Lyons GE, Kim H-RC, Lee YJ. Jumonji regulates cardiomyocyte proliferation via interaction with retinoblastoma protein. Journal of Biological Chemistry. 2005; 280(35):30916–23.
39. Clarke AR, Maandag ER, van Roon M, van der Lugt NM, van der Valk M, Hooper ML, et al. Requirement for a functional Rb-1 gene in murine development. Nature. 1992;359(6393):328–30.
40. Jacks T, Fazeli A, Schmitt EM, Bronson RT, Goodell MA, Weinberg RA. Effects of an Rb mutation in the mouse. Nature. 1992;359(6393):295–300.
41. Eva Y-HL, Chang C-Y, Hu N, Wang Y-CJ, Lai C-C, Herrup K, et al. Mice deficient for Rb are nonviable and show defects in neurogenesis and haematopoiesis. Nature. 1992; 359(6393):288–94.
42. Logan N, Graham A, Zhao X, Fisher R, Maiti B, Leone G, et al. E2F-8: an E2F family member with a similar organization of DNA-binding domains to E2F-7. Oncogene. 2005;24(31):5000–4.
43. Lees JA, Saito M, Vidal M, Valentine M, Look T, Harlow E, et al. The retinoblastoma protein binds to a family of E2F transcription factors. Mol Cell Biol. 1993;13(12):7813–25.
44. Helin K, Harlow E. Heterodimerization of the transcription factors E2F-1 and DP-1 is required for binding to the adenovirus E4 (ORF6/7) protein. J Virol. 1994;68(8):5027–35.
45. Dagnino L, Fry C, Bartley S, Farnham P, Gallie B, Phillips R. Expression patterns of the E2F family of transcription factors during mouse nervous system development. Mech Dev. 1997;66(1–2):13–25.
46. Moberg K, Starz MA, Lees JA. E2F-4 switches from p130 to p107 and pRB in response to cell cycle reentry. Mol Cell Biol. 1996;16(4):1436–49.
47. Hijmans EM, Voorhoeve PM, Beijersbergen RL, Van'T Veer L, Bernards RJM, biology c. E2F-5, a new E2F family member that interacts with p130 in vivo. Molecular and Cellular Biology. 1995; 15(6):3082–9.
48. DeGregori J, Leone G, Miron A, Jakoi L, Nevins JRJPotNAoS. Distinct roles for E2F proteins in cell growth control and apoptosis. Proceedings of the National Academy of Sciences USA. 1997; 94(14):7245–50.

49. Johnson DG, Cress WD, Jakoi L, Nevins JR. Oncogenic capacity of the E2F1 gene. Proceedings of the National Academy of Sciences USA. 1994;91(26):12823–7.
50. Qin X-Q, Livingston DM, Kaelin WG, Adams PD. Deregulated transcription factor E2F–1 expression leads to S-phase entry and p53-mediated apoptosis. Proceedings of the National Academy of Sciences USA. 1994;91(23):10918–22.
51. Strom DK, Cleveland JL, Chellappan S, Nip J, Hiebert SW. E2F–1 and E2F–3 are functionally distinct in their ability to promote myeloid cell cycle progression and block granulocyte differentiation. Cell Growth Differ. 1998;9(1):59–69.
52. Persengiev SP, Kondova II, Kilpatrick DL. E2F4 actively promotes the initiation and maintenance of nerve growth factor-induced cell differentiation. Mol Cell Biol. 1999;19(9):6048–56.
53. Furukawa Y, Iwase S, Kikuchi J, Nakamura M, Yamada H, Matsuda M. Transcriptional repression of the E2F–1 gene by interferon-α is mediated through induction of E2F–4/pRB and E2F–4/p130 complexes. Oncogene. 1999;18(11):2003–14.
54. Tommasi S, Pfeifer GP. In vivo structure of the human cdc2 promoter: release of a p130–E2F-4 complex from sequences immediately upstream of the transcription initiation site coincides with induction of cdc2 expression. Mol Cell Biol. 1995;15(12):6901–13.
55. Bennett JD, Farlie PG, Watson RJ. E2F binding is required but not sufficient for repression of B-myb transcription in quiescent fibroblasts. Oncogene. 1996;13(5):1073–82.
56. Wu L, Timmers C, Maiti B, Saavedra HI, Sang L, Chong GT, et al. The E2F1–3 transcription factors are essential for cellular proliferation. Nature. 2001;414(6862):457–62.
57. Ebelt H, Hufnagel N, Neuhaus P, Neuhaus H, Gajawada P, Simm A, et al. Divergent siblings: E2F2 and E2F4 but not E2F1 and E2F3 induce DNA synthesis in cardiomyocytes without activation of apoptosis. Circ Res. 2005;96(5):509–17.
58. Rempel RE, Saenz-Robles MT, Storms R, Morham S, Ishida S, Engel A, et al. Loss of E2F4 activity leads to abnormal development of multiple cellular lineages. Mol Cell. 2000;6(2):293–306.
59. Yamasaki L, Jacks T, Bronson R, Goillot E, Harlow E, Dyson NJ. Tumor induction and tissue atrophy in mice lacking E2F–1. Cell. 1996;85(4):537–48.
60. Cloud JE, Rogers C, Reza TL, Ziebold U, Stone JR, Picard MH, et al. Mutant mouse models reveal the relative roles of E2F1 and E2F3 in vivo. Mol Cell Biol. 2002;22(8):2663–72.
61. Evan GI, Littlewood TD. The role of c-myc in cell growth. Curr Opin Genet Dev. 1993;3(1):44–9.
62. De Alboran IM, O'Hagan RC, Gärtner F, Malynn B, Davidson L, Rickert R, et al. Analysis of C-MYC function in normal cells via conditional gene-targeted mutation. Immunity. 2001;14(1):45–55.
63. Iritani BM, Eisenman RN. c-Myc enhances protein synthesis and cell size during B lymphocyte development. Proceedings of the National Academy of Sciences USA. 1999;96(23):13180–5.
64. Trumpp A, Refaeli Y, Oskarsson T, Gasser S, Murphy M, Martin GR, et al. c-Myc regulates mammalian body size by controlling cell number but not cell size. Nature. 2001;414(6865):768–73.
65. Davis AC, Wims M, Spotts GD, Hann SR, Bradley A. A null c-myc mutation causes lethality before 10.5 days of gestation in homozygotes and reduced fertility in heterozygous female mice. Genes & Development 1993; 7(4):671–82.
66. Jackson T, Allard M, Sreenan C, Doss L, Bishop S, Swain J, et al. The c-myc proto-oncogene regulates cardiac development in transgenic mice. Mol Cell Biol. 1990;10(7):3709–16.
67. Cramer T, Yamanishi Y, Clausen BE, Förster I, Pawlinski R, Mackman N, et al. HIF-1α is essential for myeloid cell-mediated inflammation. Cell. 2003;112(5):645–57.
68. Staller P, Sulitkova J, Lisztwan J, Moch H, Oakeley EJ, Krek WJN. Chemokine receptor CXCR4 downregulated by von Hippel–Lindau tumour suppressor pVHL. Nature. 2003; 425(6955):307–11.
69. Iyer NV, Kotch LE, Agani F, Leung SW, Laughner E, Wenger RH, et al. Cellular and developmental control of O2 homeostasis by hypoxia-inducible factor 1α. Genes Dev. 1998;12(2):149–62.

70. Koshiji M, Kageyama Y, Pete EA, Horikawa I, Barrett JC, Huang LEJTEj. HIF-1α induces cell cycle arrest by functionally counteracting Myc. The EMBO Journal 2004; 23(9):1949–56.

71. van Slegtenhorst M, de Hoogt R, Hermans C, Nellist M, Janssen B, Verhoef S, et al. Identification of the tuberous sclerosis gene TSC1 on chromosome 9q34. Science. 1997;277(5327):805–8.

72. Eurpoean Chromosome 16 Tuberous Sclerosis Consortium. Identification and characterization of the tuberous sclerosis gene on chromosome 16. Cell. 1993; 75(7):1305–15.

73. Watson GJ. Cardiac rhabdomyomas in tuberous sclerosis. Ann N Y Acad Sci. 1991;615:50–7.

74. Liang Q, Molkentin JD. Redefining the roles of p38 and JNK signaling in cardiac hypertrophy: dichotomy between cultured myocytes and animal models. J Mol Cell Cardiol. 2003;35(12):1385–94.

75. Wang Y, Huang S, Sah VP, Ross J, Brown JH, Han J, et al. Cardiac muscle cell hypertrophy and apoptosis induced by distinct members of the p38 mitogen-activated protein kinase family. J Biol Chem. 1998;273(4):2161–8.

76. Ambrosino C, Iwata T, Scafoglio C, Mallardo M, Klein R, Nebreda AR. TEF-1 and C/EBPβ are major p38α MAPK-regulated transcription factors in proliferating cardiomyocytes. Biochemical Journal. 2006;396(1):163–72.

77. Zarubin T, Jiahuai HJ. Activation and signaling of the p38 MAP kinase pathway. Cell Res. 2005;15(1):11–8.

78. Ambrosino C, Nebreda AR. Cell cycle regulation by p38 MAP kinases. Biol Cell. 2001;93(1–2):47–51.

79. Nebreda AR, Porras AJ. p38 MAP kinases: beyond the stress response. Trends Biochem Sci. 2000;25(6):257–60.

80. Wu Z, Woodring PJ, Bhakta KS, Tamura K, Wen F, Feramisco JR, et al. p38 and extracellular signal-regulated kinases regulate the myogenic program at multiple steps. Mol Cell Biol. 2000;20(11):3951–64.

81. Lavoie JN, L'Allemain G, Brunet A, Müller R, Pouysségur J. Cyclin D1 expression is regulated positively by the p42/p44MAPK and negatively by the p38/HOGMAPK pathway. J Biol Chem. 1996;271(34):20608–16.

82. Maher PJ. p38 mitogen-activated protein kinase activation is required for fibroblast growth factor-2-stimulated cell proliferation but not differentiation. J Biol Chem. 1999;274(25):17491–8.

83. Nagata Y, Takahashi N, Davis RJ, Todokoro KJB. Activation of p38 MAP kinase and JNK but not ERK is required for erythropoietin-induced erythroid differentiation. The Journal of the American Society of Hematology. 1998;92(6):1859–69.

84. Rausch O, Marshall CJJJoBC. Cooperation of p38 and extracellular signal-regulated kinase mitogen-activated protein kinase pathways during granulocyte colony-stimulating factor-induced hemopoietic cell proliferation. Journal of Biological Chemistry. 1999; 274(7):4096–105.

85. Claycomb WC, Bradshaw HD Jr. Acquisition of multiple nuclei and the activity of DNA polymerase α and reinitiation of DNA replication in terminally differentiated adult cardiac muscle cells in culture. Dev Biol. 1983;99(2):331–7.

86. Engel FB, Hsieh PC, Lee RT, Keating MTJPotNAoS. FGF1/p38 MAP kinase inhibitor therapy induces cardiomyocyte mitosis, reduces scarring, and rescues function after myocardial infarction. Proceedings of the National Academy of Sciences USA. 2006; 103(42):15546–51.

87. Engel FB, Schebesta M, Duong MT, Lu G, Ren S, Madwed JB, et al. p38 MAP kinase inhibition enables proliferation of adult mammalian cardiomyocytes. Genes Dev. 2005;19(10):1175–87.

88. Braz JC, Bueno OF, Liang Q, Wilkins BJ, Dai Y-S, Parsons S, et al. Targeted inhibition of p38 MAPK promotes hypertrophic cardiomyopathy through upregulation of calcineurin-NFAT signaling. J Clin Investig. 2003;111(10):1475–86.

89. LeRoith D, Helman LJ. The new kid on the block (ade) of the IGF-1 receptor. Cancer Cell. 2004;5(3):201–2.

90. Reiss K, Cheng W, Ferber A, Kajstura J, Li P, Li B, et al. Overexpression of insulin-like growth factor-1 in the heart is coupled with myocyte proliferation in transgenic mice. Proceedings of the National Academy of Sciences USA. 1996;93(16):8630–5.

91. Lau M, Stewart C, Liu Z, Bhatt H, Rotwein P, Stewart CLJG, et al. Loss of the imprinted IGF2/cation-independent mannose 6-phosphate receptor results in fetal overgrowth and perinatal lethality. Genes Dev. 1994;8(24):2953–63.

92. Molkentin JD, Dorn GW II. Cytoplasmic signaling pathways that regulate cardiac hypertrophy. Annu Rev Physiol. 2001;63(1):391–426.

93. Condorelli G, Drusco A, Stassi G, Bellacosa A, Roncarati R, Iaccarino G, et al. Akt induces enhanced myocardial contractility and cell size in vivo in transgenic mice. Proceedings of the National Academy of Sciences USA. 2002;99(19):12333–8.

94. Cook SA, Matsui T, Li L, Rosenzweig AJ. Transcriptional effects of chronic Akt activation in the heart. J Biol Chem. 2002;277(25):22528–33.

95. Rota M, Boni A, Urbanek K, Padin-Iruegas ME, Kajstura TJ, Fiore G, et al. Nuclear targeting of Akt enhances ventricular function and myocyte contractility. Circ Res. 2005;97(12):1332–41.

96. Shioi T, McMullen JR, Kang PM, Douglas PS, Obata T, Franke TF, et al. Akt/protein kinase B promotes organ growth in transgenic mice. Mol Cell Biol. 2002;22(8):2799–809.

97. Shioi T, Kang PM, Douglas PS, Hampe J, Yballe CM, Lawitts J, et al. The conserved phosphoinositide 3-kinase pathway determines heart size in mice. EMBO J. 2000;19(11):2537–48.

98. Mima T, Ueno H, Fischman D, Williams L, Mikawa TJ. Fibroblast growth factor receptor is required for in vivo cardiac myocyte proliferation at early embryonic stages of heart development. Proceedings of the National Academy of Sciences USA. 1995;92(2):467–71.

99. Pasumarthi KB, Kardami E, Cattini PA. High and low molecular weight fibroblast growth factor-2 increase proliferation of neonatal rat cardiac myocytes but have differential effects on binucleation and nuclear morphology: evidence for both paracrine and intracrine actions of fibroblast growth factor-2. Circ Res. 1996;78(1):126–36.

100. Beltrami AP, Barlucchi L, Torella D, Baker M, Limana F, Chimenti S, et al. Adult cardiac stem cells are multipotent and support myocardial regeneration. Cell. 2003;114(6):763–76.

101. Eswarakumar V, Lax I, Schlessinger J. Cellular signaling by fibroblast growth factor receptors. Cytokine Growth Factor Rev. 2005;16(2):139–49.

102. Kardami E, Liu L, Kishore S, Pasumarthi B, Doble BW, Cattini PA. Regulation of basic fibroblast growth factor (bFGF) and FGF receptors in the heart. Ann N Y Acad Sci. 1995;752:353–69.

103. Lavine KJ, Yu K, White AC, Zhang X, Smith C, Partanen J, et al. Endocardial and epicardial derived FGF signals regulate myocardial proliferation and differentiation in vivo. Dev Cell. 2005;8(1):85–95.

104. Kopf M, Baumann H, Freer G, Freudenberg M, Lamers M, Kishimoto T, et al. Impaired immune and acute-phase responses in interleukin-6-deficient mice. Nature. 1994;368(6469):339–42.

105. Escary J-L, Perreau J, Duménil D, Ezine S, Brûlet PJN. Leukaemia inhibitory factor is necessary for maintenance of haematopoietic stem cells and thymocyte stimulation. Nature. 1993;363(6427):361–4.

106. Stewart CL, Kaspar P, Brunet LJ, Bhatt H, Gadi I, Köntgen F, et al. Blastocyst implantation depends on maternal expression of leukaemia inhibitory factor. Nature. 1992;359(6390):76–9.

107. Masu Y, Wolf E, Holtmann B, Sendtner M, Brem G, Thoenen HJN. Disruption of the CNTF gene results in motor neuron degeneration. Nature. 1993;365(6441):27–32.

108. Robb L, Li R, Hartley L, Nandurkar HH, Koentgen F, Begley CGJ. Infertility in female mice lacking the receptor for interleukin 11 is due to a defective uterine response to implantation. Nat Med. 1998;4(3):303–8.

109. Yoshida K, Taga T, Saito M, Suematsu S, Kumanogoh A, Tanaka T, et al. Targeted disruption of gp130, a common signal transducer for the interleukin 6 family of cytokines, leads to myocardial and hematological disorders. Proceedings of the National Academy of Sciences USA. 1996;93(1):407–11.

110. Betz UA, Bloch W, Van Den Broek M, Yoshida K, Taga T, Kishimoto T, et al. Postnatally induced inactivation of gp130 in mice results in neurological, cardiac, hematopoietic, immunological, hepatic, and pulmonary defects. J Exp Med. 1998;188(10):1955–65.
111. Hirota H, Chen J, Betz UA, Rajewsky K, Gu Y, Ross J Jr, et al. Loss of a gp130 cardiac muscle cell survival pathway is a critical event in the onset of heart failure during biomechanical stress. Cell. 1999;97(2):189–98.
112. Hyams JS. Cytokinesis: the great divide. Trends Cell Biol. 2005;15(1):1.
113. Glotzer M. The molecular requirements for cytokinesis. Science. 2005;307(5716):1735–9.
114. Engel FB, Schebesta M, Keating MT. Anillin localization defect in cardiomyocyte binucleation. J Mol Cell Cardiol. 2006;41(4):601–12.
115. Ahuja P, Perriard E, Trimble W, Perriard J-C, Ehler EJ. Probing the role of septins in cardiomyocytes. Exp Cell Res. 2006;312(9):1598–609.

Chapter 4
Cardiac Fibroblast and Cardiomyocyte Growth

Abstract Fibroblasts in the heart (cardiac fibroblasts) aid to maintain structural, mechanical, and electrical functions of the organ. They are also integral to the response of the heart to stressors and injury. These cells interact with cardiomyocytes via paracrine mechanisms, alterations in extracellular matrix homeostasis, and direct cell—cell interactions. Thus, they promote embryonic cardiomyocyte proliferation and contribute to cardiomyocyte growth during development of the heart. How the bidirectional signalling is coordinated during development and promote proliferation of cardiomyocyte progenitors is only partially clear. The indications are that the mechanisms are important for cardiac repair and regeneration.

Keywords Cardiac fibroblasts · Cardiomyocytes · Heart development · Paracrine signalling · Extracellular matrix · Fibroblast growth factors

Introduction

Fibroblasts are cells of mesenchymal origin. They synthesize collagen types I, III, and VI and fibronectin, which are the main components of the extracellular matrix (ECM) [1]. ECM preserves the structural integrity of various organs in the body of vertebrates. Fibroblasts are important in wound healing and repair as well [2].

There are different phenotypes of fibroblasts among various tissues. Within the same tissue under different physiological conditions, fibroblast phenotypes may also vary [2, 3].

Fibroblasts in the heart (cardiac fibroblasts) aid to maintain structural, mechanical, and electrical functions of the organ [2, 4]. They are also integral to the response of the heart to stressors and injury. These cells interact with cardiomyocytes (CMs) via paracrine mechanisms, alterations in extracellular matrix homeostasis, and direct cell—cell interactions. Thus, they contribute to cardiomyocyte growth during development of the heart.

Origin of Cardiac Fibroblasts

Cardiac fibroblasts in the embryo are derived from three different pools of progenitor cells. The main source of cardiac fibroblasts in the embryo is the pro-epicardial organ, a transient structure located near the base of the developing heart [3, 5–8]. They also arise from the progenitor cell populations of the endocardium and a small population is derived from the neural crest [6, 9, 10]. Neither bone marrow-derived or hematopoietic cell-derived fibroblasts have been identified among the fibroblast population of the heart [6, 9]. During cardiac development, cardiac fibroblasts form a network of cell processes and collagen fibres to guide organization of CMs.

Pro-epicardial cells by E9.5 in the mouse, migrate and cover the entire embryonic heart to form the embryonic epicardium [8]. A subset of cells undergoes epithelial-to-mesenchymal transition (EMT) and migrates to the myocardium. These epicardial-derived cells (EPDCs) migrate deeper and differentiate into interstitial and adventitial cardiac fibroblasts [3, 8]. Interstitial cardiac fibroblasts are first detected on E12.5 in the mouse embryo [3, 11]. Transcription factor 21 (Tcf21) is essential for epicardial cells to differentiate into cardiac fibroblasts before EMT [12]. EMT and migration to the myocardium, require interactions with other growth and transcription factors, such as Ets factors, fibroblast growth factors (FGFs), platelet-derived growth factor-β, Sox9, Tbx5, thymosin β4, and transforming growth factors (TGFs) [3]. FGF10 and its receptor FGFR2β regulate the migration of EPDCs to the compact myocardium [8].

Cardiac Fibroblasts and Heart Development

During the first weeks after birth in the mouse, cardiac fibroblasts increase in number to promote ventricular thickness and tensile strength, required to compensate the increase in systolic pressure after birth [3]. Whether a similar mechanism is present in the human neonatal heart is unclear.

Cell–cell interactions between cardiac fibroblasts and embryonic cardiomyocytes are considered to promote proliferation of embryonic CMs and are thus critical for heart development. The lack of a single definition of a fibroblast, as well as the paucity of specific molecular markers and enhancers of cardiac fibroblast have however impaired investigations on interactions between cardiac fibroblasts and CMs early in development.

Fibroblast–Cardiomyocyte Paracrine Interactions

Cardiac fibroblasts modulate CMs via paracrine hormonal pathways (Figs. 4.1 and 4.2). CMs regulate fibroblast as well. Several factors have been implicated in the

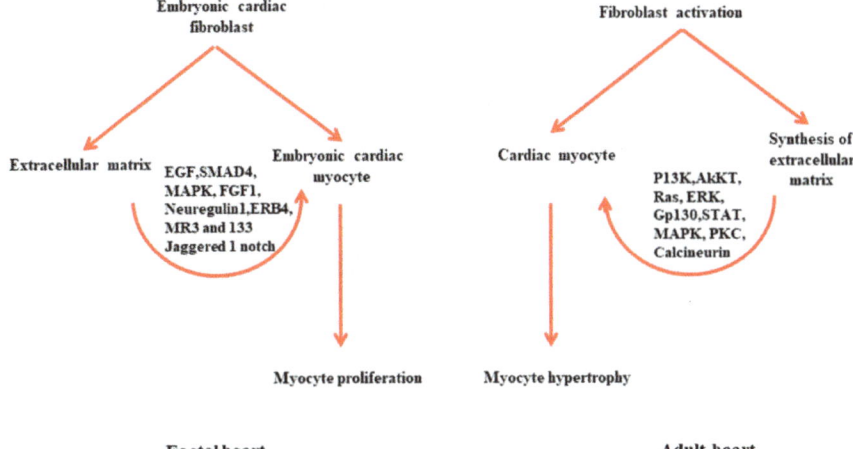

Fig. 4.1 Paracrine signalling mechanisms among fibroblasts, extracellular matrix and cardiomy-ocytes in the developing and adult heart

crosstalk between fibroblast and CMs (Figs. 4.1 and 4.2). The intercellular communication is mediated by proteins such as TGF1, FGF2, members of the IL-6 family of proteins, and the cytokine IL-33 released by the cells into the surrounding microenvironment.

Ieda et al. have demonstrated the effects of embryonic cardiac fibroblasts on developing CMs [11]. Embryonic cardiac fibroblasts express fibronectin, collagens, periostin, hyaluronan, and proteoglycan link protein 1, all of which are components of the ECM. Embryonic CMs proliferate when grown on plates enriched with fibronectin, collagen type III, periostin, or laminin. This effect is mediated by 1-integrin signalling and HB-EGF (heparin-binding epidermal growth factor–like growth factor) and induction of downstream extracellular signal-regulated kinase and p38 mitogen-activated protein kinase signalling.

Fibroblast Growth Factors

Fibroblast growth factors FGF2 and FGF4 induce expression of early cardiac transcription factors as well as ventricular (but not atrial) specific markers in the developing chick embryo [13]. The loss of FGF1 arrests differentiation of multipotent precursor cells into a cardiomyocyte lineage [14]. Other FGFs are possibly involved in Wnt/-catenin signalling and formation of the anterior heart field [15]. Specific members of the FGF family may influence regional development of the heart.

CMs and interstitial cells of both the foetal and adult heart express and release TGF [16–19]. TGF1 is induced and released from CMs in response to mechanical stretch (Fig. 4.2) [20]. TGF expression is upregulated during pressure overload and after

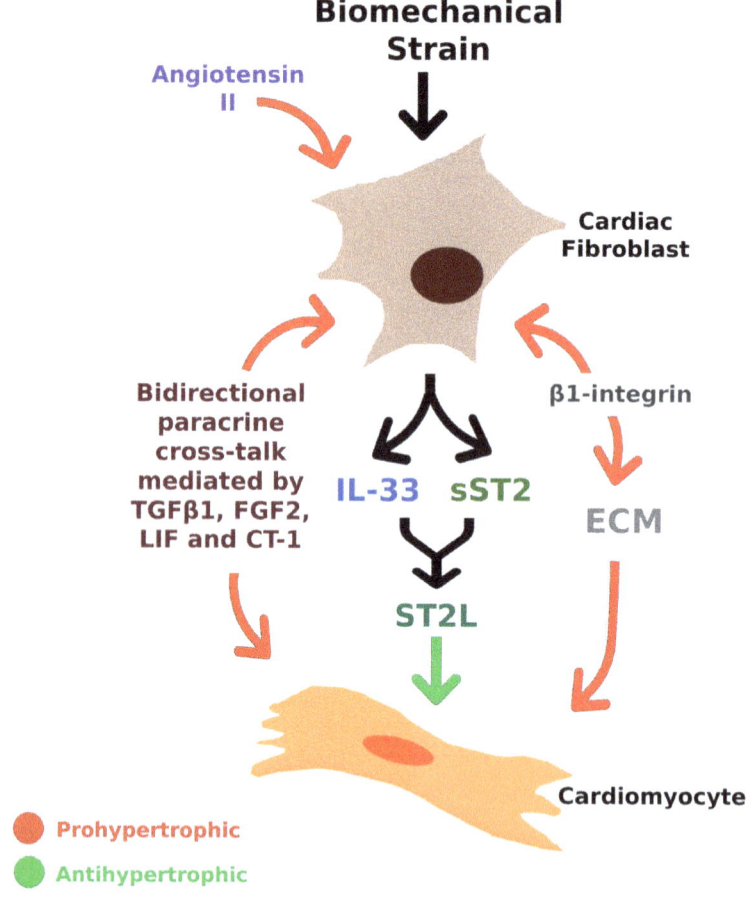

Fig. 4.2 Paracrine factors that mediate the effects of biomechanical strain and angiotensin on fibroblast-cardiomyocyte cross-talk

myocardial infarction [17, 21, 22]. Both ventricular myocytes and fibroblasts have receptors for TGF [23]. Angiotensin II induced myocyte hypertrophy is possibly mediated by TGF1 and may require cardiac fibroblasts [24]. The primary target of angiotensin is the cardiac fibroblast and the effect of angiotensin on CMs may occur via TGF1 secreted by the fibroblast (Fig. 4.2). Angiotensin II stimulation of fibroblasts induces TGF1 through the angiotensin type 1 receptor [25, 26]. It can also alter TGF1 receptor expression and increase matrix protein synthesis [27–30].

TGF has also been found to induce pluripotent cells towards a transcriptional and morphological phenotype of cardiomyocyte [31–34]. TGF upregulates cardiac transcription factors such as Nkx2.5 and MEF2C in mouse embryonic stem cells and enhances the formation of rhythmically contractile embryoid bodies [31]. These observations indicate that fibroblast–myocyte cross-talk may occur at the level of both terminally differentiated CMs and cardiomyogenic progenitor cells.

Fibroblast growth factor-2 (FGF2) is secreted primarily by cardiac fibroblasts and mediates a paracrine pro-hypertrophic response in CMs. Embryonic and adult cardiac fibroblasts mostly express the higher-molecular weight form of FGF2, which can induce a foetal gene program and promote hypertrophy of CMs [35, 36]. Secreted FGF2 binds to heparin sulfate proteoglycans and basement membrane [37]. Adrenergic stimulation, angiotensin II stimulation and ischemia or hypoxia can induce (FGF2) in CMs (Fig. 4.2) [38–41]. FGF2 acts on receptors on the surface of target CMs as well as on fibroblasts which release other pro-hypertrophic factors [42, 43].

Low-molecular-weight FGF2 may induce stem cell factor (the ligand for the c-kit receptor) and can also cause accumulation of c-kit cells at the site of ischemic injury [37]. FGF2 is possibly one of the several fibroblast-derived signals for the homing and differentiation of circulating progenitors and can also induce resident stem cells for myocyte differentiation [44]. FGF2 may also promote differentiation of embryonic stem cells into CMs [45].

Interleukins

Cardiac fibroblasts can secrete several of the peptides of the IL-6 family. Both CMs and fibroblasts synthesize leukemia inhibitory factor (LIF) and CT-1, members of the IL-6 family. Both are mediators of fibroblast– myocyte crosstalk (Fig. 4.2) [46]. LIF induces hypertrophy of CMs and proliferation of fibroblasts but inhibits myofibroblast transition and collagen deposition [47]. CT-1 also promotes hypertrophy of CMs and proliferation of fibroblasts [48, 49]. LIF and CT-1 derived from cardiac fibroblasts are mediators of the pro-hypertrophic effects of angiotensin II (Fig. 4.2) [50].

Gp130 is a transmembrane protein subunit that transfer signals for members of the interleukin (IL)-6 family, has a key role in cardiac development. The Gp130 knockout mouse do not develop compact myocardium and dies in mid to late gestation [51]. Deletion of several members of the IL-6 family does not however result in death of the embryo in utero. Cardiotrophin (CT)-1 and leukemia inhibitory factor (LIF), members of the IL-6 family that are known to be produced by cardiac fibroblasts do not seem to be required for heart development [51–53]. The above findings suggest a functional redundancy in this signalling pathway.

Interleukin-33 (IL-33 or IL-1F11) is produced primarily by cardiac fibroblasts and is involved in paracrine signalling between fibroblast and cardiomyocyte [54]. In the extracellular milieu, IL-33 binds to a transmembrane form of ST2 (ST2L) [55]. (55). IL-33 exerts a dose-dependent antihypertrophic effect in the presence of phenylephrine or angiotensin II and also decreases cardiomyocyte hypertrophy in pressure overloaded heart [54]. The IL-33/ST2 system signalling pathway constitutes an antihypertrophic and cardioprotective mechanism during mechanical overload (Fig. 4.2).

Extracellular Matrix–Based Fibroblast–Cardiomyocyte Interactions

Fibroblasts are the major source of non-basement membrane collagen and other proteins of the ECM [56–58]. Fibrillar collagen is the primary protein of the ECM. ECM of the heart (cardiac interstitium) contains mostly types I and type III collagen; type I is more abundant than type III [59]. ECM proteins, which signal through integrin receptors on the cell membrane are intermediaries in the cross-talk between cardiac fibroblasts and CMs. Fibroblast mediated regulation of ECM has been found to affect cardiomyocyte development and regeneration [60]. When collagen synthesis is inhibited, embryonic cardiomyocyte differentiation is disrupted in vitro.

Fibronectin produced by cardiac fibroblasts interacts with cardiomyocyte surface integrins to mediate cellular hypertrophy [61–63]. Fibronectin also promotes proliferation of embryonic CMs via 1-integrin signalling [8].

Both CMs and fibroblasts produce and secrete matrix metalloproteinases (MMP) and their tissue inhibitors (TIMPs), which regulate ECM homeostasis [64, 65]. Continuous interaction between CMs and fibroblasts in matrix synthesis and degradation including feedback mechanisms regulate the matrix composition in both physiological and pathological conditions.

Direct Fibroblast–Cardiomyocyte Communication and Electrical Coupling

Interestingly, there is also evidence for direct cell-to-cell communication between CMs and fibroblasts [66]. There seems to be chemi-electric communication between fibroblast and cardiomyocyte as among CMs. Cardiac fibroblasts express connexin 43 within gap junctions. Connexin 43 is also the major protein of gap junction-based connections among CMs [67–69]. Cardiac fibroblasts use connexin 43 for communication with CMs, in contrast to connexin 40 for communication with fibroblasts [70–72]. Electric conductance between fibroblasts and myocytes has features of a hemichannel which suggests a mixed pool of gap junctions [73].

There are also evidences to suggest direct cell-to-cell connections between fibroblasts and CMs, which permits transfer of molecular and ionic signals between these cells in response to environmental changes. Flow of membrane-impermeant dyes and calcium fluxes have been found between myocytes and adjacent fibroblasts [74]. Cardiac fibroblasts in culture acquire the rhythmic depolarization of nearby CMs. They also seem to alter the depolarization properties of the cardiomyocyte, mediate cardiomyocyte electric synchrony and also contribute to cardiomyocyte automaticity [71, 75–81].

Fibroblast and Cardiomyocyte Dedifferentiation

Myocardial hibernation, characterized by sarcomere depletion and loss of structure of the sarcoplasmic reticulum and T-tubules, glycogen accumulation, nuclear heterochromatin redistribution, mitochondrial redistribution and switch in mitochondrial metabolism is seen in conditions of myocardial ischemia, pressure overload, and atrial fibrillation [82, 83]. The changes of hibernation mimic an embryonic cellular phenotype and hence considered a form of cardiomyocyte 'dedifferentiation'. A dedifferentiated phenotype can be induced in CMs when they are cocultured with cardiac fibroblasts. Cardiomyocyte dedifferentiation does not occur in monoculture with fibroblast conditioned medium [84, 85]. Activation of the transcription factor GATA4 is considered to be involved in this effect of fibroblasts on CMs [86].

Reprogramming of Fibroblasts into Cardiomyocytes

Fibroblasts have been successfully reprogrammed or transdifferentiated directly into 'functional' CMs without passing through a pluripotent state [87–91, Chap. 19 in this book].

Cardiac Fibroblast–Myocyte Interactions in Tissue Engineering

There have been numerous attempts to engineer ex vivo, functional myocardial tissue for possible engraftment on to injured heart. Recent studies indicate that when cardiac fibroblasts are included in 3D cultures of myocytes and endothelial cells, growth, vascularity, stability and function of myocardial structures are enhanced [92–96].

Role of Cardiac Fibroblasts in Regeneration of Cardiomyocytes in the Adult Heart

Cardiac fibroblasts are present in the adult mammalian heart in almost the same numbers as CMs. They are normally in a quiescent state [1, 97, 98]. They can proliferate in response to pathological stimuli [1, 6, 98–100]. The role of cardiac fibroblasts in cardiomyocyte regeneration in the adult heart is however unclear. A generally held view is that fibroblast activity impairs regeneration [15]. Interestingly, the effects of cardiac fibroblasts derived paracrine factors on embryonic and adult CMs differ (Fig. 4.1). When adult cardiac fibroblasts are cocultured with embryonic mouse CMs, the result is myocyte hypertrophy rather than proliferation.

Conclusions

The essential role of cardiac fibroblasts in cardiac development, cardiac homeostasis, and several disease states is presently well recognized. It is however unclear whether they have functional variability and differing responses during mechanical stress and injury to the heart, and during cardiac repair and regeneration. There is considerable progress in our understanding of the origins of cardiac fibroblasts, the dynamic nature of these cells, and their responses during time of mechanical stress or injury to the heart. Several paracrine factors involved in the cross-talk between fibroblasts and cardiomyocytes have been delineated. How the bidirectional signalling coordinate during development and promote proliferation of cardiomyocyte progenitors is only partially clear. The indications are that the mechanisms are important for cardiac repair and regeneration.

References

1. Camelliti P, Borg TK, Kohl P. Structural and functional characterisation of cardiac fibroblasts. Cardiovasc Res. 2005;65(1):40–51.
2. Souders CA, Bowers SL, Baudino TA. Cardiac fibroblast: the renaissance cell. Circ Res. 2009;105(12):1164–76.
3. Lajiness JD, Conway SJ. Origin, development, and differentiation of cardiac fibroblasts. J Mol Cell Cardiol. 2014;70:2–8.
4. Zeisberg EM, Kalluri R. Origins of cardiac fibroblasts. Circ Res. 2010;107(11):1304–12.
5. Harvey RP. Patterning the vertebrate heart. Nat Rev Genet. 2002;3(7):544–56.
6. Moore-Morris T, Guimarães-Camboa N, Banerjee I, Zambon AC, Kisseleva T, Velayoudon A, et al. Resident fibroblast lineages mediate pressure overload–induced cardiac fibrosis. J Clin Investig. 2014;124(7):2921–34.
7. Deb A, Ubil E. Cardiac fibroblast in development and wound healing. J Mol Cell Cardiol. 2014;70:47–55.
8. Vega-Hernández M, Kovacs A, De Langhe S, Ornitz DM. FGF10/FGFR2b signaling is essential for cardiac fibroblast development and growth of the myocardium. Development. 2011;138(15):3331–40.
9. Moore-Morris T, Guimarães-Camboa N, Yutzey KE, Pucéat M, Evans SM. Cardiac fibroblasts: from development to heart failure. J Mol Med (Berl). 2015;93(8):823–30.
10. Lighthouse JK, Small EM. Transcriptional control of cardiac fibroblast plasticity. J Mol Cell Cardiol. 2016;91:52–60.
11. Ieda M, Tsuchihashi T, Ivey KN, Ross RS, Hong T-T, Shaw RM, et al. Cardiac fibroblasts regulate myocardial proliferation through β1 integrin signaling. Dev Cell. 2009;16(2):233–44.
12. Acharya A, Baek ST, Huang G, Eskiocak B, Goetsch S, Sung CY, et al. The bHLH transcription factor Tcf21 is required for lineage-specific EMT of cardiac fibroblast progenitors. Development. 2012;139(12):2139–49.
13. Lopez-Sanchez C, Climent V, Schoenwolf GC, Alvarez IS, Garcia-Martinez V. Induction of cardiogenesis by Hensen's node and fibroblast growth factors. Cell Tissue Res. 2002;309(2):237–49.
14. Dell'Era P, Ronca R, Coco L, Nicoli S, Metra M, Presta M. Fibroblast growth factor receptor-1 is essential for in vitro cardiomyocyte development. Circ Res. 2003;93(5):414–20.
15. Cohen ED, Wang Z, Lepore JJ, Lu MM, Taketo MM, Epstein DJ, et al. Wnt/β-catenin signaling promotes expansion of Isl-1–positive cardiac progenitor cells through regulation of FGF signaling. J Clin Investig. 2007;117(7):1794–804.

16. Thompson NL, Flanders KC, Smith JM, Ellingsworth LR, Roberts AB, Sporn MB. Expression of transforming growth factor-beta 1 in specific cells and tissues of adult and neonatal mice. J Cell Biol. 1989;108(2):661–9.
17. Eghbali MJC, research t. Cellular origin and distribution of transforming growth factor-β 1 in the normal rat myocardium. Cell Tissue Res. 1989; 256(3):553–558.
18. Lee AA, Delhaas T, McCulloch AD, Villarreal FJ. Differential responses of adult cardiac fibroblasts to in vitro biaxial strain patterns. J Mol Cell Cardiol. 1999;31(10):1833–43.
19. Takahashi N, Calderone A, Izzo NJ, Mäki T, Marsh JD, Colucci WS. Hypertrophic stimuli induce transforming growth factor-beta 1 expression in rat ventricular myocytes. J Clin Investig. 1994;94(4):1470–6.
20. van Wamel AJ, Ruwhof C, van der Valk-Kokshoorn LJ, Schrier PI, Van Der Laarse A. Stretch-induced paracrine hypertrophic stimuli increase TGF-β 1 expression in cardiomyocytes. Mol Cell Biochem. 2002;236(1):147–53.
21. Bujak M, Frangogiannis NG. The role of TGF-β signaling in myocardial infarction and cardiac remodeling. Cardiovasc Res. 2007;74(2):184–95.
22. Thompson NL, Bazoberry F, Speir EH, Casscells W, Ferrans VJ, Flanders KC, et al. Transforming growth factor beta-1 in acute myocardial infarction in rats. Growth Factors. 1988;1(1):91–9.
23. Engelmann G, Grutkoski P. Coordinate TGF-beta receptor gene expression during rat heart development. Cell Mol Biol Res. 1994;40(2):93–104.
24. Gray MO, Long CS, Kalinyak JE, Li H-T, Karliner JS. Angiotensin II stimulates cardiac myocyte hypertrophy via paracrine release of TGF-β1 and endothelin-1 from fibroblasts. Cardiovasc Res. 1998;40(2):352–63.
25. Schultz JEJ, Witt SA, Glascock BJ, Nieman ML, Reiser PJ, Nix SL, et al. TGF-β1 mediates the hypertrophic cardiomyocyte growth induced by angiotensin II. J Clin Investig. 2002;109(6):787–96.
26. Tsybouleva N, Zhang L, Chen S, Patel R, Lutucuta S, Nemoto S, et al. Aldosterone, through novel signaling proteins, is a fundamental molecular bridge between the genetic defect and the cardiac phenotype of hypertrophic cardiomyopathy. Circulation. 2004;109(10):1284–91.
27. Campbell SE, Katwa LC. Angiotensin II stimulated expression of transforming growth factor-β1 in cardiac fibroblasts and myofibroblasts. J Mol Cell Cardiol. 1997;29(7):1947–58.
28. Lee AA, Dillmann WH, McCulloch AD, Villarreal FJ. Angiotensin II stimulates the autocrine production of transforming growth factor-β1 in adult rat cardiac fibroblasts. J Mol Cell Cardiol. 1995;27(10):2347–57.
29. Chen K, Mehta JL, Li D, Joseph L, Joseph J. Transforming growth factor β receptor endoglin is expressed in cardiac fibroblasts and modulates profibrogenic actions of angiotensin II. Circ Res. 2004;95(12):1167–73.
30. Sadoshima J-I, Izumo SJCR. Molecular characterization of angiotensin II–induced hypertrophy of cardiac myocytes and hyperplasia of cardiac fibroblasts. Critical role of the AT1 receptor subtype. Circ Res. 1993; 73(3):413–423.
31. Behfar A, Zingman LV, Hodgson DM, Rauzier JM, Kane GC, Terzic A, et al. Stem cell differentiation requires a paracrine pathway in the heart. FASEB J. 2002;16(12):1558–66.
32. Li T-S, Hayashi M, Ito H, Furutani A, Murata T, Matsuzaki M, et al. Regeneration of infarcted myocardium by intramyocardial implantation of ex vivo transforming growth factor-β–preprogrammed bone marrow stem cells. Circulation. 2005;111(19):2438–45.
33. Gwak SJ, Bhang SH, Yang HS, Kim SS, Lee DH, Lee SH, et al. In vitro cardiomyogenic differentiation of adipose-derived stromal cells using transforming growth factor-β1. Cell Biochem Funct. 2009;27(3):148–54.
34. Lim J-Y, Kim WH, Kim J, Park SI. Involvement of TGF-β1 Signaling in Cardiomyocyte Differentiation from P19CL6 Cells. Mol Cells. 2007; 24(3).
35. Schneider MD, McLellan WR, Black FM, Parker TG, editors. Growth factors, growth factor response elements, and the cardiac phenotype. Basic Res Cardiol. 1992; 87(suppl 2):33–48.
36. Jiang Z-S, Jeyaraman M, Wen G-B, Fandrich RR, Dixon IM, Cattini PA, et al. High-but not low-molecular weight FGF-2 causes cardiac hypertrophy in vivo; possible involvement of cardiotrophin-1. J Mol Cell Cardiol. 2007;42(1):222–33.

37. Kardami E, Detillieux K, Ma X, Jiang Z, Santiago J-J, Jimenez SK, et al. Fibroblast growth factor-2 and cardioprotection. Heart Fail Rev. 2007;12(3):267–77.
38. Jimenez SK, Sheikh F, Jin Y, Detillieux KA, Dhaliwal J, Kardami E, et al. Transcriptional regulation of FGF-2 gene expression in cardiac myocytes. Cardiovasc Res. 2004;62(3):548–57.
39. Detillieux KA, Meij JT, Kardami E, Cattini PA. α1-Adrenergic stimulation of FGF-2 promoter in cardiac myocytes and in adult transgenic mouse hearts. Am J Physiol. 1999;276(3):H826–33.
40. Peifley KA, Winkles JA. Angiotensin II and endothelin-1 increase fibroblast growth factor-2 mRNA expression in vascular smooth muscle cells. Biochem Biophys Res Commun. 1998;242(1):202–8.
41. Endoh M, Pulsinelli WA, Wagner JA. Transient global ischemia induces dynamic changes in the expression of bFGF and the FGF receptor. Mol Brain Res. 1994;22(1–4):76–88.
42. Schultz JEJ, Witt SA, Nieman ML, Reiser PJ, Engle SJ, Zhou M, et al. Fibroblast growth factor-2 mediates pressure-induced hypertrophic response. J Clin Investig. 1999;104(6):709–19.
43. Pellieux C, Foletti A, Peduto G, Aubert J-F, Nussberger J, Beermann F, et al. Dilated cardiomyopathy and impaired cardiac hypertrophic response to angiotensin II in mice lacking FGF-2. J Clin Investig. 2001;108(12):1843–51.
44. Rosenblatt-Velin N, Lepore MG, Cartoni C, Beermann F, Pedrazzini T. FGF-2 controls the differentiation of resident cardiac precursors into functional cardiomyocytes. J Clin Investig. 2005;115(7):1724–33.
45. Kawai T, Takahashi T, Esaki M, Ushikoshi H, Nagano S, Fujiwara H, et al. Efficient cardiomyogenic differentiation of embryonic stem cell by fibroblast growth factor 2 and bone morphogenetic protein 2. Circ J. 2004;68(7):691–702.
46. Matsui H, Fujio Y, Kunisada K, Hirota H, Yamauchi-Takihara K. Leukemia inhibitory factor induces a hypertrophic response mediated by gp130 in murine cardiac myocytes. Res Commun Mol Pathol Pharmacol. 1996;93(2):149–62.
47. Wang F, Trial J, Diwan A, Gao F, Birdsall H, Entman M, et al. Regulation of cardiac fibroblast cellular function by leukemia inhibitory factor. J Mol Cell Cardiol. 2002;34(10):1309–16.
48. Wollert KC, Taga T, Saito M, Narazaki M, Kishimoto T, Glembotski CC, et al. Cardiotrophin-1 activates a distinct form of cardiac muscle cell hypertrophy: assembly of sarcomeric units in series via gp130/leukemia inhibitory factor receptor-dependent pathways. J Biol Chem. 1996;271(16):9535–45.
49. Tsuruda T, Jougasaki M, Boerrigter G, Huntley BK, Chen HH, D'Assoro AB, et al. Cardiotrophin-1 stimulation of cardiac fibroblast growth: roles for glycoprotein 130/leukemia inhibitory factor receptor and the endothelin type A receptor. Circ Res. 2002;90(2):128–34.
50. Sano M, Fukuda K, Kodama H, Pan J, Saito M, Matsuzaki J, et al. Interleukin-6 family of cytokines mediate angiotensin II-induced cardiac hypertrophy in rodent cardiomyocytes. J Biol Chem. 2000;275(38):29717–23.
51. Wollert KC, Chien KR. Cardiotrophin-1 and the role of gp130-dependent signaling pathways in cardiac growth and development. J Mol Med. 1997;75(7):492–501.
52. Holtmann B, Wiese S, Samsam M, Grohmann K, Pennica D, Martini R, et al. Triple knock-out of CNTF, LIF, and CT-1 defines cooperative and distinct roles of these neurotrophic factors for motoneuron maintenance and function. J Neurosci. 2005;25(7):1778–87.
53. Oppenheim RW, Wiese S, Prevette D, Armanini M, Wang S, Houenou LJ, et al. Cardiotrophin-1, a muscle-derived cytokine, is required for the survival of subpopulations of developing motoneurons. J Neurosci. 2001;21(4):1283–91.
54. Sanada S, Hakuno D, Higgins LJ, Schreiter ER, McKenzie AN, Lee RT. IL-33 and ST2 comprise a critical biomechanically induced and cardioprotective signaling system. J Clin Investig. 2007;117(6):1538–49.
55. Kakkar R, Lee RT. The IL-33/ST2 pathway: therapeutic target and novel biomarker. Nat Rev Drug Discov. 2008;7(10):827–40.
56. Bashey RI, Donnelly M, Insinga F, Jimenez SA. Growth properties and biochemical characterization of collagens synthesized by adult rat heart fibroblasts in culture. J Mol Cell Cardiol. 1992;24(7):691–700.

57. Eghbali M, Blumenfeld O, Seifter S, Buttrick P, Leinwand L, Robinson T, et al. Localization of types I, III and IV collagen mRNAs in rat heart cells by in situ hybridization. J Mol Cell Cardiol. 1989;21(1):103–13.

58. Eghbali M, Czaja MJ, Zeydel M, Weiner FR, Zern MA, Seifter S, et al. Collagen chain mRNAs in isolated heart cells from young and adult rats. J Mol Cell Cardiol. 1988;20(3):267–76.

59. MacKenna D, Summerour SR, Villarreal F. Role of mechanical factors in modulating cardiac fibroblast function and extracellular matrix synthesis. Cardiovasc Res. 2000;46(2):257–63.

60. Fisher SA, Periasamy M. Collagen synthesis inhibitors disrupt embryonic cardiocyte myofibrillogenesis and alter the expression of cardiac specific genes in vitro. J Mol Cell Cardiol. 1994;26(6):721–31.

61. Laser M, Willey CD, Jiang W, Cooper G 4th, Menick DR, Zile MR, et al. Integrin activation and focal complex formation in cardiac hypertrophy. J Biol Chem. 2000;275(45):35624–30.

62. Hynes RO. Fibronectins. Sci Am. 1986;254:42–51.

63. Farhadian F, Contard F, Corbier A, Barrieux A, Rappaport L, Samuel JL, et al. Fibronectin expression during physiological and pathological cardiac growth. J Mol Cell Cardiol. 1995;27(4):981–90.

64. Manso AM, Elsherif L, Kang S-M, Ross RS. Integrins, membrane-type matrix metalloproteinases and ADAMs: potential implications for cardiac remodeling. Cardiovasc Res. 2006;69(3):574–84.

65. Li YY, McTiernan CF, Feldman AM. Interplay of matrix metalloproteinases, tissue inhibitors of metalloproteinases and their regulators in cardiac matrix remodeling. Cardiovasc Res. 2000;46(2):214–24.

66. Doppler SA, Carvalho C, Lahm H, Deutsch M-A, Dreßen M, Puluca N, et al. Cardiac fibroblasts: more than mechanical support. J Thorac Dis. 2017;9(Suppl 1):S36–51.

67. Beyer EC, Kistler J, Paul DL, Goodenough DA. Anti sera directed against connexin43 peptides react with a 43-kD protein localized to gap junctions in myocardium and other tissues. J Cell Biol. 1989;108(2):595–605.

68. Beyer EC, Paul DL, Goodenough DA. Connexin43: a protein from rat heart homologous to a gap junction protein from liver. J Cell Biol. 1987;105(6):2621–9.

69. Luke R, Beyer E, Hoyt R, Saffitz J. Quantitative analysis of intercellular connections by immunohistochemistry of the cardiac gap junction protein connexin43. Circ Res. 1989;65(5):1450–7.

70. Zhang Y, Kanter EM, Laing JG, Aprhys C, Johns DC, Kardami E, et al. Connexin43 expression levels influence intercellular coupling and cell proliferation of native murine cardiac fibroblasts. Cell Commun Adhes. 2008;15(3):289–303.

71. Camelliti P, Green CR, LeGrice I, Kohl P. Fibroblast network in rabbit sinoatrial node: structural and functional identification of homogeneous and heterogeneous cell coupling. Circ Res. 2004;94(6):828–35.

72. Oyamada M, Kimura H, Oyamada Y, Miyamoto A, Ohshika H, Mori M. The expression, phosphorylation, and localization of connexin 43 and gap-junctional intercellular communication during the establishment of a synchronized contraction of cultured neonatal rat cardiac myocytes. Exp Cell Res. 1994;212(2):351–8.

73. Rook M, Jongsma H, De Jonge B. Single channel currents of homo-and heterologous gap junctions between cardiac fibroblasts and myocytes. Pflugers Arch. 1989;414(1):95–8.

74. Chilton L, Giles WR, Smith GL. Evidence of intercellular coupling between co-cultured adult rabbit ventricular myocytes and myofibroblasts. J Physiol. 2007;583(1):225–36.

75. Mark GE, Strasser F. Pacemaker activity and mitosis in cultures of newborn rat heart ventricle cells. Exp Cell Res. 1966;44(2–3):217–33.

76. Goshima K. Synchronized beating of and electrotonic transmission between myocardial cells mediated by heterotypic strain cells in monolayer culture. Exp Cell Res. 1969;58(2–3):420–6.

77. Goshima K. Formation of nexuses and electrotonic transmission between myocardial and FL cells in monolayer culture. Exp Cell Res. 1970;63(1):124–30.

78. Miragoli M, Salvarani N, Rohr S. Myofibroblasts induce ectopic activity in cardiac tissue. Circ Res. 2007;101(8):755–8.

79. Kohl P, Kamkin A, Kiseleva I, Noble D. Mechanosensitive fibroblasts in the sino-atrial node region of rat heart: interaction with cardiomyocytes and possible role. Exp Physiol Trans Integr. 1994;79(6):943–56.
80. Shiraishi I, Takamatsu T, Minamikawa T, Onouchi Z, Fujita SJC. Quantitative histological analysis of the human sinoatrial node during growth and aging. Circulation. 1992;85(6):2176–84.
81. De Maziere A, Van Ginneken A, Wilders R, Jongsma H, Bouman L. Spatial and functional relationship between myocytes and fibroblasts in the rabbit sinoatrial node. J Mol Cell Cardiol. 1992;24(6):567–78.
82. Vanoverschelde J-LJ, Wijns W, Borgers M, Heyndrickx G, Depré C, Flameng W, et al. Chronic myocardial hibernation in humans: from bedside to bench. Circulation. 1997; 95(7):1961–1971.
83. Ausma J, Wijffels M, Van Eys G, Koide M, Ramaekers F, Allessie M, et al. Dedifferentiation of atrial cardiomyocytes as a result of chronic atrial fibrillation. Am J Pathol. 1997;151(4):985–97.
84. Driesen RB, Verheyen FK, Dispersyn GD, Thoné F, Lenders M-H, Ramaekers FC, et al. Structural adaptation in adult rabbit ventricular myocytes. Cell Biochem Biophys. 2006;44(1):119–28.
85. Dispersyn GD, Geuens E, Ver Donck L, Ramaekers FC, Borgers M. Adult rabbit cardiomyocytes undergo hibernation-like dedifferentiation when co-cultured with cardiac fibroblasts. Cardiovasc Res. 2001;51:230–40.
86. Zaglia T, Dedja A, Candiotto C, Cozzi E, Schiaffino S, Ausoni S, et al. Cardiac interstitial cells express GATA4 and control dedifferentiation and cell cycle re-entry of adult cardiomyocytes. J Mol Cell Cardiol. 2009;46(5):653–62.
87. Song K, Nam Y-J, Luo X, Qi X, Tan W, Huang GN, et al. Heart repair by reprogramming non-myocytes with cardiac transcription factors. Nature. 2012;485(7400):599–604.
88. Qian L, Huang Y, Spencer CI, Foley A, Vedantham V, Liu L, et al. In vivo reprogramming of murine cardiac fibroblasts into induced cardiomyocytes. Nature. 2012;485(7400):593–8.
89. Jayawardena TM, Egemnazarov B, Finch EA, Zhang L, Payne JA, Pandya K, et al. MicroRNA-mediated in vitro and in vivo direct reprogramming of cardiac fibroblasts to cardiomyocytes. Cardiovasc Res. 2012;110(11):1465–73.
90. Ieda M, Fu J-D, Delgado-Olguin P, Vedantham V, Hayashi Y, Bruneau BG, et al. Direct reprogramming of fibroblasts into functional cardiomyocytes by defined factors. Cell. 2010;142(3):375–86.
91. Lalit PA, Salick MR, Nelson DO, Squirrell JM, Shafer CM, Patel NG, et al. Lineage reprogramming of fibroblasts into proliferative induced cardiac progenitor cells by defined factors. Cell Stem Cell. 2016;18(3):354–67.
92. Levenberg S, Rouwkema J, Macdonald M, Garfein ES, Kohane DS, Darland DC, et al. Engineering vascularized skeletal muscle tissue. Nat Biotechnol. 2005;23(7):879–84.
93. Caspi O, Lesman A, Basevitch Y, Gepstein A, Arbel G, Habib IHM, et al. Tissue engineering of vascularized cardiac muscle from human embryonic stem cells. Circ Res. 2007;100(2):263–72.
94. Naito H, Melnychenko I, Didié M, Schneiderbanger K, Schubert P, Rosenkranz S, et al. Optimizing engineered heart tissue for therapeutic applications as surrogate heart muscle. Circulation. 2006; 114 (1_supplement):I-72-I-8.
95. Nichol JW, Engelmayr GC Jr, Cheng M, Freed LE. Co-culture induces alignment in engineered cardiac constructs via MMP-2 expression. Biochem Biophys Res Commun. 2008;373(3):360–5.
96. Radisic M, Park H, Martens TP, Salazar-Lazaro JE, Geng W, Wang Y, et al. Pre-treatment of synthetic elastomeric scaffolds by cardiac fibroblasts improves engineered heart tissue. J Biomed Mater Res Part A. 2008;86(3):713–24.
97. Pinto AR, Ilinykh A, Ivey MJ, Kuwabara JT, D'antoni ML, Debuque R, et al. Revisiting cardiac cellular composition. Circ Res. 2016; 118(3):400–409.

98. Banerjee I, Fuseler JW, Price RL, Borg TK, Baudino TA. Determination of cell types and numbers during cardiac development in the neonatal and adult rat and mouse. Am J Physiol Heart Circ Physiol. 2007;293(3):H1883–91.
99. Travers JG, Kamal FA, Robbins J, Yutzey KE, Blaxall BC. Cardiac fibrosis: the fibroblast awakens. Circ Res. 2016;118(6):1021–40.
100. Ali SR, Ranjbarvaziri S, Talkhabi M, Zhao P, Subat A, Hojjat A, et al. Developmental heterogeneity of cardiac fibroblasts does not predict pathological proliferation and activation. Circ Res. 2014;115(7):625–35.

Chapter 5
Role of Endocardium and Epicardium in Generation of Cardiomyocytes

Abstract Endocardial cells form the innermost layer of the heart. Epicardium is the inner layer of the pericardium, the sac of fibrous tissue that encloses the heart. A sub population of cardiomyocytes are possibly derived from endocardial cells, which are structurally and functionally distinct endothelial cells. Earlier studies had found that a subset of Flk1$^+$ cells which are progenitors of endocardial cells are also progenitors for cardiomyocytes. The observation of ectopic islands of beating cardiogenic cells emerging from hemogenic endocardial endothelium in stem cell leukemia null ($Scl^{-/-}$) embryos, indicates the capacity of endocardial cells to transdifferentiate into cardiomyocytes if critical genes are manipulated. Congenital heart defects can result from either abnormal development of the endocardium or inappropriate communication between endocardium and myocardium. Recent evidence suggests that endothelial cells are also important players in post-infarction remodeling and cardiac regeneration. The sub epicardial region has several cell types of stem cells, which could also be a potential source of cardiac progenitors for cardiac regenerative therapy.

Keyword Endocardium. · Endothelial cell · Epicardium · Cardiomyocyte · Transdifferentiation · Stem cell · Cardiac progenitors · Cardiac remodelling

Introduction

Endocardial cells which form the innermost layer of the heart are structurally and functionally distinct endothelial cells [1]. Endocardium not only defines cardiac chambers but has pluripotent potential in the spatiotemporal context of the development of the heart. Endocardium generates diverse cell types during development of the heart [2]. Mesenchymal cells of the cardiac endothelial cushion expand, migrate along the ventricular walls, and contribute to pericytes, smooth muscle cells, and fibroblasts as well as intramyocardial adipocytes and hematopoietic cells; a sub population of cardiomyocytes (CMs) are also derived from endocardial cells [3].

Endocardium

Endocardial cells originate from the ventral surface of the early cardiac mesoderm. They are located between anterior visceral endoderm and myocardial cells [4]. The single layer of pluripotent cells arises from a population of Flk1$^+$ progenitor cells of the late primitive streak [4–6]. Cross talk between endocardial cells and the outer layer of CMs mediates both maturation of the heart chambers as well as formation of cardiac valves [7, 8]. Congenital heart defects can result from either abnormal development of the endocardium or inappropriate communication between endocardium and myocardium [9–14].

Endocardial Contribution to Cardiomyocytes

Cardiomyocytes are not derived from endothelial cells during development of the normal heart. Regulatory mechanisms intrinsic to the endocardial cell along with those of the surrounding microenvironment regulate the developmental fate of myocardial cells towards a cardiomyocyte lineage [15].

Earlier studies had found that a subset of Flk1$^+$ cells which are progenitors of endocardial cells are also progenitors for CMs [4, 16]. Flk1$^+$ progenitors differentiate into endothelium and CMs at early embryonic stages [16]. Whether there are intermediates of Flk1$^+$ endothelial and myocardial cell lineages and whether transcription factors have a role in determining the myogenic lineage of Flk1$^+$ progenitor cells, are unclear [3].

Ectopic islands of beating cardiogenic cells have been seen to emerge from hemogenic endocardial endothelium in the stem cell leukemia null ($Scl^{-/-}$ embryos) [17]. (Scl) is a candidate bHLH (basic helix–loop–helix) transcription factor which determines the differentiation of hemogenic endothelial cells from the mesoderm. Expression of Isl1 (ISL LIM homeobox 1), Gata6 (GATA-binding protein 6), Tbx5 (T-box 5), Nkx2-5 (NK2 homeobox 5), and Tnnt2 (troponin T2) are induced in endothelial cells in the absence of Scl [17]. Thus, endocardium may transdifferentiate into cardiomyocytes if critical genes are manipulated [3].

A recent study identified that endothelial derived CMs first appear two weeks after birth. This finding raises the possibility that endothelial cells may contribute to the generation of adult ventricular myocytes during cardiac homeostasis [18].

Sub Epicardial Cells and Cardiomyocytes

Epicardium is the inner layer of the pericardium, the sac of fibrous tissue that encloses the heart. The sub epicardial region has several cell types other than endothelial cells and fibroblasts. They include cells with mesenchymal stem cell (MSC) characteristics or those expressing other markers of potential progenitor cells [19–21]. Adult stem cell niches have also been reported in interstitial spaces around intramyocardial vessels in the atria and apical regions of the heart [22]. Whether MSCs have self-renewal properties and multipotency in vivo is unsettled [23]. MSC-like stem cells resident in proepicardial regions of the adult heart have been described as colony-forming units-fibroblasts (CFU-Fs) and can give rise to many cell fates but predominantly cardiac fibroblasts [24].

A subset of multipotent proepicardial progenitors which express Tbx18 have been found to differentiate into CMs [25]. Epicardium-derived progenitors of an injured mouse heart can be induced to re-express the embryonic epicardial marker Wt1 and differentiate into cardiomyocytes, using thymosin β4 peptide [26].

Sub epicardial cells have been suggested to provide paracrine benefits for CMs after injury to the heart [27].

Paracrine Signalling Between Endothelial Cells and Cardiomyocytes

Several endothelium derived factors which act on CMs have been identified (Fig. 5.1). Nitric oxide can regulate contractile responses of CMs [28]. Neuregulin-1 (NRG-1), which belongs to the epidermal growth factor family binds and activates ErbB receptors in CMs and thus promotes CM proliferation and growth [29, 30]. Apelin, highly EC-specific secreted peptide, regulates cardiomyocyte hypertrophy through its receptor APJ [31–33]. CMs express receptors for endothelin-1 (ET-1). ET-1 regulates contractility of CMs and maladaptive cardiac remodeling [34]. There are other potential cardiac endothelial cell secreted factors (angiocrines) such as prostacyclin, periostin, thrombospondins, follistatin/follistatin-like 1, angiotensin II, and connective tissue growth factor [35, 36]. The role of these molecules in communication between endothelial cells and CMs is unclear. In addition to the paracrine mediators, direct physical cell–cell contacts may also participate in endothelial cell-cardiomyocyte cross-talk. Cx43, Cx40, and Cx37 expressed in both the cells [37]. may mediate communication between them. The role of exosomes and other extracellular vesicles in endothelial cell- cardiomyocyte cross-talk has not been characterized.

Cardiomyocytes produce and secrete 30–60 different proteins or peptides, which are called cardiokines or cardiomyokines [38]. Cardiokines include growth factors, endocrine hormones, cytokines, extracellular matrix proteins, and peptides. Natriuretic peptides A (ANP) and B (BNP), the stress-induced cardiokines exert effects on endothelial cells [39]. The cardiokine that modulates activation and proliferation of endothelial cells is vascular endothelial growth factor (VEGF) that binds to and activates the VEGF receptor 2 (VEGFR2) [40]. VEGFR2 signalling induces angiogenesis. Other cardiokines that induce angiogenesis include VEGF-B [41], VEGF-C

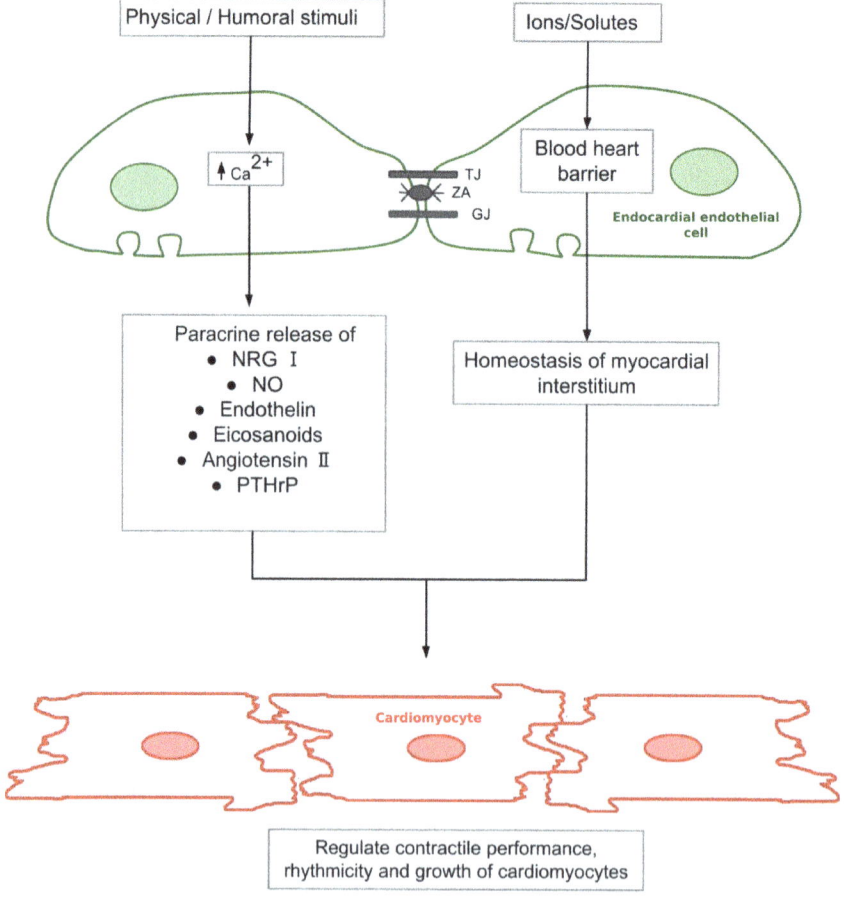

Fig. 5.1 Endocardial endothelium derived factors and ions which influence cardiomyocyte function and growth. GJ- Gap junction. NO- Nitric oxide. NRG–Neuregulin. PTHrP–Parathyroid hormone related protein. TJ–Tight junction. ZA-Zona adherens

[42], placental growth factor [43], fibroblast growth factors [44], hepatocyte growth factor [45], and angiopoietin-1 [46]. CMs also produce and secrete several members of the TGF-β superfamily. Among them, follistatin like-1 affects both CMs and endothelial cells [47].

Delineating the mechanisms of the cross-talk between endothelial cells and CMs could aid in developing regenerative therapies for heart diseases [48].

Conclusion

Most of the studies on endocardium have focused on its role during early embryonic stages. The lineage differentiation potential of endocardium indicates its heterogenous and critical role in development of the heart. Function of the endocardium in the adult heart especially under different disease conditions has not received serious attention. Endocardium and sub epicardium could be a potential source of cardiac progenitors useful for generating cells or bioengineered tissues for cardiac regenerative therapy.

References

1. Kuruvilla L, Kartha CC. Molecular mechanisms in endothelial regulation of cardiac function. Mol Cell Biochem. 2003;253(1):113–23.
2. von Gise A, Pu WT. Endocardial and epicardial epithelial to mesenchymal transitions in heart development and disease. Circ Res. 2012;110(12):1628–45.
3. Zhang H, Lui KO, Zhou B. Endocardial cell plasticity in cardiac development, diseases and regeneration. Circ Res. 2018;122(5):774–89.
4. Misfeldt AM, Boyle SC, Tompkins KL, Bautch VL, Labosky PA, Baldwin HS. Endocardial cells are a distinct endothelial lineage derived from Flk1+ multipotent cardiovascular progenitors. Dev Biol. 2009;333(1):78–89.
5. Harris IS, Black BL. Development of the endocardium. Pediatr Cardiol. 2010;31:391–9.
6. Nakano A, Nakano H, Smith KA, Palpant NJ. The developmental origins and lineage contributions of endocardial endothelium. Biochimica et Biophysica Acta. 2016; 1863(7 Pt B):1937–47.
7. Bressan M, Yang PB, Louie JD, Navetta AM, Garriock RJ, Mikawa T. Reciprocal myocardial-endocardial interactions pattern the delay in atrioventricular junction conduction. Development. 2014;141(21):4149–57.
8. D'Amato G, Luxán G, del Monte-Nieto G, Martínez-Poveda B, Torroja C, Walter W, et al. Sequential Notch activation regulates ventricular chamber development. Nat Cell Biol. 2016;18(1):7–20.
9. de la Pompa JL. Notch signaling in cardiac development and disease. Pediatr Cardiol. 2009;30(5):643–50.
10. Del Monte G, Grego-Bessa J, González-Rajal A, Bolós V, De La Pompa JL. Monitoring Notch1 activity in development: evidence for a feedback regulatory loop. Dev Dyn Off Publ Am Assoc Anat. 2007;236(9):2594–614.
11. D'Amato G, Luxán G, de la Pompa JL. Notch signalling in ventricular chamber development and cardiomyopathy. FEBS J. 2016;283(23):4223–37.
12. Luxán G, D'Amato G, MacGrogan D, de la Pompa JL. Endocardial notch signaling in cardiac development and disease. Circ Res. 2016;118(1):e1–18.
13. MacGrogan D, D'Amato G, Travisano S, Martinez-Poveda B, Luxán G, del Monte-Nieto G, et al. Sequential ligand-dependent notch signaling activation regulates valve primordium formation and morphogenesis. Circ Res. 2016;118(10):1480–97.
14. Luxán G, Casanova JC, Martínez-Poveda B, Prados B, D'amato G, MacGrogan D, et al. Mutations in the NOTCH pathway regulator MIB1 cause left ventricular noncompaction cardiomyopathy. Nat Med. 2013; 19(2):193–201.
15. Nostro MC, Cheng X, Keller GM, Gadue P. Wnt, activin, and BMP signaling regulate distinct stages in the developmental pathway from embryonic stem cells to blood. Cell Stem Cell. 2008;2(1):60–71.

16. Motoike T, Markham DW, Rossant J, Sato TN. Evidence for novel fate of Flk1+ progenitor: contribution to muscle lineage. Genesis. 2003;35(3):153–9.
17. Van Handel B, Montel-Hagen A, Sasidharan R, Nakano H, Ferrari R, Boogerd CJ, et al. Scl represses cardiomyogenesis in prospective hemogenic endothelium and endocardium. Cell. 2012;150(3):590–605.
18. Fioret BA, Heimfeld JD, Paik DT, Hatzopoulos AK. Endothelial cells contribute to generation of adult ventricular myocytes during cardiac homeostasis. Cell Rep. 2014;8(1):229–41.
19. Stevens SM, von Gise A, VanDusen N, Zhou B, Pu WT. Epicardium is required for cardiac seeding by yolk sac macrophages, precursors of resident macrophages of the adult heart. Dev Biol. 2016;413(2):153–9.
20. Popescu L, Gherghiceanu M, Manole C, Faussone-Pellegrini MS. Cardiac renewing: interstitial Cajal-like cells nurse cardiomyocyte progenitors in epicardial stem cell niches. J Cell Mol Med. 2009;13(5):866–86.
21. Cencioni C, Atlante S, Savoia M, Martelli F, Farsetti A, Capogrossi MC, et al. The double life of cardiac mesenchymal cells: epimetabolic sensors and therapeutic assets for heart regeneration. Pharmacol Ther. 2017;171:43–55.
22. Urbanek K, Cesselli D, Rota M, Nascimbene A, De Angelis A, Hosoda T, et al. Stem cell niches in the adult mouse heart. Proc Natl Acad Sci USA. 2006;103(24):9226–31.
23. Guimarães-Camboa N, Cattaneo P, Sun Y, Moore-Morris T, Gu Y, Dalton ND, et al. Pericytes of multiple organs do not behave as mesenchymal stem cells in vivo. Cell Stem Cell. 2017; 20(3):345–59. (e5).
24. Chong JJ, Chandrakanthan V, Xaymardan M, Asli NS, Li J, Ahmed I, et al. Adult cardiac-resident MSC-like stem cells with a proepicardial origin. Cell Stem Cell. 2011;9(6):527–40.
25. Cai C-L, Martin JC, Sun Y, Cui L, Wang L, Ouyang K, et al. A myocardial lineage derives from Tbx18 epicardial cells. Nature. 2008;454(7200):104–8.
26. Smart N, Bollini S, Dubé KN, Vieira JM, Zhou B, Davidson S, et al. De novo cardiomyocytes from within the activated adult heart after injury. Nature. 2011;474(7353):640–4.
27. Zhou B, Honor LB, He H, Ma Q, Oh J-H, Butterfield C, et al. Adult mouse epicardium modulates myocardial injury by secreting paracrine factors. J Clin Investig. 2011;121(5):1894–904.
28. Mohan P, Brutsaert DL, Paulus WJ, Sys SU. Myocardial contractile response to nitric oxide and cGMP. Circulation. 1996;93:1223–9.
29. Gemberling M, Karra R, Dickson AL, Poss KD. Nrg1 is an injury-induced cardiomyocyte mitogen for the endogenous heart regeneration program in zebrafish. Elife. 2015; 4:e.05871. https://doi.org/10.7554/eLife.05871.
30. D'Uva G, Aharonov A, Lauriola M, Kain D, Yahalom-Ronen Y, Carvalho S, et al. ERBB2 triggers mammalian heart regeneration by promoting cardiomyocyte dedifferentiation and proliferation. Nat Cell Biol. 2015;17:627–38.
31. Liu Q, Hu T, He L, Huang X, Tian X, Zhang H, et al. Genetic targeting of sprouting angiogenesis using Apln-CreER. Nat Commun. 2015;6:6020. https://doi.org/10.1038/ncomms7020.
32. Scimia MC, Hurtado C, Ray S, Metzler S, Wei K, Wang J, et al. APJ acts as a dual receptor in cardiac hypertrophy. Nature. 2012;488:394–8.
33. Wang W, McKinnie SM, Patel VB, Haddad G, Wang Z, Zhabyeyev P, et al. Loss of Apelin exacerbates myocardial infarction adverse remodeling and ischemia-reperfusion injury: therapeutic potential of synthetic Apelin analogues. J Am Heart Assoc. 2013;2:e000249. https://doi.org/10.1161/JAHA.113.000249.
34. Drawnel FM, Archer CR, Roderick HL. The role of the paracrine/autocrine mediator endothelin-1 in regulation of cardiac contractility and growth. Br J Pharmacol. 2013;168:296–317.
35. Lim SL, Lam CS, Segers VF, Brutsaert DL, De Keulenaer GW. Cardiac endothelium-myocyte interaction: clinical opportunities for new heart failure therapies regardless of ejection fraction. Eur Heart J. 2015;36:2050–60.
36. Kamo T, Akazawa H, Komuro I. Cardiac nonmyocytes in the hub of cardiac hypertrophy. Circ Res. 2015;117:89–98.

37. Johnson RD, Camelliti P. Role of non-myocyte gap junctions and connexin hemichannels in cardiovascular health and disease: novel therapeutic targets? Int J Mol Sci. 2018;19:E866. https://doi.org/10.3390/ijms19030866.
38. Doroudgar S, Glembotski CC. The cardiokine story unfolds: ischemic stress induced protein secretion in the heart. Trends Mol Med. 2011;17:207–14.
39. Kuhn M. Endothelial actions of atrial and B-type natriuretic peptides. Br J Pharmacol. 2012;166:522–31.
40. Olsson AK, Dimberg A, Kreuger J, Claesson-Welsh L. VEGF receptor signalling-in control of vascular function. Nat Rev Mol Cell Biol. 2006;7:59–371.
41. Bry M, Kivelä R, Leppänen VM, Alitalo K. Vascular endothelial growth factor-B in physiology and disease. Physiol Rev. 2014;94:779–84.
42. Chen HI, Poduri A, Numi H, Kivelä R, Saharinen P, McKay AS, et al. VEGFC and aortic cardiomyocytes guide coronary artery stem development. J Clin Invest. 2014;124:4899–914.
43. Accornero F, Molkentin JD. Placental growth factor as a protective paracrine effector in the heart. Trends Cardiovasc Med. 2011;21:220–4.
44. Itoh N, Ohta H, Nakayama Y, Konishi M. Roles of FGF signals in heart development, health, disease. Front Cell Dev Biol. 2016;4:110. https://doi.org/10.3389/fcell.2016.00110.
45. Gallo S, Sala V, Gatti S, Crepaldi T. Cellular and molecular mechanisms of HGF/Met in the cardiovascular system. Clin Sci. 2015;129:1173–93.
46. Arita Y, Nakaoka Y, Matsunaga T, Kidoya H, Yamamizu K, Arima Y, et al. Myocardium-derived angiopoietin-1 is essential for coronary vein formation in the developing heart. Nat Commun. 2014;5:4552. https://doi.org/10.1038/ncomms5552.
47. Shimano M, Ouchi N, Walsh K. Cardiokines: recent progress in elucidating the cardiac secretome. Circulation. 2012;126:e327–32.
48. Talman V, Kivela R. Cardiomyocyte—endothelial cell Interactions in cardiac remodeling and regeneration. Front Cardiovasc Med. 2018;5:101. https://doi.org/10.3389/fcvm.2018.00101.

Chapter 6
Cardiomyocytes in the Mammalian Adult Heart

Abstract Cardiomyocytes of the mammalian adult heart differ significantly from cardiomyocytes in the foetal heart, both in structure and biochemical properties. Adult cardiomyocytes are terminally differentiated and the rate of DNA synthesis is extremely low. The cells exit the cell cycle resulting in a postnatal decline in cell proliferation. The mechanisms for the cell cycle exit and permanent growth arrest after birth in cardiomyocytes of the mammalian heart are unclear. The arrest is known to be associated with downregulation of positive cell cycle regulators, centrosome disassembly and absence of cytokinesis. Though resident progenitors are present in the adult heart, cell renewal from stem cell populations in the normal mammalian adult heart is also low compared to the developing heart.

Keyword Mammalian heart · Adult cardiomyocytes · Cell cycle regulation · Cell cycle exit · Cytokinesis · Cardiac progenitors · Cardiac reprogramming

Introduction

Cardiomyocytes (CMs) of the mammalian adult heart differ significantly from foetal CMs, both in structure and in biochemical properties. In the mammalian adult heart, there is a postnatal decline in cell proliferation as well as cell renewal from stem cell populations. The rate of DNA synthesis in CMs is also low in the normal mammalian adult heart compared to the developing heart [1]. The loss in regenerative potential in the adult heart is associated with changes that occur in the microenvironment of the heart during maturation. The structure and environment of the adult heart seems to be limiting for proliferation of CMs.

Post-natal Growth of Cardiomyocytes

In amphibians and lower vertebrates, the mitotic machinery is preserved in the adult organisms [2–8]. Newt has the capacity for myocardial regeneration and in them, mononucleated cardiomyocytes undergo cytokinesis [5]. In zebrafish, as part of the

© The Author(s), under exclusive license to Springer Nature Switzerland AG 2021   63
C. C. Kartha, *Cardiomyocytes in Health and Disease*,
https://doi.org/10.1007/978-3-030-85536-9_6

response to injury, limited dedifferentiation of a small population of cells and prolif-eration of CMs do occur [6–8]. In contrast, adult CMs in mammals have very limited potential for self-renewal and growth is achieved by cell enlargement [9]. CMs in the mammalian adult heart synthesize DNA without undergoing cell division [10]. Be that as it may, a limited renewal of adult CMs possibly balances cell loss through apoptosis [11, 12].

There is evidence for cell turnover in the mammalian heart albeit at a very low rate [13]. The estimated rate of DNA synthesis in CMs in the adult heart is however less than 1% per year [14, 15]. Less than 50% of CMs are possibly replaced during the lifespan of an individual [15]. Contrarily, the renewal rates of non-myocytes in the heart are much higher than those of CMs; endothelial cells have a renewal rate of >15% per year and mesenchymal cells have a rate about 4% per year [14].

Division of resident CMs in the adult heart may in a small measure contribute to replacement of the heart tissue in mammals [16]. Adjacent to areas of injury, increased proliferation of pre-existing CMs at a low rate has been observed; it is decreased by oxidative DNA damage mediated by aerobic respiration [16, 17]. Stress signalling in the myocardium also switches on genes which encode foetal isoforms of proteins assisting cardiomyocyte proliferation [18]. While the cell turnover may aid structural maintenance, it is inadequate to repair the heart after an injury or for maintaining cardiac function in disease conditions.

Regulation of Cell Cycle Exit in Cardiomyocytes

Adult cardiomyocytes are terminally differentiated. Terminal differentiation of CMs during the post-natal period involves permanent exit from the cell cycle and cell type-specific differentiation marked by the upregulation of tissue-specific genes. As foetal CMs undergo terminal differentiation, skeletal actin is downregulated and myosin heavy chain (MHC) and cardiac actin are upregulated [19]. When terminally differentiated, cells do not re-enter the cell cycle in response to mitogens or physi-ological stress. The potential for cell cycle re-entry is however different in various species. There is also no evidence for the formation of a contractile ring required for cytokinesis in adult CMs of any species.

Cell cycle re-entry however occurs to a limited extent in the adult human heart in response to stress or injury [20]. Increased DNA content per nuclei as well as increased number of nuclei per myocyte have been reported in cardiomyopathic hearts of humans [21–23]. Beltrami et al. found mitotic nuclei in a small population of CMs in the failing heart and opined that this finding suggests division of CMs [20]. Later studies have suggested that re-entry of human CMs into the cell cycle after myocardial infarction is transient and limited and that it results in endoreduplication and not in cytokinesis and proliferation [24]. The critical factor that determines proliferation of adult CMs is its cellular size. In the adult heart, cells smaller in size have higher potential to re-enter into the cell cycle in response to stress or injury [20, 25, 26].

There is another explanation for the presence of cycling CMs in the adult myocardium [27, 28]. Several cardiac progenitor cells capable of proliferation have been identified in the adult heart [29, 30]. Be that as it may, the progeny of a stem cell after commitment to the cardiomyocyte lineage can only undergo three to four rounds of cell division; after that they permanently exit from the cell cycle [29].

Thus, in adult hearts, terminally differentiated CMs that do not re-enter the cell cycle predominate; a small number of CMs or resident stem cells have a limited potential for re-entry into the cell cycle. Neither adult CMs nor cardiac stem cells seem to have the potential for proliferation, adequate to regenerate the heart after ischemic injury.

Mechanisms for Cell Cycle Exit and Terminal Differentiation in Cardiomyocytes

The mechanisms for the cell cycle exit and permanent growth arrest after birth in CMs of the mammalian heart are unclear. The arrest is known to be linked with downregulation of positive cell cycle regulators, as well as centrosome disassembly [10, 31]. Several intrinsic and extrinsic factors regulate the cell cycle in CMs (Fig. 6.1). Factors which promote cell division are downregulated and there is increased expression of those that inhibit cell division in postnatal CMs when they transit from a proliferative to hypertrophic mode of growth [32]. Cell cycle arrest is also accompanied by decrease in the expression of cell cycle genes and increase in reactive oxygen species activity as well as DNA damage (Fig. 6.2) [33].

Cell cycle exit of CMs during the perinatal period is also associated with downregulation of the expression of small Rho GTPases such as RhoA, Cdc42, Rac1, ROCK-I,

Fig. 6.1 Factors that promote and inhibit cardiomyocyte proliferation in post-natal heart during transition from a proliferative to hypertrophic mode

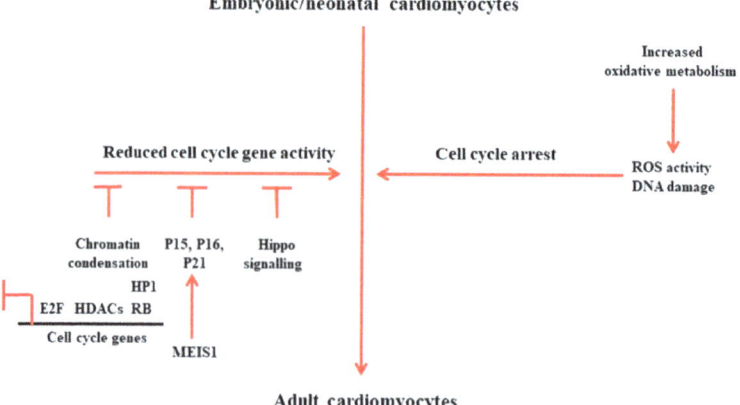

Fig. 6.2 Mechanisms of cell cycle arrest in adult cardiomyocytes

ROCK-II, and p-cofilin, which are important for the formation of actomyosin ring proteins [34].

The retinoblastoma protein (Rb) is suspected to mediate both the cell cycle exit as well as irreversible cell cycle arrest with terminal differentiation in skeletal muscle, adipocytes and macrophages [35–37]. Rb has been found to be specifically required for the control of the normal myogenic cell cycle and complete differentiation [38–40]. When Rb is deleted, before myogenic differentiation, both differentiation and apoptosis are defective. If Rb is deleted after differentiation, the result is multinucleated myotubes which when stimulated, do not enter S phase. Rb is thought to play a key role in the switch from proliferation to differentiation in skeletal myocytes [41].

There may be tissue specificity in the role of Rb in maintaining terminal differentiation. Mice in which Rb was deleted in differentiated CMs, employing an–MHC-driven Cre transgene, the cells did not have cell cycle defects [42]. CMs may have a developmental specific dependence on Rb as observed in the skeletal muscle [43]. CMs may require Rb for commitment and differentiation, but Rb is not necessary after differentiation. CMs of Rb-p130-null mice have defects in cell cycle exit and differentiation [42]. This finding suggests that Rb and p130 have coinciding roles in mediation of cell cycle exit of CMs. The role of Rb and p130 for maintaining quiescence in CMs is yet to be discovered. Interestingly, in the hearts of adult transgenic mice with increased Cdk4 or Cdk2 activity, which are expected to inactivate all pocket proteins including Rb, an increase in CMs and continued synthesis of DNA have been observed [44–46].

Causes for the Absence of Cytokinesis

A likely cause for the uncoupling of karyokinetic and cytokinetic events in postnatal CMs is their complex myofibrillar cytoarchitecture coupled with downregulation in the expression of proteins which regulate cell cycle and cytokinesis. The highly organized myofibrils in adult CMs are implicated in the physical prevention of cell division. Foetal CMs completely disassemble their cytoskeletal filaments before entering cell division [34]. In adult CMs, disassembly of myofibrils may impact contractile function and hence detrimental.

During progressive differentiation from the embryonic to adult stage, ventricular CMs increase in size and number as well as in the complexity of organization of myofibrils [47–49]. Atrial CMs in contrast are smaller in size, lesser in myofibrils and are less differentiated. Interestingly, they can regenerate both in vivo and in vitro [50, 51]. Though sarcomeric structure could be an important factor, the exact reasons for the inability of adult CMs to undergo cytokinesis are at present uncertain.

The contractile ring composed of actin, myosin and several other proteins is important for cytokinesis in eukaryotic cells. Formation of the contractile ring has been studied in postnatal rat CMs [52, 53]. Defective localization of anillin in the midbody region is reported to result in failure of cytoplasmic division and lead to binucleation [54]. Anillin is a known regulator of the cleavage furrow formation. In adult CMs, there is also downregulation of Polo-like kinase, a protein involved in spindle formation and chromosome segregation during mitosis [55].

Cardiac Progenitors in the Adult Heart

Several groups have independently identified in the adult mammalian heart, resident stem cell or progenitor cell populations with the capacity to differentiate into CMs [29, 30, 56]. Such cells have been found in the human heart as well [26, 28].

Many studies have searched the adult heart for resident, self-renewing multipotent cells with capacity to differentiate into the important cardiac lineages [57, 58]. Cardiac resident side population cells, ISL1+ progenitors, c-Kit+ cells, Sca1+ cells, epicardial progenitors, and mesenchymal cells are some of the cell types which have been reported. Resident c-Kit+ cells with regenerative capacity have been found in the adult mammalian heart and they are similar to the multipotent, clonogenic and self-renewing c-Kit+ cells isolated from bone marrow [29, 59]. These cells were found in niches that modulate the migration and differentiation of the cells [60]. While some suggested a role for these cells in cardiac repair, others reported that c-Kit+ cells in adults do not differentiate into CMs [61, 62]. They may turn at a low rate into CMs through cell fusion [63].

c-Kit+ cells could may also in very small numbers generate new CMs during ageing and after injury [64]. An endothelial population of cells in mouse hearts have been found to express c-Kit. This finding has led to the view that c-Kit+ cells are endothelial cells that can rarely de-differentiate into cardiac stem cells [64, 65]. Lineage tracing studies have confirmed that in the adult heart, ∼0.03% of CMs express c-Kit [38, 66]. In mammals, c-Kit + cardiovascular progenitors are capable of differentiating into the major cardiac lineages until the neonatal stage of development. How their capacity for cardiac myogenesis is largely lost later is not known [61, 67].

Quiescent stem cells and progenitors in the adult heart are re-activated in response to injury but, generate mostly non-cardiomyocyte cells. They do not contribute significantly to the turnover of CMs. Progenitor cells that can differentiate into endothelial cells, may provide pro-survival and angiogenic benefits and thus aid tissue maintenance [68].

Generation of Adult Cardiomyocytes Through Reprogramming of Other Cell Types

Adult CMs can be derived in vitro from embryonic stem (ES) or induced pluripotent stem (iPS) cells [69]. The differentiation process is a repetition of the stages of differentiation during in vivo development and is associated with expression of transcription factors [70]. Efficiency of cardiomyocyte induction from iPS cells can be enhanced by inhibition or activation of specific signalling pathways or through the effective use of signalling molecules and several microRNAs which regulate specification of the cardiac mesoderm [71–74].

Adult CMs can also be derived from other somatic cells, such as fibroblasts through the process of trans differentiation [57, 69, 75]. Genes that are specific for CMs, such as ryanodine receptor 2 (Ryr2), connexin 43 (Gja1), cTnT, and α-MHC are upregulated in the transdifferentiated CMs [75]. Transcription factors such as Gata4, Mef2c, and Tbx5, are sufficient for cardiomyocyte induction in vitro and have been used in vivo in mice to reprogramme cardiac fibroblasts to CMs [76]. Addition of Hand2 or enhancing Mef2c expression relative to Gata4 and Tbx5 are seen to significantly enhance the transformation of fibroblasts to CMs [77, 78]. After initiation of reprogramming with iPS factors, Oct4, Klf4, and Sox2, cardiomyogenesis has been achieved with the use of BMP4 and inhibition of Janus kinase (JAK1) [79, 80].

A better understanding of cardiac stem cell niches and mechanisms by which cardiac microenvironment support developmental progenitors could pave way for identifying effective therapies using cardiac stem cells.

Conclusions

In recent years, there has been considerable advances in our understanding of cell cycle control in cardiomyocytes (CMs). Be that as it may, the mechanisms for the cell cycle exit and permanent growth arrest after birth in CMs of the mammalian adult heart are unclear. A better understanding of the regulation of the cell cycle exit in adult CMs could be advantageous for therapeutically reactivating cell cycle progression in existing adult CMs. Several approaches to genetically manipulate key cell cycle regulators to promote cell cycle progression in adult CMs for use in cardiac repair, are under investigation. In the adult heart, there are resident stem cell or progenitor cell populations with the capacity to differentiate into CMs. Clinical use of these progenitor cells necessitates a better understanding of the signals that activate their proliferation and migration to sites of injury as well as identification of methods to isolate and expand them in vitro, and deliver them to sites of myocardial injury.

References

1. Soonpaa MH, Field LJ. Survey of studies examining mammalian cardiomyocyte DNA synthesis. Circ Res. 1998;83(1):15–26.
2. Bettencourt-Dias M, Mittnacht S, Brockes JP. Heterogeneous proliferative potential in regenerative adult newt cardiomyocytes. J Cell Sci. 2003;116(19):4001–9.
3. Nag AC, Healy CJ, Cheng M. DNA synthesis and mitosis in adult amphibian cardiac muscle cells in vitro. Science. 1979;205(4412):1281–2.
4. Oberpriller JO, Oberpriller JC, Matz DG, Soonpaa MH. Stimulation of proliferative events in the adult amphibian cardiac myocyte. Ann N Y Acad Sci. 1995;752:30–46.
5. Matz D, Oberpriller JO, Oberpriller JC. Comparison of mitosis in binucleated and mononucleated newt cardiac myocytes. Anat Rec. 1998;251(2):245–55.
6. Jopling C, Sleep E, Raya M, Martí M, Raya A, Belmonte JCI. Zebrafish heart regeneration occurs by cardiomyocyte dedifferentiation and proliferation. Nature. 2010;464(7288):606–9.
7. Matrone G, Tucker CS, Denvir MA. Cardiomyocyte proliferation in zebrafish and mammals: lessons for human disease. Cell Mol Life Sci. 2017;74(8):1367–78.
8. Poss KD, Wilson LG, Keating MT. Heart regeneration in zebrafish. Science. 2002;298(5601):2188–90.
9. Li F, Wang X, Capasso JM, Gerdes AM. Rapid transition of cardiac myocytes from hyperplasia to hypertrophy during postnatal development. J Mol Cell Cardiol. 1996;28(8):1737–46.
10. Walsh S, Ponten A, Fleischmann BK, Jovinge S. Cardiomyocyte cell cycle control and growth estimation in vivo—an analysis based on cardiomyocyte nuclei. Cardiovasc Res. 2010;86(3):365–73.
11. Leri A, Rota M, Hosoda T, Goichberg P, Anversa P. Cardiac stem cell niches. Stem Cell Res. 2014;13(3):631–46.
12. Anversa P, Kajstura J. Ventricular myocytes are not terminally differentiated in the adult mammalian heart. Circ Res. 1998;83(1):1–14.
13. Eschenhagen T, Bolli R, Braun T, Field LJ, Fleischmann BK, Frisén J, et al. Cardiomyocyte regeneration: a consensus statement. Circulation. 2017;136(7):680–6.
14. Bergmann O, Zdunek S, Felker A, Salehpour M, Alkass K, Bernard S, et al. Dynamics of cell generation and turnover in the human heart. Cell. 2015;161(7):1566–75.
15. Bergmann O, Bhardwaj RD, Bernard S, Zdunek S, Barnabé-Heider F, Walsh S, et al. Evidence for cardiomyocyte renewal in humans. Science. 2009;324(5923):98–102.
16. Senyo SE, Steinhauser ML, Pizzimenti CL, Yang VK, Cai L, Wang M, et al. Mammalian heart renewal by pre-existing cardiomyocytes. Nature. 2013;493(7432):433–6.
17. Nakada Y, Canseco DC, Thet S, Abdisalaam S, Asaithamby A, Santos CX, et al. Hypoxia induces heart regeneration in adult mice. Nature. 2017;541(7636):222–7.

18. Depre C, Shipley GL, Chen W, Han Q, Doenst T, Moore ML, et al. Unloaded heart in vivo replicates fetal gene expression of cardiac hypertrophy. Nat Med. 1998;4(11):1269–75.

19. Subramaniam A, Jones WK, Gulick J, Wert S, Neumann J, Robbins J. Tissue-specific regulation of the alpha-myosin heavy chain gene promoter in transgenic mice. J Biol Chem. 1991;266(36):24613–20.

20. Beltrami AP, Urbanek K, Kajstura J, Yan S-M, Finato N, Bussani R, et al. Evidence that human cardiac myocytes divide after myocardial infarction. N Engl J Med. 2001;344(23):1750–7.

21. Beltrami C, Di Loreto C, Finato N, Yan SM. DNA content in end-stage heart failure. Adv Clin Pathol Off J Adriat Soc Pathol. 1997;1(1):59–73.

22. Goodman L, Epling S, Kelly S, Lee S, Fishbein MJ. DNA flow cytometry of myocardial cell nuclei in paraffin-embedded, human autopsy, cardiac tissue. Am J Cardiovasc Pathol. 1990;3(1):55–9.

23. Herget G, Neuburger M, Plagwitz R, Adler C. DNA content, ploidy level and number of nuclei in the human heart after myocardial infarction. Cardiovasc Res. 1997;36(1):45–51.

24. Meckert PC, Rivello HG, Vigliano C, González P, Favaloro R, Laguens R. Endomitosis and polyploidization of myocardial cells in the periphery of human acute myocardial infarction. Cardiovasc Res. 2005;67(1):116–23.

25. Kajstura J, Leri A, Finato N, Di Loreto C, Beltrami CA, Anversa P. Myocyte proliferation in end-stage cardiac failure in humans. Proc Natl Acad Sci USA. 1998;95(15):8801–5.

26. Urbanek K, Torella D, Sheikh F, De Angelis A, Nurzynska D, Silvestri F, et al. Myocardial regeneration by activation of multipotent cardiac stem cells in ischemic heart failure. Proc Natl Acad Sci USA. 2005;102(24):8692–7.

27. Rosenthal N. High hopes for the heart. N Engl J Med. 2001;344:1785–7.

28. Urbanek K, Quaini F, Tasca G, Torella D, Castaldo C, Nadal-Ginard B, et al. Intense myocyte formation from cardiac stem cells in human cardiac hypertrophy. Proc Natl Acad Sci USA. 2003;100(18):10440–5.

29. Beltrami AP, Barlucchi L, Torella D, Baker M, Limana F, Chimenti S, et al. Adult cardiac stem cells are multipotent and support myocardial regeneration. Cell. 2003;114(6):763–76.

30. Oh H, Bradfute SB, Gallardo TD, Nakamura T, Gaussin V, Mishina Y, et al. Cardiac progenitor cells from adult myocardium: homing, differentiation, and fusion after infarction. Proc Natl Acad Sci USA. 2003;100(21):12313–8.

31. Zebrowski DC, Vergarajauregui S, Wu C-C, Piatkowski T, Becker R, Leone M, et al. Developmental alterations in centrosome integrity contribute to the post-mitotic state of mammalian cardiomyocytes. eLife. 2015; 4:e05563.

32. Foglia MJ, Poss KD. Building and re-building the heart by cardiomyocyte proliferation. Development. 2016;143(5):729–40.

33. Cui M, Wang Z, Bassel-Duby R, Olson EN. Genetic and epigenetic regulation of cardiomyocytes in development, regeneration and disease. Development. 2018; 145(24):dev171983.

34. Ahuja P, Perriard E, Perriard J-C, Ehler E. Sequential myofibrillar breakdown accompanies mitotic division of mammalian cardiomyocytes. J Cell Sci. 2004;117(15):3295–306.

35. Gu W, Schneider JW, Condorelli G, Kaushal S, Mahdavi V, Nadal-Ginard B. Interaction of myogenic factors and the retinoblastoma protein mediates muscle cell commitment and differentiation. Cell. 1993;72(3):309–24.

36. Chen P-L, Riley DJ, Chen Y, Lee W-H. Retinoblastoma protein positively regulates terminal adipocyte differentiation through direct interaction with C/EBPs. Genes Dev. 1996;10(21):2794–804.

37. Chen P-L, Riley DJ, Chen-Kiang S, Lee W-H. Retinoblastoma protein directly interacts with and activates the transcription factor NF-IL6. Proc Natl Acad Sci USA. 1996;93(1):465–9.

38. Huh MS, Parker MH, Scimè A, Parks R, Rudnicki MA. Rb is required for progression through myogenic differentiation but not maintenance of terminal differentiation. J Cell Biol. 2004;166(6):865–76.

39. Novitch BG, Mulligan GJ, Jacks T, Lassar AB. Skeletal muscle cells lacking the retinoblastoma protein display defects in muscle gene expression and accumulate in S and G2 phases of the cell cycle. J Cell Biol. 1996;135(2):441–56.

40. Novitch BG, Spicer DB, Kim PS, Cheung WL, Lassar AB. pRb is required for MEF2-dependent gene expression as well as cell-cycle arrest during skeletal muscle differentiation. Curr Biol. 1999;9(9):449–59.
41. Camarda G, Siepi F, Pajalunga D, Bernardini C, Rossi R, Montecucco A, et al. A pRb-independent mechanism preserves the postmitotic state in terminally differentiated skeletal muscle cells. J Cell Biol. 2004;167(3):417–23.
42. MacLellan W, Garcia A, Oh H, Frenkel P, Jordan M, Roos K, et al. Overlapping roles of pocket proteins in the myocardium are unmasked by germ line deletion of p130 plus heart-specific deletion of Rb. Mol Cell Biol. 2005;25(6):2486–97.
43. Papadimou E, Menard C, Grey C, Puceat M. Interplay between the retinoblastoma protein and LEK1 specifies stem cells toward the cardiac lineage. EMBO J. 2005;24(9):1750–61.
44. Pasumarthi KB, Nakajima H, Nakajima HO, Soonpaa MH, Field LJ. Targeted expression of cyclin D2 results in cardiomyocyte DNA synthesis and infarct regression in transgenic mice. Circ Res. 2005;96(1):110–8.
45. Soonpaa MH, Koh GY, Pajak L, Jing S, Wang H, Franklin MT, et al. Cyclin D1 overexpression promotes cardiomyocyte DNA synthesis and multinucleation in transgenic mice. J Clin Investig. 1997;99(11):2644–54.
46. Liao H-S, Kang PM, Nagashima H, Yamasaki N, Usheva A, Ding B, et al. Cardiac-specific overexpression of cyclin-dependent kinase 2 increases smaller mononuclear cardiomyocytes. Circ Res. 2001;88(4):443–50.
47. Ehler E, Rothen BM, Hammerle S, Komiyama M, Perriard J-C. Myofibrillogenesis in the developing chicken heart: assembly of Z-disk, M-line and the thick filaments. J Cell Sci. 1999;112(10):1529–39.
48. Hirschy A, Schatzmann F, Ehler E, Perriard J-C. Establishment of cardiac cytoarchitecture in the developing mouse heart. Dev Biol. 2006;289(2):430–41.
49. Rumyantsev PP. Interrelations of the proliferation and differentiation processes during cardiac myogenesis and regeneration. Int Rev Cytol. 1977; 51:187–273.
50. Rumyantsev P, Mirakjan V. Reactive synthesis of DNA and mitotic division in atrial heart muscle cells following ventricle infarction. Experientia. 1968;24(12):1234–5.
51. Steinhelper M, Lanson N Jr, Dresdner K, Delcarpio J, Wit A, Claycomb W, et al. Proliferation in vivo and in culture of differentiated adult atrial cardiomyocytes from transgenic mice. Am J Phys-Heart Circ Phys. 1990;259(6):H1826–34.
52. Li F, Wang X, Bunger PC, Gerdes AM. Formation of binucleated cardiac myocytes in rat heart: I. Role of actin–myosin contractile ring. J Mol Cell Cardiol. 1997; 29(6):1541–51.
53. Li F, Wang X, Gerdes AM. Formation of binucleated cardiac myocytes in rat heart: II. Cytoskeletal organisation. J Mol Cell Cardiol. 1997; 29(6):1553–65.
54. Engel FB, Schebesta M, Keating MT. Anillin localization defect in cardiomyocyte binucleation. J Mol Cell Cardiol. 2006;41(4):601–12.
55. Georgescu SP, Komuro I, Hiroi Y, Mizuno T, Kudoh S, Yamazaki T, et al. Downregulation of polo-like kinase correlates with loss of proliferative ability of cardiac myocytes. J Mol Cell Cardiol. 1997;29(3):929–37.
56. Laugwitz K-L, Moretti A, Lam J, Gruber P, Chen Y, Woodard S, et al. Postnatal isl1+ cardioblasts enter fully differentiated cardiomyocyte lineages. Nature. 2005;433(7026):647–53.
57. Lambers E, Kume T. Navigating the labyrinth of cardiac regeneration. Dev Dyn. 2016;245(7):751–61.
58. Aguilar-Sanchez C, Michael M, Pennings S. Cardiac stem cells in the postnatal heart: lessons from development. Stem Cells Int. 2018:1247857. https://doi.org/10.1155/2018/1247857.
59. Ellison GM, Vicinanza C, Smith AJ, Aquila I, Leone A, Waring CD, et al. Adult c-kitpos cardiac stem cells are necessary and sufficient for functional cardiac regeneration and repair. Cell. 2013;154(4):827–42.
60. Urbanek K, Cesselli D, Rota M, Nascimbene A, De Angelis A, Hosoda T, et al. Stem cell niches in the adult mouse heart. Proc Natl Acad Sci USA. 2006;103(24):9226–31.

61. Jesty SA, Steffey MA, Lee FK, Breitbach M, Hesse M, Reining S, et al. C-kit+ precursors support postinfarction myogenesis in the neonatal, but not adult, heart. Proc Natl Acad Sci USA. 2012;109(33):13380–5.
62. Murry CE, Soonpaa MH, Reinecke H, Nakajima H, Nakajima HO, Rubart M, et al. Haematopoietic stem cells do not transdifferentiate into cardiac myocytes in myocardial infarcts. Nature. 2004;428(6983):664–8.
63. Nygren JM, Jovinge S, Breitbach M, Säwén P, Röll W, Hescheler J, et al. Bone marrow–derived hematopoietic cells generate cardiomyocytes at a low frequency through cell fusion, but not transdifferentiation. Nat Med. 2004;10(5):494–501.
64. Van Berlo JH, Kanisicak O, Maillet M, Vagnozzi RJ, Karch J, Lin S-CJ, et al. C-kit+ cells minimally contribute cardiomyocytes to the heart. Nature. 2014; 509(7500):337–41.
65. Sultana N, Zhang L, Yan J, Chen J, Cai W, Razzaque S, et al. Resident c-kit+ cells in the heart are not cardiac stem cells. Nat Commun. 2015;6(1):1–10.
66. Liu Q, Yang R, Huang X, Zhang H, He L, Zhang L, et al. Genetic lineage tracing identifies in situ Kit-expressing cardiomyocytes. Cell Res. 2016;26(1):119–30.
67. Tallini YN, Greene KS, Craven M, Spealman A, Breitbach M, Smith J, et al. c-Kit expression identifies cardiovascular precursors in the neonatal heart. Proc Natl Acad Sci USA. 2009;106(6):1808–13.
68. Santini MP, Forte E, Harvey RP, Kovacic JC. Developmental origin and lineage plasticity of endogenous cardiac stem cells. Development. 2016;143(8):1242–58.
69. Burridge PW, Keller G, Gold JD, Wu JC. Production of de novo cardiomyocytes: human pluripotent stem cell differentiation and direct reprogramming. Cell Stem Cell. 2012;10(1):16–28.
70. Niwa H. How is pluripotency determined and maintained? Development. 2007;134(4):635–46.
71. Tirosh-Finkel L, Zeisel A, Brodt-Ivenshitz M, Shamai A, Yao Z, Seger R, et al. BMP-mediated inhibition of FGF signaling promotes cardiomyocyte differentiation of anterior heart field progenitors. Development. 2010;137(18):2989–3000.
72. Lian X, Hsiao C, Wilson G, Zhu K, Hazeltine LB, Azarin SM, et al. Robust cardiomyocyte differentiation from human pluripotent stem cells via temporal modulation of canonical Wnt signaling. Proc Natl Acad Sci USA. 2012;109(27):E1848-1857.
73. Kattman SJ, Witty AD, Gagliardi M, Dubois NC, Niapour M, Hotta A, et al. Stage-specific optimization of activin/nodal and BMP signaling promotes cardiac differentiation of mouse and human pluripotent stem cell lines. Cell Stem Cell. 2011;8(2):228–40.
74. Lee DS, Chen J-H, Lundy DJ, Liu C-H, Hwang S-M, Pabon L, et al. Defined microRNAs induce aspects of maturation in mouse and human embryonic-stem-cell-derived cardiomyocytes. Cell Rep. 2015;12(12):1960–7.
75. Ieda M, Fu J-D, Delgad-Olguin P, Vedantham V, Hayashi Y, Bruneau BG, et al. Direct reprogramming of fibroblasts into functional cardiomyocytes by defined factors. Cell. 2010;142(3):375–86.
76. Qian L, Huang Y, Spencer CI, Foley A, Vedantham V, Liu L, et al. In vivo reprogramming of murine cardiac fibroblasts into induced cardiomyocytes. Nature. 2012;485(7400):593–8.
77. Song K, Nam Y-J, Luo X, Qi X, Tan W, Huang GN, et al. Heart repair by reprogramming non-myocytes with cardiac transcription factors. Nature. 2012;485(7400):599–604.
78. Wang L, Liu Z, Yin C, Asfour H, Chen O, Li Y, et al. Stoichiometry of Gata4, Mef2c, and Tbx5 influences the efficiency and quality of induced cardiac myocyte reprogramming. Circ Res. 2015;116(2):237–44.
79. Efe JA, Hilcove S, Kim J, Zhou H, Ouyang K, Wang G, et al. Conversion of mouse fibroblasts into cardiomyocytes using a direct reprogramming strategy. Nat Cell Biol. 2011;13(3):215–22.
80. Snyder M, Huang X-Y, Zhang JJ. Stat3 directly controls the expression of Tbx5, Nkx2. 5, and GATA4 and is essential for cardiomyocyte differentiation of P19CL6 cells. J Biol Chem. 2010; 285(31):23639–46.

Chapter 7
Energy Metabolism in Cardiomyocyte

Abstract Development, differentiation, and postnatal growth of cardiomyocyte are accompanied by substantial changes in its energy metabolism. These changes influence cardiomyocyte proliferation during the early phases of development of the heart and also terminal differentiation of the cardiomyocyte during later stages. In the early phase, glycolysis is a major source of energy for proliferating cardiomyocytes. After birth, there is an increase in the workload of the heart. The adult heart does not have energy reserves and hence has to ceaselessly produce energy in the form of adenosine triphosphate to meet the demands of uninterrupted contractile function. Cardiomyocyte improves its contractile capacity by shifting to a more efficient energy source (lipids) and by adapting to specialized systems for energy transfer. Metabolic maturation precedes the maturation of excitation–contraction coupling. The exact interacting mechanisms are unclear. As cardiomyocytes become terminally differentiated, its mitochondrial oxidative capacity increases and fatty acid β-oxidation is the major source of energy. Energy metabolism drives the post-natal development of cardiomyocyte and its complex architecture. The cell microenvironment modulates the cross-talk between cytoskeletal architecture and metabolism in cardiomyocytes.

Keyword Cardiomyocyte · Cardiac metabolism · Heart development · Glycolysis · Mitochondrial oxidation · Fatty acid oxidation · Metabolic switch · Energy transfer · Mechano-transduction

Introduction

The adult heart has a very–high-energy demand, because of its uninterrupted contractile function. As the heart has no energy reserves, to sustain its function it has to ceaselessly produce the needed energy in the form of adenosine triphosphate (ATP) [1–5]. More than 90% of the energy demands of the adult heart is met through mitochondrial oxidative metabolism; fatty acid oxidation is the major source [3, 4]. The fully differentiated cardiomyocyte can also additionally oxidize glucose, lactate, ketones, and amino acids [3, 4]. In contrast, embryonic cardiomyocytes (CMs) are dependent on glycolysis as the source of energy [6, 7]. Mitochondrial oxidative metabolism is deficient in embryonic CMs [7–9]. Glycolytic metabolism aid proliferation of CMs [7–9].

A switch from glycolytic metabolism to mitochondrial oxidative phosphorylation is requisite to meet the post-natal increase in energy demand in the cardiomyocyte. As mitochondrial oxidative capacity increases, the cardiomyocyte appears to convert to a terminally differentiated state [6, 7].

Metabolic Switch During Differentiation of Pluripotent Embryonic Stem Cells into Cardiomyocytes

Energy substrate metabolism in proliferating pluripotent stem cells is significantly different from that in quiescent or differentiated cells. High rates of glycolysis and lactate production even in the presence of adequate oxygen are features of energy substrate metabolism in proliferating cells. This metabolic phenotype is desirable for the synthesis of lipids, amino acids, nucleotides, and other macromolecules, to result in adequate cellular mass for viable daughter cells at mitosis [10]. Cellular growth and proliferation are also associated with increase in lipid synthesis from carbon derived from glucose coupled with low rates of fatty acid β-oxidation [11, 12]. Alterations in the expression and activity of various enzymes involved in glycolysis (hexokinase and phospofructokinase-1) and fatty acid β-oxidation [carnitine palmitoyltransferase-1 (CPT1)] are features of change in the cellular metabolic phenotype.

Undifferentiated human embryonic stem cells are characterized by small mitochondria with poorly developed cristae [7–9, 13]. Mitochondrial network is not well developed in them. Mitochondria in differentiated cells are tubular in structure with elongated cristae. They are dispersed in the cytoplasm in contrast to their perinuclear disposition in embryonic cells [7–9]. Indications for the initiation of oxidative phosphorylation in mitochondria are seen in differentiated cells. The evidences include elevated mitochondrial membrane potential, increase in oxygen consumption, and increased synthesis of ATP [7–9]. The expression of glycolytic enzymes decreases and the expression of enzymes of the electron transport chain and citric acid cycle increases [7]. Metabolic switch leads to loss of pluripotency [14, 15]. These findings indicate that switch from anaerobic glycolysis to mitochondrial oxidative phosphorylation possibly is a basis for the differentiation of pluripotent embryonic cells into CMs [7].

Energy Metabolism in Cardiomyocytes

Lipids and carbohydrates are the main sources of energy for CMs [16–18]. Their respective utilization however varies greatly during organ development [19, 20]. Enzyme activity in the energy pathways and the availability of substrates in circulation dictate the metabolic orientation of the heart at a given stage of cardiac growth. Thus, the energy demand, the oxygen content and the availability of substrates essentially determine the adaptation of cardiac metabolism.

Metabolic Phenotype of Foetal Cardiomyocytes

The foetal cardiomyocyte in a hypoxic environment is dependent on glycolysis for the synthesis of ATP [19, 21]. In the foetal heart, 50% of ATP produced is through anaerobic glycolysis [21]; the remaining is from oxidative phosphorylation [22, 23]. Foetal CMs have significant oxidative capacity as well [19, 24–28]. Most of the oxygen consumption is through lactate oxidation [24, 29, 30]. In the foetus, only 15% of total ATP produced is obtained from fatty acids and fatty acid β-oxidation [31]. Lactate inhibits oxidation of lipids and fatty acids repress the pathways for utilization of carbohydrates [32]. The higher activity and control of glycolytic enzymes during development in utero also assist anaerobic ATP production [33–35].

The metabolic phenotype of foetal cardiomyocyte is possibly related to the nature of energy substrates available for the foetal heart (Fig. 7.1). The high glycolytic activity in the foetal heart is because of the low levels of circulating fatty acid and the high levels of lactate in utero [19, 36–38]. HIF-1α mediates the transcriptional changes necessary for activation of the enzymes of the glycolytic pathway in foetal CMs [19, 21, 24, 25]. In addition to being a major source of energy for the early foetal heart, glucose metabolism promotes proliferation of CMs and thus contributes to cardiac growth in utero [19, 21–23, 39].

During the late phase of heart development in the foetus, most of the oxygen consumption in the heart is because of circulating lactate [24, 29, 40]; there is less contribution from glucose and fatty acids oxidation [32, 41]. In humans, it has also been shown that the foetal period is marked by a gradual increase in the expression of genes involved in fatty acid metabolism in the heart [42]. The earliest changes of transition to a metabolism less dependent on glucose and which assist differentiation of CMs are seen in the late phase of foetal development. The events that trigger the metabolic shift are not clear.

The mature heart has almost entirely oxidative metabolism (Fig. 7.1) [31, 35, 40]. Fatty acids are the substrate of choice in low oxygen environment as is seen in the hypoxic adult heart [43, 44].

Postnatal Transition in Cardiomyocyte Metabolism

Significant changes in the blood levels of substrates trigger swift metabolic switch in CMs during the neonatal period. Maternal milk, the first food of the new born is rich in lipids [36] and milk intake leads to reversal of the lactate–fatty acid ratio in blood [31]. The high lipid content of the colostrum contributes to the switch from carbohydrate to lipid metabolism in the cardiomyocyte [45].

A highly oxidative metabolism is essential for the development of the postnatal heart because of circulatory, respiratory and nutritive changes that happen in the early period after birth [7, 46]. There is in the postnatal heart, an increase in the mitochondrial mass as well as in the expression of mitochondrial proteins which

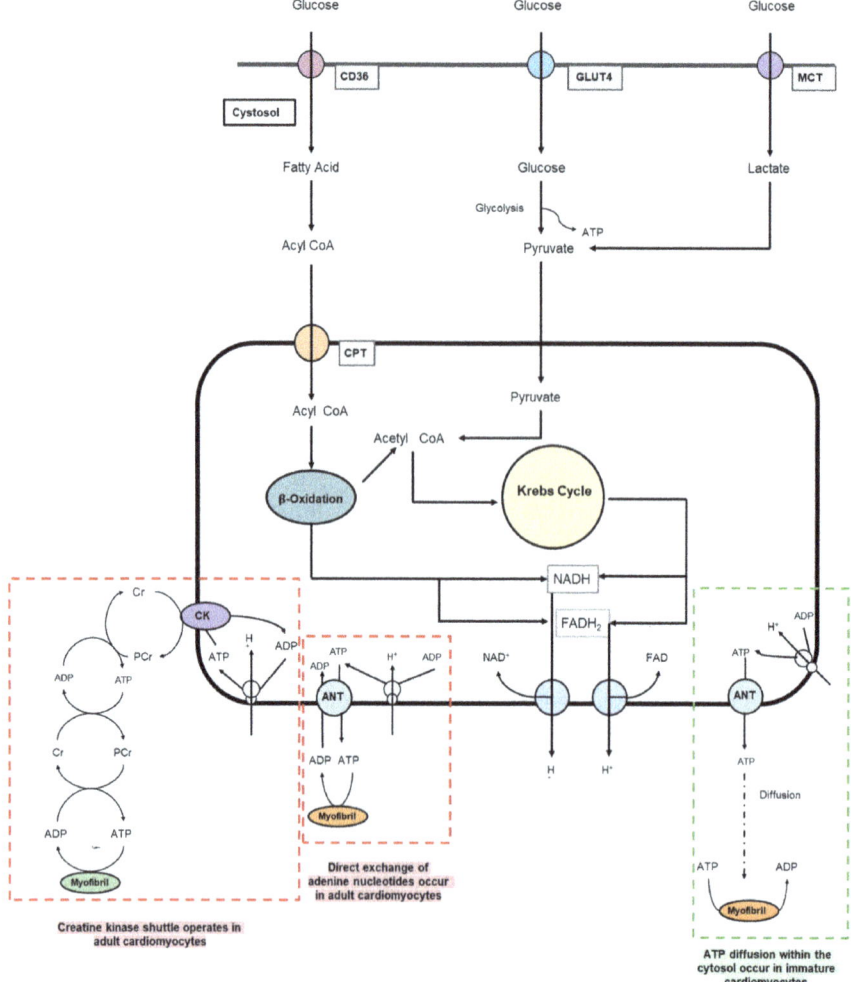

Fig. 7.1 Metabolic pathways for energy production and transfer in immature and adult cardiomyocytes. Acyl COA–Acetyl coenzyme A. ANT–Adenine nucleotide translocator. CD36–Fatty acid transclocase. CK–Creatine kinase. FAD–Flavin adenine nucleotide. GLUT–Glucose transporter. MCT–Monocarboxylate transporter. PCr–Phosphocreatine

control the fusion/fission mechanisms [47–51]. Both in animals as well as in humans during the first weeks after birth, there is increased expression of several genes linked to mitochondrial biogenesis [52, 53]. An increase in the expression of the peroxisome proliferator-activated receptor α (PPARα) and of enzymes and transporters necessary for fatty acid oxidation invigorate pathways for degradation of fatty acids [54–56]. The lipids in a few days forms the main energy source of the cardiomyocyte [21, 57]. Mechanisms for protection against possible oxidative damage are also strengthened in the post-natal heart thanks to the increase in mitochondria, a rich source of reactive oxygen species (ROS) [58–60].

The increase in mitochondrial biogenesis as well as the dramatic shift in substrate utilization in the heart during the foetal-to-adult transition, are controlled by nuclear receptor signalling which responds to developmental signals and physiologic conditions after birth. The developmental changes in energy metabolism of the post-natal heart are the result of enrichment of a sub set of nuclear receptors which can match mitochondrial fuel preferences and the ability to produce ATP according to changes in the energy demands of the heart, as well as post-translational modifications such as acetylation and succinylation of key metabolic enzymes and transcription factors in CMs [57, 61].

Maturation of Cardiac Energy Metabolism During Perinatal Development

During development, the energy produced within the cardiomyocyte is determined by work load of the heart, availability of substrate and oxygen pressure. These determinants change immensely at birth. Major hemodynamic changes in the cardiovascular system secondary to the sudden expansion of the lung alveoli at birth is associated with alterations in oxygen levels, pressure as well as volume of the two cardiac ventricles. Postnatal growth is associated with increase in workload of the heart. Maturation of the cardiomyocyte is necessary for optimal function of the heart. This need is met by cardiomyocyte hypertrophy for greater contractile ability and a greater capacity for energy production. Increased cardiac efficiency is attained by remodelling of the cardiomyocyte structure, excitation–contraction coupling and energy metabolism. Remodelling of the cardiomyocyte structure during early postnatal period results in compartmentalization of the cell, necessitating systems for energy transfer between energy producing and energy consuming compartments in the cardiomyocyte. More efficient energy production systems get organized in mature CMs [31, 50].

Metabolic Phenotype of Neonatal Cardiomyocytes

Proliferation of CMs ceases in the late stages of foetal growth [62]. Cellular growth in the peri natal period is characterized by hypertrophy and binucleation [62, 63]. Metabolism in neonatal CMs is favourable for cellular growth. In the early new born period, glycolysis contributes to almost 50% of total ATP production. By one week after birth, glycolysis supplies less than 10% of ATP production; this value is similar to that seen in the adult heart [64]. In the neonatal period, there is also a marked decrease in the ATP produced through lactate oxidation.

Fatty acid β-oxidation increases in the neonatal period attaining levels seen in adult hearts [30]. Levels of free fatty acids rise soon after birth to their levels in adults [37, 65, 66]. Energy supply from fatty acid oxidation is yet less than 15% of the needs of the heart in the new born [64, 67]. This is partly because of inhibition of fatty acid uptake by the mitochondria [21, 25, 68, 69]. Mitochondrial fatty acid uptake is a key regulator of fatty acid β-oxidation [70–73]. Changes in malonyl-CoA levels is considered to be important in the control of fatty acid β-oxidation in CMs. Malonyl-CoA levels in the heart decrease within days of birth, because of decreased synthesis as well as increased degradation [74, 75]. The decrease in malonyl-CoA contributes to the maturation of fatty acid β-oxidation during the neonatal period [74, 76].

The enzyme acetyl CoA carboxylase (ACC) is important for malonyl-CoA synthesis. In the heart, mainly the 280 kDa isoform (ACC280) of the enzyme is expressed [68, 75, 77–79]. After birth, ACC activity in the heart substantially decreases in response to activation of 5′ adenosine monophosphate–activated protein kinase (AMPK) [74, 75, 79–82]. In the adult and neonatal heart, AMPK activity is high [67, 74, 83]. Increase in AMPK activity after birth results in increased ACC phosphorylation and decreased ACC activity, promoting fatty acid oxidation [74]. Activation of AMPK may be the result of decrease in blood insulin levels in the post-natal period.

Other changes may also contribute to alterations in the production of malonyl-CoA in the neonatal heart. An age-dependent decrease in ACC expression and age dependent increase in the expression of malonyl-CoA decarboxylase (MCD), which degrades malonyl-CoA have been observed in the heart tissue of infants [84]. Increase in PPARα levels and or changes in levels of PPARβ/δ may determine MCD expression [85–89].

While fatty acid oxidation in CMs rapidly matures during the neonatal period, ATP production through glucose oxidation continues at low levels until weaning [36]. This may limit oxidative metabolism in the neonatal heart. Increase in the expression of pyruvate dehydrogenase kinase (PDK) and phosphorylation of pyruvate dehydrogenase (PDH) in neonatal cardiomyocytes may contribute to the low PDH flux and glucose oxidation [90–93].

Energy Metabolism in Adult Cardiomyocytes

Oxygen consumption and cardiac work have a linear correlation in the adult heart [94]. The adult heart consumes 1 mM ATP per second. Oxidative phosphorylation occurring in mitochondria provides this energy. Energy is stored in the form of adenosine triphosphate (ATP) and phosphocreatine (PCr). These energy stores are sufficient only for a few seconds of activity.

Metabolism of the adult heart depend on the high oxygenation and high lipid levels in blood [3, 4, 24]. A high mitochondrial density in the cardiomyocyte and the ideal activity of the Krebs cycle and enzymes of the respiratory chain secure appropriate conditions for energy production from fatty acids (Fig. 7.1) [22, 48]. In terms of energy production, though it uses more oxygen per ATP produced, lipid metabolism has a higher yield than glycolytic metabolism. Carbohydrates supply only 10–40% of mitochondria-oxidized acetyl-CoA [4]. This is augmented by a complex reciprocal inhibition between lipid oxidation and carbohydrate oxidation [95].

Transcriptional Regulation of Metabolic Phenotype in Cardiomyocytes

Energy metabolism in CMs is regulated mainly by three nuclear receptor pathways. These are (i) the hypoxia-inducible factor 1α (HIF-1α) pathway, (ii) the peroxisome proliferator activated receptor γ (PPARγ) coactivator 1α (PGC-1α)/PPARα pathway and (iii) the PGC-1α/PPARβ/δ pathway. HIFs regulate the expression of several genes which express proteins involved in the intermediary metabolism of glucose, particularly those of anaerobic glycolysis. HIF-1α regulates the hypoxia inducible expression of glucose transporter (GLUT)-1, hexokinase (HK)-1, lactate dehydrogenase (LDH)-A and pyruvate dehydrogenase kinase (PDK)-1 (Fig. 7.2) [96–100]. In the foetal heart, which has a low oxygen environment as well as in the early postnatal period, HIF-1α may play a key role in maintaining or promoting glycolysis [19, 21]. Changes in HIF-1α and its target genes may initiate the postnatal metabolic switch in CMs.

PGC-1α/PPARα regulate the expression of genes that code for the enzymes involved in fatty acid activation (fatty acyl-CoA synthase), mitochondrial fatty acid uptake (mCPT1/CPT1β)], and fatty acid oxidation (medium chain acyl-CoA dehydrogenase and long chain acyl CoA dehydrogenase) [101–107]. The varied expression of both PPARα and PGC-1α could contribute to the metabolic phenotype of neonatal CMs [108, 109]. PGC-1α is also a transcriptional coactivator for PPARβ/δ [110].

PPARβ/δ and PPARα are the major PPAR isoforms expressed in the cardiomyocyte [111]. The effects of PPARβ/δ on the pathways of fatty acid and glucose metabolism may determine the metabolic phenotype of neonatal heart [31]. PPARβ/δ regulates the expression of fatty acyl-CoA synthase, mCPT1, long chain acyl CoA dehydrogenase, and medium chain acyl-CoA dehydrogenase [85, 111, 112] and can thus control the rates of fatty acid utilization, particularly of fatty acid β-oxidation [111, 113, 114]. PPARβ/δ also modulates the rates of glucose oxidation as well via controlling the expression of the genes of GLUT4 and phosphofructokinase [86]. Interestingly, the expression of PGC-1α as well as PPARα is also influenced by PPARβ/δ [112, 115].

HIF-1α decreases the DNA-binding activity of the PPARα/RXR heterodimer. A reciprocal relationship between HIF-1α and PPARα impacts energy metabolism in CMs. [116, 117]. This is evident from the increase in fatty acid oxidation and concomitant decrease in glycolysis seen in neonatal CMs. In vivo, HIF-1α may

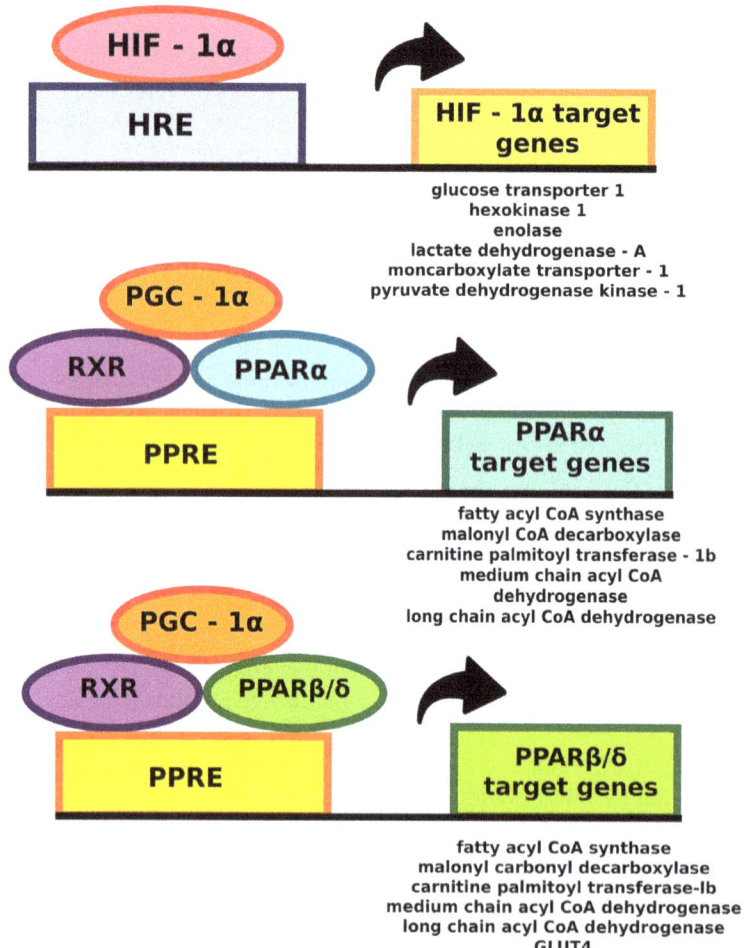

Fig. 7.2 Transcriptional nuclear receptors which modulate energy metabolism in cardiomyocytes. HIF–Hypoxia inducible factor. HRE–Hormone response element. PPAR–Peroxisome proliferator-activated receptor. PGC- Peroxisome proliferator-activated receptor-gamma coactivator. RXR–Retinoid X receptor

have a role in hypoxia associated downregulation of genes involved in fatty acid metabolism [116].

Energy Transfer Systems in Cardiomyocyte

Cardiomyocyte need a mechanism to transfer the energy produced in mitochondria to ATPases of myosin myofilaments and sarcoplasmic reticulum as diffusion is limited in the adult cardiomyocyte because of its dense structure [118, 119]. Intracellular energy transfer unit is represented in Fig. 7.3. Creatine kinase (CK) acts as the 'shuttle' for the reversible transfer of the high energy bond of ATP to creatine [120]. CK exists in different isoforms. Cytosolic CK predominantly represented by the MM-CK isoenzyme in the heart is either free in the cytoplasm or bound near ATPases of myofilaments and sarcoplasmic reticulum [121–123]. Mitochondrial CK (mi-CK) is located in the mitochondrial intermembrane space near the adenine nucleotide translocase (ANT) [121, 124]. ANT is responsible for both the import of ADP into the mitochondrial matrix and the export of the newly synthesized ATP to the intermembrane space [125]. ANT thus ensures supply of ATP to the mi-CK which then produces phosphocreatine.

In the adult heart, nearly 90% of energy is transferred out of mitochondria as PCr. The high-energy phosphate moiety of PCr is transferred via cytosolic CK and the bound MM-CK near ATPases is used for re-phosphorylation of the locally produced ADP. Enzymatic coupling between mi-CK and ANT produces a high ADP/ATP ratio, which is advantageous for the stimulation of oxidative phosphorylation. This mechanism is mainly responsible for regulating oxygen consumption in consonance

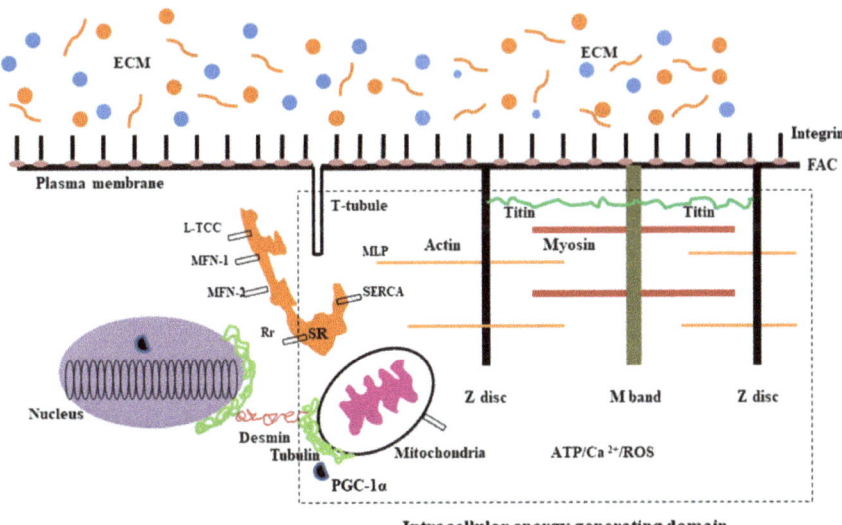

Intracellular energy generating domain

Fig. 7.3 Intracellular microdomains of energy transfer and cytoskeletal proteins involved in mechanotransduction in cardiomyocytes. ECM–Extracellular matrix. FAC-Focal adhesion kinase. LTCC–L-type calcium channel. MFN–Mitofusin. MLP–Muscle LIM protein. PGC–PPARγ coactivator. Rr–Ryanodine receptor. ROS-Reactive oxygen species. SERCA–Sarcoplasmic endoplasmic reticulum. SR-Sarcoplasmic reticulum

with cardiac performance [126]. The 'shuttle' also safeguards effective coordination of energy production and consumption [118, 127].

There are also other compartmentalized energy transfer systems in the cardiomyocyte. These include the adenylate kinase system and the nucleotide diphosphokinase (NDPK) system [128]. Another mechanism, known as direct adenine nucleotide channelling (DANC) helps to maintain a favourable ATP/ADP ratio in the proximity of ATPases and has a yield comparable to the CK system [129, 130]. The CK shuttle and DANC can compensate one another to keep up with demand [131, 132].

Postnatal Establishment of Mechanisms for Energy Transfer

Foetal CMs do not seem to have efficient systems for energy transfer. The foetal heart operates without highly organized mechanisms for energy transfer [133–135]. The organization of glycolytic enzymes in the vicinity of ATPases permits an effective control of the local ATP/ADP ratio [18, 136]. A specific transfer system is hence not required. The loose architecture of the foetal cardiomyocyte also favours diffusion of energy molecules [137, 138].

There is increase in creatine kinase (CK) activity around birth, linked to rise in the cellular content of creatine. With the surge in CK activity, phosphotransfer system gets established progressively to aid the work of ATPases and also provide an effective positive feedback on energy production [50, 133, 139, 140]. Mitochondrial permeability for ADP decreases in parallel with the increased efficiency of mi-CK. Further, the CK shuttle controls mitochondrial respiration in the post-natal heart [133]. Several studies suggest that effectiveness of the CK shuttle depends on the degree of maturity of the cardiomyocyte [134]. Energy transfer dependent on DANC only happens after the establishment of the complex cytoarchitecture, organized mitochondrial network and formation of energy micro-domains in the cardiomyocyte during the early phase of postnatal development of the heart [50]. The link between cytoarchitecture and energy metabolism is evident from studies in CK null mice in which CK deficiency is associated with reorganization of mitochondria within myofilaments and increased DANC efficiency [130].

The Link Between Metabolism and Mechano-Transduction in Cardiomyocytes

Several studies have investigated how sarcomerogenesis and myofibrillogenesis are regulated by the physicochemical characteristics of the cell microenvironment [141–146]. Microenvironmental cues provide a chemo-mechanical signal which controls the shape and organization of the contractile apparatus in the cardiomyocyte [147–149]. Conversely, mechano-transduction may influence cell metabolism as well [147, 150, 151]. CMs in a pathological stiffer cell microenvironment have been shown to remodel to preserve contractile function at the cellular level [147, 152–154].

In addition to their direct effects on excitation–contraction coupling in CMs, interactions between sarcomeres and mitochondria regulate the development and maintenance of intracellular energetic units [155].

Shortening of sarcomeres and swelling of mitochondria are sources of mechanical forces [156, 157]. During contraction of the sarcomeres, mitochondria are squeezed, thus providing an intracellular load [157]. Swollen mitochondria exert pressure against structures nearby [156]. Swollen mitochondria impact the microtubule network and severely deform nuclear shape in CMs. This effect could initiate chromatin remodelling as well as epigenetic and transcriptional events [158]. Thus, the intracellular energetic unit senses and responds to changes in the cell microenvironment (Fig. 7.3).

Mitochondria in adult CMs are more well organized than those in embryonic CMs. They have elevated mitochondrial membrane potential, higher oxygen consumption, and increased ATP production [7, 8]. Cytoskeletal maturation is associated with closure of mitochondrial permeability pores, which are open during the early stages of heart development [159]. Pore closure reduces the cytoplasmic levels of ROS and thus facilitates completion of myofibrillogenesis [159]. Mitochondrial dynamics affects myofibrillogenesis through ROS-mediated mechanisms [151]. Studies using heart tissues from patients with Barth syndrome suggest that impaired mitochondrial structure would not only result in mitochondrial dysfunction but also lead to disruptions in structure of the sarcomere and cardiomyocyte contraction [160].

Mitochondrial tethering to mechanosensing cytoskeletal proteins represents a pathway for the effects of extracellular matrix on both contractile and metabolic apparatuses. This structure facilitates nucleotide channelling and the transport of metabolites involved in oxidative phosphorylation [161–163]. Microtubules and actin are possibly involved in long- and short-distance mitochondrial trafficking respectively [164, 165]. Mitochondrial cristae acquire their distinctive invaginated structure through tethering to tubulin, actin and intermediate filaments [166, 167]. Tubulin interacts also with the mitochondrial outer membrane both physically and chemically and thus modulates the function of the permeability transition pore and of voltage dependent anion channels [168, 169].

External mechanical forces also activate protein-tyrosine kinases of the Src family, focal adhesion kinases (FAK) and mitogen-activated protein kinases (MAPK) [170, 171]. The responses are mediated by integrin. Importantly, integrin also via Rho GTPases, regulates mitochondrial function for signal transduction [172].

The impact of microenvironmental cues on cardiomyocyte metabolism can also be independent from those by the changes in cell structure or function [150]. The cell microenvironment modulates the cross-talk between cytoskeletal architecture and metabolism in CMs through a mechanosensitive feedback loop [155].

Conclusion

The studies on the metabolism of cardiomyocytes (CMs) during heart development have revealed how cellular energy production responds to growth of the heart. After birth, there is an increase in the workload of the heart and CMs improve their contractile capacity by shifting to a more efficient energy source (lipids) and by adapting to specialized systems for energy transfer. Metabolic maturation precedes the maturation of excitation–contraction coupling. Energy metabolism drives the post-natal development of CMs and their complex architecture. The exact interacting mechanisms are unclear. The period of hyperplasia of the cardiomyocyte seems to be the phase when CMs acquire metabolic and morphological features that suit post-natal physiological needs. A clear understanding of the role of energy metabolism in the development and maturation of the cardiomyocyte would be advantageous for the efforts aimed at regenerating CMs in the diseased or injured heart.

References

1. Opie LH. Metabolism of the heart in health and disease. Part I Am Heart J. 1968;76(5):685–98.
2. Opie LH. Metabolism of the heart in health and disease. Part II Am Heart J. 1969;77(1):100–22.
3. Lopaschuk GD, Ussher JR, Folmes CD, Jaswal JS, Stanley WC. Myocardial fatty acid metabolism in health and disease. Physiol Rev. 2010;90(1):207–58.
4. Stanley WC, Recchia FA, Lopaschuk GD. Myocardial substrate metabolism in the normal and failing heart. Physiol Rev. 2005;85(3):1093–129.
5. Neely JR, Morgan HE. Relationship between carbohydrate and lipid metabolism and the energy balance of heart muscle. Annu Rev Physiol. 1974;36(1):413–59.
6. Chung S, Arrell DK, Faustino RS, Terzic A, Dzeja PP. Glycolytic network restructuring integral to the energetics of embryonic stem cell cardiac differentiation. J Mol Cell Cardiol. 2010; 48(4):725–34.
7. Chung S, Dzeja PP, Faustino RS, Perez-Terzic C, Behfar A, Terzic A. Mitochondrial oxidative metabolism is required for the cardiac differentiation of stem cells. Nat Clin Pract Cardiovasc Med. 2007;4(1):S60–7.
8. Cho YM, Kwon S, Pak YK, Seol HW, Choi YM, Park DJ, et al. Dynamic changes in mitochondrial biogenesis and antioxidant enzymes during the spontaneous differentiation of human embryonic stem cells. Biochem Biophys Res Commun. 2006;348(4):1472–8.
9. St. John JC, Ramalho-Santos J, Gray HL, Petrosko P, Rawe VY, Navara CS, et al. The expression of mitochondrial DNA transcription factors during early cardiomyocyte in vitro differentiation from human embryonic stem cells. Cloning Stem Cells. 2005; 7(3):141–53.
10. Vander Heiden MG, Cantley LC, Thompson CB. Understanding the Warburg effect: the metabolic requirements of cell proliferation. Science. 2009;324(5930):1029–33.
11. DeBerardinis RJ, Lum JJ, Thompson CB. Phosphatidylinositol 3-kinase-dependent modulation of carnitine palmitoyltransferase 1A expression regulates lipid metabolism during hematopoietic cell growth. J Biol Chem. 2006;281(49):37372–80.
12. DeBerardinis RJ, Lum JJ, Hatzivassiliou G, Thompson CB. The biology of cancer: metabolic reprogramming fuels cell growth and proliferation. Cell Metab. 2008;7(1):11–20.
13. Baharvand H, Matthaei KI. The ultrastructure of mouse embryonic stem cells. Reprod Biomed Online. 2003;7(3):330–5.

14. Forristal CE, Wright KL, Hanley NA, Oreffo RO, Houghton FD. Hypoxia inducible factors regulate pluripotency and proliferation in human embryonic stem cells cultured at reduced oxygen tensions. Reproduction. 2010;139(1):85.
15. Ezashi T, Das P, Roberts RM. Low O2 tensions and the prevention of differentiation of hES cells. Proc Natl Acad Sci USA. 2005;102(13):4783–8.
16. Williamson JR, Ford C, Illingworth J, Safer B. Coordination of citric acid cycle activity with electron transport flux. Circ Res. 1976;38(5 Suppl 1):I39-51.
17. Saks V, Favier R, Guzun R, Schlattner U, Wallimann T. Molecular system bioenergetics: regulation of substrate supply in response to heart energy demands. J Physiol. 2006;577(3):769–77.
18. Ventura-Clapier R, Garnier A, Veksler V, Joubert F. Bioenergetics of the failing heart. Biochimica et Biophysica Acta (BBA)-Mol Cell Res. 2011; 1813(7):1360–72.
19. Makinde A, Kantor PF, Lopaschuk GD. Maturation of fatty acid and carbohydrate metabolism in the newborn heart. Mol Cell Biochem. 1998; 188:49–56.
20. Lopaschuk GD, Spafford MA, Marsh DR. Glycolysis is predominant source of myocardial ATP production immediately after birth. Am J Physiol-Heart Circul Physiol. 1991;261(6):H1698–705.
21. Lopaschuk GD, Collins-Nakai RL, Itoi T. Developmental changes in energy substrate use by the heart. Cardiovasc Res. 1992;26(12):1172–80.
22. Werner JC, Whitman V, Musselman J, Schuler G. Perinatal changes in mitochondrial respiration of the rabbit heart. Neonatology. 1982;42(5–6):208–16.
23. Minai L, Martinovic J, Chretien D, Dumez F, Razavi F, Munnich A, et al. Mitochondrial respiratory chain complex assembly and function during human fetal development. Mol Genet Metab. 2008;94(1):120–6.
24. Fisher DJ, Heymann MA, Rudolph AM. Myocardial oxygen and carbohydrate consumption in fetal lambs in utero and in adult sheep. Am J Physiol-Heart Circ Physiol. 1980;238(3):H399–405.
25. Rolph T, Jones C. Regulation of glycolytic flux in the heart of the fetal guinea pig. J Dev Physiol. 1983;5(1):31–49.
26. Warshaw JB. Cellular energy metabolism during fetal development: I. Oxidative phosphorylation in the fetal heart. J Cell Biology. 1969; 41(2):651.
27. Warshaw JB. Cellular energy metabolism during fetal development: IV. Fatty acid activation, acyl transfer and fatty acid oxidation during development of the chick and rat. Dev Biology. 1972; 28(4):537–44.
28. Dallman PR, Schwartz HC. Cytochrome c concentrations during rat and guinea pig development. Pediatrics. 1964;33(1):106–10.
29. Werner JC, Sicard RE. Lactate metabolism of isolated, perfused fetal, and newborn pig hearts. Pediatr Res. 1987;22(5):552–6.
30. Itoi T, Lopaschuk GD. The contribution of glycolysis, glucose oxidation, lactate oxidation, and fatty acid oxidation to ATP production in isolated biventricular working hearts from 2-week-old rabbits. Pediatr Res. 1993;34(6):735–41.
31. Lopaschuk GD, Jaswal JS. Energy metabolic phenotype of the cardiomyocyte during development, differentiation, and postnatal maturation. J Cardiovasc Pharmacol. 2010;56(2):130–40.
32. Werner JC, Sicard RE, Schuler HG. Palmitate oxidation by isolated working fetal and newborn pig hearts. Am J Physiol-Endocrinol Metab. 1989;256(2):E315–21.
33. Jones CT, Rolph TP. Metabolism during fetal life: a functional assessment of metabolic development. Physiol Rev. 1985;65(2):357–430.
34. Bristow J, Bier D, Lange L. Regulation of adult and fetal myocardial phosphofructokinase. Relief of cooperativity and competition between fructose 2, 6-bisphosphate, ATP, and citrate. J Biol Chem. 1987; 262(5):2171–175.
35. Ascuitto RJ, Ross-Ascuitto NT. Substrate metabolism in the developing heart. Semin Perinatol. 1996;20:542–63.
36. Girard J, Ferre P, Pegorier J, Duee P. Adaptations of glucose and fatty acid metabolism during perinatal period and suckling-weaning transition. Physiol Rev. 1992;72(2):507–62.

37. Portman O, Behrman R, Soltys P. Transfer of free fatty acids across the primate placenta. Am J Physiol-Legacy Content. 1969;216(1):143–7.
38. Bartelds B, Gratama J-WC, Knoester H, Takens J, Smid GB, Aarnoudse JG, et al. Perinatal changes in myocardial supply and flux of fatty acids, carbohydrates, and ketone bodies in lambs. Am J Physiol-Heart Circ Physiol. 1998; 274(6):H1962-H9.
39. Nakano H, Minami I, Braas D, Pappoe H, Wu X, Sagadevan A, et al. Glucose inhibits cardiac muscle maturation through nucleotide biosynthesis Elife. 2017; 6:e29330. https://doi.org/10.7554/eLife.29330.
40. Bartelds B, Knoester H, Smid GB, Takens J, Visser GH, Penninga L, van der Leij FR, Beaufort-Krol GC, Zijlstra WG, Heymans HS, Kuipers JR. Perinatal changes in myocardial metabolism in lambs. Circulation. 2000;102:926–31.
41. Warshaw JB, Terry ML. Cellular energy metabolism during fetal development. Dev Biol. 1976;52(1):161–6.
42. Iruretagoyena J, Davis W, Bird C, Olsen J, Radue R, Teo Broman A, et al. Metabolic gene profile in early human fetal heart development. Mol Hum Reprod. 2014;20(7):690–700.
43. Park JH, Stoffers DA, Nicholls RD, Simmons RA. Development of type 2 diabetes following intrauterine growth retardation in rats is associated with progressive epigenetic silencing of Pdx1. J Clin Investig. 2008;118(6):2316–24.
44. Voloshyna I, Littlefield MJ, Reiss AB. Atherosclerosis and interferon-γ: new insights and therapeutic targets. Trends Cardiovasc Med. 2014;24(1):45–51.
45. Bitman J, Wood L, Hamosh M, Hamosh P, Mehta NR. Comparison of the lipid composition of breast milk from mothers of term and preterm infants. Am J Clin Nutr. 1983;38(2):300–12.
46. de Carvalho AETS, Bassaneze V, Forni MF, Keusseyan AA, Kowaltowski AJ, Krieger JE. Early postnatal cardiomyocyte proliferation requires high oxidative energy metabolism. Sci Rep. 2017;7(1):1–11.
47. Hallman M. Changes in mitochondrial respiratory chain proteins during perinatal development. Evidence of the importance of environmental oxygen tension. Biochimica et Biophysica Acta (BBA)-Bioenergetics. 1971; 253(2):360–72.
48. Glatz J, Veerkamp J. Postnatal development of palmitate oxidation and mitochondrial enzyme activities in rat cardiac and skeletal muscle. Biochem Biophys Acta. 1982;711(2):327–35.
49. Taha M, Lopaschuk GD. Alterations in energy metabolism in cardiomyopathies. Ann Med. 2007;39(8):594–607.
50. Piquereau J, Novotova M, Fortin D, Garnier A, Ventura-Clapier R, Veksler V, et al. Postnatal development of mouse heart: formation of energetic microdomains. J Physiol. 2010;588(13):2443–54.
51. Martin OJ, Lai L, Soundarapandian MM, Leone TC, Zorzano A, Keller MP, et al. A role for peroxisome proliferator-activated receptor γ coactivator-1 in the control of mitochondrial dynamics during postnatal cardiac growth. Circ Res. 2014;114(4):626–36.
52. Pohjoismäki JL, Boettger T, Liu Z, Goffart S, Szibor M, Braun T. Oxidative stress during mitochondrial biogenesis compromises mtDNA integrity in growing hearts and induces a global DNA repair response. Nucleic Acids Res. 2012;40(14):6595–607.
53. Pohjoismäki JL, Krüger M, Al-Furoukh N, Lagerstedt A, Karhunen PJ, Braun T. Postnatal cardiomyocyte growth and mitochondrial reorganization cause multiple changes in the proteome of human cardiomyocytes. Mol BioSyst. 2013;9(6):1210–9.
54. Barger P, Kelly DP. PPAR signaling in the control of cardiac energy metabolism. Trends Cardiovasc Med. 2000;10:238–45.
55. Lehman JJ, Barger PM, Kovacs A, Saffitz JE, Medeiros DM, Kelly DP. Peroxisome proliferator–activated receptor γ coactivator-1 promotes cardiac mitochondrial biogenesis. J Clin Investig. 2000;106(7):847–56.
56. Warren JS, Oka S-i, Zablocki D, Sadoshima J. Metabolic reprogramming via PPAR signaling in cardiac hypertrophy and failure: From metabolomics to epigenetics. Am J Physiol Heart Circ Physiol. 2017; 313: H584–H596.
57. Fukushima A, Alrob OA, Zhang L, Wagg CS, Altamimi T, Rawat S, et al. Acetylation and succinylation contribute to maturational alterations in energy metabolism in the newborn heart. Am J Physiol-Heart Circ Physiol. 2016;311(2):H347–63.

58. Das DK, Flansaas D, Engelman RM, Rousou JA, Breyer RH, Jones R, et al. Age-related development profiles of the antioxidative defense system and the peroxidative status of the pig heart. Neonatology. 1987;51(3):156–69.
59. Bódi B, Tóth EP, Nagy L, Tóth A, Mártha L, Kovács Á, et al. Titin isoforms are increasingly protected against oxidative modifications in developing rat cardiomyocytes. Free Radical Biol Med. 2017;113:224–35.
60. Taverne YJ, Bogers AJ, Dirk JD, Merkus D. Reactive oxygen species and the cardiovascular system. Oxidative Medicine and Cellular Longevity. 2013; Article ID 862423. https://doi.org/10.1155/2013/862423.
61. Vega RB, Kelly DP. Cardiac nuclear receptors: architects of mitochondrial structure and function. J Clin Investig. 2017;127(4):1155–64.
62. Soonpaa MH, Kim KK, Pajak L, Franklin M, Field LJ. Cardiomyocyte DNA synthesis and binucleation during murine development. Am J Physiol-Heart Circ Phys. 1996;271(5):H2183–9.
63. Ahuja P, Sdek P, MacLellan WR. Cardiac myocyte cell cycle control in development, disease, and regeneration. Physiol Rev. 2007;87(2):521–44.
64. Lopaschuk GD, Spafford MA. Energy substrate utilization by isolated working hearts from newborn rabbits. Am J Physiol-Heart Circ Physiol. 1990;258(5):H1274–80.
65. Medina J. The role of lactate as an energy substrate for the brain during the early neonatal period. Neonatology. 1985;48(4):237–44.
66. Knopp RH, Warth MR, Charles D, Childs M, Li JR, Mabuchi H, et al. Lipoprotein metabolism in pregnancy, fat transport to the fetus, and the effects of diabetes. Neonatology. 1986;50(6):297–317.
67. Kudo N, Gillespie JG, Kung L, Witters LA, Schulz R, Clanachan AS, et al. Characterization of $5'$ AMP-activated protein kinase activity in the heart and its role in inhibiting acetyl-CoA carboxylase during reperfusion following ischemia. Biochimica et Biophysica Acta (BBA)-Lipids Lipid Metab. 1996; 1301(1–2):67–75.
68. Fisher DJ, Heymann MA, Rudolph AM. Regional myocardial blood flow and oxygen delivery in fetal, newborn, and adult sheep. Am J Physiol-Heart Circ Physiol. 1982;243(5):H729–31.
69. Lopaschuk GD, Spafford MA. Differences in myocardial ischemic tolerance between 1-and 7-day-old rabbits. Can J Physiol Pharmacol. 1992;70(10):1315–23.
70. Kunau W-H, Dommes V, Schulz H. β-Oxidation of fatty acids in mitochondria, peroxisomes, and bacteria: a century of continued progress. Prog Lipid Res. 1995;34(4):267–342.
71. Lopaschuk GD, Belke DD, Gamble J, Toshiyuki I, Schönekess BO. Regulation of fatty acid oxidation in the mammalian heart in health and disease. Biochimica et Biophysica Acta (BBA)-Lipids Lipid Metab. 1994; 1213(3):263–76.
72. McGarry JD, Woeltje KF, Kuwajima M, Foster DW. Regulation of ketogenesis and the renaissance of carnitine palmitoyltransferase. Diabetes Metab Rev. 1989;5(3):271–84.
73. Van der Vusse G, Glatz J, Stam H, Reneman RS. Fatty acid homeostasis in the normoxic and ischemic heart. Physiol Rev. 1992;72(4):881–940.
74. Lopaschuk GD, Witters LA, Itoi T, Barr R, Barr A. Acetyl-CoA carboxylase involvement in the rapid maturation of fatty acid oxidation in the newborn rabbit heart. J Biol Chem. 1994;269(41):25871–8.
75. Lopaschuk GD, Gamble J. The 1993 Merck Frosst Award. Acetyl-CoA carboxylase: an important regulator of fatty acid oxidation in the heart. Can J Physiol Pharmacol. 1994; 72(10):1101–109.
76. Dyck JR, Barr AJ, Barr RL, Kolattukudy PE, Lopaschuk GD. Characterization of cardiac malonyl-CoA decarboxylase and its putative role in regulating fatty acid oxidation. Am J Physiol-Heart Circ Physiol. 1998;275(6):H2122–9.
77. Saddik M, Gamble J, Witters L, Lopaschuk G. Acetyl-CoA carboxylase regulation of fatty acid oxidation in the heart. J Biol Chem. 1993;268(34):25836–45.
78. Bianchi A, Evans JL, Iverson AJ, Nordlund A-C, Watts TD, Witters L. Identification of an isozymic form of acetyl-CoA carboxylase. J Biol Chem. 1990;265(3):1502–9.

79. Hardie DG. Regulation of fatty acid synthesis via phosphorylation of acetyl-CoA carboxylase. Prog Lipid Res. 1989;28(2):117–46.
80. Hardie DG. Regulation of fatty acid and cholesterol metabolism by the AMP-activated protein kinase. Biochimica et Biophysica Acta (BBA)-Lipids Lipid Metab. 1992; 1123(3):231–38.
81. Hardie DG. An emerging role for protein kinases: the responses to nutritional and environmental stress. Cell Signal. 1994;6(8):813–21.
82. Hawley SA, Davison M, Woods A, Davies SP, Beri RK, Carling D, et al. Characterization of the AMP-activated protein kinase kinase from rat liver and identification of threonine 172 as the major site at which it phosphorylates AMP-activated protein kinase. J Biol Chem. 1996;271(44):27879–87.
83. Makinde A-O, Gamble J, Lopaschuk GD. Upregulation of 5′-AMP–activated protein kinase is responsible for the increase in myocardial fatty acid oxidation rates following birth in the newborn rabbit. Circ Res. 1997;80(4):482–9.
84. Yatscoff MA, Jaswal JS, Grant MR, Greenwood R, Lukat T, Beker DL, et al. Myocardial hypertrophy and the maturation of fatty acid oxidation in the newborn human heart. Pediatr Res. 2008;64(6):643–7.
85. Cheng L, Ding G, Qin Q, Huang Y, Lewis W, He N, et al. Cardiomyocyte-restricted peroxisome proliferator-activated receptor-δ deletion perturbs myocardial fatty acid oxidation and leads to cardiomyopathy. Nat Med. 2004;10(11):1245–50.
86. Schuler M, Ali F, Chambon C, Duteil D, Bornert J-M, Tardivel A, et al. PGC1α expression is controlled in skeletal muscles by PPARβ, whose ablation results in fiber-type switching, obesity, and type 2 diabetes. Cell Metab. 2006;4(5):407–14.
87. Yoda-Murakami M, Taniguchi M, Takahashi K, Kawamata S, Saito K, Choi-Miura N-H, et al. Change in expression of GBP28/adiponectin in carbon tetrachloride-administrated mouse liver. Biochem Biophys Res Commun. 2001;285(2):372–7.
88. Lee GY, Kim NH, Zhao Z-S, Cha BS, KIM YS. Peroxisomal-proliferator-activated receptor alpha activates transcription of the rat hepatic malonyl-CoA decarboxylase gene: a key regulation of malonyl-CoA level. Biochem J. 2004; 378(3):983–90.
89. Li Y, Cheng L, Qin Q, Liu J, Lo W-k, Brako LA, et al. High-fat feeding in cardiomyocyte-restricted PPARδ knockout mice leads to cardiac overexpression of lipid metabolic genes but fails to rescue cardiac phenotypes. J Mol Cell Cardiol. 2009; 47(4):536–43.
90. Holness M, Sugden M. Regulation of pyruvate dehydrogenase complex activity by reversible phosphorylation. Biochem Soc Trans. 2003;31(6):1143–51.
91. Sugden MC, Holness MJ. Recent advances in mechanisms regulating glucose oxidation at the level of the pyruvate dehydrogenase complex by PDKs. Am J Physiol-Endocrinol Metab. 2003;284(5):E855–62.
92. McCORMACK JG, Denton RM. The activation of pyruvate dehydrogenase in the perfused rat heart by adrenaline and other inotropic agents. Biochem J. 1981;194(2):639–43.
93. Unitt JF, McCormack JG, Reid D, MacLachlan L, England P. Direct evidence for a role of intramitochondrial Ca2+ in the regulation of oxidative phosphorylation in the stimulated rat heart. Studies using31P nmr and ruthenium red. Biochem J. 1989; 262(1):293–301.
94. Stepanov V, Mateo P, Gillet B, Beloeil J, Lechene P, Hoerter J. Kinetics of creatine kinase in an experimental model of low phosphocreatine and ATP in the normoxic heart. Am J Physiol Cell Physiol. 1997;273(4):C1397–408.
95. Hue L, Taegtmeyer H. The Randle cycle revisited: a new head for an old hat. Am J Physiol-Endocrinol Metab. 2009;297(3):E578–91.
96. Iyer NV, Kotch LE, Agani F, Leung SW, Laughner E, Wenger RH, et al. Cellular and developmental control of O2 homeostasis by hypoxia-inducible factor 1α. Genes Dev. 1998;12(2):149–62.
97. Ryan HE, Lo J, Johnson RS. HIF-1α is required for solid tumor formation and embryonic vascularization. EMBO J. 1998;17(11):3005–15.
98. Wood SM, Wiesener MS, Yeates KM, Okada N, Pugh CW, Maxwell PH, et al. Selection and analysis of a mutant cell line defective in the hypoxia-inducible factor-1 α-subunit (HIF-1α): Characterization of HIF-1α-dependent and-independent hypoxia-inducible gene expression. J Biol Chem. 1998;273(14):8360–8.

99. Kim J-W, Tchernyshyov I, Semenza GL, Dang CV. HIF-1-mediated expression of pyruvate dehydrogenase kinase: a metabolic switch required for cellular adaptation to hypoxia. Cell Metab. 2006; 3(3):177–85.

100. Papandreou I, Cairns RA, Fontana L, Lim AL, Denko NC. HIF-1 mediates adaptation to hypoxia by actively downregulating mitochondrial oxygen consumption. Cell Metab. 2006;3(3):187–97.

101. Huss JM, Kelly DP. Nuclear receptor signaling and cardiac energetics. Circ Res. 2004;95(6):568–78.

102. Finck BN. The PPAR regulatory system in cardiac physiology and disease. Cardiovasc Res. 2007;73(2):269–77.

103. van der Lee KA, Vork MM, De Vries JE, Willemsen PH, Glatz JF, Reneman RS, et al. Long-chain fatty acid-induced changes in gene expression in neonatal cardiac myocytes. J Lipid Res. 2000;41(1):41–7.

104. Brandt JM, Djouadi F, Kelly DP. Fatty acids activate transcription of the muscle carnitine palmitoyltransferase I gene in cardiac myocytes via the peroxisome proliferator-activated receptor α. J Biol Chem. 1998;273(37):23786–92.

105. Mascaró C, Acosta E, Ortiz JA, Marrero PF, Hegardt FG, Haro D. Control of human muscle-type carnitine palmitoyltransferase I gene transcription by peroxisome proliferator-activated receptor. J Biol Chem. 1998;273(15):8560–3.

106. Gulick T, Cresci S, Caira T, Moore DD, Kelly DP. The peroxisome proliferator-activated receptor regulates mitochondrial fatty acid oxidative enzyme gene expression. Proc Natl Acad Sci USA. 1994;91(23):11012–6.

107. Leone TC, Weinheimer CJ, Kelly DP. A critical role for the peroxisome proliferator-activated receptor α (PPARα) in the cellular fasting response: the PPARα-null mouse as a model of fatty acid oxidation disorders. Proc Natl Acad Sci USA. 1999;96(13):7473–8.

108. Panadero M, Herrera E, Bocos C. Peroxisome proliferator-activated receptor-α expression in rat liver during postnatal development. Biochimie. 2000;82(8):723–6.

109. Škárka L, Bardová K, Brauner P, Flachs P, Jarkovská D, Kopecký J, et al. Expression of mitochondrial uncoupling protein 3 and adenine nucleotide translocase 1 genes in developing rat heart: putative involvement in control of mitochondrial membrane potential. J Mol Cell Cardiol. 2003;35(3):321–30.

110. Wang Y-X, Lee C-H, Tiep S, Ruth TY, Ham J, Kang H, et al. Peroxisome-proliferator-activated receptor δ activates fat metabolism to prevent obesity. Cell. 2003;113(2):159–70.

111. Gilde AJ, van der Lee KA, Willemsen PH, Chinetti G, van der Leij FR, van der Vusse GJ, et al. Peroxisome proliferator-activated receptor (PPAR) α and PPARβ/δ, but not PPARγ, modulate the expression of genes involved in cardiac lipid metabolism. Circ Res. 2003;92(5):518–24.

112. Hondares E, Pineda-Torra I, Iglesias R, Staels B, Villarroya F, Giralt M. PPARδ, but not PPARα, activates PGC-1α gene transcription in muscle. Biochem Biophys Res Commun. 2007;354(4):1021–7.

113. Pellieux C, Montessuit C, Papageorgiou I, Lerch R. Angiotensin II downregulates the fatty acid oxidation pathway in adult rat cardiomyocytes via release of tumour necrosis factor-α. Cardiovasc Res. 2009;82(2):341–50.

114. Planavila A, Laguna JC, Vázquez-Carrera M. Nuclear factor-κB activation leads to down-regulation of fatty acid oxidation during cardiac hypertrophy. J Biol Chem. 2005;280(17):17464–71.

115. Burkart EM, Sambandam N, Han X, Gross RW, Courtois M, Gierasch CM, et al. Nuclear receptors PPARβ/δ and PPARα direct distinct metabolic regulatory programs in the mouse heart. J Clin Investig. 2007;117(12):3930–9.

116. Razeghi P, Young ME, Abbasi S, Taegtmeyer H. Hypoxia in vivo decreases peroxisome proliferator-activated receptor α-regulated gene expression in rat heart. Biochem Biophys Res Commun. 2001;287(1):5–10.

117. Belanger AJ, Luo Z, Vincent KA, Akita GY, Cheng SH, Gregory RJ, et al. Hypoxia-inducible factor 1 mediates hypoxia-induced cardiomyocyte lipid accumulation by reducing the DNA binding activity of peroxisome proliferator-activated receptor α/retinoid X receptor. Biochem Biophys Res Commun. 2007;364(3):567–72.

118. Ventura-Clapier R, Kuznetsov A, Veksler V, Boehm E, Anflous K. Functional coupling of creatine kinases in muscles: Species and tissue specificity. Bioenergetics of the Cell: Quantitative Aspects: Springer; 1998. p. 231–47.

119. Saks V, Beraud N, Wallimann T. Metabolic compartmentation–a system level property of muscle cells. Int J Mol Sci. 2008;9(5):751–67.

120. Bessman SP, Yang WC, Geiger PJ, Erickson-Viitanen S. Intimate coupling of creatine phosphokinase and myofibrillar adenosinetriphosphatase. Biochem Biophys Res Commun. 1980;96(3):1414–20.

121. Wallimann T, Wyss M, Brdiczka D, Nicolay K, Eppenberger H. Intracellular compartmentation, structure and function of creatine kinase isoenzymes in tissues with high and fluctuating energy demands: the'phosphocreatine circuit'for cellular energy homeostasis. Biochem J. 1992;281(Pt 1):21.

122. Wallimann T, Eppenberger HM. Localization and function of M-line-bound creatine kinase. In: Shay JW, editor.Cell and Muscle Motility. Boston: Springer;1985. pp. 239–85. https://doi.org/10.1007/978-1-4757-4723-2_8.

123. Rossi A, Eppenberger H, Volpe P, Cotrufo R, Wallimann T. Muscle-type MM creatine kinase is specifically bound to sarcoplasmic reticulum and can support Ca2+ uptake and regulate local ATP/ADP ratios. J Biol Chem. 1990;265(9):5258–66.

124. Schnyder T, Rojo M, Furter R, Wallimann T. The structure of mitochondrial creatine kinase and its membrane binding properties. Mol Cell Biochem. 1994;133:115–23.

125. Saks V, Khuchua Z, Vasilyeva E, Belikova OY, Kuznetsov A. Metabolic compartmentation and substrate channelling in muscle cells. Mol Cell Biochem. 1994;133(1):155–92.

126. Saks V, Dzeja P, Schlattner U, Vendelin M, Terzic A, Wallimann T. Cardiac system bioenergetics: metabolic basis of the Frank-Starling law. J Physiol. 2006;571(2):253–73.

127. Saks VA, Belikova YO, Kuznetsov AV. In vivo regulation of mitochondrial respiration in cardiomyocytes: specific restrictions for intracellular diffusion of ADP. Biochimica et Biophysica Acta (BBA)-Gen Subj. 1991; 1074(2):302–11.

128. Dzeja PP, Terzic A. Phosphotransfer networks and cellular energetics. J Exp Biol. 2003;206(12):2039–47.

129. Jacobus WE. Theoretical support for the heart phosphocreatine energy transport shuttle based on the intracellular diffusion limited mobility of ADP. Biochem Biophys Res Ccommun. 1985;133(3):1035–41.

130. Kaasik A, Veksler V, Boehm E, Novotova M, Minajeva A, Ventura-Clapier R. Energetic crosstalk between organelles: architectural integration of energy production and utilization. Circ Res. 2001;89:153–9.

131. Ventura-Clapier R, Kaasik A, Veksler V. Structural and functional adaptations of striated muscles to CK deficiency. Mol Cell Biochem. 2004;256:29–41.

132. Tylková L. Architectural and functional remodeling of cardiac and skeletal muscle cells in mice lacking specific isoenzymes of creatine kinase. Gen Physiol Biophys. 2009;28(3):219.

133. Hoerter JA, Kuznetsov A, Ventura-Clapier R. Functional development of the creatine kinase system in perinatal rabbit heart. Circ Res. 1991;69(3):665–76.

134. Hoerter JA, Ventura-Clapier R, Kuznetsov A. Compartmentation of creatine kinases during perinatal development of mammalian heart. 1994; 133:277–-86.

135. Tiivel T, Kadaya L, Kuznetsov A, Käämbre T, Peet N, Sikk P, et al. Developmental changes in regulation of mitochondrial respiration by ADP and creatine in rat heart in vivo. Mol Cell Biochem. 2000;208(1):119–28.

136. Brooks S, Storey KB. Where is the glycolytic complex? A critical evaluation of present data from muscle tissue. FEBS Lett. 1991;278(2):135–8.

137. Hirschy A, Schatzmann F, Ehler E, Perriard J-C. Establishment of cardiac cytoarchitecture in the developing mouse heart. Dev Biol. 2006;289(2):430–41.

138. Lozyk MD, Papp S, Zhang X, Nakamura K, Michalak M, Opas M. Ultrastructural analysis of development of myocardium in calreticulin-deficient mice. BMC Dev Biol. 2006;6(1):1–16.

139. Fischer A, Ten Hove M, Sebag-Montefiore L, Wagner H, Clarke K, Watkins H, et al. Changes in creatine transporter function during cardiac maturation in the rat. BMC Dev Biol. 2010;10(1):1–9.

140. Anmann T, Varikmaa M, Timohhina N, Tepp K, Shevchuk I, Chekulayev V, et al. Formation of highly organized intracellular structure and energy metabolism in cardiac muscle cells during postnatal development of rat heart. Biochimica et Biophysica Acta (BBA)-Bioenergetics. 2014; 1837(8):1350–61.

141. Chien KR, Domian IJ, Parker KK. Cardiogenesis and the complex biology of regenerative cardiovascular medicine. Science. 2008;322(5907):1494–7.

142. Dabiri GA, Turnacioglu KK, Sanger JM, Sanger JW. Myofibrillogenesis visualized in living embryonic cardiomyocytes. Proc Natl Acad Sci USA. 1997;94(17):9493–8.

143. Grosberg A, Kuo P-L, Guo C-L, Geisse NA, Bray M-A, Adams WJ, et al. Self-organization of muscle cell structure and function. Plos Comput Biol. 2011; 7(2):e1001088.

144. Pasqualini FS, Sheehy SP, Agarwal A, Aratyn-Schaus Y, Parker KK. Structural phenotyping of stem cell-derived cardiomyocytes. Stem Cell Reports. 2015;4(3):340–7.

145. Parker KK, Tan J, Chen CS, Tung L. Myofibrillar architecture in engineered cardiac myocytes. Circ Res. 2008;103(4):340–2.

146. Dasbiswas K, Majkut S, Discher D, Safran SA. Substrate stiffness-modulated registry phase correlations in cardiomyocytes map structural order to coherent beating. Nat Commun. 2015;6(1):1–8.

147. McCain ML, Yuan H, Pasqualini FS, Campbell PH, Parker KK. Matrix elasticity regulates the optimal cardiac myocyte shape for contractility. Am J Physiol-Heart Circ Physiol. 2014;306(11):H1525–39.

148. McCain ML, Lee H, Aratyn-Schaus Y, Kléber AG, Parker KK. Cooperative coupling of cell-matrix and cell–cell adhesions in cardiac muscle. Proc Natl Acad Sci USA. 2012;109(25):9881–6.

149. McCain ML, Sheehy SP, Grosberg A, Goss JA, Parker KK. Recapitulating maladaptive, multiscale remodeling of failing myocardium on a chip. Proc Natl Acad Sci USA. 2013;110(24):9770–5.

150. McCain ML, Agarwal A, Nesmith HW, Nesmith AP, Parker KK. Micromolded gelatin hydrogels for extended culture of engineered cardiac tissues. Biomaterials. 2014;35(21):5462–71.

151. Wang G, McCain ML, Yang L, He A, Pasqualini FS, Agarwal A, et al. Modeling the mitochondrial cardiomyopathy of Barth syndrome with induced pluripotent stem cell and heart-on-chip technologies. Nat Med. 2014;20(6):616–23.

152. McCain ML, Parker KK. Mechanotransduction: the role of mechanical stress, myocyte shape, and cytoskeletal architecture on cardiac function. Pflügers Archiv-Eur J Physiol. 2011;462(1):89–104.

153. Horton RE, Yadid M, McCain ML, Sheehy SP, Pasqualini FS, Park S-J, et al. Angiotensin II induced cardiac dysfunction on a chip. PLoS One. 2016; 11(1):e0146415.

154. Gerdes AM. Cardiac myocyte remodeling in hypertrophy and progression to failure. J Cardiac Fail. 2002;8(6):S264–8.

155. Pasqualini FS, Nesmith AP, Horton RE, Sheehy SP, Parker KK. Mechanotransduction and metabolism in cardiomyocyte microdomains. BioMed Res Int. 2016; Article ID 4081638. https://doi.org/10.1155/2016/4081638.

156. Kaasik A, Kuum M, Joubert F, Wilding J, Ventura-Clapier R, Veksler V. Mitochondria as a source of mechanical signals in cardiomyocytes. Cardiovasc Res. 2010;87(1):83–91.

157. Yaniv Y, Juhaszova M, Wang S, Fishbein KW, Zorov DB, Sollott SJ. Analysis of mitochondrial 3D-deformation in cardiomyocytes during active contraction reveals passive structural anisotropy of orthogonal short axes. PLoS One. 2011; 6(7):e21985.

158. Ramdas NM, Shivashankar G. Cytoskeletal control of nuclear morphology and chromatin organization. J Mol Biol. 2015;427(3):695–706.

159. Hom JR, Quintanilla RA, Hoffman DL, de Mesy Bentley KL, Molkentin JD, Sheu S-S, et al. The permeability transition pore controls cardiac mitochondrial maturation and myocyte differentiation. Dev Cell. 2011;21(3):469–78.

160. Zangi L, Lui KO, Von Gise A, Ma Q, Ebina W, Ptaszek LM, et al. Modified mRNA directs the fate of heart progenitor cells and induces vascular regeneration after myocardial infarction. Nat Biotechnol. 2013;31(10):898–907.

161. Saks VA, Kaambre T, Sikk P, Eimre M, Orlova E, Paju K, et al. Intracellular energetic units in red muscle cells. Biochem J. 2001;356(2):643–57.
162. Appaix F, Kuznetsov AV, Usson Y, Kay L, Andrienko T, Olivares J, et al. Possible role of cytoskeleton in intracellular arrangement and regulation of mitochondria. Exp Physiol. 2003;88(1):175–90.
163. Anmann T, Guzun R, Beraud N, Pelloux S, Kuznetsov AV, Kogerman L, et al. Different kinetics of the regulation of respiration in permeabilized cardiomyocytes and in HL-1 cardiac cells: importance of cell structure/organization for respiration regulation. Biochimica et Biophysica Acta (BBA)-Bioenergetics. 2006; 1757(12):1597–606.
164. Elhanany-Tamir H, Yu YV, Shnayder M, Jain A, Welte M, Volk T. Organelle positioning in muscles requires cooperation between two KASH proteins and microtubules. J Cell Biol. 2012;198(5):833–46.
165. Boldogh IR, Pon LA. Interactions of mitochondria with the actin cytoskeleton. Biochimica et Biophysica Acta (BBA)-Molecular Cell Research. 2006; 1763(5–6):450–62.
166. Voeltz GK, Prinz WA. Sheets, ribbons and tubules—how organelles get their shape. Nat Rev Mol Cell Biol. 2007;8(3):258–64.
167. Tang HL, Lung HL, Wu KC, Le A-HP, Tang HM, Fung MC. Vimentin supports mitochondrial morphology and organization. Biochem J. 2008; 410(1):141–6.
168. Kumazawa A, Katoh H, Nonaka D, Watanabe T, Saotome M, Urushida T, et al. Microtubule disorganization affects the mitochondrial permeability transition pore in cardiac myocytes. Circ J. 2014;78(5):1206–15.
169. Rostovtseva TK, Sheldon KL, Hassanzadeh E, Monge C, Saks V, Bezrukov SM, et al. Tubulin binding blocks mitochondrial voltage-dependent anion channel and regulates respiration. Proc Natl Acad Sci USA. 2008;105(48):18746–51.
170. de Cavanagh EM, Ferder M, Inserra F, Ferder L. Angiotensin II, mitochondria, cytoskeletal, and extracellular matrix connections: an integrating viewpoint. Am J Physiol-Heart Circ Physiol. 2009;296(3):H550–8.
171. Schlaepfer DD, Hunter T. Signal transduction from the extracellular matrix-a role for the focal adhesion protein-tyrosine kinase FAK. Cell Struct Funct. 1996;21(5):445–50.
172. Werner E, Werb Z. Integrins engage mitochondrial function for signal transduction by a mechanism dependent on Rho GTPases. J Cell Biol. 2002;158(2):357–68.

Part II
Cardiomyocyte Responses

Chapter 8
Response of Cardiomyocytes to Mechanical Stress

Abstract Mechanical stretch activates several intracellular signalling networks and modulates gene expressions which initiate structural and functional remodelling in cardiomyocytes. Various durations, loads and frequencies of mechanical forces induce differing response mechanisms. Cardiomyocytes remodel in response to mechanical stress both during cardiac development as well as in disease conditions. In both of these situations, the cytoskeleton has a key role in sensing mechanical stress and in mediating adaptive or maladaptive changes within the cardiomyocyte. Cytoskeleton which is highly sensitive and adaptive to mechanical forces acts as a signal integrator for mechanical and structural inputs. The information is transmitted throughout the cardiomyocyte to ensure that the cell respond and adapt appropriately. Mechanical stretch induces secretion or synthesis of molecules with autocrine and paracrine effects. Energy metabolism is also altered in adult cardiomyocytes, which hypertrophy in response to mechanical stress.

Keyword Heart · Cardiomyocyte · Cytoskeleton · Haemodynamics · Mechanical stress · Mechanosensitivity · Mechanotransduction · Cardiac hypertrophy

Introduction

Sustained haemodynamic stress induced by pressure or volume overload is known to cause structural and functional alterations in cardiomyocytes (CMs) and thus contribute to hypertrophy of cardiac ventricles. Mechanical stretch induced hypertrophy of CMs is accompanied by an increase in cell size, enhanced organization of sarcomeres and induction of the 'foetal' gene programme [1, 2].

Studies subjecting isolated CMs and isolated heart to mechanical stress have revealed the cellular and molecular mechanisms of gene expression and signal transduction induced solely by mechanical stress. Three models of mechanical stress were used for such studies: mechanical stretch, aortocaval shunt and balloon dilation of the left ventricle in the isolated heart [3–5]. In vitro studies have indicated that mechanical stretch regulates growth, apoptosis, electric remodelling, alterations in gene expression, and autocrine and paracrine effects.

© The Author(s), under exclusive license to Springer Nature Switzerland AG 2021   95
C. C. Kartha, *Cardiomyocytes in Health and Disease*,
https://doi.org/10.1007/978-3-030-85536-9_8

Role of Mechanical Forces in the Development of Heart

In vivo studies indicate that mechanical stimuli dictate cardiac form and chamber formation in the developing heart [6]. There is a direct link between hemodynamic pressure and cardiac morphogenesis. Blood forms vortices inside the ventricle, creating gradients of shear stress which induce localized differences in cellular and cytoskeletal remodelling [7–9]. Change in the shape of the cardiomyocyte is an important contributor to cardiac looping, an early step in the formation of the four-chambered heart [10, 11]. Looping occurs because of asymmetric changes in the muscular layer. One side of the heart tube becomes thin because of flattening of CMs in focal regions. Mechanical factors such as hemodynamic stress are considered to regulate regional cell shape changes in the heart tube [9].

Cytoskeleton has a key role in linking extracellular mechanical forces with remodelling of the shape of cardiomyocyte during development, which impacts cardiac morphogenesis. There are evidences to suggest that localized remodelling of cell shape and looping of the heart tube are modulated by actin dynamics [12–18]. Cross-talk between the extracellular matrix (ECM) and CMs is assisted by transmembrane integrin receptors that attach to ECM ligands and to the actomyosin cytoskeleton via proteins such as talin, vinculin, and α-actinin [19–21]. Complex interactions between the ECM components and the diverse integrin subunits coregulate cytoskeletal remodelling during development of the heart by providing structural stability to the cardiomyocyte and regulating spatiotemporal proliferation and migration of the cell.

Effects of Mechanical Stress on the Cytoskeleton

Cells convert mechanical forces to chemical and electrical responses by a process known as mechanotransduction. Mechanotransduction aid balancing the structure and function of the cell. The mechanosensory integrator of mechanical inputs from the extracellular microenvironment and a wide range of physiological responses needed for functional homeostasis in cells is the cytoskeletal network [22]. The cytoskeleton is linked to mechanosensing protein complexes on the cell surface. Cytoskeletal reorganization is one among the several responses provoked by mechanical stimuli. The cytoskeletal structure of CMs is capable of responding to mechanical forces based on their direction, source, and temporal frequency. Cytoskeleton also propagates mechanical signals throughout the cell and activates multiple functional responses.

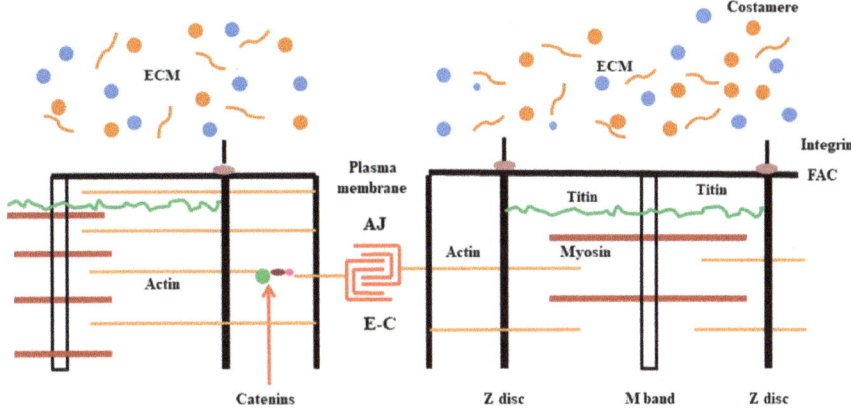

Fig. 8.1 Mechanosensors in the cardiomyocyte. ECM-Extracellular matrix. AJ—Adherens junction. E-C—E cadherin

Mechanosensitivity of Cytoskeleton

Several mechanosensory strategies aid the specific and sensitive responses of the cytoskeleton to mechanical stress. Mechanosensing complexes which can detect both the source and direction of mechanical stress are directly linked to distinctly oriented myofibril components (Fig. 8.1). This feature permits the cytoskeleton to initiate requisite responses within the cardiomyocyte [22]. Integrins attach Z-discs to the ECM at costameres. Cadherins link actin myofilament of adjacent cells at adherens junctions. Titin extends between Z-discs and M-line in the sarcomere. These structures function respectively as mechanotransducers for extracellular, intercellular, and intracellular mechanical stimuli (Fig. 8.2).

Extracellular Mechanics

Integrins and costameres are involved in the response of CMs to alterations in extracellular mechanics. Integrins localized to costameres in the adult CMs (Fig. 8.1) are the physical links between ECM and intracellular cytoskeleton. They are mechanotransducers [23, 24]. Focal adhesion proteins and costameres sense mechanical stress from the ECM [25]. They interact with signalling molecules such as focal adhesion kinase and integrin-linked kinase [26]. The costameric mechanosensing protein melusin, binds directly to the cytoplasmic tail of β1-integrin at costameres [27]. Melusin is involved in stretch-activated signalling pathways for protein synthesis and hypertrophy induced by mechanical stress [28]. Muscle LIM protein (MLP), yet another mechanosensor of the costameric complex [29] is considered to permit titin to sense and respond to mechanical stresses relayed from the ECM [30, 31].

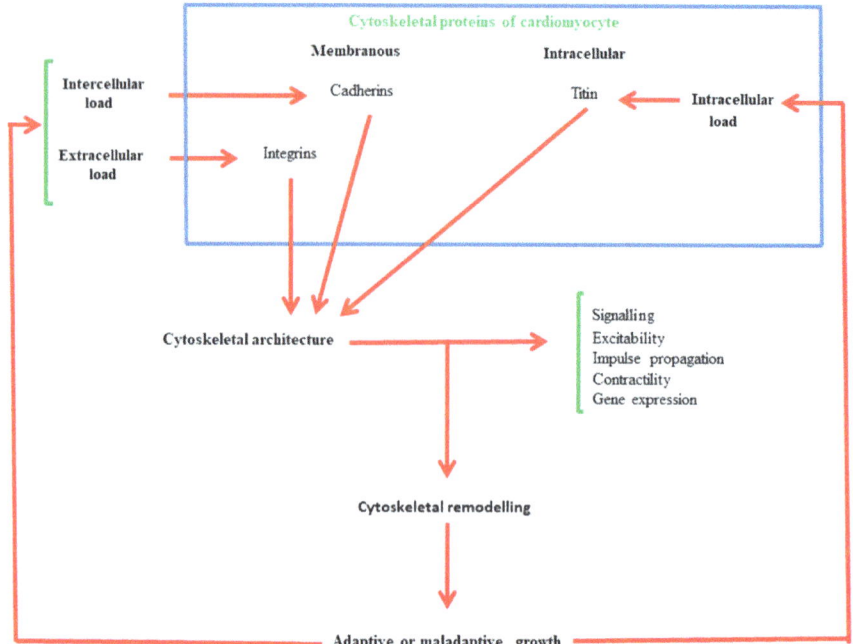

Fig. 8.2 Functional responses of cytoskeleton to intercellular, intra cellular and extracellular loads and the sensors associated with the cytoskeleton

Intracellular Mechanics

Titin is a highly elastic protein that extends between Z-discs and M-bands within sarcomeres and functions as a molecular spring which supports sarcomere recoiling after contraction (Fig. 8.1) [32–34]. Titin is closely associated with both myosin and actin filaments [35]. In addition to its role in cardiomyocyte mechanics, titin also regulates gene expression by interacting with several molecules involved in various signalling pathways [17, 36, 37]. Thus, titin can function as a mechanosensor sensitive to sarcomere length and intracellular strains as well as regulate mechanotransduction.

Several mechanisms for the mechano sensing function of titin have been proposed [37, 38]. Muscle ankyrin repeat proteins (MARPs) are activated after binding between specific domains of the titin molecule in the elastic I-band region [39]. MARPs activate signalling and transcription. When myocytes are stretched, titin–MARP binding is enhanced [39]. The observation that MARP expression is upregulated during both pressure overload hypertrophy and in dilated cardiomyopathy suggest that MARPs are sensitive to increased mechanical stress [40, 41]. Four-and-a-half-LIM-domain (FHL) proteins, which activate transcription and signalling also bind to the elastic I-band region of titin [36]. They are also thought to be involved in mechanosensing [42–44]. The elasticity of titin allows it to function as an intracellular stress sensor depending on the level of tension applied to the molecule, by either exposing or hiding different domains to bind with signalling molecules [22].

Intercellular Mechanics

Cadherins are transmembrane receptors that bind together adjacent cells and link intracellularly to actin filaments via catenins (Fig. 8.1) [45]. In the foetal and neonatal heart, cadherins are distributed around all borders and is also seen in costameric Z disc structures [46–51]. Cadherins form lateral cell–cell junctions in foetal and neonatal CMs and hence detect mechanical stresses along the transverse axis. Cadherins can transmit mechanical stress from adjacent cells directly to the actomyosin cytoskeleton. Cadherins assist the bi-directional transmission of cytoskeletal tension between CMs. Cadherins also strengthen cell–cell junctions in response to pulling forces and act as mechanosensors [52, 53]. The mechanism of mechanosensing is unclear. The observation that myofibrillogenesis is disrupted on inhibiting cadherin function indicates that mechanotransduction at cadherins is essential for integrated development of the heart [49, 54–58]. Cadherins may actively remodel the cytoskeleton to strengthen the adherens junction, localized to intercalated discs, where they bind ends of myofibrils in the adult CMs [46].

Adherens junctions have directional specificity. They sense mechanical stress transmitted primarily from the longitudinal direction. Formation of the adherens junction mediates cytoskeletal structure and is necessary for myofibrillogenesis and the formation of a cohesive, functional tissue. In adult CMs, adherens junctions localize to longitudinal borders in intercalated discs. They are absent in lateral borders [59]. Polarization of adherens junctions coincides temporally with birth when CMs elongate and hypertrophy and is coincident with increased force generation [60, 61]. In vitro studies indicate that adherens junctions positively respond to mechanical stress [52, 53, 62–64]. Lateral borders subjected to high shear forces may prevent the formation and maintenance of adherens junctions. Since adherens junctions are responsive to tension, maturation of intrinsic contractility after birth could induce polarization of adherens junctions to intercalated discs.

Microtubules as Mechanosensors

Microtubules may also participate in sensing both exogenous and intrinsic mechanical stress. Microtubules bear compressive loads and the compressive elements collapse when mechanical load exceeds a specific intensity [65]. The microtubule network within the cardiomyocyte is known to undergo topological alterations during both heart development and in disease conditions. Pressure overload hypertrophy is associated with hyper polymerization of the microtubule, which increases the viscosity of the cytoplasm and deter sarcomere motion [66–68]. Microtubules in beating CMs buckle during systole and unbuckle during diastole [69]. Compressive loads borne by the microtubule are transmitted throughout the cytoskeletal network resulting in integration of responses of several intracellular compartments to mechanical stress.

Mechanical Stress and Cellular Excitability

Pressure or volume overload results in a stretched cardiac ventricle. Shortening of action potential duration, slight depolarization of resting membrane potential, premature ventricular excitations and arrhythmias are associated with ventricular stretch [70–72]. External mechanical forces are considered to regulate cellular excitability through mechanisms mediated by the cytoskeleton. Studies indicate that transmission of mechanical forces by the cytoskeleton modulate activity and excitability of the ion channels [73–76]. Ion channel kinetics are altered by changes in the cytoskeleton. Distinct structures of the cytoskeleton have dissimilar electrophysiological properties [77, 78]. Interestingly, the cytoskeletal topography as seen in healthy adult CMs is ideal for normal expression and function of ion channels.

Mechanical Stress Induced Structural Changes in Cardiomyocytes

CMs in vitro respond to cyclic mechanical strain by elongation, altered orientation, significant enlargement in size, increase in overall protein expression and the activation of foetal gene programs [1, 79–82]. Cell alignment and cell polarity are also determined by mechanical stretch [83]. Rac1 activity which is necessary for cardiomyocyte alignment in response to mechanical stretch is considered a common downstream signal transducer of integrins as mechanoreceptors [84, 85].

The changes in gene expression may equip the cardiomyocyte to endure mechanical stress. Some of the stretch mediated mechanisms are governed by the direction of stretch relative to the axis of cellular alignment. Transversely applied stretch in contrast to longitudinally applied stretch results in increase in protein turnover and upregulation of expression of hypertrophy markers such as Cx43, and N-cadherin [82, 86] and rise in protein kinase C activity [87]. The directional dependency may be determined by specific localization and orientation of mechanosensors such as integrin and titin relative to the cytoskeletion. The differential activation of these mechanosensors could permit cytoskeleton to detect the direction of mechanical stress and activate gene expression, suited for better myocyte adaptation. There is evidence that an elongated morphology is an intrinsic cue for cardiomyocyte differentiation of mesenchymal cells or progenitor cells [88–90]. Cardiac-specific genes are upregulated in cyclically stretched embryonic stem cell-derived CMs [91, 92]. These studies suggest that microenvironmental mechanical cues transmitted through the cytoskeleton modulate gene expression in differentiating stem cells and progenitor cells.

Role of Extracellular Rigidity

Extracellular rigidity is thought to contribute to cytoskeletal remodelling. The inter-relationship has been explored through in vitro studies. CMs remodel their acto-myosin cytoskeleton in response to stiffness [93]. Soft substrates provide little resistance to focal adhesion attachments, and thus limit myocyte spreading, cytoskeletal strengthening and myofibril formation [94–97]. Excessive rigidity is also unfavorable for myofibril assembly. Substrates with moderate elastic moduli of 10 kPa appear ideal for myofibril maturation and force generation [94–97]. Intra cellular calcium handling is also dependent on substrate stiffness [97]. This observation suggests that rigidity can also modulate the expression and function of non-cytoskeletal proteins.

Mechanical Stretch and Hypertrophy of Cardiomyocytes

Hypertrophy is seen in both atrial and ventricular myocytes subjected to mechanical stress [98, 99]. The hypertrophic response is accompanied by increase in cell size, sarcomeric organization and expression of the genes of ANF (atrial natriuretic factor), BNP (brain natriuretic peptide) and β-MHC (myosin heavy chain) [1]. RhoA and FAK (focal adhesion kinase) have been found to be important factors which control the hypertrophic genetic programme in CMs subjected to mechanical stretch [100]. The stimulatory effect of RhoA on cardiac hypertrophy is mediated by ERK1/2 through controlling activation of GATA4. PKC-α and -δ are key regulators which mediate activation of Rho GTPases and MAPKs in the stretch-induced hypertrophic response [101]. The hypertrophy response to mechanical stretch also includes activation of c-jun, c-fos, c-myc and skeletal α-actin [102, 103].

Mechanical Stress and Heart Diseases

The organization of the heart from the alignment of individual sarcomeres to whole heart construct has physiological significance. In the adult heart, elongated cardiomy-ocytes aligned along a common axis assist speedy electrical propagation and the uniaxial alignment of sarcomeres aid precise cardiac contraction [104].

Structural changes of the ventricle and abnormalities in ventricular cardiomy-ocyte shape are features of several cardiomyopathies and are induced by increased mechanical loads [105–108]. Pressure overload results in concentric hypertrophy of the ventricle associated with parallel addition of myofibrils and increase in cross-sectional area of ventricular CMs [2, 52, 109–112]. In contrast, volume overload results in dilatation of the ventricle with eccentric hypertrophy, which is associated with the addition of sarcomeres in series and lengthening of the ventricular CMs with cross sectional area remaining unaltered [110, 113–115]. The increased wall stress causes mechanical load on CMs and thus contribute to maladaptive remodelling of the cytoskeleton [52, 110, 114].

Mutations in genes of sarcomeric proteins required for force generation or force transmission to ECM are associated with abnormalities in cardiomyocyte shape and chamber morphology [116–118]. Genetic mutations which compromise contractility or mechanotransduction may prevent the cytoskeleton to respond to mechanical loads effectively to preserve normal function of the heart.

Effect of Mechanical Stretch on Apoptosis in Cardiomyocytes

Stretch elongation of more than 20% as may occur in pathological conditions may induce apoptosis in CMs. Apoptosis of CMs is an important factor in the transition from pathological hypertrophy to cardiac failure. A low amplitude 5% stretch has been found to cause hypertrophy, whereas a high amplitude 25% stretch induces apoptosis in CMs [119]. High amplitude stretch result in significant increase in the expression of the cell death signal gene Bax, through ROS production and MAPK activation. Liao et al. found that mechanical stretch induces a marked elevation in intracellular NO levels in cardiomyocytes through activation of Ca2 + -dependent eNOS (endothelial nitric oxide synthase) and iNOS (inducible nitric oxide synthase) [120]. Nitric oxide (NO) is a known bidirectional regulator of apoptosis.

Mechanical Stretch and Electric Remodelling in Cardiomyocytes

Cyclic stretch as may occur in pathological conditions may result in remodelling of conduction pathways and arrhythmogenesis. Mechanical stretch of CMs alters gap junctions. In atrial myocytes, mechanical stretch shortens the duration of the action potential and increases the expression of genes encoding IK1 (inward rectifier K + current) and IKur (ultra-rapid delayed rectifier K + current) [98]. Gap junction proteins such as Cx43 (connexin43) are also upregulated in cardiomyocyte after mechanical stretch [121, 122]. There is evidence that Ang II, VEGF and TGF-β may mediate the stretch-induced Cx expression [121, 123]. Mechanical stretch also activates ROS and decreases the expression of Kv4.3 transient outward current channel [124].

Mechanical Stress Induced Alterations in Gene Expression in Cardiomyocytes

DNA microarray studies on neonatal cardiomyocytes have revealed the specific genes that are induced by mechanical stretch [1]. The genes that are upregulated are ANF, BNP, skeletal α-actin, HSP70 (70 kDa heat-shock protein), the proto-oncogene c-myc, CKS-2 (cyclin dependent kinase regulatory subunit-2), intoxicative and cardio-protective genes, such as metallothionein-1 and HO-1 (haem oxidase-1), and cytokine growth and differentiation factors. Lipocalin, Cx40, cell-adhesion molecules and phospholipase are down-regulated. Among more than 28 000 genes analysed, Frank et al. found only 185 genes that were significantly regulated by mechanical stretch [1].

Ang II has a key role in the induction of gene expression by mechanical stretch [125]. Mechanical stretch of short duration (5– 8 min) activates the AT1 receptor without the participation of Ang II (104). Ang II involvement has been observed in studies where mechanical stretch has been employed for a longer period (more than 4 h).

Varying durations and degree of mechanical stress activate different mechanisms and also differently influence the pattern and types of gene expression. A 20% mechanical stretch of cardiomyocytes increases myostatin gene expression through the action of IGF-1 and increases serotonin 2B receptor expression to modulate BNP function [126, 127].

Mechanical stretch also enhances the expression of resistin gene in CMs suggesting a metabolic link between mechanical stretch and cardiac hypertrophy [128]. Naka et al. observed that 20% mechanical stretch induces IL-18 expression in CMs through the AT1 receptor and the ETA receptor [129].

Mechanical Stretch Induced Autocrine and Paracrine Effects

Mechanical stretch is also known to induce secretion or synthesis of molecules with autocrine and paracrine effects [102, 130]. AngII, IGF-1, TNF-α and VEGF are known to mediate the autocrine and paracrine effects [102, 123, 126, 128]. The autocrine and paracrine effects could include activation of the expression of other growth factors or cytokines as well. The secreted proteins interacting with their receptors can activate intracellular signalling pathways and thus initiate cellular and molecular changes in CMs (Fig. 8.3).

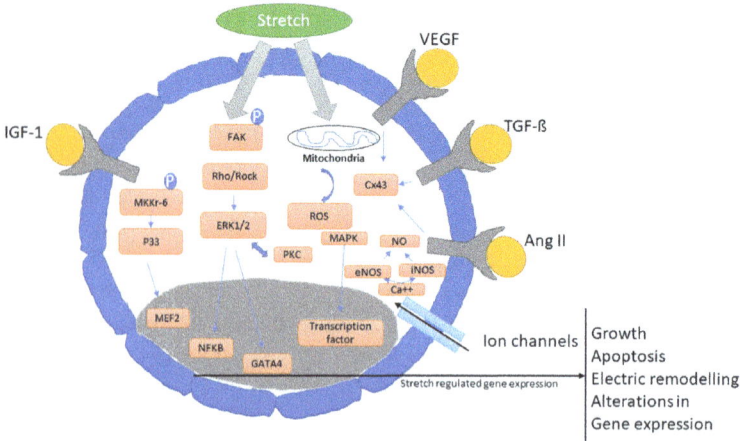

Fig. 8.3 Intracellular signalling and autocrine or paracrine cytokine secretion linked to mechanical stretch induced responses in cardiomyocytes. Ang—Angiotensin. Cx—Connexin. FAK—Focal adhesion kinase. IGF—Insulin growth factor. TGF—Transforming growth factor. VEGF—Vascular endothelial growth factor. NO—Nitric oxide. NOS—Nitric oxide synthase. ROS—Reactive oxygen species

Metabolic Alterations in Hypertrophic Cardiomyocytes

Adult CMs, which hypertrophy in response to mechanical stress return to a 'foetal' pattern of energy substrate metabolism [131, 134].

In pressure overload hypertrophy, there is increase in glycolysis [135–137]. Glucose uptake during hypertrophy is dependent more on GLUT1 9glucose transporter 1) translocation, which is upregulated in hypertrophied CMs [138]. Activity of the glycolytic enzyme enolase is also increased [139]. There is decrease in oxygen tension per unit area in hypertrophied cardiomyocyte. HIF-1α may mediate at least partly the alterations in carbohydrate metabolism. HIF-1α expression is increased in pressure overload hypertrophy [140]. Hypoxia in a HIF-1α–dependent manner increases the expression of 6-phosphofructo-2- kinase (PFK2) [141]. PFK2 generates fructose-2,6-bisphosphate, an activator of PFK1, the rate limiting enzyme of glycolysis. The expression of the lactic dehydrogenase isoform also shifts to a more foetal form in hypertrophic CMs [142]. Hypoxia also increases the expression of monocarboxylate transporter (MCT)-4 expression in a HIF-1a– dependent manner [143]. In hypertrophy there is increased expression of monocarboxylate transporter-1 [144]. These alterations may promote lactate efflux, maintain increased rates of glycolysis and thus aid hypertrophy. Rates of glucose and lactate oxidation are low in hypertrophied CMs though the rate of glycolysis is increased [135, 145]. The decreased carbohydrate oxidation may be because of the increase in PDK1 expression mediated by HIF-1α [146].

In both pressure- and volume-overload hypertrophy there is decrease in fatty acid oxidation [135, 147, 148]. There are parallel changes in the activity and or expression of enzymes of these metabolic pathways [149–151]. The rates of fatty acid oxidation are decreased during pathological hypertrophy [141].

Both PPARα and PPARβ/δ may have a role in the metabolic phenotype [152]. PPARα and PGC-1α levels are reduced during hypertrophy [153]. Upregulation of PPARα in neonatal rat CMs has been found to suppress hypertrophy-induced increases in protein synthesis as well as hypertrophy and lessen the reduction in mCPT1/CPT1β expression [154, 155]. The precise role of PPARα in cardiomyocyte hypertrophy is unclear. In mice with cardiac specific ablation of PPARβ/δ also, fatty acid oxidation is reduced in the heart along with hypertrophy [156]. Activation of PPARβ/δ with selective ligands prevents the downregulation of genes involved in fatty acid metabolism in CMs subjected to hypertrophic stimuli [157, 158]. The rates of fatty acid oxidation are however preserved [157, 159]. In the whole heart, cardiac-specific overexpression of PPARβb/δ does not increase myocardial fatty acid oxidation. Cardiac glucose oxidation rates are increased and this is associated with an increased expression of GLUT4 [160].

A summary of differences in the features of physiological and metabolic changes in heart during physiological and pathological response of cardiomyocytes to mechanical stress are given in Table 8.1.

Conclusions

Mechanical stretch activates several intracellular signalling networks and modulates gene expressions and functional responses in cardiomyocytes. Various durations, loads and frequencies of mechanical forces induce differing response mechanisms. In vitro studies have helped to decipher how cardiomyocytes sense and respond to mechanical inputs and how mechanical stress influence structure and function of cardiomyocytes. These studies indicate a feedback loop between mechanosensing and cytoskeletal adaptation. Cytoskeleton transmits mechanical forces to regulate other processes within the cell. Cytoskeletal remodelling links the extracellular microenvironment to a wide range of cellular functions. Microenvironmental cues are sensed by costameres, adherens junctions, and titin embedded within the cytoskeleton. After sensing mechanical inputs, cytoskeleton undergoes structural remodelling to adapt to external mechanical stress. Cytoskeleton also propagates mechanical inputs and aid to mediate signalling, excitability, impulse propagation, contractility and gene expression in cardiomyocytes. Even as in vitro models have benefited to understand how individual aspects of the microenvironment directly influence cytoskeletal remodelling, observations in these studies are to be cautiously interpreted because mechanical stress secondary to pressure or volume overload in vivo is more complex than the mechanical stretch used in models in vitro.

Table 8.1 Physiological and metabolic changes in heart during physiological hypertrophy and pathological hypertrophy

Characteristics	Physiological hypertrophy	Pathological hypertrophy/ Heart failure
Stimuli	Pregnancy, Exercise	Hypertension, Coronary artery disease, Diabetes
Reversible	Yes	No
Cardiomyocyte size	Increase	Increase
Contractility	Increase	Increase initially and then decrease
Ejection fraction	Increase	Decrease or no change
Signaling mechanisms	Thyroid receptor α/β (TRα/β), Insulin/IGF-1, platelet derived growth factor (PDGF), vascular endothelial growth factor (VEGF) and nitric oxide (NO)	Angiotensin II and endothelin mediated calcineurin-NFAT signaling, Catecholamine mediated PKA-ERK signaling, Mechanical forces such as transient receptor protein channel (TRPC), stromal interaction molecule 1 (STIM1), atrial or brain natriuretic peptide (ANP or BNP) mediated cGMP-PKG signaling
Collagen (Type 1) deposition	No	Yes (Myocytes replaced with collagen)
Metabolic changes	Normal utilization of glycolysis and fatty acid oxidation pathway Normal iron metabolism Increase in ATP supply No other significant change in metabolism of heart	Accumulation of fatty acid/lipid intermediates Impaired iron metabolism Decrease in fatty acid oxidation, Increase in glycolysis Uncoupling of glycolysis with glucose oxidation Increase in ATP demand/supply Decrease in phosphocreatine levels
Mitochondria	Dense cristae and well-organized matrix Packed ETC super complex assembly at mitochondrial membrane Normal ROS levels Normal mitochondrial dynamics (fission and fusion) and quality control pathway (Mitophagy)	Loosely arranged cristae and loss of matrix components Lack of well-organized super complex assembly Increased levels of reactive oxygen species Abnormal mitochondrial dynamics (fission and fusion) and quality control pathway (Mitophagy)
Fetal gene reprogramming	No	Increased expression of MHCβ, MYH 11, ANP, BNP genes
Cell death	No	Yes

References

1. Frank D, Kuhn C, Brors B, Hanselmann C, Lüdde M, Katus HA, et al. Gene expression pattern in biomechanically stretched cardiomyocytes: Evidence for a stretch-specific gene program. Hypertension. 2008;51(2):309–18.
2. Smith SH, Bishop SP. Regional myocyte size in compensated right ventricular hypertrophy in the ferret. J Mol Cell Cardiol. 1985;17(10):1005–11.
3. Shyu K, Chen J, Shih N, Wang D, Chang H, Lien W, et al. Regulation of human cardiac myosin heavy chain genes by cyclical mechanical stretch in cultured cardiocytes. Biochem Biophys Res Commun. 1995;210(2):567–73.
4. McNicholas-Bevensee CM, DeAndrade KB, Bradley WE, Dell'Italia LJ, Lucchesi PA, Bevensee MO. Activation of gadolinium-sensitive ion channels in cardiomyocytes in early adaptive stages of volume overload-induced heart failure. Cardiovasc Res. 2006;72(2):262–70.
5. Otani H, Matsuhisa S, Akita Y, Kyoi S, Enoki C, Tatsumi K, et al. Role of mechanical stress in the form of cardiomyocyte death during the early phase of reperfusion. Circ J. 2006;70(10):1344–55.
6. Taber LA. Biomechanics of cardiovascular development. Annu Rev Biomed Eng. 2001;3(1):1–25.
7. Auman HJ, Coleman H, Riley HE, Olale F, Tsai H-J, Yelon D. Functional modulation of cardiac form through regionally confined cell shape changes. PLoS Biol. 2007;5(3):e53.
8. Berdougo E, Coleman H, Lee DH, Stainier DY, Yelon D. Mutation of weak atrium/atrial myosin heavy chain disrupts atrial function and influences ventricular morphogenesis in zebrafish. Development. 2003;130(24):6121–9.
9. Hove JR, Köster RW, Forouhar AS, Acevedo-Bolton G, Fraser SE, Gharib M. Intracardiac fluid forces are an essential epigenetic factor for embryonic cardiogenesis. Nature. 2003;421(6919):172–7.
10. Price R, Chintanowonges C, Shiraishi I, Borg T, Terracio L. Local and regional variations in myofibrillar patterns in looping rat hearts. Anatom Record: Official Publ Am Assoc Anatom. 1996;245(1):83–93.
11. Taber LA, Lin IE, Clark EB. Mechanics of cardiac looping. Dev Dyn. 1995;203(1):42–50.
12. Icardo J, Ojeda J. Effects of colchicine on the formation and looping of the tubular heart of the embryonic chick. Cells Tissues Organs. 1984;119(1):1–9.
13. Itasaki N, Nakamura H, Sumida H, Yasuda M. Actin bundles on the right side in the caudal part of the heart tube play a role in dextro-looping in the embryonic chick heart. Anat Embryol. 1991;183(1):29–39.
14. Latacha KS, Rémond MC, Ramasubramanian A, Chen AY, Elson EL, Taber LA. Role of actin polymerization in bending of the early heart tube. Develop Dyn: Official Publ Am Assoc Anatom. 2005;233(4):1272–86.
15. Manasek FJ, Burnside MB, Waterman RE. Myocardial cell shape change as a mechanism of embryonic heart looping. Dev Biol. 1972;29(4):349–71.
16. Manasek FJ, Monroe RG. Early cardiac morphogenesis is independent of function. Dev Biol. 1972;27(4):584–8.
17. Manning A, McLachlan J. Looping of chick embryo hearts in vitro. J Anat. 1990;168:257.
18. Rémond MC, Fee JA, Elson EL, Taber LA. Myosin-based contraction is not necessary for cardiac c-looping in the chick embryo. Anat Embryol. 2006;211(5):443–54.
19. Baker EL, Zaman MH. The biomechanical integrin. J Biomech. 2010;43(1):38–44.
20. Humphries M. Integrin structure. Biochem Soc Trans. 2000;28(4):311–40.
21. Ross RS, Borg TK. Integrins and the myocardium. Circ Res. 2001;88(11):1112–9.
22. McCain ML, Parker KK. Mechanotransduction: the role of mechanical stress, myocyte shape, and cytoskeletal architecture on cardiac function. Pflügers Archiv-Europ J Physiol. 2011;462(1):89–104.
23. Brancaccio M, Hirsch E, Notte A, Selvetella G, Lembo G, Tarone G. Integrin signalling: the tug-of-war in heart hypertrophy. Cardiovasc Res. 2006;70(3):422–33.

24. Meyer CJ, Alenghat FJ, Rim P, Fong JH-J, Fabry B, Ingber DE. Mechanical control of cyclic AMP signalling and gene transcription through integrins. Nat Cell Biol. 2000;2(9):666–8.
25. Hoshijima M. Mechanical stress-strain sensors embedded in cardiac cytoskeleton: Z disk, titin, and associated structures. Am J Physiol-Heart Circul Physiol. 2006;290(4):H1313–25.
26. Bendig G, Grimmler M, Huttner IG, Wessels G, Dahme T, Just S, et al. Integrin-linked kinase, a novel component of the cardiac mechanical stretch sensor, controls contractility in the zebrafish heart. Genes Dev. 2006;20(17):2361–72.
27. Brancaccio M, Guazzone S, Menini N, Sibona E, Hirsch E, De Andrea M, et al. Melusin is a new muscle-specific interactor for β1integrin cytoplasmic domain. J Biol Chem. 1999;274(41):29282–8.
28. Brancaccio M, Fratta L, Notte A, Hirsch E, Poulet R, Guazzone S, et al. Melusin, a muscle-specific integrin β 1–interacting protein, is required to prevent cardiac failure in response to chronic pressure overload. Nat Med. 2003;9(1):68–75.
29. Flick MJ, Konieczny SF. The muscle regulatory and structural protein MLP is a cytoskeletal binding partner of betaI-spectrin. J Cell Sci. 2000;113(9):1553–64.
30. Boateng SY, Belin RJ, Geenen DL, Margulies KB, Martin JL, Hoshijima M, et al. Cardiac dysfunction and heart failure are associated with abnormalities in the subcellular distribution and amounts of oligomeric muscle LIM protein. Am J Physiol-Heart Circul Physiol. 2007;292(1):H259–69.
31. Knöll R, Hoshijima M, Hoffman HM, Person V, Lorenzen-Schmidt I, Bang M-L, et al. The cardiac mechanical stretch sensor machinery involves a Z disc complex that is defective in a subset of human dilated cardiomyopathy. Cell. 2002;111(7):943–55.
32. Fürst DO, Osborn M, Nave R, Weber K. The organization of titin filaments in the half-sarcomere revealed by monoclonal antibodies in immunoelectron microscopy: a map of ten nonrepetitive epitopes starting at the Z line extends close to the M line. J Cell Biol. 1988;106(5):1563–72.
33. Helmes M, Trombitas K, Centner T, Kellermayer M, Labeit S, Linke W, et al. Mechanically driven contour-length adjustment in rat cardiac titin's unique N2B sequence: titin is an adjustable spring. Circ Res. 1999;84(11):1339–52.
34. Helmes M, Trombita´s Kr, Granzier H. Titin develops restoring force in rat cardiac myocytes. Circul Res. 1996;79(3):619–26.
35. de Tombe PP, Mateja RD, Tachampa K, Mou YA, Farman GP, Irving TC. Myofilament length dependent activation. J Mol Cell Cardiol. 2010;48(5):851–8.
36. Krüger M, Linke WA. Titin-based mechanical signalling in normal and failing myocardium. J Mol Cell Cardiol. 2009;46(4):490–8.
37. Linke WA. Sense and stretchability: the role of titin and titin-associated proteins in myocardial stress-sensing and mechanical dysfunction. Cardiovasc Res. 2008;77(4):637–48.
38. LeWinter MM, Granzier H. Cardiac titin: a multifunctional giant. Circulation. 2010;121(19):2137–45.
39. Miller MK, Bang M-L, Witt CC, Labeit D, Trombitas C, Watanabe K, et al. The muscle ankyrin repeat proteins: CARP, ankrd2/Arpp and DARP as a family of titin filament-based stress response molecules. J Mol Biol. 2003;333(5):951–64.
40. Aihara Y, Kurabayashi M, Saito Y, Ohyama Y, Tanaka T, Takeda S-i, et al. Cardiac ankyrin repeat protein is a novel marker of cardiac hypertrophy: role of M-CAT element within the promoter. Hypertension. 2000;36(1):48–53.
41. Nagueh SF, Shah G, Wu Y, Torre-Amione G, King NM, Lahmers S, et al. Altered titin expression, myocardial stiffness, and left ventricular function in patients with dilated cardiomyopathy. Circulation. 2004;110(2):155–62.
42. Granzier HL, Radke MH, Peng J, Westermann D, Nelson OL, Rost K, et al. Truncation of titin's elastic PEVK region leads to cardiomyopathy with diastolic dysfunction. Circ Res. 2009;105(6):557–64.
43. Sheikh F, Raskin A, Chu P-H, Lange S, Domenighetti AA, Zheng M, et al. An FHL1-containing complex within the cardiomyocyte sarcomere mediates hypertrophic biomechanical stress responses in mice. J Clin Investig. 2008;118(12):3870–80.

44. Lange S, Auerbach D, McLoughlin P, Perriard E, Schäfer BW, Perriard J-C, et al. Subcellular targeting of metabolic enzymes to titin in heart muscle may be mediated by DRAL/FHL-2. J Cell Sci. 2002;115(24):4925–36.

45. Harris TJ, Tepass U. Adherens junctions: from molecules to morphogenesis. Nat Rev Mol Cell Biol. 2010;11(7):502–14.

46. Angst BD, Khan LU, Severs NJ, Whitely K, Rothery S, Thompson RP, et al. Dissociated spatial patterning of gap junctions and cell adhesion junctions during postnatal differentiation of ventricular myocardium. Circ Res. 1997;80(1):88–94.

47. Gourdie RG, Green CR, Severs NJ, Thompson RP. Immunolabelling patterns of gap junction connexins in the developing and mature rat heart. Anat Embryol. 1992;185(4):363–78.

48. Peters NS, Severs NJ, Rothery SM, Lincoln C, Yacoub MH, Green CR. Spatiotemporal relation between gap junctions and fascia adherens junctions during postnatal development of human ventricular myocardium. Circulation. 1994;90(2):713–25.

49. Goncharova EJ, Kam Z, Geiger B. The involvement of adherens junction components in myofibrillogenesis in cultured cardiac myocytes. Development. 1992;114(1):173–83.

50. Wu JC, Chung TH, Tseng YZ, Wang SM. N-cadherin/catenin–based costameres in cultured chicken cardiomyocytes. J Cell Biochem. 1999;75(1):93–104.

51. Wu JC, Sung HC, Chung TH, DePhilip RM. Role of N-cadherin-and integrin-based costameres in the development of rat cardiomyocytes. J Cell Biochem. 2002;84(4):717–24.

52. Le Duc Q, Shi Q, Blonk I, Sonnenberg A, Wang N, Leckband D, et al. Vinculin potentiates E-cadherin mechanosensing and is recruited to actin-anchored sites within adherens junctions in a myosin II–dependent manner. J Cell Biol. 2010;189(7):1107–15.

53. Liu Z, Tan JL, Cohen DM, Yang MT, Sniadecki NJ, Ruiz SA, et al. Mechanical tugging force regulates the size of cell–cell junctions. Proceedings of the National Academy of Sciences USA. 2010;107(22):9944–9.

54. Hertig CM, Eppenberger-Eberhardt M, Koch S, Eppenberger HM. N-cadherin in adult rat cardiomyocytes in culture. I. Functional role of N-cadherin and impairment of cell-cell contact by a truncated N-cadherin mutant. J Cell Sci. 1996;109(1):1–10.

55. Imanaka-Yoshida K, Knudsen KA, Linask KK. N-cadherin is required for the differentiation and initial myofibrillogenesis of chick cardiomyocytes. Cell Motil Cytoskelet. 1998;39(1):52–62.

56. Luo Y, Radice GL. Cadherin-mediated adhesion is essential for myofibril continuity across the plasma membrane but not for assembly of the contractile apparatus. J Cell Sci. 2003;116(8):1471–9.

57. Radice GL, Rayburn H, Matsunami H, Knudsen KA, Takeichi M, Hynes RO. Developmental defects in mouse embryos lacking N-cadherin. Dev Biol. 1997;181(1):64–78.

58. Soler AP, Knudsen KA. N-cadherin involvement in cardiac myocyte interaction and myofibrillogenesis. Dev Biol. 1994;162(1):9–17.

59. Gourdie R, Green C, Severs N. Gap junction distribution in adult mammalian myocardium revealed by an anti-peptide antibody and laser scanning confocal microscopy. J Cell Sci. 1991;99(1):41–55.

60. Grant DA. Ventricular constraint in the fetus and newborn. Can J Cardiol. 1999;15(1):95–104.

61. Hirschy A, Schatzmann F, Ehler E, Perriard J-C. Establishment of cardiac cytoarchitecture in the developing mouse heart. Dev Biol. 2006;289(2):430–41.

62. Salameh A, Wustmann A, Karl S, Blanke K, Apel D, Rojas-Gomez D, et al. Novelty and significance. Circ Res. 2010;106(10):1592–602.

63. Simpson DG, Decker ML, Clark WA, Decker RS. Contractile activity and cell-cell contact regulate myofibrillar organization in cultured cardiac myocytes. J Cell Biol. 1993;123(2):323–36.

64. Zhuang J, Yamada KA, Saffitz JE, Kléber AG. Pulsatile stretch remodels cell-to-cell communication in cultured myocytes. Circ Res. 2000;87(4):316–22.

65. Ingber DE, Tensegrity I. Cell structure and hierarchical systems biology. J Cell Sci. 2003;116(7):1157–73.

66. Tsutsui H, Ishihara K, Cooper G. Cytoskeletal role in the contractile dysfunction of hypertrophied myocardium. Science. 1993;260(5108):682–7.
67. Tagawa H, Wang N, Narishige T, Ingber DE, Zile MR, Cooper G IV. Cytoskeletal mechanics in pressure-overload cardiac hypertrophy. Circ Res. 1997;80(2):281–9.
68. Tsutsui H, Tagawa H, Kent RL, McCollam PL, Ishihara K, Nagatsu M, et al. Role of microtubules in contractile dysfunction of hypertrophied cardiocytes. Circulation. 1994;90(1):533–55.
69. Brangwynne CP, MacKintosh FC, Kumar S, Geisse NA, Talbot J, Mahadevan L, et al. Microtubules can bear enhanced compressive loads in living cells because of lateral reinforcement. J Cell Biol. 2006;173(5):733–41.
70. Franz MR, Burkhoff D, Yue DT, Sagawa K. Mechanically induced action potential changes and arrhythmia in isolated and in situ canine hearts. Cardiovasc Res. 1989;23(3):213–23.
71. Boland J, Troquet J. Intracellular action potential changes induced in both ventricles of the rat by an acute right ventricular pressure overload. Cardiovasc Res. 1980;14(12):735–40.
72. Franz MR, Cima R, Wang D, Profitt D, Kurz R. Electrophysiological effects of myocardial stretch and mechanical determinants of stretch-activated arrhythmias. Circulation. 1992;86(3):968–78.
73. Furukawa T, Yamane T-i, Terai T, Katayama Y, Hiraoka M. Functional linkage of the cardiac ATP-sensitive K+ channel to the actin cytoskeleton. Pflügers Archiv. 1996;431(4):504–12.
74. Galli A, DeFelice LJ. Inactivation of L-type Ca channels in embryonic chick ventricle cells: Dependence on the cytoskeletal agents colchicine and taxol. Biophys J. 1994;67(6):2296–304.
75. Parker KK, Taylor LK, Atkinson B, Hansen DE, Wikswo JP. The effects of tubulin-binding agents on stretch-induced ventricular arrhythmias. Eur J Pharmacol. 2001;417(1–2):131–40.
76. Terzic A, Kurachi Y. Actin microfilament disrupters enhance K (ATP) channel opening in patches from guinea-pig cardiomyocytes. J Physiol. 1996;492(2):395–404.
77. Walsh KB, Parks GE. Changes in cardiac myocyte morphology alter the properties of voltage-gated ion channels. Cardiovasc Res. 2002;55(1):64–75.
78. Yin L, Bien H, Entcheva E. Scaffold topography alters intracellular calcium dynamics in cultured cardiomyocyte networks. Am J Physiol-Heart Circul Physiol. 2004;287(3):H1276–85.
79. Blaauw E, van Nieuwenhoven FA, Willemsen P, Delhaas T, Prinzen FW, Snoeckx LH, et al. Stretch-induced hypertrophy of isolated adult rabbit cardiomyocytes. Am J Physiol-Heart Circul Physiol. 2010;299(3):H780–7.
80. Cadre BM, Qi M, Eble DM, Shannon TR, Bers DM, Samarel AM. Cyclic stretch down-regulates calcium transporter gene expression in neonatal rat ventricular myocytes. J Mol Cell Cardiol. 1998;30(11):2247–59.
81. De Jonge HW, Dekkers DH, Houtsmuller AB, Sharma HS, Lamers JM. Differential signaling and hypertrophic responses in cyclically stretched vs endothelin-1 stimulated neonatal rat cardiomyocytes. Cell Biochem Biophys. 2007;47(1):21–32.
82. Simpson D, Majeski M, Borg T, Terracio L. Regulation of cardiac myocyte protein turnover and myofibrillar structure in vitro by specific directions of stretch. Circ Res. 1999;85(10):e59–69.
83. Matsuda T, Takahashi K, Nariai T, Ito T, Takatani T, Fujio Y, et al. N-cadherin-mediated cell adhesion determines the plasticity for cell alignment in response to mechanical stretch in cultured cardiomyocytes. Biochem Biophys Res Commun. 2004;326(1):228–32.
84. Yamane M, Matsuda T, Ito T, Fujio Y, Takahashi K, Azuma J. Rac1 activity is required for cardiac myocyte alignment in response to mechanical stress. Biochem Biophys Res Commun. 2007;353(4):1023–7.
85. Aikawa R, Nagai T, Kudoh S, Zou Y, Tanaka M, Tamura M, et al. Integrins play a critical role in mechanical stress–induced p38 MAPK activation. Hypertension. 2002;39(2):233–8.
86. Gopalan SM, Flaim C, Bhatia SN, Hoshijima M, Knoell R, Chien KR, et al. Anisotropic stretch-induced hypertrophy in neonatal ventricular myocytes micropatterned on deformable elastomers. Biotechnol Bioeng. 2003;81(5):578–87.

87. Bullard TA, Hastings JL, Davis JM, Borg TK, Price RL. Altered PKC expression and phosphorylation in response to the nature, direction, and magnitude of mechanical stretch. Can J Physiol Pharmacol. 2007;85(2):243–50.

88. Domian IJ, Chiravuri M, Van Der Meer P, Feinberg AW, Shi X, Shao Y, et al. Generation of functional ventricular heart muscle from mouse ventricular progenitor cells. Science. 2009;326(5951):426–9.

89. Pijnappels DA, Schalij MJ, Ramkisoensing AA, van Tuyn J, de Vries AA, van der Laarse A, et al. Forced alignment of mesenchymal stem cells undergoing cardiomyogenic differentiation affects functional integration with cardiomyocyte cultures. Circ Res. 2008;103(2):167–76.

90. Tay CY, Yu H, Pal M, Leong WS, Tan NS, Ng KW, et al. Micropatterned matrix directs differentiation of human mesenchymal stem cells towards myocardial lineage. Exp Cell Res. 2010;316(7):1159–68.

91. Gwak S-J, Bhang SH, Kim I-K, Kim S-S, Cho S-W, Jeon O, et al. The effect of cyclic strain on embryonic stem cell-derived cardiomyocytes. Biomaterials. 2008;29(7):844–56.

92. Shimko VF, Claycomb WC. Effect of mechanical loading on three-dimensional cultures of embryonic stem cell-derived cardiomyocytes. Tissue Eng Part A. 2008;14(1):49–58.

93. Pelham RJ, Wang Y-l. Cell locomotion and focal adhesions are regulated by substrate flexibility. Proc Natl Acad Sciences USA. 1997;94(25):13661–5.

94. Bajaj P, Tang X, Saif TA, Bashir R. Stiffness of the substrate influences the phenotype of embryonic chicken cardiac myocytes. J Biomed Mater Res, Part A. 2010;95(4):1261–9.

95. Bhana B, Iyer RK, Chen WLK, Zhao R, Sider KL, Likhitpanichkul M, et al. Influence of substrate stiffness on the phenotype of heart cells. Biotechnol Bioeng. 2010;105(6):1148–60.

96. Engler AJ, Carag-Krieger C, Johnson CP, Raab M, Tang H-Y, Speicher DW, et al. Embryonic cardiomyocytes beat best on a matrix with heart-like elasticity: scar-like rigidity inhibits beating. J Cell Sci. 2008;121(22):3794–802.

97. Jacot JG, McCulloch AD, Omens JH. Substrate stiffness affects the functional maturation of neonatal rat ventricular myocytes. Biophys J . 2008;95(7):3479–87.

98. Yamazaki T, Komuro I, Kudoh S, Zou Y, Shiojima I, Hiroi Y, et al. Endothelin-1 is involved in mechanical stress-induced cardiomyocyte hypertrophy. J Biol Chem. 1996;271(6):3221–8.

99. Saygili E, Rana OR, Saygili E, Reuter H, Frank K, Schwinger RH, et al. Losartan prevents stretch-induced electrical remodeling in cultured atrial neonatal myocytes. Am J Physiol-Heart Circul Physiol. 2007;292(6):H2898–905.

100. Torsoni AS, Marin TM, Velloso LA, Franchini KG. RhoA/ROCK signaling is critical to FAK activation by cyclic stretch in cardiac myocytes. Am J Physiol-Heart Circul Physiol. 2005;289(4):H1488–96.

101. Pan J, Singh US, Takahashi T, Oka Y, Palm-Leis A, Herbelin BS, et al. PKC mediates cyclic stretch-induced cardiac hypertrophy through Rho family GTPases and mitogen-activated protein kinases in cardiomyocytes. J Cell Physiol. 2005;202(2):536–53.

102. Sadoshima J-i, Izumo S. Mechanical stretch rapidly activates multiple signal transduction pathways in cardiac myocytes: potential involvement of an autocrine/paracrine mechanism. EMBO J. 1993;12(4):1681–92.

103. Liang F, Gardner DG. Mechanical strain activates BNP gene transcription through a p38/NF-κB–dependent mechanism. J Clin Investig. 1999;104(11):1603–12.

104. Parker KK, Ingber DE. Extracellular matrix, mechanotransduction and structural hierarchies in heart tissue engineering. Philos Trans R Soc B: Biol Sci. 2007;362(1484):1267–79.

105. Anversa P, Ricci R, Olivetti G. Quantitative structural analysis of the myocardium during physiologic growth and induced cardiac hypertrophy: a review. J Am Coll Cardiol. 1986;7(5):1140–9.

106. Gerdes A. Remodeling of ventricular myocytes during cardiac hypertrophy and heart failure. J Fla Med Assoc. 1992;79(4):253–5.

107. Gerdes AM. Cardiac myocyte remodeling in hypertrophy and progression to failure. J Cardiac Fail. 2002;8(6):S264–8.

108. Gerdes AM, Capasso JM. Structural remodeling and mechanical dysfunction of cardiac myocytes in heart failure. J Mol Cell Cardiol. 1995;27(3):849–56.

109. Katz AM. Maladaptive growth in the failing heart: the cardiomyopathy of overload. Cardiovasc Drugs Ther. 2002;16(3):245–9.
110. Grossman W, Jones D, McLaurin L. Wall stress and patterns of hypertrophy in the human left ventricle. J Clin Investig. 1975;56(1):56–64.
111. Werchan PM, Summer WR, Gerdes AM, McDonough K. Right ventricular performance after monocrotaline-induced pulmonary hypertension. Am J Physiol-Heart Circul Physiol. 1989;256(5):H1328–36.
112. Zierhut W, Zimmer H, Gerdes A. Effect of angiotensin converting enzyme inhibition on pressure-induced left ventricular hypertrophy in rats. Circ Res. 1991;69(3):609–17.
113. Capasso JM, Fitzpatrick D, Anversa P. Cellular mechanisms of ventricular failure: myocyte kinetics and geometry with age. Am J Physiol-Heart Circul Physiol. 1992;262(6):H1770–81.
114. Gerdes AM, Kellerman SE, Moore JA, Muffly KE, Clark LC, Reaves PY, et al. Structural remodeling of cardiac myocytes in patients with ischemic cardiomyopathy. Circulation. 1992;86(2):426–30.
115. Luk A, Ahn E, Soor GS, Butany J. Dilated cardiomyopathy: a review. J Clin Pathol. 2009;62(3):219–25.
116. Kamisago M, Sharma SD, DePalma SR, Solomon S, Sharma P, McDonough B, et al. Mutations in sarcomere protein genes as a cause of dilated cardiomyopathy. N Engl J Med. 2000;343(23):1688–96.
117. Lapidos KA, Kakkar R, McNally EM. The dystrophin glycoprotein complex: signaling strength and integrity for the sarcolemma. Circ Res. 2004;94(8):1023–31.
118. Niimura H, Patton KK, McKenna WJ, Soults J, Maron BJ, Seidman J, et al. Sarcomere protein gene mutations in hypertrophic cardiomyopathy of the elderly. Circulation. 2002;105(4):446–51.
119. Pimentel DR, Amin JK, Xiao L, Miller T, Viereck J, Oliver-Krasinski J, et al. Reactive oxygen species mediate amplitude-dependent hypertrophic and apoptotic responses to mechanical stretch in cardiac myocytes. Circ Res. 2001;89(5):453–60.
120. Liao X, Liu JM, Du L, Tang A, Shang Y, Wang SQ, et al. Nitric oxide signaling in stretch-induced apoptosis of neonatal rat cardiomyocytes. FASEB J. 2006;20(11):1883–5.
121. Shyu K-G, Chen C-C, Wang B-W, Kuan P. Angiotensin II receptor antagonist blocks the expression of connexin43 induced by cyclical mechanical stretch in cultured neonatal rat cardiac myocytes. J Mol Cell Cardiol. 2001;33(4):691–8.
122. Saffitz JE, Kléber AG. Effects of mechanical forces and mediators of hypertrophy on remodeling of gap junctions in the heart. Circ Res. 2004;94(5):585–91.
123. Pimentel RC, Yamada KA, Kléber AG, Saffitz JE. Autocrine regulation of myocyte Cx43 expression by VEGF. Circ Res. 2002;90(6):671–7.
124. Zhou C, Ziegler C, Birder LA, Stewart AF, Levitan ES. Angiotensin II and stretch activate NADPH oxidase to destabilize cardiac Kv4. 3 channel mRNA. Circul Res. 2006;98(8):1040–7.
125. Zou Y, Akazawa H, Qin Y, Sano M, Takano H, Minamino T, et al. Mechanical stress activates angiotensin II type 1 receptor without the involvement of angiotensin II. Nat Cell Biol. 2004;6(6):499–506.
126. Shyu K-G, Ko W-H, Yang W-S, Wang B-W, Kuan P. Insulin-like growth factor-1 mediates stretch-induced upregulation of myostatin expression in neonatal rat cardiomyocytes. Cardiovasc Res. 2005;68(3):405–14.
127. Liang Y-J, Lai L-P, Wang B-W, Juang S-J, Chang C-M, Leu J-G, et al. Mechanical stress enhances serotonin 2B receptor modulating brain natriuretic peptide through nuclear factor-κB in cardiomyocytes. Cardiovasc Res. 2006;72(2):303–12.
128. Wang B-W, Hung H-F, Chang H, Kuan P, Shyu K-G. Mechanical stretch enhances the expression of resistin gene in cultured cardiomyocytes via tumor necrosis factor-α. Am J Physiol-Heart Circul Physiol. 2007;293(4):H2305–12.
129. Naka T, Sakoda T, Doi T, Akagami T, Tsujino T, Masuyama T, et al. Mechanical stretch induced interleukin-18 (IL-18) expression through angiotensin subtype 1 receptor (AT1R) and endothelin-1 in cardiomyocytes. Preparat Biochem Biotechnol. 2008; 38(2):201–12.

130. Sadoshima J-i, Xu Y, Slayter HS, Izumo S. Autocrine release of angiotensin II mediates stretch-induced hypertrophy of cardiac myocytes in vitro. Cell. 1993; 75(5):977–84.
131. Lorell BH, Grossman W. Cardiac hypertrophy: The consequences for diastol. J Am Coll Cardiol. 1987;9(5):1189–93.
132. Clerk A, Cullingford TE, Fuller SJ, Giraldo A, Markou T, Pikkarainen S, et al. Signaling pathways mediating cardiac myocyte gene expression in physiological and stress responses. J Cell Physiol. 2007;212(2):311–22.
133. Razeghi P, Young ME, Alcorn JL, Moravec CS, Frazier O, Taegtmeyer H. Metabolic gene expression in fetal and failing human heart. Circulation. 2001;104(24):2923–31.
134. Sambandam N, Lopaschuk GD, Brownsey RW, Allard MF. Energy metabolism in the hypertrophied heart. Heart Fail Rev. 2002;7(2):161–73.
135. Allard M, Schonekess B, Henning S, English D, Lopaschuk GD. Contribution of oxidative metabolism and glycolysis to ATP production in hypertrophied hearts. Am J Physiol-Heart Circul Physiol. 1994;267(2):H742–50.
136. Schönekess BO, Allard MF, Henning SL, Wambolt RB, Lopaschuk GD. Contribution of glycogen and exogenous glucose to glucose metabolism during ischemia in the hypertrophied rat heart. Circ Res. 1997;81(4):540–9.
137. Wambolt RB, Henning SL, English DR, Dyachkova Y, Lopaschuk GD, Allard MF. Glucose utilization and glycogen turnover are accelerated in hypertrophied rat hearts during severe low-flow ischemia. J Mol Cell Cardiol. 1999;31(3):493–502.
138. Morissette MR, Howes AL, Zhang T, Brown JH. Upregulation of GLUT1 expression is necessary for hypertrophy and survival of neonatal rat cardiomyocytes. J Mol Cell Cardiol. 2003;35(10):1217–27.
139. Keller A, Rouzeau J-D, Farhadian F, Wisnewsky C, Marotte F, Lamande N, et al. Differential expression of alpha-and beta-enolase genes during rat heart development and hypertrophy. Am J Physiol-Heart Circul Physiol. 1995;269(6):H1843–51.
140. Choi Y-H, Cowan DB, Nathan M, Poutias D, Stamm C, Pedro J, et al. Myocardial hypertrophy overrides the angiogenic response to hypoxia. PLoS One. 2008;3(12):e4042.
141. Minchenko O, Opentanova I, Caro J. Hypoxic regulation of the 6-phosphofructo-2-kinase/fructose-2, 6-bisphosphatase gene family (PFKFB-1–4) expression in vivo. FEBS Lett. 2003;554(3):264–70.
142. Bishop SP, Altschuld RA. Increased glycolytic metabolism in cardiac hypertrophy and congestive failure. Am J Physiol-Legacy Content. 1970;218(1):153–9.
143. Ullah MS, Davies AJ, Halestrap AP. The plasma membrane lactate transporter MCT4, but not MCT1, is up-regulated by hypoxia through a HIF-1α-dependent mechanism. J Biol Chem. 2006;281(14):9030–7.
144. Evans RK, Schwartz DD, Gladden LB. Effect of myocardial volume overload and heart failure on lactate transport into isolated cardiac myocytes. J Appl Physiol. 2003;94(3):1169–76.
145. Schonekess B, Allard M, Lopaschuk G. Recovery of glycolysis and oxidative metabolism during postischemic reperfusion of hypertrophied rat hearts. Am J Physiol-Heart Circul Physiol. 1996;271(2):H798–805.
146. Kim J-w, Tchernyshyov I, Semenza GL, Dang CV. HIF-1-mediated expression of pyruvate dehydrogenase kinase: a metabolic switch required for cellular adaptation to hypoxia. Cell Metab. 2006;3(3):177–85.
147. El Alaoui-Talibi Z, Guendouz A, Moravec M, Moravec J. Control of oxidative metabolism in volume-overloaded rat hearts: effect of propionyl-L-carnitine. Am J Physiol-Heart Circul Physiol. 1997;272(4):H1615–24.
148. el Alaoui-Talibi Z, Landormy S, Loireau A, Moravec J. Fatty acid oxidation and mechanical performance of volume-overloaded rat hearts. Am J Physiol-Heart Circul Physiol. 1992;262(4):H1068–74.
149. van Bilsen M, Smeets PJ, Gilde AJ, van der Vusse GJ. Metabolic remodelling of the failing heart: the cardiac burn-out syndrome? Cardiovasc Res. 2004;61(2):218–26.
150. van der Vusse GJ, van Bilsen M, Glatz JF. Cardiac fatty acid uptake and transport in health and disease. Cardiovasc Res. 2000;45(2):279–93.

151. Aitman TJ, Glazier AM, Wallace CA, Cooper LD, Norsworthy PJ, Wahid FN, et al. Identification of Cd36 (Fat) as an insulin-resistance gene causing defective fatty acid and glucose metabolism in hypertensive rats. Nat Genet. 1999;21(1):76–83.

152. Campbell FM, Kozak R, Wagner A, Altarejos JY, Dyck JR, Belke DD, et al. A role for peroxisome proliferator-activated receptor α (PPARα) in the control of cardiac malonyl-CoA levels: reduced fatty acid oxidation rates and increased glucose oxidation rates in the hearts of mice lacking PPARα are associated with higher concentrations of malonyl-CoA and reduced expression of malonyl-CoA decarboxylase. J Biol Chem. 2002;277(6):4098–103.

153. Young ME, Patil S, Ying J, Depre C, Singh AH, Shipley GL, et al. Uncoupling protein 3 transcription is regulated by peroxisome proliferator-activated receptor α in the adult rodent heart. FASEB J. 2001;15(3):833–45.

154. Liang F, Wang F, Zhang S, Gardner DG. Peroxisome proliferator activated receptor (PPAR) α agonists inhibit hypertrophy of neonatal rat cardiac myocytes. Endocrinology. 2003;144(9):4187–94.

155. Smeets PJ, Teunissen BE, Planavila A, de Vogel-van den Bosch H, Willemsen PH, van der Vusse GJ, et al. Inflammatory pathways are activated during cardiomyocyte hypertrophy and attenuated by peroxisome proliferator-activated receptors PPARα and PPARδ. J Biol Chem. 2008;283(43):29109–18.

156. Cheng L, Ding G, Qin Q, Huang Y, Lewis W, He N, et al. Cardiomyocyte-restricted peroxisome proliferator-activated receptor-δ deletion perturbs myocardial fatty acid oxidation and leads to cardiomyopathy. Nat Med. 2004;10(11):1245–50.

157. Planavila A, Laguna JC, Vázquez-Carrera M. Nuclear factor-κB activation leads to down-regulation of fatty acid oxidation during cardiac hypertrophy. J Biol Chem. 2005;280(17):17464–71.

158. Planavila A, Rodríguez-Calvo R, Jové M, Michalik L, Wahli W, Laguna JC, et al. Peroxisome proliferator-activated receptor β/δ activation inhibits hypertrophy in neonatal rat cardiomyocytes. Cardiovasc Res. 2005;65(4):832–41.

159. Pellieux C, Montessuit C, Papageorgiou I, Lerch R. Angiotensin II downregulates the fatty acid oxidation pathway in adult rat cardiomyocytes via release of tumour necrosis factor-α. Cardiovasc Res. 2009;82(2):341–50.

160. Burkart EM, Sambandam N, Han X, Gross RW, Courtois M, Gierasch CM, et al. Nuclear receptors PPARβ/δ and PPARα direct distinct metabolic regulatory programs in the mouse heart. J Clin Investig. 2007;117(12):3930–9.

Chapter 9
Cardiomyocyte Responses to Hormones

Abstract Several hormones and their receptors have a critical role in the home-ostasis of cardiomyocytes. They also modulate pathophysiological alterations in the cells and are therapeutic targets in heart diseases. Insulin controls substrate utilization in cardiomyocytes. Insulin promotes glucose uptake and its utilization via glycolysis and also regulates uptake of long-chain fatty acid and protein synthesis. There is evidence that varied thyroid hormone levels alter the growth patterns of both neonatal and adult cardiomyocytes. Oestrogen and its receptors influence the complex network of genomic and nongenomic pathways that govern cardiac metabolism, cytoprotection, cardiomyocyte regeneration and the electrophysiological and contractile function of the heart. Growth hormone is a participant in stimulating cardiomyocyte growth during development of the heart as well as in the maintenance of the structure and function of the normal adult heart. Aldosterone-mediated effects on the heart include increased oxidative stress, apoptosis, cardiac fibrosis, as well as left-ventricular hypertrophy.

Keyword Cardiomyocyte · Insulin · Thyroid hormone · Oestrogen · Growth hormone · Aldosterone · Glucocorticoids · Leptin

Introduction

Several hormones and their receptors have a critical role in the homeostasis as well as in pathophysiological alterations of cardiomyocytes (CMs) in heart diseases. They govern growth and metabolism of CMs as well as alterations in the cells in pathological conditions. Several of the cellular and molecular mechanisms of hormonal action on CMs have been elucidated.

Insulin

The key role of insulin in CMs is the regulation of substrate utilization. Insulin promotes glucose uptake and its utilization via glycolysis. Insulin participates in the regulation of long-chain fatty acid uptake and protein synthesis as well. Significant progress has been made in elucidating the signal transduction mechanisms underlying the effects of insulin [1–3]. In disease conditions, insulin signal transduction pathways and action are altered.

Insulin Signalling

Insulin binds to the extracellular region of insulin receptor (IR), resulting in the activation of intrinsic tyrosine kinase of the β-subunits of the receptor. Activated and phosphorylated IR phosphorylates several downstream elements [4–7]. Phosphatidylinositol 3-kinase (PI3K) and the mitogen-activated protein kinase (MAPK) are the main pathways, which are activated. The metabolic action of insulin is via PI3K (especially class Ia); The MAPK pathway is mainly involved in cell growth and differentiation.

Regulation of glucose uptake in CMs seems to require phosphoinositide—dependent kinase-1 (PDK1) and the PKBβ/Akt2 isoform [8–10]. Insulin is a potent activator of PKB/Akt in CMs [11]. PDK1 is necessary for activation of PKB/Akt by insulin [12].

Insulin signalling pathways involved in the regulation of cardiomyocyte metabolism is depicted in Fig. 9.1.

Regulation of Glucose Metabolism

The entry of glucose into the cardiomyocyte is dependent on glucose transporters at the plasma membrane. Glut1 and Glut4 are the two glucose transporters expressed in CMs. Glut4 is mainly involved in insulin regulated glucose uptake [13, 14]. Glut4 translocation from their intracellular storage sites to sarcolemmal membrane is caused by insulin [13, 15]. Insulin-induced Glut4 translocation involves the PI3K/PKB/Akt-signalling pathway [16]. Several other mechanisms which involve substrates of PKB/Akt such as the Akt substrate protein AS160 and Tre2/BUB2/cdc 1 (TBC1D1), the GTP-hydrolysis activating proteins (GAP) of the small G proteins Rabs, as well as the phosphoinositide 5-kinase and Synip, the SNARE-associated protein also regulate insulin-induced glucose Glut-4 translocation [16–24]. Along with PKB/Akt, the atypical PKC l/z family, another PDK1 substrate has also been suggested to have a role in Glut4 translocation to the sarcolemmal membrane [25].

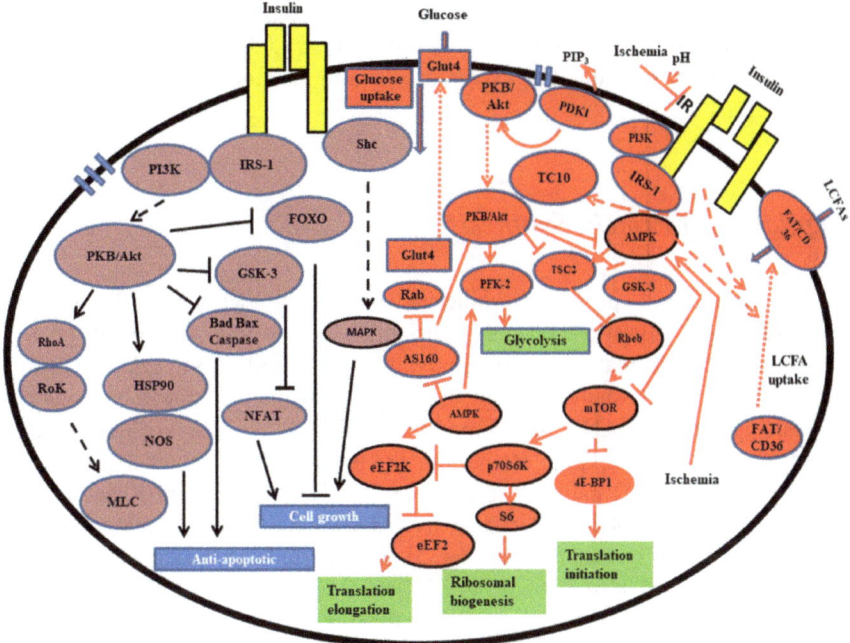

Fig. 9.1 Effects of insulin on substrate metabolism, protein translation, cell growth and apoptosis in cardiomyocyte and related signalling pathways. Solid lines indicate direct interactions, dashed lines denote indirect regulation and dotted lines represent translocation

In CMs, insulin increases the phosphorylation of the G protein TC10 (member of the Rho family of GTPases) coupled to IR and the IRS Cbl located in lipid raft microdomains [26]. This pathway also may contribute to the control of Glut4 transfer. The roles of TBC1D1, PIK five, synip, PKC l/z, or Cbl/TC10 in the insulin-induced stimulation of glucose uptake in CMs have not yet been completely elucidated.

Glucose after entry into CMs undergoes glycolysis and is also metabolized for glycogen synthesis. Both these are insulin-dependent. Insulin stimulates glycolysis and thus assists the use of glucose by CMs [27–29]. Glycogen level in CMs is considered to be regulated by a mechanism that involves allosteric activation of glucose-6-phosphate [12]. Insulin-induced and PKB/Akt-dependent stimulation of glycogen synthesis do not seem to be involved [12].

Regulation of Long-Chain Fatty Acid Metabolism

Insulin also induces the uptake of long chain fatty acids (LCFA) in CMs [30, 31]. The uptake is facilitated by insulin-dependent translocation of the LCFA transporter

FAT/CD36 to the sarcolemmal membrane (Fig. 9.1) and possibly involves insulin-induced PI3K activation [32]. Increase in intracellular LCFA levels does not lead to increased LCFA oxidation. The excess is added to the stored intracellular pool of lipids [33].

Regulation of Protein Synthesis

Insulin promotes protein synthesis and cell growth. Phosphorylation/dephosphorylation of several translation factors and ribosomal proteins contribute to the regulation of protein synthesis by insulin [34, 35]. PKB/Akt is a key player of this regulation and its action is mediated through the tuberous sclerosis factor 2 (TSC2), the G protein Rheb and the mammalian target of rapamycin (mTOR) (Fig. 9.1) [36–38]. Insulin is now clearly known to regulate the PKB/Akt/TSC2/mTOR pathway in CMs [39]. mTOR on activation regulates the 4E-binding protein-1 (4E-BP1) and the p70 ribosomal S6 protein kinase (p70S6K) which participate in protein translation [35]. Several studies have shown that insulin regulates 4E-BP1, p70S6K/ S6, and eEF2K/eEF2 in the heart or CMs [40–46].

GSK-3 (glycogen synthase kinase 3) and the fork head transcription factor FOXO family, which control protein translation and atrophy respectively are two other substrates regulated by PKB/Akt [35]. GSK-3 is known to regulate cardiomyocyte hypertrophy by inactivating the nuclear factor of activated T (NFAT), responsible for the expression of pro-hypertrophic genes [47]. FOXO family members, comprising FOXO1, FOXO3a, and FOXO4, promote the atrogene transcriptional program in skeletal muscle [48]. Skurk et al. reported that insulin inhibits FOXO3a through a PKB/Akt dependent pathway and thus restricts atrophy of cardiomyocytes [49]. FOXO1 in CMs is also regulated in the same way [50].

Insulin Action in Ischaemic Conditions

Acidosis that results from ischaemia inhibits tyrosine kinase activity of IR which correlates with the decrease in phosphorylation of PKB/Akt, p70S6K, and GSK-3 (Fig. 9.1) [41]. Ischaemia significantly alters insulin signalling and action mainly via a pathway dependent on AMP-activated protein kinase (AMPK) [28, 51–55]. Ischaemia, activates AMPK and also affects production of oxidative ATP and increases AMP levels [56].

Activated AMPK may regulate several components of the insulin-signalling pathway which control glucose metabolism and protein synthesis (Fig. 9.1) [28, 51–54, 56–58]. AMPK stimulates glucose uptake and glycolysis in CMs by regulating Glut-4 translocation and PFK-2 activation [59, 60]. AMPK also stimulates the uptake of LCFA [61]. AMPK inhibits eEF2 and the TSC2/mTOR/p70S6K pathways [44, 62–66]. Thus, AMPK antagonizes the stimulatory effect of insulin on protein synthesis. Insulin antagonizes AMPK signalling in ischemic conditions [67, 68]. Ischaemia-induced AMPK activation is reduced by phosphorylation of AMPK by activated PKB/Akt [69–71].

Though insulin-signalling pathway is blunted during ischaemia, insulin can still act during reperfusion. Insulin mediated activation of PKB/Akt has been proposed as a mechanism for the protective effect of a cocktail of glucose/ insulin/potassium (GIK) in reducing the infarct size after myocardial infarction [52, 72–75]. Metabolic reasons suggested are: (i) shift from LCFAs to glucose oxidation which is more oxygen efficient and prevents the production of toxic LCFA intermediates and (ii) activation of PKB/Akt resulting in several protective mechanisms including phosphorylation, sequestration, and/or inactivation of many pro-apoptotic proteins (Fig. 9.1) [76, 77]. PKB/Akt also phosphorylates and activates eNOS, leading to activation of NO-dependent PKG and inhibition of the mitochondrial transition pore [77]. PKB/Akt/mTOR/p70S6K is considered to promote the synthesis of contractile proteins during the post-ischaemic period.

IGF-1 acts on the same targets of insulin and exerts a protective action during ischaemia/reperfusion [78, 79].

Insulin Signalling and Cardiac Hypertrophy

Insulin, insulin growth factor-1 (IGF-1) and the neurohormone angiotensin II have common signalling pathways for stimulation of cell growth and protein synthesis as well as for inhibition of protein breakdown [47, 80]. All of them activate the MAPK and/or the PI3K/PKB/Akt signalling.

PI3K/PKB/Akt pathway participates in the growth of CMs during both embryonic and postnatal periods [81, 82]. In the adult heart, physiological (exercise) and patho-logical (hypertension, valvular dysfunction) conditions can cause chronic stimula-tion of PI3K/PKB/ Akt and/or MAPK pathways and thus induce cardiac hypertrophy [47, 80, 81]. Though the signalling pathways involved in physiological and patho-logical hypertrophy share some common signalling elements, they are dissimilar as well. The PI3K/PKB/Akt axis seems more linked to physiological hypertrophy. The mTOR/P70S6K pathway is necessary for PKB/Akt action [83]. MAPK signalling, in concert with PKC and calcineurin/NFAT pathways, participates in the development of pathological hypertrophy specifically induced by angiotensin II [81]. Sustained hyper insulinemia stimulates angiotensin II signalling that is involved in pathological hypertrophy [84]. There is also remodelling of substrate utilization in pathological hypertrophy and this is accompanied by a shift to carbohydrate usage simulating the insulin-stimulated phenotype [85, 86]. Reactive oxygen species production that accompanies metabolic shift also seems to cause hypertrophy [86].

Insulin Resistance in Type 2 Diabetes

Insulin-induced Glut4 translocation and glucose uptake are impaired in diabetic hearts. Glucose oxidation is also reduced in the diabetic heart, while LCFA uptake and oxidation are increased. The enhanced LCFA oxidation does not limit lipid accumulation in CMs. The metabolic disturbance has an important role in the development of insulin resistance in CMs. The molecular mechanisms for insulin resistance in CMs are not completely understood. The mechanisms have been studied in adipocytes and skeletal muscle. Insulin resistance is characterized by an alteration of the insulin-induced activation of the PI3K/PKB/Akt-signalling pathway [87–89].

Thyroid Hormones

Thyroid hormone (TH) is important for the development of several organs including the heart. The level of TH in the foetus is determined by the level of the hormone in the plasma of the mother. Thyroid function in the mother is known to impact the growth and maturation of foetal organs [90]. Abnormal levels of TH in the mother are known to cause abnormal development of the heart as well as other organs in the foetus [91–95].

   Most of our knowledge on the influence of thyroid hormone on the development of the foetal heart and the mechanisms that regulate cardiomyocyte growth have come from studies in sheep. Sheep is considered a suitable model because the thyroid endocrine system in the sheep is similar to that in humans and several organs including the heart in both sheep and humans have similar path for growth [96].

Thyroid Hormone Dynamics During Development

Circulating levels of 3,5,3'-triiodothyronine (T3) are low over most of the gestation period in foetuses of both sheep and humans. Serum T3 cannot be measured until about 30-week of pregnancy in the human foetus. Serum T3 level in the sheep foetus increases slowly during the last third of gestation. During the week immediately before birth, there is a prenatal surge in the serum levels of T3. The plateau phase coincides with the final stage of prenatal developmental of various organs including the heart. Serum T3 rises to a mean level of about 50 ng/dL at birth [97]. After birth, during the first 4–6 h, there is a 3–sixfold increase in the serum level of T3 [98, 99]. Studies indicate that plasma thyroid hormone levels during the 2–3 weeks prior to delivery and the increase in levels in the prepartum and postnatal periods are important for cardiovascular and metabolic adaptations requisite in the new born for extrauterine life [100].

Thyroid Hormone and Cardiomyocyte Differentiation

The terminal differentiation in CMs starts nearly thirty 30 days before the prepartum rise in T3 levels. T3 is an important contributor to the maturation process [101]. The effect of T3 on proliferation of isolated CMs from sheep foetus has been studied at two developmental ages viz., 100 days and 135 days (~10 days prior to birth). The proliferation rates are different in the two ages. T3 inhibits proliferation of CMs in the presence of IGF1, which is an important stimulant of cardiomyocyte proliferation throughout gestation [102].

In vivo studies on the effect of T3 on the proliferation of foetal CMs have revealed that when T3 (54 μg/day) is infused into foetal sheep about 15 days prior to the normal T3 surge, Ki-67 expression is lower and the cell cycle inhibitor p21 is elevated, indicating reduced proliferation of CMs. Cardiomyocytes were enlarged and several hypertrophy and maturation related genes such as mammalian target of rapamycin (mTOR), atrial natriuretic peptide and sarcoplasmic reticulum Ca^{2+}—ATPase 2A (SERCA2A) were upregulated [102].

These observations are consonant with the findings in studies in vitro. p21 protein levels were found increased and the cell cycle promoter, cyclin D1 levels were found lowered in CMs which were treated with varying doses of T3 for 24 h [103, 104].

T3- Related Signalling Pathways

T3 is known to stimulate the mitogen-activated protein kinase/extracellular signal-regulated kinase (MAPK/ERK) and Akt/PKB pathways [105–109]. In neonatal rats, angiotensin type 1 receptor (AT1R) mediates T3-induced hypertrophy of CMs via Akt/glycogen synthase kinase-3β/mammalian target of rapamycin (Akt/GSK-3β/mTOR) pathway [110].

The role of T3 in signalling and survival of CMs has been investigated using neonatal rat [111]. T3 rescues the loss of sarcomeric structure and apoptosis seen in neonatal CMs under serum-starved conditions [111]. The T3 effect is via the phosphatidylinositol-3 kinase/Akt/GSK-3β (PI3K/Akt/GSK-3β) pathway. The effect is attenuated by LY294002 (a specific PI3K inhibitor). These studies were carried out in postnatal rodent CMs. Be that as it may, the studies provide evidence that T3 is necessary for signalling in immature CMs through several pathways. Whether prenatal CMs in sheep and humans use the same pathways is unclear; T3 has different effects on CMs from sheep foetus.

T3 Interaction with IGF1

IGF1 stimulates the proliferation rates of CMs from 135-day gestation age (dGA) sheep by nearly four times in vitro. When T3 is administered in vivo, it increases the cardiac mass by about 30% [106]. Both T3 and IGF1 stimulates phosphorylation of both ERK and Akt in cultured cells. In younger cells, the combination of T3 and IGF1 suppresses both cell proliferation and phosphorylation of ERK and Akt [109]. Thus, when T3 levels are high, IGF-1 induced proliferation stops.

In CMs from more than 135-dGA sheep, with a mixed cell population in terms of maturation, T3 + IGF1 leads to suppression of BrdU uptake but a super stimulation of ERK and Akt, well above the levels induced by only T3 or IGF1 [109]. Cell signal patterning is altered as the cells mature and IGF1 and T3 stimulation results in different phosphorylation patterns and signalling outcomes than in cells of an earlier age [102].

When cells are treated with T3 and IGF1 together, both the ERK and Akt pathways are important in bringing about a cellular response. These pathways that are required for the growth of cells at a younger age may stimulate hypertrophy or metabolic functions in more mature cells [112–114]. In contrast to exogenous T3 administration, thyroidectomy does not cause changes in the phosphorylation of ERK, Akt and mTOR in sheep foetal CMs [114].

Studies suggest that at lower physiological levels of T3, cells proliferate under the effect of IGF1, which is preponderant all through development. When T3 levels increase, older CMs require stimulation of both pathways for their maturation. Segar and colleagues have shown that the foetal heart during late gestation requires thyroid hormone for adaptive growth of the heart in response to pressure overload [115].

T3 and FOXO1–p21 Interaction

T3, similar to IGF-1 stimulates not only the phosphorylation of Akt and ERK, but also p21, the cell cycle inhibitor [109]. The Forkhead box O (FOXO) protein is an important intermediate in this action. FOXO has a key role in cell cycle arrest, oxidative stress resistance, cell survival, energy metabolism and cell death in the heart [116–118]. FOXO1 is known to induce the transcription of several downstream target genes that collectively inhibit proliferation and induce cell cycle exit [119]. Direct downstream targets of FOXO1 include the cyclin kinase inhibitors p21 and p27, implicated in postnatal cell cycle exit of CMs [120–122]. The regulation of FOXO1 during cardiomyocyte cell cycle exit is not well characterized. FOXO1 activity is thought to be the determinant of the different actions of T3 and IGF1. Irrcher and colleagues have opined that AMPK phosphorylates FOXO1 disrupting its nuclear export and causes T3 to arrest the cell cycle through p21 [123]. One of the mechanisms by which T3 inhibits FOXO1 inactivation might be through sirtuins [124].

Actions of Thyroid Hormone Receptors

Signalling pathways through which TH molecules stimulate cellular actions are presently known. The classical or 'genomic' pathways involve TH binding to specific thyroid receptors (TR) in the nucleus, affecting gene expression via thyroid hormone response elements (TRE). In most species separate genes express two thyroid receptors (THRA (TRα) and THRB (TRβ)) and their isoforms, which bind T3 and thyroxine (T4) [125–128].

Most of the effects of T3 on CMs are mediated via TRα [126]. The affinity of T4 for TRα1 and TRβ1 in the foetal heart is considerably (10–15 times) less than those of T3 [127–130]. The foetal sheep heart expresses both receptor types [131, 132]. TRα1 is the predominant isoform which increases over gestation and in adulthood [133]. Among the different alpha isoforms, TRα1 is the predominant isoform expressed in the adult heart, known to bind T3. The other isoforms may have mostly inhibitory actions.

Phenotype analysis in mutant mice have identified the functional characteristics of the TR isoforms [134, 135]. The loss of TRα1 and TRα2 leads to decrease in the expression of the pacemaker current gene. Receptor-mediated effects of T3 have been revealed in thyroid hormone receptors alpha- and beta-knockout mice [131]. The sodium–calcium exchanger, the L/T type calcium channels, the ryanodine receptor and the sarcoplasmic reticulum calcium ATPase (SERCA2) are all targets of T3 action through its receptor. The hormone also influences the expression and contractile state of the contractile proteins in the muscle [136].

The genomic and non-genomic actions of thyroid hormones are summarized in Fig. 9.2.

Estrogen

Among estrone (E1), 17β-estradiol (E2) and estriol (E3), the three naturally occurring human estrogens in circulation E2 is the most potent and biologically relevant and hence the most commonly studied oestrogen.

A large number of studies indicate that estrogen has anti-inflammatory, anti-atherogenic and anti-arrhythmogenic effects [137–141]. In addition to the well-known role of estrogen in maintaining the functional integrity of endothelium and blood vessels, there are evidences for intricate oestrogen signalling and action on CMs as well.

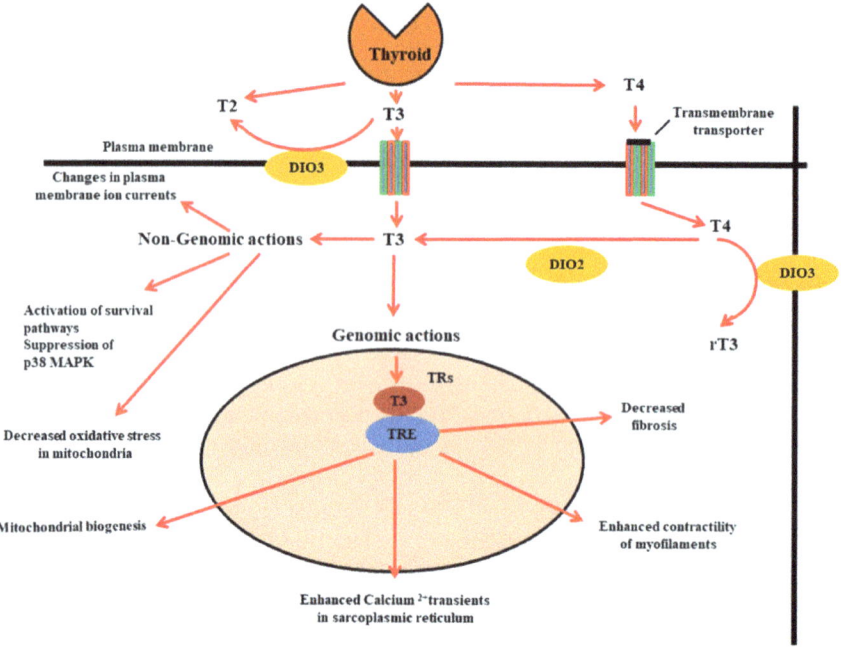

Fig. 9.2 Genomic and non-genomic actions of thyroid hormones (T2, T3, T4) on cardiomyocyte. DIO3—Iodothyronine deiodinase 3. TR—Thyroid hormone receptor

Actions of Estrogen on Cardiomyocytes

Estrogen biosynthesis is dependent on cytochrome P450 aromatase (CYP19A1). The enzyme metabolizes testosterone to E2 and androstenedione to E1 [142]. Expression of aromatase in the heart is evidence for production of estrogen in the organ [143, 144]. E2 regulates several genes linked to mitochondrial function, redox homeostasis, carbohydrate metabolism and lipogenesis [145, 146]. Thus, aromatase may have a role in estrogen-mediated cardiac protection.

Estrogen receptors (ER) are expressed and functional in CMs. There are two receptor subtypes: ERα and Erβ. Both ERα and ERβ are expressed in the cytoplasm of CMs of atria and ventricles of neonatal and adult hearts in both genders [147–150]. ER regulates gene expression and posttranslational modifications. Genomic and nongenomic mechanisms for ER action are depicted in Fig. 9.3. There is evidence for antiapoptotic, pro-hypertrophic and anti-inflammatory actions of estrogen-activated ER (E2/ER) in the heart [151].

Each receptor subtype, depending on their subcellular localization, contributes to specific forms of estrogen signalling. Each ER subtype has dissimilar target molecules and different degree of regulation [145, 152–154].

G-protein coupled oestrogen receptor (GPER) is involved in rapid, nongenomic E2 signalling at the plasma membrane [155]. GPER is present in the hearts of both genders and is cardioprotective during ischemic and hypertensive stress [156, 157]. Studies indicate that the receptor also inhibits the mitochondria permeability transition pore (mPTP) opening and activates pro-survival pathway [158].

Estrogen and Estrogen Receptors in Cardiac Metabolism

Eestrogen regulates energy balance and glucose homeostasis via different mechanisms [159]. Among the three known estrogen receptors (ERs), ERα is the major factor in E2-specific regulation of energy homeostasis. Several studies have explored the role of estrogen in the regulation of cardiac metabolism. E2 supplementation substantially improves the levels of myocardial ATP and mitochondrial respiratory function in the heart [160]. ERα seems necessary for normal glucose uptake in the murine heart [161]. The discovery of a link between ERα and GLUT4 suggests a molecular mechanism for E2-mediated regulation of glucose metabolism in CMs.

Eestrogen via its receptors has a key role in maintaining mitochondrial integrity and function, and thus protects the cardiomyocyte from oxidative stress and metabolic derangement during ischemic or hypoxic conditions [153, 162–164].

Eestrogen receptor-related receptor alpha (ERRα) together with peroxisome proliferator-activated receptor-γ (PPAR γ) coactivator-1 alpha (PGC-1α) seems requisite for cardiomyocyte metabolism [165]. Given the evidence for cross-talk between the ERα and ERRα subunits, it is possible that ERRα modulates E2/ERα signalling in cardiomyocyte metabolism or that E2/ERα signalling could affect ERRα activity [166]. While ERα activity in energy metabolism is evident, the role of ERβ in metabolic regulation is less obvious. There is however evidence to suggest a role for ERβ in the regulation of metabolism in mitochondria [154, 167–169].

Our knowledge on the E2/ER input in cardiomyocyte energetics are based on experiments in animals or cultured cells and are thus limited in understanding the role of E2/ER in cardiomyocyte metabolism in vivo.

Estrogen and Estrogen Receptors in Cardiac Hypertrophy

Estrogen is involved in a wide array of signalling linked to ventricular hypertrophy. The spectrum of its action includes regulation of normal postnatal growth of CMs as well as of physiological response to exercise and reduction of maladaptive ventricular hypertrophy induced by pressure overload. In experiments, estrogen has been found to antagonize angiotensin II induced hypertrophy of CMs. In humans, administration of the hormone improves the geometry of left ventricle [170, 171].

In the E2-mediated anti-hypertrophic effects, both ER isoforms are implicated [172, 173]. The molecular mechanism of estrogen action on cardiomyocyte hypertrophy is supposed to involve autocrine and or paracrine effects which are mediated by atrial natriuretic factor (ANF), calcineurin degradation, mTOR signalling, p38

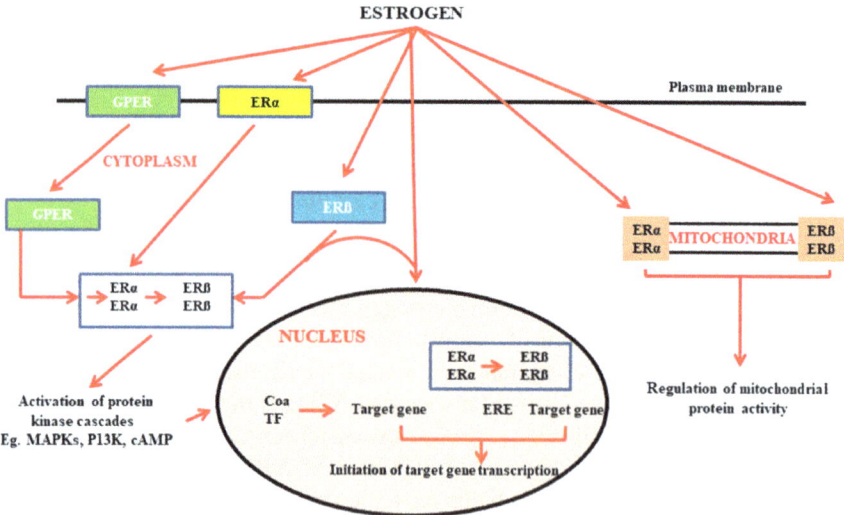

Fig. 9.3 Mechanisms of actions of oestrogen and estrogen receptors on cardiomyocyte. ER—Estrogen receptor. ERE—Estrogen response element. GPER—G protein-coupled estrogen receptor. TF—Transcription factor

MAPK pathways, and regulation of histone deacetylases (Fig. 9.3). Exercise-induced ventricular hypertrophy seems to be dependent on Erβ [170, 174–178].

ERα mediates oestrogenic regulation for normal myocardial development and the action involves upregulation of genes for insulin-like growth factor 1 and myocardin [179, 180].

Endonuclease G (endo G), a mitochondrial-localized nuclease expressed in CMs is a central link between maladaptive hypertrophy and mitochondrial processes, which are however unrelated to apoptosis [181]. Endo G is a known transcriptional target of ERR-α and PGC-1α, which are master regulators of mitochondrial function and cardiomyocyte metabolism. Inhibition of endo G expression directly induces hypertrophy in cultured CMs. Endo G loss-of-function in vivo results in cardiomyocyte hypertrophy and increase in left ventricular mass after angiotensin II stimulation. It has been suggested that exploring the link between the E2/ER and endo G may aid to improve our knowledge on estrogen mediated hypertrophic response in CMs [141].

Estrogen and Estrogen Receptors in Apoptosis of Cardiomyocytes

E2 inhibits both the intrinsic and extrinsic apoptotic pathways and thus reduces apoptosis in CMs [182]. Cytoprotective or pro-survival oestrogen signalling pathways involved in the antiapoptotic effect of oestrogen on CMs include inhibition of apoptosis signal-regulating kinase 1 (ASK1) activity, downregulation of NF-κB, activation of PI3K/Akt signalling and upregulation of corticotropin-releasing hormone receptor type 2 [165, 183–185]. It also involves stimulation of p38β activity leading to inhibition of p53 and attenuation of mitochondrial redox response [153, 164, 186]. Both ERα and ERβ present in mitochondria, a major site of apoptotic signalling, are involved in E2-specific inhibition of apoptosis in CMs (147, 152, 154). GPER also inhibits cardiomyocyte apoptosis [156, 158, 187]. There is no consensus on which ER receptor has the principal role in the anti-apoptotic effect [188, 189].

Estrogen, Estrogen Receptors and Excitation–Contraction Coupling

The cellular and molecular mechanisms of oestrogen's impact on action potential of the cardiomyocyte have been elucidated. Release of endogenous nitric oxide and inhibition of Na^{++}/H^{++} exchanger (NHE1) are components of the E2 signalling pathways [190, 191].

Estrogen is important to maintain normal excitation–contraction coupling in the heart of females [192, 193]. The gender differences in E-C coupling seen at the organ and cellular level are thought to be determined by the control of calcium homeostasis by E2/ER in the heart. E2 modulates L-type Ca^{2+} channel activity in CMs via alterations in the density and expression of both L-type Ca^{2+} channels and low-voltage-activated Ca(V)3.2 T-type calcium channels in the membrane [194–196]. While one study did not find evidence for role of ERα or ERβ in E2-mediated control of E-C coupling in the heart, others suggest the involvement of the classic ER subtypes [197–199]. The rapid effects by the hormone are attributed to the membrane-associated ER [200]. Studies using ER subtype-selective agonists and genetic deletion models indicate that only ERβ has positive effects on E-C coupling and myocyte contractility of the heart in females [201, 202]. The precise role of each ER subtype in E-C coupling in CMs is yet to be identified.

Estrogen and Estrogen Receptors in Cardiac Regeneration

There is now evidence that oestrogen can stimulate regeneration of CMs. When human resident cardiac cells that are positive for c-kit (stem cell-related surface antigen), were treated with E2 and infused into isolated mouse hearts subjected to acute ischemia, significant paracrine production of cardiac stem cell derived protective factors, improvement in cardiac function and better cardiomyocyte survival have been observed [203]. Similar effects were also observed when E2 pre-treated mesenchymal stem cells (MSC) were infused to the heart [204].

E2 stimulates proliferation of mouse embryonic stem cells as well. This action is via p44/42 mitogen-activated protein kinases (MAPKs), cyclin dependent kinase (CDK)2 and CDK4, and is associated with upregulated expression of c-fos, c-jun, and c-myc proto-oncogene [205]. E2 signalling also seems to increase the regenerative capability of stem cells after ischemic injury [206]. ERα is upregulated in c-kit+ progenitor cells homing in the peri-infarct regions of the heart [207].

Growth Hormone

Growth hormone (GH) has a vital role in stimulating cardiomyocyte growth during development of a normal heart as well as in the maintenance of the structure and function of the normal adult heart [78, 208, 209]. In response to GH, production of insulin growth factor −1 (IGF-1) in the heart is up-regulated [210]. The heart has functional receptors for both GH [211, 212] and IGF-1 [213, 214]. Thus, GH has direct actions and also affects the endocrine or autocrine/paracrine effects of IGF-1 on the heart (Fig. 9.4); the relative importance of these is however unclear.

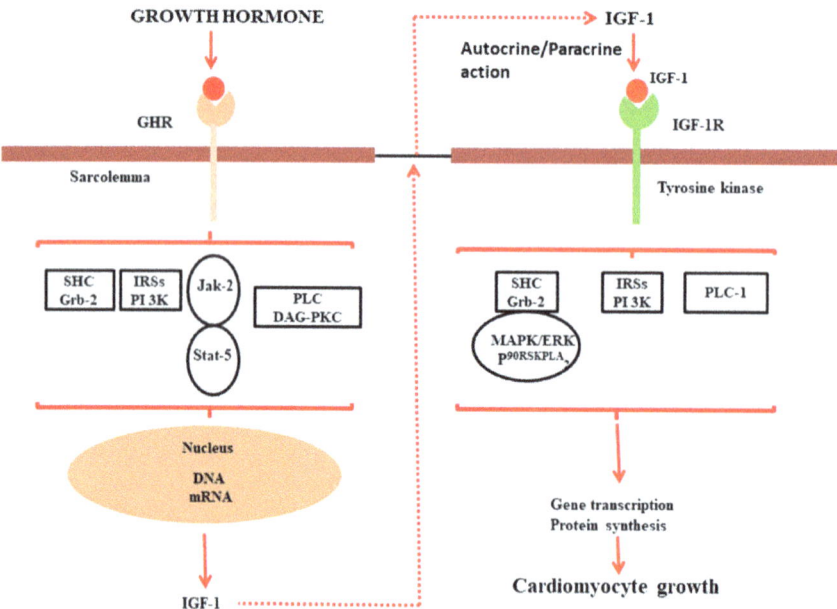

Fig. 9.4 Insulin growth factor (IGF)—1 mediated growth hormone (GH) action on cardiomyocyte and related signalling pathways. DAG—Diacylglycerol. IR—Insulin receptor. IRSs—Insulin receptor substrates. Jak-Janus kinase. MAPK/ERK—Mitogen activated protein kinase/extracellular signal regulated kinase. PKC—Protein kinase C. PLC—Phospholipase C. Stat—Signalling transducers and activators of transcription

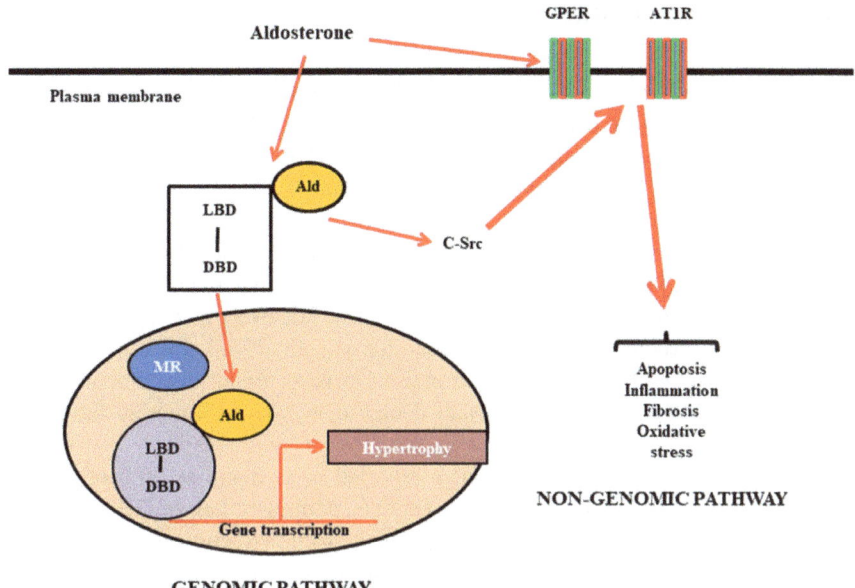

Fig. 9.5 Genomic and non-genomic pathways of action of aldosterone on cardiomyocytes. AT1R—Angiotensin II type1 receptor. DBD—DNA binding domain. GPER—G protein-coupled estrogen receptor. LBD—Ligand binding domain. MR—Mineralocorticoid receptor

Effects of GH and IGF-1 on Cardiomyocytes

It has been suggested that GH may also alter cardiomyocyte metabolism and stimulate cardiomyocyte growth independently of IGF-1 [215, 216]. The direct hypertrophic effects of GH on CMs, independent of IGF-1 have however not been validated in many studies [217, 218].

Several target genes of GH and IGF-1 have been identified. IGF-1 stimulates the expression of muscle-specific genes in rat CMs [218]. In vitro studies have indicated that IGF-1 increases protein synthesis and the size of CMs [217–219]. Both GH and IGF-1 stimulate the expression of skeletal α-actin and atrial natriuretic factor genes in the heart of hypoxic rats [220]. Apart from their growth promoting effects on CMs, GH and IGF-1 may protect the myocardium by preventing loss of CMs through apoptosis. There is evidence that GH/IGF-1 can inhibit apoptosis [221, 222].

Several in vitro studies have demonstrated the direct effects of IGF-1 on intrinsic cardiomyocyte contractility [223–225]. There is no evidence for direct, IGF-1-independent effects of GH on cardiomyocyte contractility. Three different mechanisms have been proposed for the induction of increased cardiomyocyte contractility by GH/IGF-1. These are: (i) altered intracellular calcium transients, (ii) increased sensitivity of myofilaments to calcium and (iii) a shift in myosin isoforms [224–232]. Whether gene regulations are involved in the alteration of calcium handling by GH/IGF-1 is not clear. GH/ IGF-I may also regulate the expression of ion channels.

Cardiac Abnormalities in GH Deficiency and Excess

A balanced GH/IGF-I system with neither deficiency nor excess is needed for normal cardiac function. Severe cardiac dysfunction is seen in GH deficient patients with childhood-onset disease because of the lack of GH/ IGF-I during growth and development of the heart. The severity is more in them than in in patients with adult-onset disease [233, 234].

Patients with acromegaly where there is increased production of GH have cardiomyopathy characterized by cardiomegaly, ventricular hypertrophy, replacement fibrosis, and degeneration of CMs [235, 236]. The exact mechanism for alteration in cardiac structure in acromegalic cardiomyopathy is unknown. In transgenic mice overexpressing bovine GH, deterioration in myocardial bioenergetics, ultrastructural changes in mitochondria and decrease in systolic function have been observed [237].

Growth hormone deficiency has been reported in as many as 40% of patients with chronic heart failure [238]. In these patients, administration of GH results in a moderate but significant increase in left ventricular ejection fraction and a reduction in circulating levels of N-terminal pro-brain natriuretic peptide [238]. Cardiac cachexia in severe heart failure could also lead to endocrine disturbances including acquired GH resistance [239].

Aldosterone

The mineralocorticoid aldosterone is synthesized in the heart as well and has paracrine effects on CMs [240, 241]. Genomic and non-genomic actions of aldosterone in the heart is via binding to cytoplasmic and transmembrane mineralocorticoid receptors [242]. There is evidence that aldosterone specifically regulates the expression of the mineralocorticoid receptor (MR) gene in CMs [243]. Aldosterone can cause increased oxidative stress, induce apoptosis, as well as myocardial fibrosis and ventricular hypertrophy [244]. Connective tissue growth factor (CTGF) expression is known to be increased in various heart diseases [245]. This increase may be induced (at least partly) by aldosterone [244].

Mineralocorticoid receptor (MR) expression in the heart is increased in hypertension, myocardial infarction, and diastolic heart failure [246–248]. Genetic deletion of cardiomyocyte-MR decreases adverse remodelling and improves ventricular function in mice with myocardial infarction [249]. Overexpression of cardiomyocyte-MR leads to arrhythmias [250].

Aldosterone effects in the heart are shown in Fig. 9.5.

Glucocorticoids

Stress is a contributor for heart diseases [251, 252]. Stress causes activation of the hypothalamic–pituitary–adrenal axis and release of glucocorticoids (cortisol) from adrenal gland. Glucocorticoid receptor (GR), the mediator of the actions of glucocorticoids is expressed in CMs; GR has a direct role for glucocorticoid signalling in the heart [253–256]. Glucocorticoids-GR complex regulates the expression of numerous genes.

RH. Oakley and colleagues have evaluated stress hormone signalling in the heart and CMs using genetically engineered mice [257]. Their findings reveal that insufficient GR signalling and unopposed MR signalling are both pathogenic in CMs. They also indicate that both receptor signalling pathways are crucial for a balanced stress response and a healthy heart.

Leptin

Leptin is a hormone secreted by adipose tissue. It reduces appetite and increases metabolic activity, thus regulating body weight. Leptin also increases fatty acid oxidation in the heart and increases myocardial glucose utilization [258–260]. The beneficial metabolic effects of leptin are thought to be mediated by elevated AMP-kinase activity [261, 262]. Cardiac-specific deletion of the leptin receptor aggravates ischemic myocardial injury in animals and worsens heart failure resulting from cardiotoxicity [261, 262].

Conclusions

Hormones have an important role in the homeostatic mechanisms in cardiomyocytes (CMs) and are recognized to be potential therapeutic targets in patients with heart failure. A large part of the insulin-signalling components involved in the regulation of processes such as glucose and LCFA utilization and protein translation in CMs has been identified. The partial elucidation of these signal transduction pathways has aided the development of efficient strategies based on the modulation of some of these signalling elements to treat heart diseases. Thyroid hormone and its related signalling pathways regulate the growth of CMs during development of the heart and postnatal terminal differentiation of CMs. There are several estrogen-specific genomic and nongenomic responses in CMs. Insights gained on the effects of estrogen on cardiomyocytes and the heart are likely to aid the design of strategies for prevention and gender-specific treatment of heart diseases.

References

1. Brownsey RW, Boone AN, Allard MF. Actions of insulin on the mammalian heart: metabolism, pathology and biochemical mechanisms. Cardiovasc Res. 1997;34(1):3–24.
2. Sano H, Kane S, Sano E, Mîinea CP, Asara JM, Lane WS, et al. Insulin-stimulated phosphorylation of a Rab GTPase-activating protein regulates GLUT4 translocation. J. Biol. Chem. 2003;278(17):14599–602.
3. Bertrand L, Horman S, Beauloye C, Vanoverschelde J-L. Insulin signalling in the heart. Cardiovasc Res. 2008;79(2):238–48.
4. Muniyappa R, Montagnani M, Koh KK, Quon MJ. Cardiovascular actions of insulin. Endocr Rev. 2007;28:463–91.
5. Avruch J. Insulin signal transduction through protein kinase cascades. Mol Cell Biochem. 1998;182:31–48.
6. Saltiel AR, Kahn CR. Insulin signalling and the regulation of glucose and lipid metabolism. Nature. 2001;414(6865):799–806.
7. Thirone AC, Huang C, Klip A. Tissue-specific roles of IRS proteins in insulin signaling and glucose transport. Trends Endocrinol Metab. 2006;17(2):72–8.
8. Dummler B, Hemmings B. Physiological roles of PKB/Akt isoforms in development and disease. Biochem Soc Trans. 2007;35:231–5.
9. DeBosch B, Sambandam N, Weinheimer C, Courtois M, Muslin AJ. Akt2 regulates cardiac metabolism and cardiomyocyte survival. J Biol Chem. 2006;281(43):32841–51.
10. Mora A, Sakamoto K, McManus EJ, Alessi DR. Role of the PDK1–PKB–GSK3 pathway in regulating glycogen synthase and glucose uptake in the heart. FEBS Lett. 2005;579(17):3632–8.
11. Deprez J, Bertrand L, Alessi DR, Krause U, Hue L, Rider MH. Partial purification and characterization of a wortmannin-sensitive and insulin-stimulated protein kinase that activates heart 6-phosphofructo-2-kinase. Biochem J. 2000;347(1):305–12.
12. Mora A, Davies AM, Bertrand L, Sharif I, Budas GR, Jovanović S, et al. Deficiency of PDK1 in cardiac muscle results in heart failure and increased sensitivity to hypoxia. EMBO J. 2003;22(18):4666–76.
13. Zorzano A, Sevilla L, Camps M, Becker C, Meyer J, Kammermeier H, et al. Regulation of glucose transport, and glucose transporters expression and trafficking in the heart: studies in cardiac myocytes. Am J Cardiol. 1997;80(3):65A-76A.
14. Abel ED. Glucose transport in the heart. Front Biosci. 2004;9:201–15.
15. Slot JW, Geuze HJ, Gigengack S, James DE, Lienhard GE. Translocation of the glucose transporter GLUT4 in cardiac myocytes of the rat. Proc Natl Acad Sci USA. 1991;88(17):7815–9.
16. Watson RT, Pessin JE. Bridging the GAP between insulin signaling and GLUT4 translocation. Trends Biochem Sci. 2006;31(4):215–22.
17. Watson RT, Pessin JE. GLUT4 translocation: the last 200 nanometers. Cell Signal. 2007;19(11):2209–17.
18. Huang S, Czech MP. The GLUT4 glucose transporter. Cell Metab. 2007;5(4):237–52.
19. He A, Liu X, Liu L, Chang Y, Fang F. How many signals impinge on GLUT4 activation by insulin? Cell Signal. 2007;19(1):1–7.
20. Kramer HF, Witczak CA, Taylor EB, Fujii N, Hirshman MF, Goodyear LJ. AS160 regulates insulin-and contraction-stimulated glucose uptake in mouse skeletal muscle. J Biol Chem. 2006;281(42):31478–85.
21. Roach WG, Chavez JA, Mîinea CP, Lienhard GE. Substrate specificity and effect on GLUT4 translocation of the Rab GTPase-activating protein Tbc1d1. Biochem J. 2007;403(2):353–8.
22. Chen S, Murphy J, Toth R, Campbell DG, Morrice NA, Mackintosh C. Complementary regulation of TBC1D1 and AS160 by growth factors, insulin and AMPK activators. Biochem J. 2008;409(2):449–59.
23. Berwick DC, Dell GC, Welsh GI, Heesom KJ, Hers I, Fletcher LM, et al. Protein kinase B phosphorylation of PIKfyve regulates the trafficking of GLUT4 vesicles. J Cell Sci. 2004;117(25):5985–93.

24. Yamada E, Okada S, Saito T, Ohshima K, Sato M, Tsuchiya T, et al. Akt2 phosphorylates Synip to regulate docking and fusion of GLUT4-containing vesicles. J Cell Biol. 2005;168(6):921–8.
25. Farese R, Sajan M, Standaert M. Atypical protein kinase C in insulin action and insulin resistance. Biochem Soc Trans. 2005;33(2):350–3.
26. Gupte A, Mora S. Activation of the Cbl insulin signaling pathway in cardiac muscle; dysregulation in obesity and diabetes. Biochem Biophys Res Commun. 2006;342(3):751–7.
27. Lefebvre V, Máchin MC, Louckx MP, Rider MH, Hue L. Signaling pathway involved in the activation of heart 6-phosphofructo-2-kinase by insulin. J Biol Chem. 1996;271(37):22289–92.
28. Hue L, Beauloye C, Marsin A-S, Bertrand L, Horman S, Rider MH. Insulin and ischemia stimulate glycolysis by acting on the same targets through different and opposing signaling pathways. J Mol Cell Cardiol. 2002;34(9):1091–7.
29. Rider MH, Bertrand L, Vertommen D, Michels PA, Rousseau GG, Hue L. 6-phosphofructo-2-kinase/fructose-2, 6-bisphosphatase: head-to-head with a bifunctional enzyme that controls glycolysis. Biochem J. 2004;381(3):561–79.
30. Coort SL, Bonen A, van der Vusse GJ, Glatz JF, Luiken JJ. Cardiac substrate uptake and metabolism in obesity and type-2 diabetes: role of sarcolemmal substrate transporters. Mol Cell Biochem. 2007;299(1):5–18.
31. Glatz JF, Bonen A, Ouwens DM, Luiken JJ. Regulation of sarcolemmal transport of substrates in the healthy and diseased heart. Cardiovasc Drugs Ther. 2006;20(6):471–6.
32. Luiken JJ, Koonen DP, Willems J, Zorzano A, Becker C, Fischer Y, et al. Insulin stimulates long-chain fatty acid utilization by rat cardiac myocytes through cellular redistribution of FAT/CD36. Diabetes. 2002;51(10):3113–9.
33. Dyck D, Steinberg G, Bonen A. Insulin increases FA uptake and esterification but reduces lipid utilization in isolated contracting muscle. Am J Physiol-Endocrinol Metab. 2001;281(3):E600–7.
34. Hedhli N, Pelat M, Depre C. Protein turnover in cardiac cell growth and survival. Cardiovasc Res. 2005;68(2):186–96.
35. Proud CG. Signalling to translation: how signal transduction pathways control the protein synthetic machinery. Biochem J. 2007;403(2):217–34.
36. Potter CJ, Pedraza LG, Xu T. Akt regulates growth by directly phosphorylating Tsc2. Nat Cell Biol. 2002;4(9):658–65.
37. Inoki K, Li Y, Zhu T, Wu J, Guan K-L. TSC2 is phosphorylated and inhibited by Akt and suppresses mTOR signalling. Nat Cell Biol. 2002;4(9):648–57.
38. Manning BD, Cantley LC. Rheb fills a GAP between TSC and TOR. Trends Biochem Sci. 2003;28(11):573–6.
39. Rolfe M, McLeod LE, Pratt PF, Proud CG. Activation of protein synthesis in cardiomyocytes by the hypertrophic agent phenylephrine requires the activation of ERK and involves phosphorylation of tuberous sclerosis complex 2 (TSC2). Biochem J. 2005;388(3):973–84.
40. Wang L, Wang X, Proud CG. Activation of mRNA translation in rat cardiac myocytes by insulin involves multiple rapamycin-sensitive steps. Am J Physiol-Heart Circulatory Physiol. 2000;278(4):H1056–68.
41. Beauloye C, Bertrand L, Krause U, Marsin A-S, Dresselaers T, Vanstapel F, et al. No-flow ischemia inhibits insulin signaling in heart by decreasing intracellular pH. Circ Res. 2001;88(5):513–9.
42. Pham FH, Sugden PH, Clerk A. Regulation of protein kinase B and 4E-BP1 by oxidative stress in cardiac myocytes. Circ Res. 2000;86(12):1252–8.
43. Sharma S, Guthrie PH, Chan SS, Haq S, Taegtmeyer H. Glucose phosphorylation is required for insulin-dependent mTOR signalling in the heart. Cardiovasc Res. 2007;76(1):71–80.
44. Horman S, Beauloye C, Vertommen D, Vanoverschelde J-L, Hue L, Rider MH. Myocardial ischemia and increased heart work modulate the phosphorylation state of eukaryotic elongation factor-2. J Biol Chem. 2003;278(43):41970–6.
45. Longnus S, Segalen C, Giudicelli J, Sajan M, Farese R, Van Obberghen E. Insulin signalling downstream of protein kinase B is potentiated by 5′ AMP-activated protein kinase in rat hearts in vivo. Diabetologia. 2005;48(12):2591–601.

46. Jonassen AK, Sack MN, Mjøs OD, Yellon DM. Myocardial protection by insulin at reper-
    fusion requires early administration and is mediated via Akt and p70s6 kinase cell-survival
    signaling. Circ Res. 2001;89(12):1191–8.
47. Heineke J, Molkentin JD. Regulation of cardiac hypertrophy by intracellular signalling
    pathways. Nat Rev Mol Cell Biol. 2006;7(8):589–600.
48. Sandri M, Sandri C, Gilbert A, Skurk C, Calabria E, Picard A, et al. Foxo transcription factors
    induce the atrophy-related ubiquitin ligase atrogin-1 and cause skeletal muscle atrophy. Cell.
    2004;117(3):399–412.
49. Skurk C, Izumiya Y, Maatz H, Razeghi P, Shiojima I, Sandri M, et al. The FOXO3a
    transcription factor regulates cardiac myocyte size downstream of AKT signaling. J Biol
    Chem. 2005;280(21):20814–23.
50. Liu T-J, Lai H-C, Ting C-T, Wang PH. Bidirectional regulation of upstream IGF-
    I/insulin receptor signaling and downstream FOXO1 in cardiomyocytes. J Endocrinol.
    2007;192(1):149–58.
51. Dolinsky VW, Dyck JR. Role of AMP-activated protein kinase in healthy and diseased
    hearts. Am J Physiol-Heart Circulatory Physiol. 2006;291(6):H2557–69.
52. Tian R, Balschi JA. Interaction of insulin and AMPK in the ischemic heart: another chapter
    in the book of metabolic therapy? Circ Res. 2006;99(1):3–5.
53. Dyck JR, Lopaschuk GD. AMPK alterations in cardiac physiology and pathology: enemy or
    ally? J Physiol. 2006;574(1):95–112.
54. Arad M, Seidman CE, Seidman J. AMP-activated protein kinase in the heart: role during
    health and disease. Circ Res. 2007;100(4):474–88.
55. Young LH, Li J, Baron SJ, Russell RR. AMP-activated protein kinase: a key stress signaling
    pathway in the heart. Trends Cardiovasc Med. 2005;15(3):110–8.
56. Kudo N, Gillespie JG, Kung L, Witters LA, Schulz R, Clanachan AS, et al. Characterization
    of 5′ AMP-activated protein kinase activity in the heart and its role in inhibiting acetyl-
    CoA carboxylase during reperfusion following ischemia. Biochimica et Biophysica Acta
    (BBA)-Lipids and Lipid Metabolism. 1996;1301(1–2):67–75.
57. Towler MC, Hardie DG. AMP-activated protein kinase in metabolic control and insulin
    signaling. Circ Res. 2007;100(3):328–41.
58. Hue L, Beauloye C, Bertrand L, Horman S, Krause U, Marsin A-S, et al. New targets of
    AMP-activated protein kinase. Biochem Soc Trans. 2003;31(1):213–5.
59. Russell RR III, Bergeron R, Shulman GI, Young LH. Translocation of myocardial GLUT-4
    and increased glucose uptake through activation of AMPK by AICAR. Am J Physiol-Heart
    Circulatory Physiol. 1999;277(2):H643–9.
60. Marsin A, Bertrand L, Rider M, Deprez J, Beauloye C, Vincent MF, et al. Phosphorylation
    and activation of heart PFK-2 by AMPK has a role in the stimulation of glycolysis during
    ischaemia. Curr Biol. 2000;10(20):1247–55.
61. Luiken JJ, Coort SL, Willems J, Coumans WA, Bonen A, van der Vusse GJ, et al. Contraction-
    induced fatty acid translocase/CD36 translocation in rat cardiac myocytes is mediated through
    AMP-activated protein kinase signaling. Diabetes. 2003;52(7):1627–34.
62. Krause U, Bertrand L, Hue L. Control of p70 ribosomal protein S6 kinase and acetyl-
    CoA carboxylase by AMP-activated protein kinase and protein phosphatases in isolated
    hepatocytes. Eur J Biochem. 2002;269(15):3751–9.
63. Bolster DR, Crozier SJ, Kimball SR, Jefferson LS. AMP-activated protein kinase suppresses
    protein synthesis in rat skeletal muscle through down-regulated mammalian target of
    rapamycin (mTOR) signaling. J Biol Chem. 2002;277(27):23977–80.
64. Cheng SW, Fryer LG, Carling D, Shepherd PR. Thr2446 is a novel mammalian target
    of rapamycin (mTOR) phosphorylation site regulated by nutrient status. J Biol Chem.
    2004;279(16):15719–22.
65. Inoki K, Zhu T, Guan K-L. TSC2 mediates cellular energy response to control cell growth
    and survival. Cell. 2003;115(5):577–90.
66. Horman S, Browne GJ, Krause U, Patel JV, Vertommen D, Bertrand L, et al. Activation of
    AMP-activated protein kinase leads to the phosphorylation of elongation factor 2 and an
    inhibition of protein synthesis. Curr Biol. 2002;12(16):1419–23.

67. Beauloye C, Marsin A-S, Bertrand L, Krause U, Hardie DG, Vanoverschelde J-L, et al. Insulin antagonizes AMP-activated protein kinase activation by ischemia or anoxia in rat hearts, without affecting total adenine nucleotides. FEBS Lett. 2001;505(3):348–52.
68. Gamble J, Lopaschuk GD. Insulin inhibition of 5′ adenosine monophosphate—activated protein kinase in the heart results in activation of acetyl coenzyme A carboxylase and inhibition of fatty acid oxidation. Metabolism. 1997;46(11):1270–4.
69. Kovacic S, Soltys C-LM, Barr AJ, Shiojima I, Walsh K, Dyck JR. Akt activity negatively regulates phosphorylation of AMP-activated protein kinase in the heart. J Biol Chem. 2003;278(41):39422–7.
70. Horman S, Vertommen D, Heath R, Neumann D, Mouton V, Woods A, et al. Insulin antagonizes ischemia-induced Thr172 phosphorylation of AMP-activated protein kinase α-subunits in heart via hierarchical phosphorylation of Ser485/491. J Biol Chem. 2006;281(9):5335–40.
71. Soltys C-LM, Kovacic S, Dyck JR. Activation of cardiac AMP-activated protein kinase by LKB1 expression or chemical hypoxia is blunted by increased Akt activity. Am J Physiol-Heart Circulatory Physiol. 2006;290(6):H2472-H9.
72. Diaz R, Paolasso E, Piegas L, Tajer C, Moreno M, Corvalan R, et al. Metabolic modulation of acute myocardial infarction. The ECLA (Estudios Cardiologicos Latinoamerica) Collaborative Group. Circulation. 1998;98(21):2227–34.
73. Schipke JD, Friebe R, Gams E. Forty years of glucose–insulin–potassium (GIK) in cardiac surgery: a review of randomized, controlled trials. Eur J Cardiothorac Surg. 2006;29(4):479–85.
74. Hausenloy DJ, Yellon DM. Reperfusion injury salvage kinase signalling: taking a RISK for cardioprotection. Heart Fail Rev. 2007;12(3–4):217–34.
75. Apstein CS. The benefits of glucose-insulin-potassium for acute myocardial infarction (and some concerns). J Am Coll Cardiol. 2003;42(5):792–5.
76. Hausenloy DJ, Yellon DM. Survival kinases in ischemic preconditioning and postconditioning. Cardiovasc Res. 2006;70(2):240–53.
77. Hausenloy DJ, Yellon DM. New directions for protecting the heart against ischaemia–reperfusion injury: targeting the Reperfusion Injury Salvage Kinase (RISK)-pathway. Cardiovasc Res. 2004;61(3):448–60.
78. Ren J, Samson WK, Sowers JR. Insulin-like growth factor I as aÈCardiac hormone: physiological and pathophysiological implications in heart disease. J Mol Cell Cardiol. 1999;31(11):2049–61.
79. Opgaard OS, Wang PH. IGF-I is a matter of heart. Growth Hormon IGF Res. 2005;15(2):89–94.
80. Proud CG. Ras, PI3-kinase and mTOR signaling in cardiac hypertrophy. Cardiovasc Res. 2004;63(3):403–13.
81. McMullen JR, Jennings GL. Differences between pathological and physiological cardiac hypertrophy: novel therapeutic strategies to treat heart failure. Clin Exp Pharmacol Physiol. 2007;34(4):255–62.
82. Shiojima I, Walsh K. Regulation of cardiac growth and coronary angiogenesis by the Akt/PKB signaling pathway. Genes Dev. 2006;20(24):3347–65.
83. Shioi T, McMullen JR, Kang PM, Douglas PS, Obata T, Franke TF, et al. Akt/protein kinase B promotes organ growth in transgenic mice. Mol Cell Biol. 2002;22(8):2799–809.
84. Samuelsson A-M, Bollano E, Mobini R, Larsson B-M, Omerovic E, Fu M, et al. Hyperinsulinemia: effect on cardiac mass/function, angiotensin II receptor expression, and insulin signaling pathways. Am J Physiol-Heart Circulatory Physiol. 2006;291(2):H787–96.
85. Sharma N, Okere IC, Duda MK, Chess DJ, O'Shea KM, Stanley WC. Potential impact of carbohydrate and fat intake on pathological left ventricular hypertrophy. Cardiovasc Res. 2007;73(2):257–68.
86. Ritchie RH, Delbridge LM, editors. Cardiac hypertrophy, substrate utilisation and metabolic remodelling: cause or effect? Proc Austr Physiol Soc. 2005;36:35–43.
87. Desrois M, Sidell RJ, Gauguier D, King LM, Radda GK, Clarke K. Initial steps of insulin signaling and glucose transport are defective in the type 2 diabetic rat heart. Cardiovasc Res. 2004;61(2):288–96.

88. Ouwens D, Boer C, Fodor M, De Galan P, Heine R, Maassen J, et al. Cardiac dysfunction induced by high-fat diet is associated with altered myocardial insulin signalling in rats. Diabetologia. 2005;48(6):1229–37.
89. Huisamen B. Protein kinase B in the diabetic heart. Biochem Diab Atherosclerosis. 2003:31–8.
90. Alemu A, Terefe B, Abebe M, Biadgo B. Thyroid hormone dysfunction during pregnancy: a review. Int J Reproduct BioMed. 2016;14(11):677.
91. Abalovich M, Gutierrez S, Alcaraz G, Maccallini G, Garcia A, Levalle O. Overt and subclinical hypothyroidism complicating pregnancy. Thyroid. 2002;12(1):63–8.
92. Calvo RM, Jauniaux E, Gulbis B, Asunción M, Gervy C, Contempré B, et al. Fetal tissues are exposed to biologically relevant free thyroxine concentrations during early phases of development. J Clin Endocrinol Metab. 2002;87(4):1768–77.
93. van Tuyl M, Blommaart PE, de Boer PA, Wert SE, Ruijter JM, Islam S, et al. Prenatal exposure to thyroid hormone is necessary for normal postnatal development of murine heart and lungs. Dev Biol. 2004;272(1):104–17.
94. Casey BM, Dashe JS, Spong CY, McIntire DD, Leveno KJ, Cunningham GF. Perinatal significance of isolated maternal hypothyroxinemia identified in the first half of pregnancy. Obstet Gynecol. 2007;109(5):1129–35.
95. Harris SE, De Blasio MJ, Davis MA, Kelly AC, Davenport HM, Wooding FP, et al. Hypothyroidism in utero stimulates pancreatic beta cell proliferation and hyperinsulinaemia in the ovine fetus during late gestation. J Physiol. 2017;595(11):3331–43.
96. Wu S-Y, Polk DH, Huang W-S, Green WL, Thai B, Fisher DA. Fetal-to-maternal transfer of thyroid hormone metabolites in late gestation in sheep. Pediatr Res. 2006;59(1):102–6.
97. Fisher DA. Thyroid function in the premature infant. Am J Diseases Children 1977;131(8):842–4.
98. Abuid J, Stinson D, Larsen P. Serum triiodothyronine and thyroxine in the neonate and the acute increases in these hormones following delivery. J Clin Investig. 1973;52(5):1195–9.
99. Fisher D, Dussault J, Hobel C, Lam R. Serum and thyroid gland triiodothyronine in the human fetus. J Clin Endocrinol Metab. 1973;36(2):397–400.
100. Breall JA, Rudolph AM, Heymann MA. Role of thyroid hormone in postnatal circulatory and metabolic adjustments. J Clin Investig. 1984;73(5):1418–24.
101. Fraser M, Liggins G. Thyroid hormone kinetics during late pregnancy in the ovine fetus. J Dev Physiol. 1988;10(5):461–71.
102. Chattergoon NN. Thyroid hormone signaling and consequences for cardiac development. J Endocrinol. 2019;242(1):T145–60.
103. Chattergoon N, Giraud G, Thornburg K. Thyroid hormone inhibits proliferation of fetal cardiac myocytes in vitro. J Endocrinol. 2007;192(2):R1–8.
104. Chattergoon NN, Louey S, Stork P, Giraud GD, Thornburg KL. Mid-gestation ovine cardiomyocytes are vulnerable to mitotic suppression by thyroid hormone. Reprod Sci. 2012;19(6):642–9.
105. Hu LW, Benvenuti LA, Liberti EA, Carneiro-Ramos MS, Barreto-Chaves MLM. Thyroxine-induced cardiac hypertrophy: influence of adrenergic nervous system versus renin-angiotensin system on myocyte remodeling. Am J Physiol-Regulat Integr Comparat Physiol. 2003;285(6):R1473–80.
106. Sundgren NC, Giraud GD, Schultz JM, Lasarev MR, Stork PJ, Thornburg KL. Extracellular signal-regulated kinase and phosphoinositol-3 kinase mediate IGF-1 induced proliferation of fetal sheep cardiomyocytes. Am J Physiol-Regulat Integrat Comparat Physiol. 2003;285(6):R1481–9.
107. Sundgren N, Giraud GD, Stork PJ, Maylie JG, Thornburg KL. Angiotensin II stimulates hyperplasia but not hypertrophy in immature ovine cardiomyocytes J Physiol. 2003;548:881–91.
108. Kuzman JA, Vogelsang KA, Thomas TA, Gerdes AM. L-Thyroxine activates Akt signaling in the heart. J Mol Cell Cardiol. 2005;39(2):251–8.
109. Chattergoon NN, Louey S, Stork P, Giraud GD, Thornburg KL. Unexpected maturation of PI3K and MAPK-ERK signaling in fetal ovine cardiomyocytes. Am J Physiol-Heart Circulatory Physiol. 2014;307(8):H1216–25.

110. Diniz GP, Carneiro-Ramos MS, Barreto-Chaves MLM. Angiotensin type 1 receptor mediates thyroid hormone-induced cardiomyocyte hypertrophy through the Akt/GSK-3β/mTOR signaling pathway. Basic Res Cardiol. 2009;104(6):653–67.
111. Kuzman JA, Gerdes AM, Kobayashi S, Liang Q. Thyroid hormone activates Akt and prevents serum starvation-induced cell death in neonatal rat cardiomyocytes. J Mol Cell Cardiol. 2005;39(5):841–4.
112. Matsui T, Nagoshi T, Rosenzweig A. Akt and PI 3-kinase signaling in cardiomyocyte hypertrophy and survival. Cell Cycle. 2003;2(3):219–22.
113. Kehat I, Molkentin JD. Extracellular signal-regulated kinases 1/2 (ERK1/2) signaling in cardiac hypertrophy. Ann NY Acad Sci. 2010;1188:96.
114. Chattergoon NN, Giraud GD, Louey S, Stork P, Fowden AL, Thornburg KL. Thyroid hormone drives fetal cardiomyocyte maturation. FASEB J. 2012;26(1):397–408.
115. Segar JL, Volk KA, Lipman MH, Scholz TD. Thyroid hormone is required for growth adaptation to pressure load in the ovine fetal heart. Exp Physiol. 2013;98(3):722–33.
116. Maiese K, Chong ZZ, Shang YC, Hou J. FoxO proteins: cunning concepts and considerations for the cardiovascular system. Clin Sci. 2009;116(3):191–203.
117. Puthanveetil P, Wang Y, Wang F, Kim MS, Abrahani A, Rodrigues B. The increase in cardiac pyruvate dehydrogenase kinase-4 after short-term dexamethasone is controlled by an Akt-p38-forkhead box other factor-1 signaling axis. Endocrinology. 2010;151(5):2306–18.
118. Puthanveetil P, Wang Y, Zhang D, Wang F, Kim MS, Innis S, et al. Cardiac triglyceride accumulation following acute lipid excess occurs through activation of a FoxO1–iNOS–CD36 pathway. Free Radical Biol Med. 2011;51(2):352–63.
119. Huang H, Tindall DJ. Dynamic FoxO transcription factors. J Cell Sci. 2007;120(15):2479–87.
120. Nakamura N, Ramaswamy S, Vazquez F, Signoretti S, Loda M, Sellers WR. Forkhead transcription factors are critical effectors of cell death and cell cycle arrest downstream of PTEN. Mol Cell Biol. 2000;20(23):8969–82.
121. Seoane J, Le H-V, Shen L, Anderson SA, Massagué J. Integration of Smad and fork-head pathways in the control of neuroepithelial and glioblastoma cell proliferation. Cell. 2004;117(2):211–23.
122. Bicknell KA, Coxon CH, Brooks G. Can the cardiomyocyte cell cycle be reprogrammed? J Mol Cell Cardiol. 2007;42(4):706–21.
123. Irrcher I, Walkinshaw DR, Sheehan TE, Hood DA. Thyroid hormone (T3) rapidly activates p38 and AMPK in skeletal muscle in vivo. J Appl Physiol. 2008;104(1):178–85.
124. Singh BK, Sinha RA, Zhou J, Xie SY, You S-H, Gauthier K, et al. FoxO1 deacetylation regulates thyroid hormone-induced transcription of key hepatic gluconeogenic genes. J Biol Chem. 2013;288(42):30365–72.
125. Schueler P, Schwartz H, Strait K, Mariash C, Oppenheimer J. Binding of 3, 5, 3′-triiodothyronine (T3) and its analogs to the in vitro translational products of c-erbA protoonco-genes: differences in the affinity of the α-and β-forms for the acetic acid analog and failure of the human testis and kidney α-2 products to bind T3. Mol Endocrinol. 1990;4(2):227–34.
126. Yen PM. Physiological and molecular basis of thyroid hormone action. Physiol Rev. 2001;81(3):1097–142.
127. Mai W, Janier MF, Alliloi N, Quignodon L, Chuzel T, Flamant F, Samarut J. Thyroid hormone receptor α is a molecular switch of cardiac function between fetal and postnatal life. Proc Natl Acad Sci USA. 2004;101(28):10332–7.
128. Kahaly GJ, Dillmann WH. Thyroid hormone action in the heart. Endocr Rev. 2005;26(5):704–28.
129. Kinugawa K, Jeong MY, Bristow MR, Long CS. Thyroid hormone induces cardiac myocyte hypertrophy in a thyroid hormone receptor α1-specific manner that requires TAK1 and p38 mitogen-activated protein kinase. Mol Endocrinol. 2005;19(6):1618–28.
130. Chopra IJ, Carlson HE, Solomon DH. Comparison of Inhibitory Effects of 3, 5, 3′-Triiodothyronine (T3), Thyroxine (T4), 3, 3′, 5-Triiodothyronine (rT3), and 3, 3′-Diiodothyronine (T2) on Thyrotropin-Releasing Hormone-Induced Release of Thyrotropin in the Rat in vitro. Endocrinology. 1978;103(2):393–402.

131. Kreft H, Jetz W. Global patterns and determinants of vascular plant diversity. Proc Natl Acad Sci USA. 2007;104(14):5925–30.
132. White P, Burton KA, Fowden A, Dauncey M. Developmental expression analysis of thyroid hormone receptor isoforms reveals new insights into their essential functions in cardiac and skeletal muscles. FASEB J. 2001;15(8):1367–76.
133. Chattergoon NN, Louey S, Scanlan T, Lindgren I, Giraud GD, Thornburg KL. Thyroid hormone receptor function in maturing ovine cardiomyocytes. J Physiol. 2019;597(8):2163–76.
134. Gloss B, Trost SU, Bluhm WF, Swanson EA, Clark R, Winkfein R, et al. Cardiac ion channel expression and contractile function in mice with deletion of thyroid hormone receptor α or β. Endocrinology. 2001;142(2):544–50.
135. Flamant F, Samarut J. Thyroid hormone receptors: lessons from knockout and knock-in mutant mice. Trends Endocrinol Metab. 2003;14(2):85–90.
136. Sayen M, Rohrer D, Dillmann W. Thyroid hormone response of slow and fast sarcoplasmic reticulum Ca^{2+} ATPase mRNA in striated muscle. Mol Cell Endocrinol. 1992;87(1–3):87–93.
137. Booth EA, Lucchesi BR. Medroxyprogesterone acetate prevents the cardioprotective and anti-inflammatory effects of 17β-estradiol in an in vivo model of myocardial ischemia and reperfusion. Am J Physiol-Heart Circulatory Physiol. 2007;293(3):H1408–15.
138. Levine RL, Chen S-J, Durand J, Chen Y-F, Oparil S. Medroxyprogesterone attenuates estrogen-mediated inhibition of neointima formation after balloon injury of the rat carotid artery. Circulation. 1996;94(9):2221–7.
139. McHugh N, Cook S, Schairer J, Bidgoli M, Merrill G. Ischemia-and reperfusion-induced ventricular arrhythmias in dogs: effects of estrogen. Am J Physiol-Heart Circulatory Physiol. 1995;268(6):H2569–73.
140. Resanovic I, Rizzo M, Zafirovic S, Bjelogrlic P, Perovic M, Savic K, et al. Anti-atherogenic effects of 17β-estradiol. Horm Metab Res. 2013;45(10):701–8.
141. Kim JK. Estrogen: Impact on cardiomyocytes and the heart. In: Legato MJ, editor. Principles of gender-specific medicine. Elsevier; 2017, pp. 363–79.
142. Hong Y, Li H, Yuan Y-C, Chen S. Molecular characterization of aromatase. Ann NY Acad Sci USA. 2009;1155:112.
143. Grohe C, Kahlert S, Lobbert K, Vetter H. Expression of oestrogen receptor alpha and beta in rat heart: role of local oestrogen synthesis. J Endocrinol. 1998;156(2):R1–7.
144. Jazbutyte V, Stumpner J, Redel A, Lorenzen JM, Roewer N, Thum T, et al. Aromatase inhibition attenuates desflurane-induced preconditioning against acute myocardial infarction in male mouse heart in vivo. PLoS One. 2012;7(8):e42032.
145. O'Lone R, Knorr K, Jaffe IZ, Schaffer ME, Martini PG, Karas RH, et al. Estrogen receptors α and β mediate distinct pathways of vascular gene expression, including genes involved in mitochondrial electron transport and generation of reactive oxygen species. Mol Endocrinol. 2007;21(6):1281–96.
146. Devanathan S, Whitehead T, Schweitzer GG, Fettig N, Kovacs A, Korach KS, et al. An animal model with a cardiomyocyte-specific deletion of estrogen receptor alpha: functional, metabolic, and differential network analysis. PloS One. 2014;9(7):e101900.
147. Grohé C, Kahlert S, Löbbert K, Stimpel M, Karas RH, Vetter H, et al. Cardiac myocytes and fibroblasts contain functional estrogen receptors. FEBS Lett. 1997;416(1):107–12.
148. Mahmoodzadeh S, Eder S, Nordmeyer J, Ehler E, Huber O, Martus P, et al. Estrogen receptor alpha up-regulation and redistribution in human heart failure. FASEB J. 2006;20(7):926–34.
149. Taylor A, Al-Azzawi F. Immunolocalisation of oestrogen receptor beta in human tissues. J Mol Endocrinol. 2000;24(1):145.
150. Lizotte E, Grandy SA, Tremblay A, Allen BG, Fiset C. Expression, distribution and regulation of sex steroid hormone receptors in mouse heart. Cell Physiol Biochem. 2009;23(1–3):075–86.
151. Knowlton AA, Lee A. Estrogen and the cardiovascular system. Pharmacol Ther. 2012;135(1):54–70.
152. Jazbutyte V, Kehl F, Neyses L, Pelzer T. Estrogen receptor alpha interacts with 17β-hydroxysteroid dehydrogenase type 10 in mitochondria. Biochem Biophys Res Commun. 2009;384(4):450–4.

153. Liu H, Yanamandala M, Lee TC, Kim JK. Mitochondrial p38β and manganese superoxide dismutase interaction mediated by estrogen in cardiomyocytes. PloS One. 2014;9(1):e85272.
154. Yang S-H, Liu R, Perez EJ, Wen Y, Stevens SM, Valencia T, et al. Mitochondrial localization of estrogen receptor β. Proceedings of the National Academy of Sciences USA. 2004;101(12):4130–5.
155. Funakoshi T, Yanai A, Shinoda K, Kawano MM, Mizukami Y. G protein-coupled receptor 30 is an estrogen receptor in the plasma membrane. Biochem Biophys Res Commun. 2006;346(3):904–10.
156. Deschamps AM, Murphy E. Activation of a novel estrogen receptor, GPER, is cardioprotective in male and female rats. Am J Physiol-Heart Circulatory Physiol. 2009;297(5):H1806–13.
157. Jessup JA, Lindsey SH, Wang H, Chappell MC, Groban L. Attenuation of salt-induced cardiac remodeling and diastolic dysfunction by the GPER agonist G-1 in female mRen2. Lewis rats. PloS One. 2010;5(11):e15433.
158. Bopassa JC, Eghbali M, Toro L, Stefani E. A novel estrogen receptor GPER inhibits mitochondria permeability transition pore opening and protects the heart against ischemia-reperfusion injury. Am J Physiol-Heart Circulatory Physiol. 2010;298(1):H16–23.
159. Gupte AA, Pownall HJ, Hamilton DJ. Estrogen: an emerging regulator of insulin action and mitochondrial function. J Diab Res. 2015;2015.
160. Chen Y, Zhang Z, Hu F, Yang W, Yuan J, Cui J, et al. 17β-estradiol prevents cardiac diastolic dysfunction by stimulating mitochondrial function: a preclinical study in a mouse model of a human hypertrophic cardiomyopathy mutation. J Steroid Biochem Mol Biol. 2015;147:92–102.
161. Arias-Loza P-A, Kreissl MC, Kneitz S, Kaiser FR, Israel I, Hu K, et al. The estrogen receptor-α is required and sufficient to maintain physiological glucose uptake in the mouse heart. Hypertension. 2012;60(4):1070–7.
162. Rattanasopa C, Phungphong S, Wattanapermpool J, Bupha-Intr T. Significant role of estrogen in maintaining cardiac mitochondrial functions. J Steroid Biochem Mol Biol. 2015;147:1–9.
163. Liu H, Pedram A, Kim JK. Oestrogen prevents cardiomyocyte apoptosis by suppressing p38α-mediated activation of p53 and by down-regulating p53 inhibition on p38β. Cardiovasc Res. 2011;89(1):119–28.
164. Satoh M, Matter CM, Ogita H, Takeshita K, Wang C-Y, Dorn GW, et al. Inhibition of apoptosis-regulated signaling kinase-1 and prevention of congestive heart failure by estrogen. Circulation. 2007;115(25):3197–204.
165. Ramjiawan A, Bagchi RA, Albak L, Czubryt MP. Mechanism of cardiomyocyte PGC-1α gene regulation by ERRα. Biochem Cell Biol. 2013;91(3):148–54.
166. Giguère V, Yang N, Segui P, Evans RM. Identification of a new class of steroid hormone receptors. Nature. 1988;331(6151):91–4.
167. Simpkins JW, Yang S-H, Sarkar SN, Pearce V. Estrogen actions on mitochondria—physiological and pathological implications. Mol Cell Endocrinol. 2008;290(1–2):51–9.
168. Gabel SA, Walker VR, London RE, Steenbergen C, Korach KS, Murphy E. Estrogen receptor beta mediates gender differences in ischemia/reperfusion injury. J Mol Cell Cardiol. 2005;38(2):289–97.
169. Hsieh Y-C, Yu H-P, Suzuki T, Choudhry MA, Schwacha MG, Bland KI, et al. Upregulation of mitochondrial respiratory complex IV by estrogen receptor-β is critical for inhibiting mitochondrial apoptotic signaling and restoring cardiac functions following trauma–hemorrhage. J Mol Cell Cardiol. 2006;41(3):511–21.
170. Babiker FA, De Windt LJ, van Eickels M, Thijssen V, Bronsaer RJ, Grohé C, et al. 17β-estradiol antagonizes cardiomyocyte hypertrophy by autocrine/paracrine stimulation of a guanylyl cyclase A receptor-cyclic guanosine monophosphate-dependent protein kinase pathway. Circulation. 2004;109(2):269–76.
171. Light KC, Hinderliter AL, West SG, Grewen KM, Steege JF, Sherwood A, et al. Hormone replacement improves hemodynamic profile and left ventricular geometry in hypertensive and normotensive postmenopausal women. J Hypertens. 2001;19(2):269–78.

172. Pelzer T, Jazbutyte V, Hu K, Segerer S, Nahrendorf M, Nordbeck P, et al. The estrogen receptor-α agonist 16α-LE2 inhibits cardiac hypertrophy and improves hemodynamic function in estrogen-deficient spontaneously hypertensive rats. Cardiovasc Res. 2005;67(4):604–12.
173. Jazbutyte V, Arias-Loza PA, Hu K, Widder J, Govindaraj V, von Poser-Klein C, et al. Ligand-dependent activation of ERβ lowers blood pressure and attenuates cardiac hypertrophy in ovariectomized spontaneously hypertensive rats. Cardiovasc Res. 2008;77(4):774–81.
174. Donaldson C, Eder S, Baker C, Aronovitz MJ, Weiss AD, Hall-Porter M, et al. Estrogen attenuates left ventricular and cardiomyocyte hypertrophy by an estrogen receptor–dependent pathway that increases calcineurin degradation. Circ Res. 2009;104(2):265–75.
175. Gürgen D, Kusch A, Klewitz R, Hoff U, Catar R, Hegner B, et al. Sex-specific mTOR signaling determines sexual dimorphism in myocardial adaptation in normotensive DOCA-salt model. Hypertension. 2013;61(3):730–6.
176. Van Eickels M, Grohé C, Cleutjens JP, Janssen BJ, Wellens HJ, Doevendans PA. 17β-Estradiol attenuates the development of pressure-overload hypertrophy. Circulation. 2001;104(12):1419–23.
177. Pedram A, Razandi M, Narayanan R, Dalton JT, McKinsey TA, Levin ER. Estrogen regulates histone deacetylases to prevent cardiac hypertrophy. Mol Biol Cell. 2013;24(24):3805–18.
178. Dworatzek E, Mahmoodzadeh S, Schubert C, Westphal C, Leber J, Kusch A, et al. Sex differences in exercise-induced physiological myocardial hypertrophy are modulated by oestrogen receptor beta. Cardiovasc Res. 2014;102(3):418–28.
179. Kararigas G, Nguyen BT, Jarry H. Estrogen modulates cardiac growth through an estrogen receptor α-dependent mechanism in healthy ovariectomized mice. Mol Cell Endocrinol. 2014;382(2):909–14.
180. Huang J, Elicker J, Bowens N, Liu X, Cheng L, Cappola TP, et al. Myocardin regulates BMP10 expression and is required for heart development. J Clin Investig. 2012;122(10):3678–91.
181. McDermott-Roe C, Ye J, Ahmed R, Sun X-M, Serafín A, Ware J, et al. Endonuclease G is a novel determinant of cardiac hypertrophy and mitochondrial function. Nature. 2011;478(7367):114–8.
182. Liou CM, Yang AL, Kuo CH, Tin H, Huang CY, Lee SD. Effects of 17beta-estradiol on cardiac apoptosis in ovariectomized rats. Cell Biochem Funct. 2010;28(6):521–8.
183. Pelzer T, Neumann M, de Jager T, Jazbutyte V, Neyses L. Estrogen effects in the myocardium: inhibition of NF-κB DNA binding by estrogen receptor-α and-β. Biochem Biophys Res Commun. 2001;286(5):1153–7.
184. Patten RD, Pourati I, Aronovitz MJ, Baur J, Celestin F, Chen X, et al. 17β-Estradiol reduces cardiomyocyte apoptosis in vivo and in vitro via activation of phospho-inositide-3 kinase/Akt signaling. Circ Res. 2004;95(7):692–9.
185. Cong B, Zhu X, Cao B, Xiao J, Wang Z, Ni X. Estrogens protect myocardium against ischemia/reperfusion insult by up-regulation of CRH receptor type 2 in female rats. Int J Cardiol. 2013;168(5):4755–60.
186. Kim JK, Pedram A, Razandi M, Levin ER. Estrogen prevents cardiomyocyte apoptosis through inhibition of reactive oxygen species and differential regulation of p38 kinase isoforms. J Biol Chem. 2006;281(10):6760–7.
187. Li WL, Xiang W, Ping Y. Activation of novel estrogen receptor GPER results in inhibition of cardiocyte apoptosis and cardioprotection. Mol Med Rep. 2015;12(2):2425–30.
188. Liu CJ, Lo JF, Kuo CH, Chu CH, Chen LM, Tsai FJ, et al. Akt mediates 17β-estradiol and/or estrogen receptor-α inhibition of LPS-induced tumor necrosis factor-α expression and myocardial cell apoptosis by suppressing the JNK1/2-NFκB pathway. J Cell Mol Med. 2009;13(9b):3655–67.
189. Cao J, Zhu T, Lu L, Geng L, Wang L, Zhang Q, et al. Estrogen induces cardioprotection in male C57BL/6J mice after acute myocardial infarction via decreased activity of matrix metalloproteinase-9 and increased Akt-Bcl-2 anti-apoptotic signaling. Int J Mol Med. 2011;28(2):231–7.
190. Node K, Kitakaze M, Kosaka H, Minamino T, Funaya H, Hori M. Amelioration of ischemia-and reperfusion-induced myocardial injury by 17β-estradiol: role of nitric oxide and calcium-activated potassium channels. Circulation. 1997;96(6):1953–63.

191. Anderson SE, Kirkland DM, Beyschau A, Cala PM. Acute effects of 17β-estradiol on myocardial pH, Na$^+$, and Ca^{2+} and ischemia-reperfusion injury. Am J Physiol Cell Physiol. 2005;288(1):C57–64.

192. Özdemir K, Çelik Ç, Altunkeser BB, İçli A, Albeni H, Düzenli A, et al. Effect of postmenopausal hormone replacement therapy on cardiovascular performance. Maturitas. 2004;47(2):107–13.

193. Ribeiro RF Jr, Pavan BM, Potratz FF, Fiorim J, Simoes MR, Dias FMV, et al. Myocardial contractile dysfunction induced by ovariectomy requires at1receptor activation in female rats. Cell Physiol Biochem. 2012;30(1):1–12.

194. Johnson BD, Zheng W, Korach KS, Scheuer T, Catterall WA, Rubanyi GM. Increased expression of the cardiac L-type calcium channel in estrogen receptor–deficient mice. J Gen Physiol. 1997;110(2):135–40.

195. Marni F, Wang Y, Morishima M, Shimaoka T, Uchino T, Zheng M, et al. 17β-Estradiol modulates expression of low-voltage-activated Cav3. 2 T-type calcium channel via extracellularly regulated kinase pathway in cardiomyocytes. Endocrinology. 2009;150(2):879–88.

196. Nakajima T, Iwasawa K, Oonuma H, Morita T, Goto A, Wang Y, et al. Antiarrhythmic effect and its underlying ionic mechanism of 17β-estradiol in cardiac myocytes. Br J Pharmacol. 1999;127(2):429–40.

197. El Gebeily G, El Khoury N, Mathieu S, Brouillette J, Fiset C. Estrogen regulation of the transient outward K$^+$ current involves estrogen receptor α in mouse heart. J Mol Cell Cardiol. 2015;86:85–94.

198. Kulpa J, Chinnappareddy N, Pyle WG. Rapid changes in cardiac myofilament function following the acute activation of estrogen receptor-alpha. PLoS One. 2012;7(7):e41076.

199. Ullrich ND, Krust A, Collins P, MacLeod KT. Genomic deletion of estrogen receptors ERα and ERβ does not alter estrogen-mediated inhibition of Ca2+ influx and contraction in murine cardiomyocytes. Am J Physiol-Heart Circulatory Physiol. 2008;294(6):H2421–7.

200. Belcher SM, Chen Y, Yan S, Wang H-S. Rapid estrogen receptor-mediated mechanisms determine the sexually dimorphic sensitivity of ventricular myocytes to 17β-estradiol and the environmental endocrine disruptor bisphenol A. Endocrinology. 2012;153(2):712–20.

201. Korte T, Fuchs M, Arkudas A, Geertz S, Meyer R, Gardiwal A, et al. Female mice lacking estrogen receptor β display prolonged ventricular repolarization and reduced ventricular automaticity after myocardial infarction. Circulation. 2005;111(18):2282–90.

202. Wang Y, Wang Q, Zhao Y, Gong D, Wang D, Li C, et al. Protective effects of estrogen against reperfusion arrhythmias following severe myocardial ischemia in rats. Circ J. 2010;74(4):634–43.

203. Wang L, Gu H, Turrentine M, Wang M. Estradiol treatment promotes cardiac stem cell (CSC)-derived growth factors, thus improving CSC-mediated cardioprotection after acute ischemia/reperfusion. Surgery. 2014;156(2):243–52.

204. Erwin GS, Crisostomo PR, Wang Y, Wang M, Markel TA, Guzman M, et al. Estradiol-treated mesenchymal stem cells improve myocardial recovery after ischemia. J Surg Res. 2009;152(2):319–24.

205. Han HJ, Heo JS, Lee YJ. Estradiol-17β stimulates proliferation of mouse embryonic stem cells: involvement of MAPKs and CDKs as well as protooncogenes. Am J Physiol Cell Physiol. 2006;290(4):C1067–75.

206. Wickman A, Friberg P, Adams MA, Matejka GrL, Brantsing C, Guron G, et al. Induction of growth hormone receptor and insulin-like growth factor-I mRNA in aorta and caval vein during hemodynamic challenge. Hypertension. 1997;29(1):123–30.

207. Brinckmann M, Kaschina E, Altarche-Xifró W, Curato C, Timm M, Grzesiak A, et al. Estrogen receptor α supports cardiomyocytes indirectly through post-infarct cardiac c-kit+ cells. J Mol Cell Cardiol. 2009;47(1):66–75.

208. Saccà L, Cittadini A, Fazio S. Growth hormone and the heart. Endocr Rev. 1994;15(5):555–73.

209. Isgaard J, Arcopinto M, Karason K, Cittadini A. GH and the cardiovascular system: an update on a topic at heart. Endocrine. 2015;48(1):25–35.

210. Isgaard J, Nilsson A, Vikman K, Isaksson O. Growth hormone regulates the level of insulin-like growth factor-I mRNA in rat skeletal muscle. J Endocrinol. 1989;120(1):107–12.
211. Mathews L, Enberg B, Norstedt G. Regulation of rat growth hormone receptor gene expression. J Biol Chem. 1989;264(17):9905–10.
212. Isgaard J, Wåhlander H, Adams MA, Friberg P. Increased expression of growth hormone receptor mRNA and insulin-like growth factor-I mRNA in volume-overloaded hearts. Hypertension. 1994;23(6_pt_2):884–8.
213. Guler H-P, Zapf J, Scheiwiller E, Froesch ER. Recombinant human insulin-like growth factor I stimulates growth and has distinct effects on organ size in hypophysectomized rats. Proc Natl Acad Sci USA. 1988;85(13):4889–93.
214. Wickman A, Isgaard J, Adams MA, Friberg P. Inhibition of nitric oxide in rats. Regulation of cardiovascular structure and expression of insulin-like growth factor I and its receptor messenger RNA. J Hypertension. 1997;15(7):751–9.
215. Schnabel P, Mies F, Nohr T, Geisler M, Böhm M. Differential regulation of phospholipase C-β isozymes in cardiomyocyte hypertrophy. Biochem Biophys Res Commun. 2000;275(1):1–6.
216. Lu C, Schwartzbauer G, Sperling MA, Devaskar SU, Thamotharan S, Robbins PD, et al. Demonstration of direct effects of growth hormone on neonatal cardiomyocytes. J Biol Chem. 2001;276(25):22892–900.
217. Donath MY, Zapf J, Eppenberger-Eberhardt M, Froesch ER, Eppenberger HM. Insulin-like growth factor I stimulates myofibril development and decreases smooth muscle alpha-actin of adult cardiomyocytes. Proc Natl Acad Sci USA. 1994;91(5):1686–90.
218. Ito H, Hiroe M, Hirata Y, Tsujino M, Adachi S, Shichiri M, et al. Insulin-like growth factor-I induces hypertrophy with enhanced expression of muscle specific genes in cultured rat cardiomyocytes. Circulation. 1993;87(5):1715–21.
219. Fuller SJ, Mynett JR, Sugden PH. Stimulation of cardiac protein synthesis by insulin-like growth factors. Biochem J. 1992;282(1):85–90.
220. Donath MY, Gosteli-Peter MA, Hauri C, Froesch ER, Zapf J. Insulin-like growth factor-I stimulates myofibrillar genes and modulates atrial natriuretic factor mRNA in rat heart. Eur J Endocrinol. 1997;137(3):309.
221. Li Q, Li B, Wang X, Leri A, Jana KP, Liu Y, et al. Overexpression of insulin-like growth factor-1 in mice protects from myocyte death after infarction, attenuating ventricular dilation, wall stress, and cardiac hypertrophy. J Clin Investig. 1997;100(8):1991–9.
222. Buerke M, Murohara T, Skurk C, Nuss C, Tomaselli K, Lefer AM. Cardioprotective effect of insulin-like growth factor I in myocardial ischemia followed by reperfusion. Proc Natl Acad Sci USA. 1995;92(17):8031–5.
223. Vetter U, Kupferschmid C, Lang D, Pentz S. Insulin-like growth factors and insulin increase the contractility of neonatal rat cardiocytes in vitro. Basic Res Cardiol. 1988;83(6):647–54.
224. Freestone NS, Ribaric S, Mason WT. The effect of insulin-like growth factor-1 on adult rat cardiac contractility. Mol Cell Biochem. 1996;163(1):223–9.
225. Cittadini A, Ishiguro Y, Strömer H, Spindler M, Moses AC, Clark R, et al. Insulin-like growth factor-1 but not growth hormone augments mammalian myocardial contractility by sensitizing the myofilament to Ca^{2+} through a wortmannin-sensitive pathway: studies in rat and ferret isolated muscles. Circ Res. 1998;83(1):50–9.
226. Strömer H, Cittadini A, Douglas PS, Morgan JP. Exogenously administered growth hormone and insulin-like growth factor-I alter intracellular Ca^{2+} handling and enhance cardiac performance: in vitro evaluation in the isolated isovolumic buffer-perfused rat heart. Circ Res. 1996;79(2):227–36.
227. Tajima M, Weinberg EO, Bartunek J, Jin H, Yang R, Paoni NF, et al. Treatment with growth hormone enhances contractile reserve and intracellular calcium transients in myocytes from rats with postinfarction heart failure. Circulation. 1999;99(1):127–34.
228. Ren J, Brown-Borg HM. Impaired cardiac excitation–contraction coupling in ventricular myocytes from Ames dwarf mice with IGF-I deficiency. Growth Hormon IGF Res. 2002;12(2):99–105.

229. Kinugawa S, Tsutsui H, Ide T, Nakamura R, Arimura K-I, Egashira K, et al. Positive inotropic effect of insulin-like growth factor-1 on normal and failing cardiac myocytes. Cardiovasc Res. 1999;43(1):157–64.

230. Solem ML, Thomas AP. Modulation of cardiac Ca^{2+} channels by IGF1. Biochem Biophys Res Commun. 1998;252(1):151–5.

231. Mayoux E, Ventura-Clapier R, Timsit J, Béhar-Cohen F, Hoffmann C, Mercadier J-J. Mechanical properties of rat cardiac skinned fibers are altered by chronic growth hormone hypersecretion. Circ Res. 1993;72(1):57–64.

232. Timsit J, Riou B, Bertherat J, Wisnewsky C, Kato N, Weisberg A, et al. Effects of chronic growth hormone hypersecretion on intrinsic contractility, energetics, isomyosin pattern, and myosin adenosine triphosphatase activity of rat left ventricle. J Clin Investig. 1990;86(2):507–15.

233. Cittadini A, Cuocolo A, Merola B, Fazio S, Sabatini D, Nicolai E, et al. Impaired cardiac performance in GH-deficient adults and its improvement after GH replacement. Am J Physiol-Endocrinol Metab. 1994;267(2):E219–25.

234. Longobardi S, Cuocolo A, Merola B, Di Rella F, Colao A, Nicolai E, et al. Left ventricular function in young adults with childhood and adulthood onset growth hormone deficiency. Clin Endocrinol. 1998;48(2):137–44.

235. Cittadini A, Douglas PS. In: Giustina A, editor. Growth hormone and the heart. Springer; 2000, pp. 1–11.

236. Saccà L, Napoli R, Cittadini A. Growth hormone, acromegaly, and heart failure: an intricate triangulation. Clin Endocrinol. 2003;59(6):660–71.

237. Bollano E, Omerovic E, Bohlooly-Y M, Kujacic V, Madhu B, Tornell J, et al. Impairment of cardiac function and bioenergetics in adult transgenic mice overexpressing the bovine growth hormone gene. Endocrinology. 2000;141(6):2229–35.

238. Cittadini A, Saldamarco L, Marra AM, Arcopinto M, Carlomagno G, Imbriaco M, et al. Growth hormone deficiency in patients with chronic heart failure and beneficial effects of its correction. J Clin Endocrinol Metab. 2009;94(9):3329–36.

239. Mangner N, Matsuo Y, Schuler G, Adams V. Cachexia in chronic heart failure: endocrine determinants and treatment perspectives. Endocrine. 2013;43(2):253–65.

240. Yoshimura M, Nakamura S, Ito T, Nakayama M, Harada E, Mizuno Y, et al. Expression of aldosterone synthase gene in failing human heart: quantitative analysis using modified real-time polymerase chain reaction. J Clin Endocrinol Metab. 2002;87(8):3936–40.

241. Mizuno Y, Yoshimura M, Yasue H, Sakamoto T, Ogawa H, Kugiyama K, et al. Aldosterone production is activated in failing ventricle in humans. Circulation. 2001;103(1):72–7.

242. Lösel RM, Feuring M, Falkenstein E, Wehling M. Nongenomic effects of aldosterone: cellular aspects and clinical implications. Steroids. 2002;67(6):493–8.

243. Messaoudi S, Gravez B, Tarjus A, Pelloux V, Ouvrard-Pascaud A, Delcayre C, et al. Aldosterone-specific activation of cardiomyocyte mineralocorticoid receptor in vivo. Hypertension. 2013;61(2):361–7.

244. Hayashi H, Kobara M, Abe M, Tanaka N, Gouda E, Toba H, et al. Aldosterone nongenomically produces nadph oxidase-dependent reactive oxygen species and induces myocyte apoptosis. Hypertens Res. 2008;31(2):363–75.

245. Daniels A, Van Bilsen M, Goldschmeding R, Van Der Vusse G, Van Nieuwenhoven F. Connective tissue growth factor and cardiac fibrosis. Acta Physiol. 2009;195(3):321–38.

246. Konishi A, Tazawa C, Miki Y, Darnel AD, Suzuki T, Ohta Y, et al. The possible roles of mineralocorticoid receptor and 11β-hydroxysteroid dehydrogenase type 2 in cardiac fibrosis in the spontaneously hypertensive rat. J Steroid Biochem Mol Biol. 2003;85(2–5):439–42.

247. Milik E, Szczepanska-Sadowska E, Linski W, Cudnoch-Jedrzejewska A. Enhanced expression of mineralocorticoid receptors. J Physiol Pharmacol. 2007;58(4):745–55.

248. Ohtani T, Ohta M, Yamamoto K, Mano T, Sakata Y, Nishio M, et al. Elevated cardiac tissue level of aldosterone and mineralocorticoid receptor in diastolic heart failure: beneficial effects of mineralocorticoid receptor blocker. Am J Physiol-Regulatory Integr Comparat Physiol. 2007;292(2):R946–54.

249. Fraccarollo D, Berger S, Galuppo P, Kneitz S, Hein L, Schütz G, et al. Deletion of cardiomyocyte mineralocorticoid receptor ameliorates adverse remodeling after myocardial infarction. Circulation. 2011;123(4):400–8.
250. Ouvrard-Pascaud A, Sainte-Marie Y, Bénitah J-P, Perrier R, Soukaseum C, Cat AND, et al. Conditional mineralocorticoid receptor expression in the heart leads to life-threatening arrhythmias. Circulation. 2005;111(23):3025–33.
251. Brotman DJ, Golden SH, Wittstein IS. The cardiovascular toll of stress. Lancet. 2007;370(9592):1089–100.
252. Steptoe A, Kivimäki M. Stress and cardiovascular disease. Nat Rev Cardiol. 2012;9(6):360–70.
253. Oakley RH, Cidlowski JA. Glucocorticoid signaling in the heart: a cardiomyocyte perspective. J Steroid Biochem Mol Biol. 2015;153:27–34.
254. Oakley RH, Ren R, Cruz-Topete D, Bird GS, Myers PH, Boyle MC, et al. Essential role of stress hormone signaling in cardiomyocytes for the prevention of heart disease. Proc Natl Acad Sci USA. 2013;110(42):17035–40.
255. Rog-Zielinska EA, Thomson A, Kenyon CJ, Brownstein DG, Moran CM, Szumska D, et al. Glucocorticoid receptor is required for foetal heart maturation. Hum Mol Genet. 2013;22(16):3269–82.
256. Richardson RV, Batchen EJ, Thomson AJ, Darroch R, Pan X, Rog-Zielinska EA, et al. Glucocorticoid receptor alters isovolumetric contraction and restrains cardiac fibrosis. J Endocrinol. 2017;232(3):437.
257. Oakley RH, Cruz-Topete D, He B, Foley JF, Myers PH, Xu X, et al. Cardiomyocyte glucocorticoid and mineralocorticoid receptors directly and antagonistically regulate heart disease in mice. Scie Signaling, 2019;12(577).
258. Unger RH, Zhou Y-T, Orci L. Regulation of fatty acid homeostasis in cells: novel role of leptin. Proc Natl Acad Sci USA. 1999;96(5):2327–32.
259. Hall ME, Mready MW, Hall JE, Stec DE. Rescue of cardiac leptin receptors in db/db mice prevents myocardial triglyceride accumulation. Am J Physiol-Endocrinol Metab. 2014;307(3):E316–25.
260. Witham W, Yester K, O'Donnell CP, McGaffin KR. Restoration of glucose metabolism in leptin-resistant mouse hearts after acute myocardial infarction through the activation of survival kinase pathways. J Mol Cell Cardiol. 2012;53(1):91–100.
261. Hall ME, Smith G, Hall JE, Stec DE. Cardiomyocyte-specific deletion of leptin receptors causes lethal heart failure in Cre-recombinase-mediated cardiotoxicity. Am J Physiol-Regulatory Integrat Comparat Physiol. 2012;303(12):R1241–50.
262. McGaffin KR, Witham WG, Yester KA, Romano LC, O'Doherty RM, McTiernan CF, et al. Cardiac-specific leptin receptor deletion exacerbates ischaemic heart failure in mice. Cardiovasc Res. 2011;89(1):60–71.

Chapter 10
Sequelae of Genetic Defects in Cardiomyocytes

Abstract The consequences of genetic defects in cardiomyocytes are congenital heart defects or inherited cardiomyopathies. Linkage analysis, candidate gene screening and next generation sequencing technologies have led to identification of hundreds of mutations in genes encoding various subcellular components of the cardiomyocyte, in patients with congenital heart defects and inherited cardiomyopathies. About 400 genes have been implicated in congenital heart disease. These include those that encode structural proteins, transcription factors, cell signalling molecules, cilia related proteins, as well as cilia-transduced cell signalling pathways and chromatin modifiers. Several single-gene mutations in cardiomyocytes have been associated with unique phenotypes of cardiomyopathies. Genes linked to cardiomyopathy are generally classified considering phenotype-specific mechanistic pathways. Disease genes which do not conform to the classification are being newly discovered. Relatively few novel mechanistic insights have however been gained into how different mutations impact phenotype.

Keywords Congenital heart defects · Cardiomyopathies · Genetics of heart diseases · Gene mutations · Cardiac development · Developmental genetics

Introduction

Both genetic and environmental factors contribute to the causation of congenital heart defects and cardiomyopathies (Figs. 10.1 and 10.2). Linkage analysis, candidate gene screening and next generation sequencing technologies have led to the identification of hundreds of mutations in genes encoding various subcellular components of the cardiomyocyte, in patients with congenital heart defects and inherited cardiomyopathies. Genetic defects linked to common heart diseases are listed in Table 10.1.

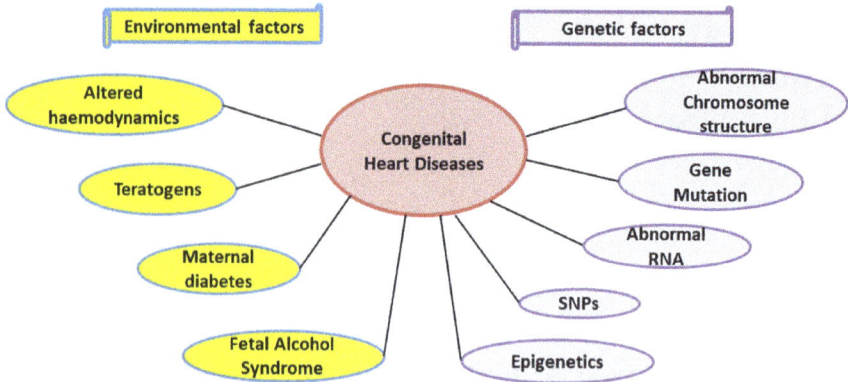

Fig. 10.1 Genetic and environmental factors which determine various congenital heart defects

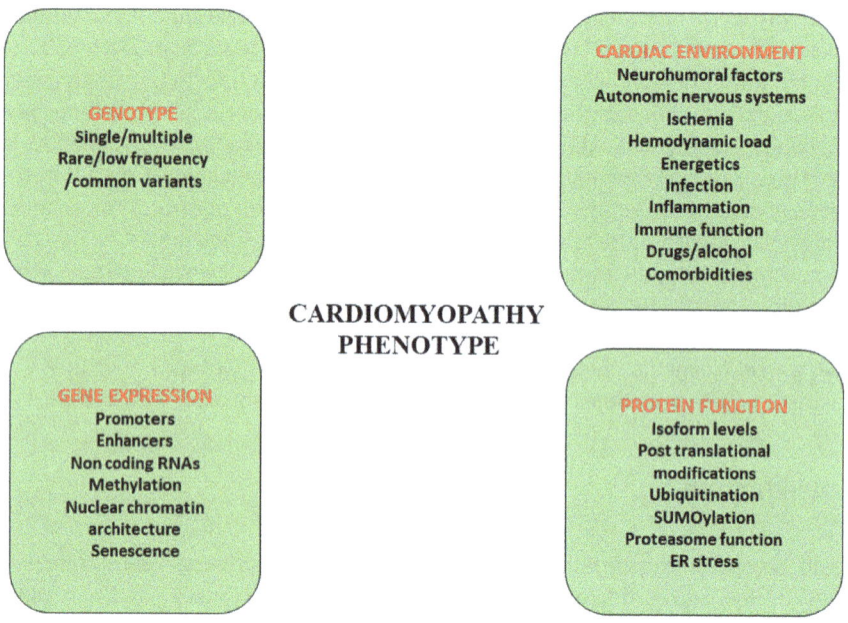

Fig. 10.2 Genetic and environmental factors which determine various cardiomyopathy phenotypes

Gene Defects and Congenital Heart Diseases

Congenital heart diseases (CHD) are classified based on the type of the structural defect, resulting blood flow patterns, risk for familial recurrence and shared susceptibility genes [1–6]. A high risk for recurrence, familial forms of the disease, and

Table 10.1 Gene defects associated with common heart diseases

Mutated genes	Cardiac defect
GATA4, Nkx2.5, Tbx20, Tbx5, CRELD1 *Nkx2.5, GATA4, Tbx20* *Tbx20* *Nkx2.5, GATA4* *Nkx2.5* *Nkx2.5, Notch1* *Nkx2.5* *GATA4, CRELD1*	**Congenital heart defects** Atrial septal defect Ventricular septal defect Patent ductus arteriosus Tetralogy of Fallot Transposition of great arteries Coarctation of aorta Double outlet right ventricle Atrioventricular septal defects
β-myosin heavy chain, α-myosin heavy chain, *α-tropomyosin, Troponin-I, Troponin-C,* *Troponin- T, Titin, Vinculin* *Titin, Desmin, cardiac actin, lamin A/C,* *Dystrophin, β-myosin heavy chain,* *α-tropomyosin, Troponin-I, Troponin-C,* *Troponin- T2, Nesprins* *Troponin I, Desmin* *Plakophilin, Plakoglobin, Desmoplakin,* *Desmoglobin-2, Desmocollin, Ryanodine* *receptor-2* *LIM domain binding protein, Lamin A/C,* *α-dystrobrevin, Tafazzin*	**Cardiomyopathies** Hypertrophic cardiomyopathy Dilated cardiomyopathy Restrictive cardiomyopathy Arrhythmogenic right ventricular dysplasia Left ventricular noncompaction
HCN4 *SCN5A* *SCN1 B* *RYR2* *CASQ2* *APAKG2* *Nkx1-5*	**Cardiac conduction diseases** Sinus bradycardia Long QT syndrome, Brugada syndrome Brugada syndrome Polymorphic ventricular tachycardia Atrio-ventricular block, Wolff-Parkinson-White syndrome Holt-Oram syndrome

association of heart defects with chromosomal anomalies suggest a genetic basis for CHD [7]. Nearly 400 genes have been implicated in CHD.

Transcription factors, cell signalling transducers, structural proteins and chromatin modifiers act synergistically during normal heart development. They have interacting functional networks [8–10]. Mutations in genes that encode them alter heart structure and function as the mutations interfere with cell type specification, differentiation and patterning.

CHD largely has non-Mendelian inheritance patterns. Studies on the genetic basis for CHD have employed targeted whole-exome or whole-genome sequencing in patients, families and cohorts. Investigations on the genetic causation of CHD are confounded by the genetic diversity of human subjects as well as genetic heterogeneity associated with CHD resulting in variable expressivity or variable penetrance [11]. Mouse models with gene knockouts, knock-ins and point mutations have also been beneficial to interrogate the genetic causes of CHD [12]. Patient-derived induced pluripotent stem cells (iPSCs) have been useful for mechanistic studies.

Mutations in Genes Coding for Myofilament and Extracellular Matrix Proteins

Mutations in ACTC1 (actin alpha cardiac muscle 1), DCHS1 (protein dachsous homolog 1 or protocadherin-16), titin (TTN), elastin (ELN), myosin heavy chain 6 (MYH6), MYH7, and MYH11 are known to cause heart defects [7]. MYH6 mutations have been associated with atrial septal defects (ASD) and with coarctation of aorta (CoA) [13]. Mutations in the cytoskeletal protein ACTC1 is associated with ASD and is thought to be the result of apoptosis in cardiomyocytes [14, 15]. The actin-binding protein NEXN has also been linked to ASD [16].

Atrioventricular septal defects (AVSD) have been reported in mice with deficiency in the matrix protein CCN1, which regulates cell adhesion and migration, proliferation, survival, and differentiation [17]. Mutations in BVES, a cell adhesion protein, have been identified in patients with tetralogy of Fallot (TOF) [18]. Mutation in PCDHA9 (protocadherin alpha 9), encoding a protocadherin cell adhesion protein may cause aortic hypoplasia or atresia in hypoplastic left heart syndrome (HLHS) as well as bicuspid aortic valve (BAV) [19].

Role of Transcription Factors

Transcription factors and cofactors involved in pathogenesis of CHD have been identified. Large cohort studies have aided the discovery of novel variants and copy number variations (CNVs) [20–22]. Transcription factors in patients with CHD are found to be enriched for de novo and loss of function mutations [2]. Proteins with harmful mutations have alterations in transcriptional or synergistic activity. They can modify the expression of downstream targets and affect cell type specification and differentiation [20].

NKX2-5

NKX2-5 that encodes a homeobox transcription factor and expressed during the earliest stages of cardiogenesis, regulate cardiomyocyte differentiation and proliferation [23]. NKX2-5 mutations were initially discovered to cause atrioventricular (AV) block and atrial septal defect (ASD) [24, 25]. They have since been found in several forms of CHD [21, 24, 26, 27]. Heterozygous mutation in NKX2-5 associated with AV block and ASD has been seen to reduce NKX2-5 nuclear import, downregulate bone morphogenetic protein (BMP) and Notch signalling and result in dysregulation of genes that participate in early cardiomyocyte differentiation and thus decrease cardiomyogenesis [28].

GATA

Family GATA4, 5, and 6 are zinc finger transcription factors expressed in the developing heart [29]. Mutations in GATA4 that decrease transcriptional activity have been linked to BAV and ventricular septal defect (VSD) [30]. Noncoding variants in GATA4 has also been related to BAV [31].

Mutations in NEXN (codes for nexilin, a Z-disc protein), a gene that regulate GATA4, are also associated with CHD [16]. GATA4 is necessary for Hh (hedgehog)-responsive progenitors involved in the development of the outflow tract (OFT). A heterozygous GATA4 mutation has been found to result in VSD and defects of the OFT such as double outlet right ventricle and atrioventricular septal defect in mice [32].

GATA6 mutations in mice result in severe defects of the OFT through disruption of the expression of SEMA3c (semaphoring 3c) and PLXNA2 (plexin-A2) [33, 34]. Mice that are double homozygous knockouts for GATA4/GATA6 generate only SHF (second heart field) progenitor cells and have acardia (32). GATA5 sequence variants have been identified in patients with TOF, VSD, familial atrial fibrillation, and BAV [35].

T-Box Family

The TBX transcription factors are expressed throughout the developing heart and they are important in determining cardiomyocyte identity [36]. TBX5 and TBX20 activate gene expression in muscles of the heart chambers. TBX2 and TBX3 repress gene expression in the precursors of the inflow and outflow tracts. TBX18 is expressed in the cells of the venous pole. Deletion of these genes results in a wide spectrum of heart defects in mice [37].

Mutations in TBX1, expressed in precursors of the OFT have been found in patients with DiGeorge syndrome, which is commonly associated with heart defects. CNVs affecting PRODH (proline dehydrogenase) and DGCR6 (DiGeorge syndrome critical region gene 6), which are known to affect TBX1 expression, have been found in patients with DiGeorge syndrome who have conotruncal defects [38].

Mutations in TBX5 are known to cause Holt–Oram syndrome, which is associated with cardiac defects [39]. Mutations in TBX20 are associated with TOF. Loss of WNT5a, a transcriptional target of TBX1 causes severe hypoplasia of SHF-dependent structures [40].

Forkhead Box Family

Several forkhead box (FOX) transcription factors are involved in the development of heart. Mutations in them lead to heart defects and death of embryos [41]. Deletion CNVs of FOXF1, FOXC2, and FOXL1 are associated with CHD, specifically hypoplastic left heart syndrome (HLHS) [42]. Mutations in FOXC2 cause TOF [43]. A mutation in FOXF1 as well as ZIC3, both of which regulate the specification of laterality was seen in a patient with VACTERL (vertebral defects, anal atresia, cardiac defects, trachea-oesophageal fistula, renal anomalies and limb abnormalities) and heterotaxy (HTX). A mutation in FOXF1 was also identified in a patient with several anomalies suggestive of abnormal left–right patterning [44, 45].

Mutations in FOXH1, a downstream target of the Nodal pathway signalling has been reported in patients with VSD, transposition of the great arteries (TGA), and laterality defects [46, 47]. (46, 47). There is evidence from large-scale mouse mutagenesis screens that mutations in FOXJ1, which modulates ciliogenesis, may result in complex CHD with HTX [9].

HAND Family

HAND1 and 2 are helix–loop–helix transcription factors that regulate the proliferation of precursors in the embryonic ventricle [48]. In HAND1 null mice, heart development is arrested at the looping stage [49]. A mutation in HAND2 is associated with VSD. HAND2 is thought to have synergistic activation effects with GATA4 and NKX2-5 [50].

Many other transcription factors also, when mutated cause CHD.

Nuclear Receptor Family

NR2F2 (nuclear receptor subfamily 2, group F) gene encodes a ligand inducible pleiotropic transcription factor, which is required for the normal development of atria, coronary vessels and aorta [51]. Mutations that alter NR2F2 transcriptional activity sans change in repressor function have been found in patients with AVSD, TOF, aortic stenosis, CoA and HLHS [52].

A de novo mutation in the DNA binding domain of NR1D2, a nuclear receptor transcriptional repressor has been identified in a cohort of patients with AVSD [53]. This mutation alters transcriptional activity. A disease associated loss of function mutation in NR2F2 has been found in a family of patients with double outlet right ventricle (DORV) [54].

CHD Associated with Defects in Signalling Pathways

Nodal Signalling

The Nodal signalling pathway is important in the regulation of left–right patterning. There is evidence of de novo and loss-of-function mutations in NODAL in patients with CHD [2]. NODAL mutations were reported in patients with TGA (47). De novo CNVs affecting NODAL have also been recognized in a cohort of patients with conotruncal defects or HLHS [55]. Mutations in ZIC3, a transcription factor that functions upstream of NODAL have been noticed in patients with CHD and HTX [56, 57]. Mutations in GDF1, CFC1, TDGF1, FOXH1, and SMAD, which are downstream targets of NODAL have been identified in a cohort of patients with CHD.

Notch Signalling

Signalling through the Notch pathway has a key role in determining cardiomyocyte fate and development of heart chambers [58]. Analysis of NOTCH1 mutations in familial CHD has revealed individuals with left sided, right-sided and conotruncal defects [59]. Loss-of-function and intronic variants in NOTCH1 seems to increase the risk for defects of the left ventricular outflow tract (LOFT) [60]. De novo and rare variants have been seen in patients with HLHS [61, 62]. Protein-altering mutations in Notch pathway genes NOTCH1, ARHGAP31, MAML1, SMARCA4, JARID2, and JAG1 are reported to co-segregate with disease in families with defects of the LOFT; patients have an enrichment of the pathogenic variants in these genes [63]. Heterozygous rare coding mutations in MIB1, which activates the Notch pathway have been identified in a Han Chinese cohort of CHD. Two of these mutations cause decrease in JAG1 ubiquitination and the induction of Notch [64]. Mice with mutations in the Slit/Robo signalling resulting in membranous VSDs and BAV; these mice have decreased expression of Notch and its downstream targets [65].

WNT/β-Catenin Signalling

The WNT/β-catenin pathway regulates cell proliferation in the second heart field (SHF) [66]. Enrichment for de novo variants in WNT pathway genes has been noticed in patients with CHD [67]. In mice, deletion of APC, a negative regulator of canonical WNT signalling, results in hypoplasia of the ventricles [68]. BCL9 and Pygo which are regulators of the Wnt pathway are also associated with defects such as AVSD in mice and TOF in humans [69]. Hypoplasia of myocardium and epicardium as well as VSD are seen in mice that are double knockouts for DKK1 and DKK2, which regulate canonical Wnt signalling [70]. Non-canonical Wnt signalling activates the planar cell polarity (PCP) pathway, which coordinates chamber remodelling through actomyosin polarization and modulates ciliogenesis [71–73]. Several components of the PCP pathway have been linked to heart defects [9].

BMP Signalling

BMP signalling is necessary for specification and differentiation of the cardiac meso-derm. It controls NKX2-5 expression [66, 74]. BMP4 deficiency can cause septal defects, and faulty remodelling of the endocardial cushion. Common variants in BMP4 were associated with CHD in a Han Chinese cohort [75]. Nonsynonymous variants in SMAD6, an inhibitor of BMP signalling, have been discovered in patients with CHD [76]. Protein-truncating, splicing, and deleterious missense variants in SMAD2 have been recognized in a cohort of patients with complex CHD with or without laterality defects and other congenital abnormalities [77].

Sonic Hedgehog (SHH) Signalling

SHH secreted from the pharyngeal arch endoderm is received by SHF cells. SHH is important for maintaining proliferation of SHF cells [78]. Mutations in GATA4 necessary for the proliferation of SHH-receiving cells and alignment of outflow tract cause DORV in mice [79]. SHH regulates development of SIX2+ progenitor cells [80]. Ablation of Six2+ cells in mice results in a common arterial trunk.

Mutations in MEGF8 can cause transposition of great arteries (TGA) or anomalies associated with HTX [81]. MEGF8 has been identified as a negative regulator of SHH signalling [82]. MGRN1, another negative SHH regulator has been found to cause HTX with CHD including TGA [83]. Though there is evidence to suggest that SHH signalling pathway is likely be important in the pathogenesis of CHD in humans, the role of SHH in CHD in humans has not been comprehensively investigated [9].

RAS/MAPK Signalling

The RAS/MAPK pathway modulates proliferation, growth and signal transduction in cells. Disruption of the pathway is associated with several developmental defects, which are together known as RASopathies. The most common among RASopathies is Noonan syndrome. Patients with Noonan syndrome have a high incidence of congen-ital heart defects [7]. In nearly 70% of cases, mutations in the genes that code for components of the RAS/MAPK signalling pathways are considered responsible for the syndrome. Mutation in the PTPN11 gene that encodes the protein tyrosine phos-phatase SHP-2, an upstream regulator of the RAS pathway, is well known to cause Noonan syndrome [84]. In one patient with Noonan syndrome, a de novo mutation in MRAS, a participant in downstream MAPK signalling has been reported [85]. Heterozygous de novo and inherited mutations in Alpha-2-Macroglobulin Like 1 (A2ML1) gene have also been identified in patients with Noonan syndrome. A2ML1 is thought to act upstream of the RAS signalling pathway. Expression of A2ML1 does not however activate the RAS/MAPK pathway in cell lines [86].

Vascular Endothelial Growth Factor (VEGF) Signalling

Damaging variants in the VEGF-related genes such as FLT4, KDR, VEGFA, FGD5, BCAR1, IQGAP, FOXO1, and PRDM1 have been identified in a cohort of patients with TOF [87]. Variants in the VEGF-A pathway genes—COL6A1, COL6A2, CRELD1, FBLN2, FRZB, and GATA5 were found in excess and thus highly probable to be deleterious in a cohort of patients with Down-syndrome-associated atrioventricular septal defects [88].

The Role of Cilia and Cilia-Transduced Cell Signalling

Normal cardiac morphogenesis requires regulated signalling processes to establish left–right asymmetry. During early embryonic development, different types of cilia are expressed in a spatiotemporal manner in the embryonic node to generate a gradient of signalling molecules, which is requisite to govern cardiogenesis [10]. Nonmotile cilia, known as primary cilia are the cell signalling transducers or mechanosensors. Cilia and cilia-transduced cell signalling modulate planar cell polarity and orga-nization of the cytoskeleton. Cilia transduce signalling pathways such as WNT, TGFβ/BMP, and SHH which are involved in heart development [89].

The key role of cilia in the pathogenesis of CHD was discovered in mice in which point mutations were randomly induced in the germline, using N-ethyl-N-nitrosourea (ENU). Analysis of the mutations that result in CHD revealed that 50% of the recovered mutations that resulted in CHD were cilia related. Thirty mutations were in genes related to cilia and ciliogenesis. Many genes identified were those involved in cilia-transduced cell signalling and vesicular trafficking, important for ciliogenesis [9].

A later study found that mutations in DNAH11, an axonemal protein, and MKS1, a basal body protein, are associated with CHD [90].

IFT88 is an intra flagellar transport protein required for cilia formation. IFT88 null mutant mice have been found to have defects in the outflow tract (OFT) of heart ventricles [91, 92].

An enrichment of ciliary genes with damaging recessive variants have also been reported in a cohort of patients with CHD [2]. The role of ciliary defects in the pathogenesis of CHD needs further scrutiny.

Chromatin Modifiers

Disruption of chromatin modifiers can interfere with key transcriptional programs linked to development of heart. Chromatin modifiers were seen to be enriched among genes with de novo mutations in a cohort with various phenotypes of CHD (LOFT obstruction, conotruncal defects, and HTX) [2]. Several genes which regulate active H3K4me/inactive H3K27me histone marks were identified. KMT2D encodes one of these histone modifiers and is associated with Kabuki syndrome with CoA, ASDs, and VSDs [93]. A mutation of CASZ1 which encodes a zinc finger transcription factor that interacts with histones is associated with reduced transcriptional activity and VSD [94].

The HDAC repressor complex has a key role in several developmental processes. Many proteins of this complex are associated with CHD [95]. Variants in SMYD4, a protein which interacts with HDAC1 and modulates histone acetylation have been seen in patients with DORV and TOF [96].

In a mouse forward genetic screen for CHD, mutations in genes regulating chromatin in SMARCA4 and PRDM1 have been found [9]. Another find was a CHD-causing mutation in SAP130, a SIN3A associated protein that is also part of the HDAC repressor complex. Mutation in SAP130 (a subunit of the histone deacetylase complex) was shown to result in left ventricular hypoplasia [97]. Double homozygous PCDHA9 and SAP130 mutations result in HLHS [19].

Gene Defects and Cardiomyopathies

Cardiomyopathies have a prevalence of one in 500 individuals. They are associated with an increased risk of heart failure, cardiac arrhythmias, stroke and sudden death [98, 99]. The consensus definition for cardiomyopathies considers both genetic and nongenetic causes of different types of cardiomyopathies as well as 'electrical' muscle disorders with characteristic features [100].

Genetic studies of cardiomyopathies in search of associated genetic defects began with cardiomyopathy phenotypes. Variants in several genes have been discovered to be associated with each of the cardiomyopathies and have also been linked functionally to the myocardial defects [101]. Gene-based classifications have refined specificity and also indicate disease specific mechanisms. Both affected and unaffected individuals carry several novel and potentially disease-causing nonsense mutations, frameshift insertion/deletions, and splice-site-altering variants in cardiomyopathy associated genes.

Hypertrophic Cardiomyopathy (HCM)

HCM is characterized by varying degrees of left ventricular hypertrophy without associated cardiovascular or systemic diseases [102]. Some genotype-positive individuals may have a little or no hypertrophy [102–104]. Typical microscopic features are cardiomyocyte hypertrophy, myofiber disarray, and interstitial fibrosis. Individuals with HCM may have a positive family history. The pattern of inheritance is autosomal-dominant. Sporadic cases have also been reported.

More than 1400 mutations have been associated with HCM [105, 106]. Most of them are in genes that encode sarcomeric contractile proteins. A small number of mutations are in genes that encode proteins of the calcium cycle in cardiomyocytes. The precise number of causative genes for HCM are unclear because of lack of uniformity in the evidence that is considered to define pathogenicity of HCM [98, 99, 102, 107–110].

Among the genotyped cases, 30–40% had mutations in MYH7, which encodes the β-myosin heavy chain and MYBPC3, which encodes cardiac myosin-binding protein C [111]. Mutations in PRKAG2, LAMP2, and GLA genes that encode the G-regulatory subunit of the AMP-activated protein kinase, lysosome-associated membrane protein-2, and a-galactosidase respectively, are also associated with phenotypes which mimic HCM [112, 113].

Dilated Cardiomyopathy (DCM)

Dilatation and impaired systolic contraction of either one or both the ventricles of the heart are the characteristics of DCM. Nearly 50% of the cases of DCM have an infectious or toxic or metabolic or endocrine or autoimmune or nutritional basis. Familial and sporadic forms with genetic defects are also seen. Conduction-system abnormalities, congenital heart defects, valvular defects, left ventricular noncompaction (LVNC) as well as extra cardiac involvement are often associated with in patients with a genetic cause [100]. The common inheritance pattern is autosomal-dominant. In a small proportion, autosomal-recessive or X-linked inheritance has been observed [114].

Mutations in more than 40 genes have been linked with adult-onset DCM. These are in the genes which encode different proteins of the sarcomere, Z-disc, cytoskeleton, sarcolemma, and nucleus [110, 115–120].

Truncating mutations in the TTN gene that encodes titin have been identified in 25% of cases of familial DCM and 18% of sporadic DCM cases [121]. Mutations in the desmosomal gene have also been identified in 5% of subjects with a clinical diagnosis of DCM [122].

The splicing factor RBM20 controls alternative splicing of genes associated with diastolic function and ion transport, as well as sarcomere assembly. The expression of shorter isoforms of titin (TTN) is related to increased expression of RBM20 [123]. In mice, mutations in RBM20 result in TTN mutations and dilated cardiomyopathy (DCM) as well as arrhythmia [124].

Restrictive Cardiomyopathy

Restrictive cardiomyopathy (RCM) is distinguished by reduced diastolic filling of ventricles, raise in end-diastolic pressure, decrease in ventricular dimensions and dilatation of both atria [125]. RCM could have a genetic cause or may be associated with infiltrative diseases such as amyloidosis or systemic disorders such as sarcoidosis and storage diseases. Mutations that cause RCM are in the genes that code for sarcomere proteins or in those that encode the cytoskeletal protein desmin [126–128]. Mutations have been reported both in familial and sporadic cases. Homozygous mutations or compound mutations which could result in higher 'dose' of the mutant protein are associated with an early onset and or more severe disease [127–129].

The high prevalence of mutations in the gene that codes sarcomere proteins have led to the debate whether RCM and HCM belong to a same spectrum of left ventricular diastolic dysfunction. Mixed RCM and HCM phenotypes are commonly seen in the same families. In both RCM and HCM, there is also increased myofilament sensitivity to calcium with distinct effects on ATPase activity [129–131].

Arrhythmogenic Right Ventricular Cardiomyopathy (ARVC)

ARVC has progressive myocyte loss and fibrofatty replacement of the right ventricle as its features. Left ventricle is involved in up to 75% of cases. A wide range of phenotypical features have been reported in cases of ARVC [132–134].

Autosomal-dominant inheritance patterns have been commonly seen in families with ARVC. Mutations in nine genes have been correlated with ARVC with different levels of evidence Five of them encode the desmosomal proteins such as plakophilin2, plakoglobin, desmoplakin, desmocollin, and desmoglein-2 [110]. About 50% of patients with ARVC have a desmosomal gene mutation, most commonly (40%) in the plakophilin-2 gene [135, 136]. A small number of mutations in genes encoding the non-desmosomal proteins such as transmembrane protein 43, TGFβ3, cardiac ryanodine receptor, and titin are also reported to be associated with ARVC.

Left Ventricular Noncompaction (LVNC)

LVNC is considered to be a developmental defect in compaction of the trabeculae in ventricles of the heart. LVNC is seen along with other developmental defects in the heart; LVNC is also found as a single defect [137].

Both sporadic and familial forms of LVNC are reported. The defect may have an autosomal-dominant, autosomal-recessive, or X-linked inheritance. In about 50% of patients with LVNC, the genes encoding sarcomere proteins are mutated [138–141]. Most of the variants are in the MYH7 gene or the MYBPC3 gene. Mutations in genes that code for Z-disc, cytoskeletal, and mitochondrial proteins are found in a small proportion of patients. Deletion and loss-of function mutations in PRDM16, a gene of unknown function(s), could also result in LVNC [142].

Ion Channelopathies

'Ion channelopathies' comprise of long QT syndrome, short QT syndrome, Brugada syndrome, catecholaminergic polymorphic ventricular tachycardia and sudden unexplained nocturnal death syndrome [143, 144]. Ventricular arrhythmias are a feature of these syndromes. Mutations in genes encoding cardiac specific sodium, potassium and calcium channels have been linked to ion channelopathies.

Ventricular arrhythmias are also seen in patients with DCM, ARVC, and LVNC. Variants in some of the ion channel genes have been detected in them as well [145–149].

Genetic Defects and Overlapping Phenotypes

Mutations in the same gene can result in different cardiomyopathies. Mutations in the MYH7 gene are associated with HCM, DCM, LVNC as well as RCM [138, 150–153]. These observations raise the questions of how distinctive cardiomyopathies are and what mechanisms lead to the phenotypical differences and common features.

Functional characterizations have been done on a few pathogenic mutations and these have indicated possible associations with cardiac dysfunction. MYH7 mutations that result in HCM (R403Q) and DCM (F764L or S532P) have alternate (increased or decreased, respectively) effects on force production by the sarcomere [154–156]. These mutations may produce similar secondary responses such as, increased energy utilization and re-expression of foetal genes.

The abundant genetic heterogeneity of cardiomyopathies raises questions on how different mutations such as in LMNA [118] or PLN [157], which encode, respectively, a nuclear membrane protein and a physiological inhibitor of the sarcoplasmic reticulum Ca2β ATPase [118, 158] or in sarcomere protein genes such as TTN,

MYH7, and ACTC1 [158] result in the overlapping phenotype of DCM. Whether multiple mechanisms are involved or mutations in genes with dissimilar functions focus onto a final common pathway need to be clarified.

Epigenetic Factors in Cardiomyopathies

Several factors intrinsic and extrinsic to the cardiomyocyte influence gene expression in the cell. These include epigenetic mechanisms that regulate gene expression (DNA methylation, histone modifications, ATP-dependent chromatin remodelling, and noncoding RNAs, (including lincRNAs and microRNAs) [159–164]. The role of epigenetic factors, including microRNAs in the pathogenesis of inherited cardiomyopathies has not been adequately investigated.

Functional Evaluation of Genetic Variants

Relatively few novel mechanistic insights have been gained into how different mutations impact phenotype [165–170]. The functional consequences of single genetic variants are usually evaluated in transfected cells. These in vitro experiments have the limitation that they cannot assess the effects of background genetic variation or various factors that affect protein function in the living organism.

Induced pluripotent stem (iPS) cells allow patient-specific functional evaluation of genetic variants [171]. Several human cardiomyopathy and arrhythmia mutations have been modelled in iPS cells despite their limitations.

Zebrafish is popular for evaluation of the in vivo effects of human gene mutations [172–175]. New candidate genes for human cardiomyopathies and arrhythmias can be identified by studies on naturally occurring or chemically induced zebrafish mutations that result in cardiac phenotypes [173, 176]. Studies after transient or sustained knockdown of zebrafish genes and use of transgenic zebrafish are useful to elucidate the molecular mechanisms [142, 174, 177, 178].

Databases of whole-genome and whole-exome sequencing (WES) sequences obtained from large numbers of individuals in large-scale initiatives have revealed the extent and range of genetic variation within populations of different ethnicity. A large number of specific variants reported earlier to be causal for cardiomyopathy have later been identified in population databases [179–183]. There is hence a need to look beyond the single variant as a cause for pathogenesis of heart diseases. Diane Fatkin and colleagues opine that new models that incorporate multiple genetic variants that can potentially affect different aspects of cardiomyocyte structure and function need to be considered [184]. The cardiac phenotype could be determined by the collective effects of and interactions of multiple genetic, epigenetic, and environmental factors.

Conclusions

A large body of evidence points to CHD as sequelae of heterogeneous genetic defects in the cardiomyocyte. The molecular mechanisms for CHD pathogenesis are however not well understood. Cilia and chromatin modifiers are thought to be central in driving the complex genetics of CHD. Genetic defects linked to cardiomyopathy are generally classified considering phenotype-specific mechanistic pathways. HCM has been termed a 'disorder of the sarcomere', DCM a 'cytoskeletal disease' and ARVC a 'disease of the desmosome'. This grouping has limitations. A single disease gene may have several biological functions, different subcellular locations, and may lead to varying phenotypes. Disease genes which do not conform to the classification are being newly discovered.

References

1. Botto LD, Lin AE, Riehle-Colarusso T, Malik S, Correa A. Seeking causes: classifying and evaluating congenital heart defects in etiologic studies. Birth Defects Res A. 2007;79(10):714–27.
2. Jin SC, Homsy J, Zaidi S, Lu Q, Morton S, DePalma SR, et al. Contribution of rare inherited and de novo variants in 2,871 congenital heart disease probands. Nat Genet. 2017;49(11):1593.
3. Houyel L, Khoshnood B, Anderson RH, Lelong N, Thieulin A-C, Goffinet F, et al. Population-based evaluation of a suggested anatomic and clinical classification of congenital heart defects based on the International Paediatric and Congenital Cardiac Code. Orphanet J Rare Dis. 2011;6(1):1–9.
4. Ellesøe SG, Workman CT, Bouvagnet P, Loffredo CA, McBride KL, Hinton RB, et al. Familial co-occurrence of congenital heart defects follows distinct patterns. Eur Heart J. 2018;39(12):1015–22.
5. Brodwall K, Greve G, Leirgul E, Tell GS, Vollset SE, Øyen N. Recurrence of congenital heart defects among siblings—a nationwide study. Am J Med Genet A. 2017;173(6):1575–85.
6. Øyen N, Poulsen G, Boyd HA, Wohlfahrt J, Jensen P, Melbye M. Recurrence of congenital heart defects in families. Circulation. 2009;120(4):295–301.
7. Pierpont ME, Brueckner M, Chung WK, Garg V, Lacro RV, McGuire AL, et al. Genetic basis for congenital heart disease: revisited: a scientific statement from the American Heart Association. Circulation. 2018;138(21):e653–711.
8. Lage K, Greenway SC, Rosenfeld JA, Wakimoto H, Gorham JM, Segrè AV, et al. Genetic and environmental risk factors in congenital heart disease functionally converge in protein networks driving heart development. Proc Natl Acad Sci USA. 2012;109(35):14035–40.
9. Li Y, Klena NT, Gabriel GC, Liu X, Kim AJ, Lemke K, et al. Global genetic analysis in mice unveils central role for cilia in congenital heart disease. Nature. 2015;521(7553):520–4.
10. Williams K, Carson J, Lo C. Genetics of congenital heart disease. Biomolecules. 2019;9(12):879.
11. Zaidi S, Brueckner M. Genetics and genomics of congenital heart disease. Circ Res. 2017;120(6):923–40.
12. Krishnan A, Samtani R, Dhanantwari P, Lee E, Yamada S, Shiota K, et al. A detailed comparison of mouse and human cardiac development. Pediatr Res. 2014;76(6):500–7.
13. Bjornsson T, Thorolfsdottir RB, Sveinbjornsson G, Sulem P, Norddahl GL, Helgadottir A, et al. A rare missense mutation in MYH6 associates with non-syndromic coarctation of the aorta. Eur Heart J. 2018;39(34):3243–9.

14. Jiang H-K, Qiu G-R, Li-Ling J, Xin N, Sun K-L. Reduced ACTC1 expression might play a role in the onset of congenital heart disease by inducing cardiomyocyte apoptosis. Circ J. 2010;74(11):2410–8.

15. Matsson H, Eason J, Bookwalter CS, Klar J, Gustavsson P, Sunnegårdh J, et al. Alpha-cardiac actin mutations produce atrial septal defects. Hum Mol Genet. 2008;17(2):256–65.

16. Yang F, Zhou L, Wang Q, You X, Li Y, Zhao Y, et al. NEXN inhibits GATA4 and leads to atrial septal defects in mice and humans. Cardiovasc Res. 2014;103(2):228–37.

17. Mo F-E, Lau LF. The matricellular protein CCN1 is essential for cardiac development. Circ Res. 2006;99(9):961–9.

18. Wu M, Li Y, He X, Shao X, Yang F, Zhao M, et al. Mutational and functional analysis of the BVES gene coding region in Chinese patients with non-syndromic tetralogy of Fallot. Int J Mol Med. 2013;31(4):899–903.

19. Liu X, Yagi H, Saeed S, Bais AS, Gabriel GC, Chen Z, et al. The complex genetics of hypoplastic left heart syndrome. Nat Genet. 2017;49(7):1152.

20. Kodo K, Nishizawa T, Furutani M, Arai S, Ishihara K, Oda M, et al. Genetic analysis of essential cardiac transcription factors in 256 patients with non-syndromic congenital heart defects. Circ J. 2012;76(7):1703–11.

21. Granados-Riveron JT, Pope M, Bu'Lock FA, Thornborough C, Eason J, Setchfield K, et al. Combined mutation screening of NKX2-5, GATA4, and TBX5 in congenital heart disease: multiple heterozygosity and novel mutations. Congenit Heart Dis. 2012;7(2):151–9.

22. Glessner JT, Bick AG, Ito K, Homsy JG, Rodriguez-Murillo L, Fromer M, et al. Increased frequency of de novo copy number variants in congenital heart disease by integrative analysis of single nucleotide polymorphism array and exome sequence data. Circ Res. 2014;115(10):884–96.

23. Prendiville T, Jay PY, Pu WT. Insights into the genetic structure of congenital heart disease from human and murine studies on monogenic disorders. Cold Spring Harbor Perspect Med. 2014; 4(10):a013946.

24. Schott J-J, Benson DW, Basson CT, Pease W, Silberbach GM, Moak JP, et al. Congenital heart disease caused by mutations in the transcription factor NKX2-5. Science. 1998;281(5373):108–11.

25. Gutierrez-Roelens I, Sluysmans T, Gewillig M, Devriendt K, Vikkula M. Progressive AV-block and anomalous venous return among cardiac anomalies associated with two novel missense mutations in the CSX/NKX2-5 gene. Hum Mutat. 2002;20(1):75–6.

26. Winston JB, Erlich JM, Green CA, Aluko A, Kaiser KA, Takematsu M, et al. Heterogeneity of genetic modifiers ensures normal cardiac development. Circulation. 2010;121(11):1313.

27. Gifford CA, Ranade SS, Samarakoon R, Salunga HT, De Soysa TY, Huang Y, et al. Oligogenic inheritance of a human heart disease involving a genetic modifier. Science. 2019;364(6443):865–70.

28. Zakariyah AF, Rajgara RF, Horner E, Cattin ME, Blais A, Skerjanc IS, et al. In vitro modeling of congenital heart defects associated with an NKX2-5 mutation revealed a dysregulation in BMP/notch-mediated signaling. Stem Cells. 2018;36(4):514–26.

29. Pikkarainen S, Tokola H, Kerkelä R, Ruskoaho H. GATA transcription factors in the developing and adult heart. Cardiovasc Res. 2004;63(2):196–207.

30. Li R-G, Xu Y-J, Wang J, Liu X-Y, Yuan F, Huang R-T, et al. GATA4 loss-of-function mutation and the congenitally bicuspid aortic valve. Am J Cardiol. 2018;121(4):469–74.

31. Yang B, Zhou W, Jiao J, Nielsen JB, Mathis MR, Heydarpour M, et al. Protein-altering and regulatory genetic variants near GATA4 implicated in bicuspid aortic valve. Nat Commun. 2017;8(1):1–10.

32. Zhao R, Watt AJ, Battle MA, Li J, Bondow BJ, Duncan SA. Loss of both GATA4 and GATA6 blocks cardiac myocyte differentiation and results in acardia in mice. Dev Biol. 2008;317(2):614–9.

33. Kodo K, Nishizawa T, Furutani M, Arai S, Yamamura E, Joo K, et al. GATA6 mutations cause human cardiac outflow tract defects by disrupting semaphorin-plexin signaling. Proceedings of the National Academy of Sciences USA. 2009;106(33):13933–8.

34. Maitra M, Koenig SN, Srivastava D, Garg V. Identification of GATA6 sequence variants in patients with congenital heart defects. Pediatr Res. 2010;68(4):281–5.
35. Bonachea EM, Zender G, White P, Corsmeier D, Newsom D, Fitzgerald-Butt S, et al. Use of a targeted, combinatorial next-generation sequencing approach for the study of bicuspid aortic valve. BMC Med Genomics. 2014;7(1):1–10.
36. Sylva M, van den Hoff MJ, Moorman AF. Development of the human heart. Am J Med Genet A. 2014;164(6):1347–71.
37. Greulich F, Rudat C, Kispert A. Mechanisms of T-box gene function in the developing heart. Cardiovasc Res. 2011;91(2):212–22.
38. Gao W, Higaki T, Eguchi-Ishimae M, Iwabuki H, Wu Z, Yamamoto E, et al. DGCR6 at the proximal part of the DiGeorge critical region is involved in conotruncal heart defects. Hum Genome Var. 2015;2(1):1–7.
39. Reamon-Buettner SM, Borlak J. TBX5 mutations in Non-Holt-Oram Syndrome (HOS) malformed hearts. Hum Mutat. 2004; 24(1):104.
40. Chen L, Fulcoli FG, Ferrentino R, Martucciello S, Illingworth EA, Baldini A. Transcriptional control in cardiac progenitors: Tbx1 interacts with the BAF chromatin remodeling complex and regulates Wnt5a. PLoS Genet. 2012; 8(3):e1002571.
41. Zhu H. Forkhead box transcription factors in embryonic heart development and congenital heart disease. Life Sci. 2016;144:194–201.
42. Stankiewicz P, Sen P, Bhatt SS, Storer M, Xia Z, Bejjani BA, et al. Genomic and genic deletions of the FOX gene cluster on 16q24. 1 and inactivating mutations of FOXF1 cause alveolar capillary dysplasia and other malformations. Am J Hum Genet. 2009; 84(6):780–791.
43. Morgenthau A, Frishman WH. Genetic origins of tetralogy of fallot. Cardiol Rev. 2018;26(2):86–92.
44. Hilger AC, Halbritter J, Pennimpede T, van der Ven A, Sarma G, Braun DA, et al. Targeted resequencing of 29 candidate genes and mouse expression studies implicate ZIC3 and FOXF1 in human VATER/VACTERL association. Hum Mutat. 2015;36(12):1150–4.
45. Li S, Liu S, Chen W, Yuan Y, Gu R, Song Y, et al. A novel ZIC3 gene mutation identified in patients with heterotaxy and congenital heart disease. Sci Rep. 2018;8(1):1–12.
46. Wang B, Yan J, Mi R, Zhou S, Xie X, Wang J, et al. Forkhead box H1 (FOXH1) sequence variants in ventricular septal defect. Int J Cardiol. 2010;145(1):83–5.
47. De Luca A, Sarkozy A, Consoli F, Ferese R, Guida V, Dentici ML, et al. Familial transposition of the great arteries caused by multiple mutations in laterality genes. Heart. 2010;96(9):673–7.
48. McFadden DG, Barbosa AC, Richardson JA, Schneider MD, Srivastava D, Olson EN. The Hand1 and Hand2 transcription factors regulate expansion of the embryonic cardiac ventricles in a gene dosage-dependent manner. Development. 2005;132(1):189–201.
49. Firulli BA, Toolan KP, Harkin J, Millar H, Pineda S, Firulli AB. The HAND1 frameshift A126FS mutation does not cause hypoplastic left heart syndrome in mice. Cardiovasc Res. 2017;113(14):1732–42.
50. Sun Y-M, Wang J, Qiu X-B, Yuan F, Li R-G, Xu Y-J, et al. A HAND2 loss-of-function mutation causes familial ventricular septal defect and pulmonary stenosis. G3 Genes Genomes Genet. 2016; 6(4):987–992.
51. Wu S-p, Cheng C-M, Lanz RB, Wang T, Respress JL, Ather S, et al. Atrial identity is determined by a COUP-TFII regulatory network. Dev Cell. 2013; 25(4):417–426.
52. Al Turki S, Manickaraj AK, Mercer CL, Gerety SS, Hitz M-P, Lindsay S, et al. Rare variants in NR2F2 cause congenital heart defects in humans. Am J Hum Genet. 2014;94(4):574–85.
53. Priest JR, Osoegawa K, Mohammed N, Nanda V, Kundu R, Schultz K, et al. De novo and rare variants at multiple loci support the oligogenic origins of atrioventricular septal heart defects. PLoS Genet. 2016; 12(4):e1005963.
54. Qiao X-H, Wang Q, Wang J, Liu X-Y, Xu Y-J, Huang R-T, et al. A novel NR2F2 loss-of-function mutation predisposes to congenital heart defect. Eur J Med Genet. 2018;61(4):197–203.

55. Warburton D, Ronemus M, Kline J, Jobanputra V, Williams I, Anyane-Yeboa K, et al. The contribution of de novo and rare inherited copy number changes to congenital heart disease in an unselected sample of children with conotruncal defects or hypoplastic left heart disease. Hum Genet. 2014;133(1):11–27.
56. Cast AE, Gao C, Amack JD, Ware SM. An essential and highly conserved role for Zic3 in left–right patterning, gastrulation and convergent extension morphogenesis. Dev Biol. 2012;364(1):22–31.
57. Li X, Liu L, Zhou J, Wang C. Heterogeneity analysis and diagnosis of complex diseases based on deep learning method. Sci Rep. 2018;8(1):1–8.
58. Li X, Liu L, Zhou J, Wang C. Heterogeneity analysis and diagnosis of complex diseases based on deep learning method. Sci Rep. 2018; 8(1):6155.
59. Kerstjens-Frederikse WS, Van De Laar IM, Vos YJ, Verhagen JM, Berger RM, Lichtenbelt KD, et al. Cardiovascular malformations caused by NOTCH1 mutations do not keep left: data on 428 probands with left-sided CHD and their families. Genet Med. 2016;18(9):914–23.
60. Helle E, Córdova-Palomera A, Ojala T, Saha P, Potiny P, Gustafsson S, et al. Loss of function, missense, and intronic variants in NOTCH1 confer different risks for left ventricular outflow tract obstructive heart defects in two European cohorts. Genet Epidemiol. 2019;43(2):215–26.
61. Iascone M, Ciccone R, Galletti L, Marchetti D, Seddio F, Lincesso A, et al. Identification of de novo mutations and rare variants in hypoplastic left heart syndrome. Clin Genet. 2012;81(6):542–54.
62. Zahavich L, Bowdin S, Mital S. Use of clinical exome sequencing in isolated congenital heart disease. Circ Cardiovasc Genet. 2017; 10(3):e001581.
63. Preuss C, Capredon M, Wünnemann F, Chetaille P, Prince A, Godard B, et al. Family based whole exome sequencing reveals the multifaceted role of notch signaling in congenital heart disease. PLoS Genet. 2016; 12(10):e1006335.
64. Li B, Yu L, Liu D, Yang X, Zheng Y, Gui Y, et al. MIB1 mutations reduce Notch signaling activation and contribute to congenital heart disease. Clin Sci. 2018;132(23):2483–91.
65. Mommersteeg MT, Yeh ML, Parnavelas JG, Andrews WD. Disrupted Slit-Robo signalling results in membranous ventricular septum defects and bicuspid aortic valves. Cardiovasc Res. 2015;106(1):55–66.
66. Rochais F, Mesbah K, Kelly RG. Signaling pathways controlling second heart field development. Circ Res. 2009;104(8):933–42.
67. Homsy J, Zaidi S, Shen Y, Ware JS, Samocha KE, Karczewski KJ, et al. De novo mutations in congenital heart disease with neurodevelopmental and other congenital anomalies. Science. 2015;350(6265):1262–6.
68. Ye B, Hou N, Xiao L, Xu Y, Boyer J, Xu H, et al. APC controls asymmetric Wnt/β-catenin signaling and cardiomyocyte proliferation gradient in the heart. J Mol Cell Cardiol. 2015;89:287–96.
69. Cantù C, Felker A, Zimmerli D, Prummel KD, Cabello EM, Chiavacci E, et al. Mutations in Bcl9 and Pygo genes cause congenital heart defects by tissue-specific perturbation of Wnt/β-catenin signaling. Genes Dev. 2018;32(21–22):1443–58.
70. Phillips MD, Mukhopadhyay M, Poscablo C, Westphal H. Dkk1 and Dkk2 regulate epicardial specification during mouse heart development. Int J Cardiol. 2011;150(2):186–92.
71. May-Simera HL, Kelley MW. Cilia, Wnt signaling, and the cytoskeleton. Cilia. 2012;1(1):1–16.
72. Caron A, Xu X, Lin X. Wnt/β-catenin signaling directly regulates Foxj1 expression and ciliogenesis in zebrafish Kupffer's vesicle. Development. 2012;139(3):514–24.
73. Merks AM, Swinarski M, Meyer AM, Müller NV, Özcan I, Donat S, et al. Planar cell polarity signalling coordinates heart tube remodelling through tissue-scale polarisation of actomyosin activity. Nat Commun. 2018;9(1):1–15.
74. Chen H, Shi S, Acosta L, Li W, Lu J, Bao S, et al. BMP10 is essential for maintaining cardiac growth during murine cardiogenesis. Development. 2004;131(9):2219–31.
75. Qian B, Mo R, Da M, Peng W, Hu Y, Mo X. Common variations in BMP4 confer genetic susceptibility to sporadic congenital heart disease in a Han Chinese population. Pediatr Cardiol. 2014;35(8):1442–7.

76. Tan HL, Glen E, Töpf A, Hall D, O'Sullivan JJ, Sneddon L, et al. Nonsynonymous variants in the SMAD6 gene predispose to congenital cardiovascular malformation. Hum Mutat. 2012;33(4):720–7.
77. Granadillo JL, Chung WK, Hecht L, Corsten-Janssen N, Wegner D, Nij Bijvank SW, et al. Variable cardiovascular phenotypes associated with SMAD2 pathogenic variants. Hum Mutat. 2018;39(12):1875–84.
78. Dyer LA, Kirby ML. Sonic hedgehog maintains proliferation in secondary heart field progenitors and is required for normal arterial pole formation. Dev Biol. 2009;330(2):305–17.
79. Liu J, Cheng H, Xiang M, Zhou L, Wu B, Moskowitz IP, et al. Gata4 regulates hedgehog signaling and Gata6 expression for outflow tract development. PLoS Genet. 2019; 15(5):e1007711.
80. Zhou Z, Wang J, Guo C, Chang W, Zhuang J, Zhu P, et al. Temporally distinct Six2-positive second heart field progenitors regulate mammalian heart development and disease. Cell Rep. 2017;18(4):1019–32.
81. Zhang Z, Alpert D, Francis R, Chatterjee B, Yu Q, Tansey T, et al. Massively parallel sequencing identifies the gene Megf8 with ENU-induced mutation causing heterotaxy. Proc Natl Acad Sci USA. 2009;106(9):3219–24.
82. Pusapati GV, Kong JH, Patel BB, Krishnan A, Sagner A, Kinnebrew M, et al. CRISPR screens uncover genes that regulate target cell sensitivity to the morphogen sonic hedgehog. Dev Cell. 2018; 44(1):113–129. e8.
83. Liu C, Cao R, Xu Y, Li T, Li F, Chen S, et al. Rare copy number variants analysis identifies novel candidate genes in heterotaxy syndrome patients with congenital heart defects. Genome Medicine. 2018;10(1):1–13.
84. Sifrim A, Hitz M-P, Wilsdon A, Breckpot J, Al Turki SH, Thienpont B, et al. Distinct genetic architectures for syndromic and nonsyndromic congenital heart defects identified by exome sequencing. Nat Genet. 2016;48(9):1060–5.
85. Higgins EM, Bos JM, Mason-Suares H, Tester DJ, Ackerman JP, MacRae CA, et al. Elucidation of MRAS-mediated Noonan syndrome with cardiac hypertrophy. JCI Insight. 2017; 2(5): e91225.
86. Vissers LE, Bonetti M, Overman JP, Nillesen WM, Frints SG, De Ligt J, et al. Heterozygous germline mutations in A2ML1 are associated with a disorder clinically related to Noonan syndrome. Eur J Hum Genet. 2015;23(3):317–24.
87. Reuter MS, Jobling R, Chaturvedi RR, Manshaei R, Costain G, Heung T, et al. Haploinsufficiency of vascular endothelial growth factor related signaling genes is associated with tetralogy of Fallot. Genet Med. 2019;21(4):1001–7.
88. Ackerman C, Locke AE, Feingold E, Reshey B, Espana K, Thusberg J, et al. An excess of deleterious variants in VEGF-A pathway genes in Down-syndrome-associated atrioventricular septal defects. Am J Hum Genet. 2012;91(4):646–59.
89. Koefoed K, Veland IR, Pedersen LB, Larsen LA, Christensen ST. Cilia and coordination of signaling networks during heart development. Organogenesis. 2014;10(1):108–25.
90. Burnicka-Turek O, Steimle JD, Huang W, Felker L, Kamp A, Kweon J, et al. Cilia gene mutations cause atrioventricular septal defects by multiple mechanisms. Hum Mol Genet. 2016;25(14):3011–28.
91. Willaredt MA, Gorgas K, Gardner HA, Tucker KL. Multiple essential roles for primary cilia in heart development. Cilia. 2012;1(1):1–18.
92. Clement CA, Kristensen SG, Møllgård K, Pazour GJ, Yoder BK, Larsen LA, et al. The primary cilium coordinates early cardiogenesis and hedgehog signaling in cardiomyocyte differentiation. J Cell Sci. 2009;122(17):3070–82.
93. Ang S-Y, Uebersohn A, Spencer CI, Huang Y, Lee J-E, Ge K, et al. KMT2D regulates specific programs in heart development via histone H3 lysine 4 di-methylation. Development. 2016;143(5):810–21.
94. Huang R-T, Xue S, Wang J, Gu J-Y, Xu J-H, Li Y-J, et al. CASZ1 loss-of-function mutation associated with congenital heart disease. Gene. 2016;595(1):62–8.

95. Montgomery RL, Davis CA, Potthoff MJ, Haberland M, Fielitz J, Qi X, et al. Histone deacetylases 1 and 2 redundantly regulate cardiac morphogenesis, growth, and contractility. Genes Dev. 2007;21(14):1790–802.
96. Xiao D, Wang H, Hao L, Guo X, Ma X, Qian Y, et al. The roles of SMYD4 in epigenetic regulation of cardiac development in zebrafish. PLoS Genet. 2018; 14(8):e1007578.
97. Yagi H, Liu X, Gabriel GC, Wu Y, Peterson K, Murray SA, et al. The genetic landscape of hypoplastic left heart syndrome. Pediatr Cardiol. 2018;39(6):1069–81.
98. Seidman CE, Seidman J. Identifying sarcomere gene mutations in hypertrophic cardiomyopathy: a personal history. Circ Res. 2011;108(6):743–50.
99. Watkins H, Ashrafian H, Redwood C. Inherited cardiomyopathies. N Engl J Med. 2011;364(17):1643–56.
100. Maron BJ, Towbin JA, Thiene G, Antzelevitch C, Corrado D, Arnett D, et al. Contemporary definitions and classification of the cardiomyopathies: an American Heart Association scientific statement from the council on clinical cardiology, heart failure and transplantation committee; quality of care and outcomes research and functional genomics and translational biology interdisciplinary working groups; and council on epidemiology and prevention. Circulation. 2006;113(14):1807–16.
101. Yotti R, Seidman CE, Seidman JG. Advances in the genetic basis and pathogenesis of sarcomere cardiomyopathies. Annu Rev Genomics Hum Genet. 2019;20:129–53.
102. Maron BJ, Maron MS. Hypertrophic cardiomyopathy. The Lancet. 2013;381(9862):242–55.
103. Klues HG, Schiffers A, Maron BJ. Phenotypic spectrum and patterns of left ventricular hypertrophy in hypertrophic cardiomyopathy: morphologic observations and significance as assessed by two-dimensional echocardiography in 600 patients. J Am Coll Cardiol. 1995;26(7):1699–708.
104. Watkins H, McKenna WJ, Thierfelder L, Suk HJ, Anan R, O'donoghue A, et al. Mutations in the genes for cardiac troponin T and α-tropomyosin in hypertrophic cardiomyopathy. New Engl J Med. 1995;332(16):1058–1065.
105. Teekakirikul P, Zhu W, Huang HC, Fung E. Hypertrophic cardiomyopathy: an overview of genetics and management. Biomolecules. 2019;9(12):878.
106. Mazzarotto F, Olivotto I, Boschi B, Girolami F, Poggesi C, Barton PJ, et al. Contemporary insights into the genetics of hypertrophic cardiomyopathy: toward a new era in clinical testing? J Am Heart Assoc. 2020;9(8):e015473.
107. Hershberger RE, Lindenfeld J, Mestroni L, Seidman CE, Taylor MR, Towbin JA. Genetic evaluation of cardiomyopathy—a Heart Failure Society of America practice guideline. J Cardiac Fail. 2009;15(2):83–97.
108. P Landstrom A, J Ackerman M. Beyond the cardiac myofilament: hypertrophic cardiomyopathy-associated mutations in genes that encode calcium-handling proteins. Curr Mol Med. 2012;12(5):507–518.
109. Maron BJ, Maron MS, Semsarian C. Genetics of hypertrophic cardiomyopathy after 20 years: clinical perspectives. J Am Coll Cardiol. 2012;60(8):705–15.
110. Teekakirikul P, Kelly MA, Rehm HL, Lakdawala NK, Funke BH. Inherited cardiomy-opathies: molecular genetics and clinical genetic testing in the postgenomic era. J Mol Diagn. 2013;15(2):158–70.
111. Ho CY. Genetics and clinical destiny: improving care in hypertrophic cardiomyopathy. Circulation. 2010;122(23):2430–40.
112. Sachdev B, Takenaka T, Teraguchi H, Tei C, Lee P, McKenna W, et al. Prevalence of Anderson-Fabry disease in male patients with late onset hypertrophic cardiomyopathy. Circulation. 2002;105(12):1407–11.
113. Arad M, Maron BJ, Gorham JM, Johnson WH Jr, Saul JP, Perez-Atayde AR, et al. Glycogen storage diseases presenting as hypertrophic cardiomyopathy. N Engl J Med. 2005;352(4):362–72.
114. Petretta M, Pirozzi F, Sasso L, Paglia A, Bonaduce D. Review and metaanalysis of the frequency of familial dilated cardiomyopathy. Am J Cardiol. 2011;108(8):1171–6.

115. Hershberger RE, Siegfried JD. Update 2011: clinical and genetic issues in familial dilated cardiomyopathy. J Am Coll Cardiol. 2011;57(16):1641–9.

116. Millat G, Bouvagnet P, Chevalier P, Sebbag L, Dulac A, Dauphin C, et al. Clinical and mutational spectrum in a cohort of 105 unrelated patients with dilated cardiomyopathy. Eur J Med Genet. 2011;54(6):e570–5.

117. van Spaendonck-Zwarts KY, van Rijsingen IA, van den Berg MP, Lekanne Deprez RH, Post JG, van Mil AM, et al. Genetic analysis in 418 index patients with idiopathic dilated cardiomyopathy: overview of 10 years' experience. Eur J Heart Fail. 2013;15(6):628–36.

118. Fatkin D, MacRae C, Sasaki T, Wolff MR, Porcu M, Frenneaux M, et al. Missense mutations in the rod domain of the lamin A/C gene as causes of dilated cardiomyopathy and conduction-system disease. N Engl J Med. 1999;341(23):1715–24.

119. Fatkin D, Otway R, Richmond Z. Genetics of dilated cardiomyopathy. Heart Fail Clin. 2010;6(2):129–40.

120. Tayal U, Prasad S, Cook SA. Genetics and genomics of dilated cardiomyopathy and systolic heart failure. Genome Med. 2017;9(1):1–14.

121. Herman DS, Lam L, Taylor MR, Wang L, Teekakirikul P, Christodoulou D, et al. Truncations of titin causing dilated cardiomyopathy. N Engl J Med. 2012;366(7):619–28.

122. Elliott P. O, Mahony C, Syrris P, Evans A, Rivera Sorensen C, Sheppard MN, Carr-White G, Pantazis A, McKenna WJ: Prevalence of desmosomal protein gene mutations in patients with dilated cardiomyopathy. Circ Cardiovasc Genet. 2010;3:314e22.

123. Rexiati M, Sun M, Guo W. Muscle-specific mis-splicing and heart disease exemplified by RBM20. Genes. 2018;9(1):18.

124. van den Hoogenhof MM, Beqqali A, Amin AS, van der Made I, Aufiero S, Khan MA, et al. RBM20 mutations induce an arrhythmogenic dilated cardiomyopathy related to disturbed calcium handling. Circulation. 2018;138(13):1330–42.

125. Muchtar E, Blauwet LA, Gertz MA. Restrictive cardiomyopathy: genetics, pathogenesis, clinical manifestations, diagnosis, and therapy. Circ Res. 2017;121(7):819–37.

126. Parvatiyar MS, Pinto JR, Dweck D, Potter JD. Cardiac troponin mutations and restrictive cardiomyopathy. J Biomed Biotechnol. 2010;2010.

127. Fatkin D, Seidman CE, Seidman JG. Genetics and disease of ventricular muscle. Cold Spring Harbor Perspect Med. 2014;4(1):a021063.

128. Caleshu C, Sakhuja R, Nussbaum RL, Schiller NB, Ursell PC, Eng C, et al. Furthering the link between the sarcomere and primary cardiomyopathies: restrictive cardiomyopathy associated with multiple mutations in genes previously associated with hypertrophic or dilated cardiomyopathy. Am J Med Genet A. 2011;155(9):2229–35.

129. Pinto JR, Yang SW, Hitz M-P, Parvatiyar MS, Jones MA, Liang J, et al. Fetal cardiac troponin isoforms rescue the increased Ca^{2+} sensitivity produced by a novel double deletion in cardiac troponin T linked to restrictive cardiomyopathy: a clinical, genetic, and functional approach. J Biol Chem. 2011;286(23):20901–12.

130. Parvatiyar MS, Pinto JR, Liang J, Potter JD. Predicting cardiomyopathic phenotypes by altering Ca^{2+} affinity of cardiac troponin C. J Biol Chem. 2010;285(36):27785–97.

131. Willott RH, Gomes AV, Chang AN, Parvatiyar MS, Pinto JR, Potter JD. Mutations in Troponin that cause HCM, DCM AND RCM: what can we learn about thin filament function? J Mol Cell Cardiol. 2010;48(5):882–92.

132. McKenna WJ, Thiene G, Nava A, Fontaliran F, Blomstrom-Lundqvist C, Fontaine G, et al. Diagnosis of arrhythmogenic right ventricular dysplasia/cardiomyopathy. Task Force of the Working Group Myocardial and Pericardial Disease of the European Society of Cardiology and of the Scientific Council on Cardiomyopathies of the International Society and Federation of Cardiology. B Heart J. 1994;71(3):215.

133. Marcus FI, McKenna WJ, Sherrill D, Basso C, Bauce B, Bluemke DA, et al. Diagnosis of arrhythmogenic right ventricular cardiomyopathy/dysplasia: proposed modification of the task force criteria. Circulation. 2010;121(13):1533–41.

134. Gacita AM, McNally EM. Genetic spectrum of arrhythmogenic cardiomyopathy. Am Heart Assoc; 2019.

135. Van Tintelen JP, Entius MM, Bhuiyan Z, Jongbloed R, Wiesfeld AC, Van der Smagt J, et al. AB47-5: Plakophilin-2 mutations are the major determinant of familial arrhythmogenic right ventricular cardiomyopathy in the Netherlands. Heart Rhythm. 2006;3(5):S98.
136. Den Haan AD, Tan BY, Zikusoka MN, Lladó LI, Jain R, Daly A, et al. Comprehensive desmosome mutation analysis in North Americans with arrhythmogenic right ventricular dysplasia/cardiomyopathy. Circ Cardiovasc Genet. 2009;2(5):428–435.
137. Oechslin E, Jenni R. Left ventricular non-compaction revisited: a distinct phenotype with genetic heterogeneity? Eur Heart J. 2011;32(12):1446–56.
138. Hoedemaekers YM, Caliskan K, Majoor-Krakauer D, van de Laar I, Michels M, Witsenburg M, et al. Cardiac β-myosin heavy chain defects in two families with non-compaction cardiomy-opathy: linking non-compaction to hypertrophic, restrictive, and dilated cardiomyopathies. Eur Heart J. 2007;28(22):2732–7.
139. Klaassen S, Probst S, Oechslin E, Gerull B, Krings G, Schuler P, et al. Mutations in sarcomere protein genes in left ventricular noncompaction. Circulation. 2008;117(22):2893.
140. Pantazis AA, Elliott PM. Left ventricular noncompaction. Curr Opin Cardiol. 2009;24(3):209–13.
141. Probst S, Oechslin E, Schuler P, Greutmann M, Boyé P, Knirsch W, et al. Sarcomere gene mutations in isolated left ventricular noncompaction cardiomyopathy do not predict clinical phenotype. Circ Cardiovasc Genet. 2011;4(4):367–374.
142. Arndt A-K, Schafer S, Drenckhahn J-D, Sabeh MK, Plovie ER, Caliebe A, et al. Fine mapping of the 1p36 deletion syndrome identifies mutation of PRDM16 as a cause of cardiomyopathy. Am J Hum Genet. 2013;93(1):67–77.
143. Cerrone M, Napolitano C, Priori SG. Genetics of ion-channel disorders. Curr Opin Cardiol. 2012;27(3):242–52.
144. Webster G, Berul CI. An update on channelopathies: from mechanisms to management. Circulation. 2013;127(1):126–140.
145. Tiso N, Stephan DA, Nava A, Bagattin A, Devaney JM, Stanchi F, et al. Identification of mutations in the cardiac ryanodine receptor gene in families affected with arrhythmogenic right ventricular cardiomyopathy type 2 (ARVD2). Hum Mol Genet. 2001;10(3):189–94.
146. Hershberger RE, Parks SB, Kushner JD, Li D, Ludwigsen S, Jakobs P, et al. Coding sequence mutations identified in MYH7, TNNT2, SCN5A, CSRP3, LBD3, and TCAP from 313 patients with familial or idiopathic dilated cardiomyopathy. Clin Transl Sci. 2008;1(1):21–6.
147. Shan L, Makita N, Xing Y, Watanabe S, Futatani T, Ye F, et al. SCN5A variants in Japanese patients with left ventricular noncompaction and arrhythmia. Mol Genet Metab. 2008;93(4):468–74.
148. McNair WP, Sinagra G, Taylor MR, Di Lenarda A, Ferguson DA, Salcedo EE, et al. SCN5A mutations associate with arrhythmic dilated cardiomyopathy and commonly localize to the voltage-sensing mechanism. J Am Coll Cardiol. 2011;57(21):2160–8.
149. Mann SA, Castro ML, Ohanian M, Guo G, Zodgekar P, Sheu A, et al. R222Q SCN5A mutation is associated with reversible ventricular ectopy and dilated cardiomyopathy. J Am Coll Cardiol. 2012;60(16):1566–73.
150. Geisterfer-Lowrance AA, Kass S, Tanigawa G, Vosberg H-P, McKenna W, Seidman CE, et al. A molecular basis for familial hypertrophic cardiomyopathy: a β cardiac myosin heavy chain gene missense mutation. Cell. 1990;62(5):999–1006.
151. Kamisago M, Sharma SD, DePalma SR, Solomon S, Sharma P, McDonough B, et al. Mutations in sarcomere protein genes as a cause of dilated cardiomyopathy. N Engl J Med. 2000;343(23):1688–96.
152. Vermeer AM, Van Engelen K, Postma AV, Baars MJ, Christiaans I, De Haij S, et al., editors. Ebstein anomaly associated with left ventricular noncompaction: an autosomal dominant condition that can be caused by mutations in MYH7. Am J Med Genet Part C Semin Med Genet; 2013. Wiley Online Library.
153. Karam S, Raboisson MJ, Ducreux C, Chalabreysse L, Millat G, Bozio A, et al. A de novo mutation of the beta cardiac myosin heavy chain gene in an infantile restrictive cardiomyopathy. Congenit Heart Dis. 2008;3(2):138–43.

154. Schmitt JP, Debold EP, Ahmad F, Armstrong A, Frederico A, Conner DA, et al. Cardiac myosin missense mutations cause dilated cardiomyopathy in mouse models and depress molecular motor function. Proc Natl Acad Sci. 2006;103(39):14525–30.

155. Debold EP, Schmitt JP, Patlak J, Beck S, Moore J, Seidman JG, et al. Hypertrophic and dilated cardiomyopathy mutations differentially affect the molecular force generation of mouse α-cardiac myosin in the laser trap assay. Am J Physiol Heart Circ Physiol. 2007;293(1):H284–91.

156. Chuan P, Sivaramakrishnan S, Ashley EA, Spudich JA. Cell-intrinsic functional effects of the α-cardiac myosin Arg-403-Gln mutation in familial hypertrophic cardiomyopathy. Biophys J . 2012;102(12):2782–90.

157. Schmitt JP, Kamisago M, Asahi M, Li GH, Ahmad F, Mende U, et al. Dilated cardiomyopathy and heart failure caused by a mutation in phospholamban. Science. 2003;299(5611):1410–3.

158. McNally EM, Golbus JR, Puckelwartz MJ. Genetic mutations and mechanisms in dilated cardiomyopathy. J Clin Investig. 2013;123(1):19–26.

159. Leach IM, van der Harst P, de Boer RA. Pharmacoepigenetics in heart failure. Curr Heart Fail Rep. 2010;7(2):83–90.

160. Movassagh M, Choy M-K, Knowles DA, Cordeddu L, Haider S, Down T, et al. Distinct epigenomic features in end-stage failing human hearts. Circulation. 2011;124(22):2411–22.

161. Chang C-P, Bruneau BG. Epigenetics and cardiovascular development. Annu Rev Physiol. 2012;74:41–68.

162. Schonrock N, Harvey RP, Mattick JS. Long noncoding RNAs in cardiac development and pathophysiology. Circ Res. 2012;111(10):1349–62.

163. Papait R, Greco C, Kunderfranco P, Latronico MV, Condorelli G. Epigenetics: a new mechanism of regulation of heart failure? Basic Res Cardiol. 2013;108(4):361.

164. Udali S, Guarini P, Moruzzi S, Choi S-W, Friso S. Cardiovascular epigenetics: from DNA methylation to microRNAs. Mol Aspects Med. 2013;34(4):883–901.

165. Moretti A, Bellin M, Welling A, Jung CB, Lam JT, Bott-Flügel L, et al. Patient-specific induced pluripotent stem-cell models for long-QT syndrome. N Engl J Med. 2010;363(15):1397–409.

166. Itzhaki I, Maizels L, Huber I, Zwi-Dantsis L, Caspi O, Winterstern A, et al. Modelling the long QT syndrome with induced pluripotent stem cells. Nature. 2011;471(7337):225–9.

167. Sun N, Yazawa M, Liu J, Han L, Sanchez-Freire V, Abilez OJ, et al. Patient-specific induced pluripotent stem cells as a model for familial dilated cardiomyopathy. Sci Trans Med. 2012;4(130):130ra47-ra47.

168. Kim C, Wong J, Wen J, Wang S, Wang C, Spiering S, et al. Studying arrhythmogenic right ventricular dysplasia with patient-specific iPSCs. Nature. 2013;494(7435):105–10.

169. Knollmann BC. Induced pluripotent stem cell–derived cardiomyocytes: Boutique science or valuable arrhythmia model? Circ Res. 2013;112(6):969–76.

170. Priori SG, Napolitano C, Di Pasquale E, Condorelli G. Induced pluripotent stem cell–derived cardiomyocytes in studies of inherited arrhythmias. J Clin Investig. 2013;123(1):84–91.

171. Takahashi K, Yamanaka S. Induction of pluripotent stem cells from mouse embryonic and adult fibroblast cultures by defined factors. Cell. 2006;126(4):663–676.

172. Lieschke GJ, Currie PD. Animal models of human disease: zebrafish swim into view. Nat Rev Genet. 2007;8(5):353–67.

173. Dahme T, Katus H, Rottbauer W. Fishing for the genetic basis of cardiovascular disease. Dis Model Mech. 2009; 2 (1–2):18–22. Epub 2009/01/10.

174. Santoriello C, Zon LI. Hooked! Modeling human disease in zebrafish. J Clin Investig. 2012;122(7):2337–43.

175. Verkerk AO, Remme CA. Zebrafish: a novel research tool for cardiac (patho) electrophysiology and ion channel disorders. Front Physiol. 2012;3:255.

176. Bendig G, Grimmler M, Huttner IG, Wessels G, Dahme T, Just S, et al. Integrin-linked kinase, a novel component of the cardiac mechanical stretch sensor, controls contractility in the zebrafish heart. Genes Dev. 2006;20(17):2361–72.

177. Sander JD, Cade L, Khayter C, Reyon D, Peterson RT, Joung JK, et al. Targeted gene disruption in somatic zebrafish cells using engineered TALENs. Nat Biotechnol. 2011;29(8):697–8.

178. Huttner IG, Trivedi G, Jacoby A, Mann SA, Vandenberg JI, Fatkin D. A transgenic zebrafish model of a human cardiac sodium channel mutation exhibits bradycardia, conduction-system abnormalities and early death. J Mol Cell Cardiol. 2013;61:123–32.
179. Norton N, Robertson PD, Rieder MJ, Züchner S, Rampersaud E, Martin E, et al. Evaluating pathogenicity of rare variants from dilated cardiomyopathy in the exome era. Circ Cardiovasc Genet. 2012;5(2):167–174.
180. Pan S, Caleshu CA, Dunn KE, Foti MJ, Moran MK, Soyinka O, et al. Cardiac structural and sarcomere genes associated with cardiomyopathy exhibit marked intolerance of genetic variation. Circ Cardiovasc Genet. 2012;5(6):602–610.
181. Golbus JR, Puckelwartz MJ, Fahrenbach JP, Dellefave-Castillo LM, Wolfgeher D, McNally EM. Population-based variation in cardiomyopathy genes. Circ Cardiovasc Genet. 2012;5(4):391–399.
182. Milting H, Klauke B. Molecular genetics of arrhythmogenic right ventricular dysplasia/cardiomyopathy. Nat Clin Pract Cardiovasc Med. 2008;5(10):E1; author reply E2.
183. Christensen AH, Benn M, Tybjærg-Hansen A, Haunso S, Svendsen JH. Missense variants in plakophilin-2 in arrhythmogenic right ventricular cardiomyopathy patients–disease-causing or innocent bystanders? Cardiology. 2010;115(2):148–54.
184. Fatkin D, Seidman CE, Seidman JG. Genetics and disease of ventricular muscle. Cold Spring Harb Perspect Med. 2014; 4(1):a021063.

Chapter 11
Response and Effects of Cardiomyocyte Progenitors in the Infarcted Heart

Abstract Several studies indicate that endogenous cardiac progenitor cells (CPCs) play some role in homeostasis and response to ischemic injury to the myocardium. There are evidences for modulation of proliferation of CPCs by paracrine factors secreted by injured cardiomyocytes. Increased levels of signalling growth factors and cytokines released in the infarcted heart have been proposed to trigger the proliferation, differentiation and migration of CPCs to the site of injury. The beneficial effect of transplanted CPCs on cardiomyocytes in the ischemic heart is considered to be because of auto or paracrine modulation rather than differentiation and functional integration of the transplanted cells. Endogenous CPCs reside in specific niches and the changes in the microenvironment in these niches post ischemic injury can affect behaviour of the CPCs. The key components of the CPC niche which can impact CPC behaviour are (i) the supporting cells, (ii) cyclic strain resulting from heart beats, (iii) extracellular matrix, and (iv) soluble factors such as cytokines and oxygen tension. Stem cell populations also participate in the healing processes in the infarcted heart. They through paracrine signalling can directly mediate the activation, recruitment, function, and switch in phenotype of immune cells involved in wound healing.

Keywords Cardiac progenitors · Myocardial infarction · Paracrine signalling · Cytokines · Growth factors · Cardiac microenvironment · Extracellular matrix

Introduction

Several studies have shown the presence of cardiac progenitor cells (CPCs) in the heart and signalled their potential for use in the regeneration of heart [1–4]. Strategies for regenerative therapies for myocardial infarction have indicated that endogenous CPCs play some role in homeostasis and response to physiological stress and ischemic injury. The ability of CPCs to generate new cardiomyocytes (CM) after injury is however not confirmed [5, 6]. Increased levels of signalling growth factors and cytokines released in the heart after ischemic injury have been proposed to trigger the proliferation, differentiation and migration of CPCs to the site of injury [7, 8]. There are evidences to indicate that proliferation of CPCs is modulated by paracrine factors secreted by injured CMs [9]. The beneficial effect of transplanted CPCs on

© The Author(s), under exclusive license to Springer Nature Switzerland AG 2021
C. C. Kartha, *Cardiomyocytes in Health and Disease*,
https://doi.org/10.1007/978-3-030-85536-9_11

CMs in the ischemic heart is considered to be because of auto or paracrine modulation rather than differentiation and functional integration of CPCs [10–12].

Mechanisms of Activation of CPCs

Whole proteome analysis of CPCs has revealed that upon injury, and in the presence of human-induced pluripotent stem cell-derived cardiomyocytes (hiPSC-CMs), proteins involved in pathways and functions associated with cell proliferation, paracrine signalling, stress response, and regeneration processes are enriched in human CPCs (hCPCs) [13]. The enriched pathways are linked to CXCL6 secretion, recovery of mechanisms for cell cycle repair, and activation of angiogenesis and vasculogenesis. These findings support the view that the regenerative capacity of hCPCs is focussed on paracrine action, and that CXCL6 is one of the key players. Proteins of VEGF, IL-2, IL-3, IL-15, and GM-CSF signalling pathways were also found to be more represented in proteome of hCPCs in ischemic conditions. All of them have paracrine roles in ischemia- reperfusion injury (Fig. 11.1) [14–16].

MJ Sebastião and colleagues have used hCPCs and hiPSC-CMs to clarify and define the response of hCPCs to ischemia–reperfusion (I/R) injury, and its effect on CMs [13]. The in vitro co culture I/R injury model they have used, has the important features of acute myocardial infarction. They discovered that upon I/R injury, there is increased secretion of CXCL6, a known angiogenic cytokine which has a regenerative role in both mesenteric and myocardial infarction [17, 18]. CXCL6 has been identified in the secretome of hCPCs [17]. CXCL6 is upregulated in ischemic conditions via HIF-1α signalling [19]. HIF-1α transcription factor and SDF-1 chemokine

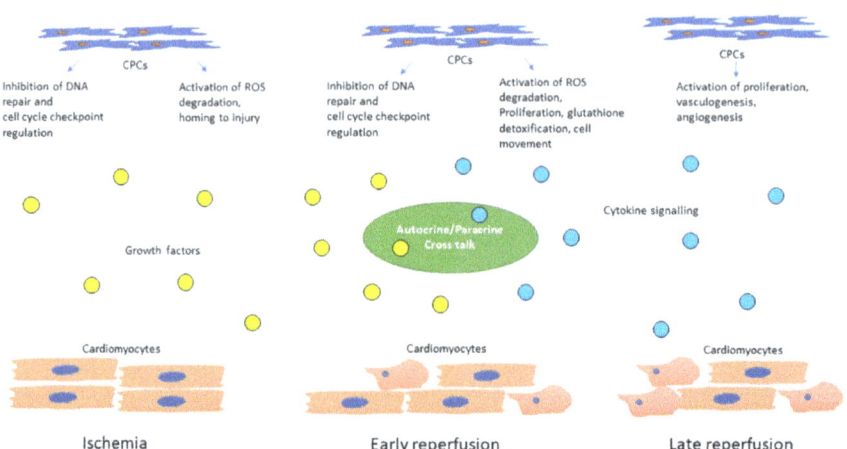

Fig. 11.1 Mechanisms proposed for human cardiac progenitor cell response to ischemia and reperfusion. CPCs—Cardiac progenitor cells. ROS—Reactive oxygen species. DNA—Deoxy ribonucleic acid

in response to ischemia are known to activate migration of cardiac stem cells to the site of injury [20]. An enrichment in proteins related to HIF-1α signalling, IGF-signalling and CCR3 signalling are important for cell homing to sites of inflammation and injury [21]. CXCR4 signalling, known to be involved in cell motility and chemotactic response has also been observed in hCPCs subjected to ischemia [22].

After reperfusion, there is also enrichment of pathways related to cell proliferation via epidermal growth factor (EGF) signalling [13]. EGF receptor is present in hCPCs [23]. EGF is one of the key signals for activation of proliferation of CPCs subjected to ischemia and reperfusion. EGF is upregulated in mice CPCs upon injury [24]. EGF has been found to have a positive effect on proliferation and migration of cardiosphere-derived hCPCs as well [25].

Other signalling pathways such as FLT3 signalling, known to be associated with cell proliferation, cytokinesis, and IGF-1 pathway were also found to be enriched after reperfusion.

A downregulation of cell cycle regulation pathways, DNA repair mechanisms, and cell repair mechanisms is also seen in hCPCs during the ischemic phase of injury (Fig. 11.1). There are previous reports that these processes are involved in activation of many types of quiescent adult stem cells [26]. Proteome analysis also suggests activation of endoplasmic reticulum stress pathways such as ERS and UPR. These pathways are related to cellular adaptation to glucose deprivation and hypoxic stimuli [27]. Proteins associated with glycolysis and oxygen consumption are also enriched, suggesting an adaptive response of hCPCs to hypoxia.

The Role of Microenvironment on CPC Response

CPCs reside in specific niches and the changes in the microenvironment in these niches post ischemic injury can affect CPC behaviour. Impaired cell division and cellular senescence associated with reduction in telomerase activity, as well as increased apoptosis have been observed in CPCs in chronic infarcts [28]. The key components of the CPC niche which can impact CPC behaviour are (i) the supporting cells, (ii) cyclic strain resulting from heart beats, (iii) extracellular matrix, and (iv) soluble factors such as cytokines and oxygen tension (Fig. 11.2A). Insight on the impact of the microenvironment on CPCs could aid in exploiting the regenerative potential of CPCs (Fig. 11.2B).

Supporting Cells

Most of our knowledge on the interactions between CPCs and the supporting cells in their neighbourhood have been obtained from in vitro studies.

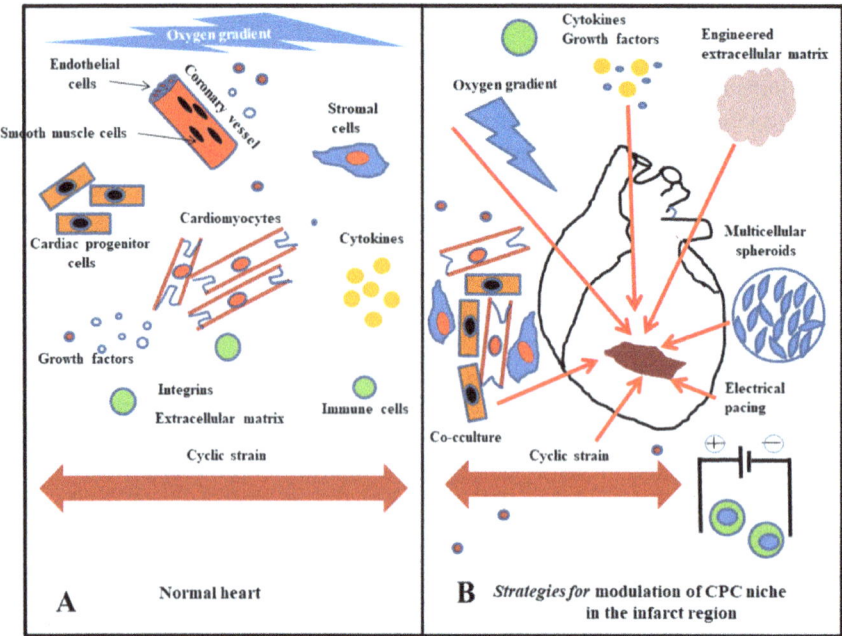

Fig. 11.2 Components of the microenvironment of cardiac progenitor cells (**A**) and strategies to modulate the niche (**B**)

Cardiomyocytes

Connexins and cadherins are present in contacts between CPCs and CMs as well as between CPCs and fibroblasts [29–31]. CPCs when co-cultured with CMs differentiate with expression of cardiomyocyte-specific proteins, develop well-organized sarcomeres, beat and also proliferate [2, 32–35]. These effects are considered to be regulated by TGF-β [36, 37] and indirectly via the Wnt/beta-catenin signalling system [38]. Coupling of CPCs with CMs is important to modulate the cardiac fate [39]. CPCs can express growth factors and cytokines [40]. Thus, they can control proliferation of cardiomyocytes and their survival [41].

Endothelial Cells

Direct interaction of CPCs with vascular endothelial cells and smooth muscle cells have not been found [29]. Indirect interactions between endothelial cells and CPCs, via the production of VEGF, may stimulate CPC migration and regulate differentiation of CPCs into endothelial or smooth muscle cells [42–44].

Immune Cells

Proliferation and differentiation of CPCs can be induced by macrophage derived growth factors such as IGF-1, VEGF, and TGF-β [45, 46]. Conversely, CPCs can polarize macrophages away from their proinflammatory phenotype, though not toward the anti-inflammatory phenotype [47]. Thus, interaction between macrophages and CPCs could have a cardioprotective effect.

Boukouaci et al. have reported that CPCs are protected from cytotoxicity from natural killer cells in an inflammatory milieu [48]. CPCs can downregulate the toxicity of natural killer cells and reverse a pro-inflammatory cytokine secretion. Retention of CPCs is improved by interaction with natural killer cells [48].

CPCs share specific features with mast cells, which express similar markers as CPCs [49]. Cell contact between these two cells has however not yet been reported. Paracrine effects are possible since mast cells produce several cytokines, growth factors, and angiogenic factors [50].

Stromal Cells

Fibroblasts are the supporting matrix of the CPC niche and may determine the differentiation potential of CPCs [38]. Fibroblasts are connected to CPCs via gap and adherens junctions. Fibroblast-conditioned medium is seen to induce differentiation via the Wnt signalling pathway [51].

Epicardium-derived cells (EPDCs) are activated following injury and promote both the migration and proliferation of CPCs [52–54]. Induction of metalloproteinases and their inhibitors has been observed during coculture of CPCs with EPDCs [54]. Paracrine stimulation from a reciprocal interaction between CPCs and EPDCs can lead to a synergistic beneficial effect [54].

Telocytes, formally known as interstitial Cajal-like cells are another type of stromal cells in the subepicardial region [55]. These cells which are in close proximity with CPCs form synapses and adherens junctions with CPCs [56, 57]. These adherens junctions could be important for division and migration of CPCs [58]. Telocytes produce growth factors and other signals, such as microRNAs [58, 59]. Thus, they may influence the activation, proliferation, and differentiation of CPCs [56, 60, 61].

Effect of Notch Signalling on CPCs

Notch signalling, which regulates survival, proliferation, and differentiation as well as development and homeostasis in tissues is an essential component of the microenvironment in the heart. Active Notch signalling is required for the differentiation of CPCs [62]. Several studies have reported that Notch, which is inactivated during maturation and after birth is reactivated in adult CMs after ischemic injury to the heart [63, 64]. Notch signalling is necessary for proliferation and expansion of the CPC pool as well as for cardiomyogenic differentiation of CPCs [62, 63, 65].

Notch has also a significant role in cardiomyocyte survival and heart repair after injury [66–69]. In the adult mouse, about 60% of c-kit + CPCs expresses the Notch1 receptor, and signalling with neighbouring cells is mediated by Boni et al. [63]. The activation of Notch1 by Jagged1 in mouse c-kit + CPCs causes the translocation of N1ICD and enhances its colocalization with Notch target gene Nkx2.5 [63, 70]. Overexpression of Notch1 in mouse c-kit + CPCs has been found to increase resistance to oxidative stress [71].

Extracellular Matrix (ECM)

The effects of the ECM environment on survival and function of CPCs have been a focus of several studies.

CPC-ECM Interactions

In vitro studies growing progenitor cells on different substrates indicate that the composition and structure of ECM has an effect on cardiomyogenic gene expression, survival, and proliferation of cardiac progenitor cells [72–75]. When CPCs were cultured on cardiac ECM, there was increased expression of cardiac genes for GATA-binding protein-4 (GATA-4), Nkx2.5, α-myosin heavy chain, and troponin C and T [73]. Fibroblast specific genes decreased and endothelial or smooth muscle cell-specific genes remained constant. The ECM components which produce the beneficial response of CPCs towards cardiac-derived ECM need to be discovered.

The advantage of fibronectin for the initial expansion of human CPCs during development and after a myocardial infarction has been demonstrated [76]. The effect may be influenced by cyclic strain and/or stiffness. ECM stiffness influences the genetic expression of titin, troponin T and f-actin in CPCs [77–80].

Integrins

CPCs sense their environment and appropriately respond via focal adhesions (FAs). FAs are transmembrane protein complexes that directly link ECM components or other cells to intracellular actin junctions, intermediate filaments, and sarcomeres [81]. Integrins are important components of the transmembrane protein complex.

The expressions of different types of integrin change in response to disease conditions [81]. Expression of FA in CPCs and interactions with specific ECM could be
important for differentiation of CPCs towards specific lineages [82].

Cyclic Strain

The effect of cyclic strain on CPCs has also been investigated [82, 83].
Mechanosensing structures such as FAs and actin stress fibres develop during
the early phase of cardiac differentiation [82]. van Marion et al. have shown that
CPCs readily turn mechanosensitive and have an increased ability for differentiation
towards the cardiomyocyte phenotype on 3D substrates [84].

Soluble Factors and Oxygen Tension

Hypoxia influence many signalling pathways, such as Notch and Oct 4, that determine
self-renewal and multipotency. They thus regulate the behaviour of several stem
and progenitor cells [80, 85, 86]. Cells express hypoxia-inducible factors (HIFs) in
conditions of low oxygen tension [87, 88]. HIF-1 controls the transcription of the
chemokine stromal cell-derived factor 1 (SDF-1) and its receptors CXCR4 which
influence the mobilization of progenitor cells [20, 89–92]. Under severe hypoxia,
expression of both CXCR4 receptor and SDF-1 are increased in murine CPCs [92].
SDF-1 induces CPC migration in a time- and dose-dependent manner [93, 94]. van
Oorschot et al. found that human Sca1 + CPCs have increased proliferation and
motility when cultured under low oxygen tension [95]. Human Sca1 + CPCs have
an increased motility directly proportional to the reduction of oxygen tension [96].

Immunomodulation by CSCs in the Infarcted Heart

Immune cells are activated following injury to the myocardium. The modulatory
role of immune cell subsets of both the innate and adaptive immune systems, in
wound healing in the heart are well-characterized [97]. Transplantation studies have
indicated that CPCs modulate the immune response and affect cardiac repair [98].

Paracrine factors secreted from stem cells can directly mediate the activation,
recruitment, function, and switch in phenotype of immune cell populations that
participate in the healing process after myocardial infarction. Stem cells secrete
cytokines, chemokines, exosomes, and miRNAs enriched within exosomes, all of
which can modulate responses of nearby immune cells. There are evidences from
in vitro studies that stem cell mediated paracrine signalling can impact both the
phenotype and function of immune cells [99].

While it is unclear how resident stem cells influence the inflammatory microenvironment in the ischemic heart, there are some insights on how paracrine signalling from engrafted stem cells regulate immune cells and promote repair of the ischemic heart [100–103]. Injected stem cells can both directly and indirectly modulate chemotaxis, activation, phenotype, and function of distinct immune cell subsets [104–108]. Stem cells derived factors have been found to directly interact with neutrophils, monocytes, T-cells, and B-cells [109–113].

The immunomodulatory effects of stem cells are seen to be dependent on the cell type [100, 101, 114–119]. Cardiac derived cells (CDCs) regulate wound healing processes via paracrine and exosome dependent signalling mechanisms [120, 121]. CDC derived exosomes mediate M1 to M2 polarization of monocytes and decrease expression of pro-inflammatory genes [122–126]. Human cardiac progenitor cells (hCPCs) seem to repress Th1/2$^+$ cell expression and promote proliferation and immunosuppressive function of Treg cells via PD-1 dependent mechanisms [127].

Conclusion

After acute myocardial infarction, CPCs in the adult heart increase in number and differentiate into mature cells. The ability of CPCs to generate new CMs after injury is however not confirmed. The mechanisms by which the CPCs exert their beneficial effects are also not completely understood. There are evidences to indicate that proliferation of CPCs is modulated by paracrine factors secreted by injured CMs. Increased levels of signalling growth factors and cytokines released in the infarcted heart are considered to trigger the proliferation, differentiation and migration of CPCs to the site of injury. Stem cell populations via paracrine signalling can in turn directly mediate the activation, recruitment, function, and switch in phenotype of immune cells that participate in the healing process after myocardial infarction.

References

1. Oh H, Bradfute SB, Gallardo TD, Nakamura T, Gaussin V, Mishina Y, et al. Cardiac progenitor cells from adult myocardium: homing, differentiation, and fusion after infarction. Proc Natl Acad Sci USA. 2003;100(21):12313–8.
2. Pfister O, Mouquet F, Jain M, Summer R, Helmes M, Fine A, et al. CD31$^-$ but not CD31$^+$ cardiac side population cells exhibit functional cardiomyogenic differentiation. Circ Res. 2005;97(1):52–61.
3. Bax NA, van Marion MH, Shah B, Goumans M-J, Bouten CV, van der Schaft DW. Matrix production and remodeling capacity of cardiomyocyte progenitor cells during in vitro differentiation. J Mol Cell Cardiol. 2012;53(4):497–508.
4. Le T, Chong J. Cardiac progenitor cells for heart repair. Cell Death Disc. 2016;2(1):1–4.
5. Van Berlo JH, Kanisicak O, Maillet M, Vagnozzi RJ, Karch J, Lin S-CJ, et al. C-kit$^+$ cells minimally contribute cardiomyocytes to the heart. Nature. 2014;509(7500):337–41.

6. Nadal-Ginard B, Ellison GM, Torella D. Absence of evidence is not evidence of absence: pitfalls of cre knock-ins in the c-Kit locus. Circul Res. 2014;115(4):415–8.
7. Torella D, Ellison GM, Karakikes I, Nadal-Ginard B. Growth-factor-mediated cardiac stem cell activation in myocardial regeneration. Nat Clin Pract Cardiovasc Med. 2007;4(1):S46–S51.
8. Li X, Ren Y, Sorokin V, Poh KK, Ho HH, Lee CN, et al. Quantitative profiling of the rat heart myoblast secretome reveals differential responses to hypoxia and re-oxygenation stress. J Proteomics. 2014;98:138–49.
9. Stastna M, Van Eyk JE. Investigating the secretome: lessons about the cells that comprise the heart. Circul: Cardiovasc Genet. 2012;5(1):o8–o18.
10. Madonna R, Van Laake LW, Davidson SM, Engel FB, Hausenloy DJ, Lecour S, et al. Position paper of the European society of cardiology working group cellular biology of the heart: cell-based therapies for myocardial repair and regeneration in ischemic heart disease and heart failure. Eur Heart J. 2016;37(23):1789–98.
11. Kawaguchi N, Smith AJ, Waring CD, Hasan MK, Miyamoto S, Matsuoka R, et al. c-kit pos GATA-4 high rat cardiac stem cells foster adult cardiomyocyte survival through IGF-1 paracrine signalling. PloS One. 2010;5(12):e14297.
12. Park C-Y, Choi S-C, Kim J-H, Choi J-H, Joo HJ, Hong SJ, et al. Cardiac stem cell secretome protects cardiomyocytes from hypoxic injury partly via monocyte chemotactic protein-1-dependent mechanism. Int J Mol Sci. 2016;17(6):800.
13. Sebastião MJ, Serra M, Pereira R, Palacios I, Gomes-Alves P, Alves PM. Human cardiac progenitor cell activation and regeneration mechanisms: exploring a novel myocardial ischemia/reperfusion in vitro model. Stem Cell Res Ther. 2019;10(1):1–16.
14. Saini HK, Xu Y-J, Zhang M, Liu PP, Kirshenbaum LA, Dhalla NS. Role of tumour necrosis factor-alpha and other cytokines in ischemia-reperfusion-induced injury in the heart. Exp Clin Cardiol. 2005;10(4):213.
15. Turillazzi E, Di Paolo M, Neri M, Riezzo I, Fineschi V. A theoretical timeline for myocardial infarction: immunohistochemical evaluation and western blot quantification for Interleukin-15 and Monocyte chemotactic protein-1 as very early markers. J Transl Med. 2014;12(1):1–10.
16. Sesti C, Hale SL, Lutzko C, Kloner RA. Granulocyte colony-stimulating factor and stem cell factor improve contractile reserve of the infarcted left ventricle XE "Ventricle" independent of restoring muscle mass. J Am Coll Cardiol. 2005;46(9):1662–9.
17. Torán JL, Aguilar S, López JA, Torroja C, Quintana JA, Santiago C, et al. CXCL6 is an important paracrine factor in the pro-angiogenic human cardiac progenitor-like cell secretome. Sci Rep. 2017;7(1):1–14.
18. Kim S-W, Lee D-W, Yu L-H, Zhang H-Z, Kim CE, Kim J-M, et al. Mesenchymal stem cells overexpressing GCP-2 improve heart function through enhanced angiogenic properties in a myocardial infarction model. Cardiovasc Res. 2012;95(4):495–506.
19. Tian H, Huang P, Zhao Z, Tang W, Xia J. HIF-1α plays a role in the chemotactic migration of hepatocarcinoma cells through the modulation of CXCL6 expression. Cell Physiol Biochem. 2014;34(5):1536–46.
20. Ceradini DJ, Kulkarni AR, Callaghan MJ, Tepper OM, Bastidas N, Kleinman ME, et al. Progenitor cell trafficking is regulated by hypoxic gradients through HIF-1 induction of SDF-1. Nat Med. 2004;10(8):858–64.
21. Bonaros N, Sondermejer H, Schuster M, Rauf R, Wang S, Seki T, et al. CCR3-and CXCR4-mediated interactions regulate migration of CD34+ human bone marrow progenitors to ischemic myocardium and subsequent tissue repair. J Thorac Cardiovasc Surg. 2008;136(4):1044–53.
22. Penn M, Pastore J, Miller T, Aras R. SDF-1 in myocardial repair. Gene Ther. 2012;19(6):583–7.
23. Gomes-Alves P, Serra M, Brito C, R.-Borlado L, López JA, Vázquez J, et al. Exploring analytical proteomics platforms toward the definition of human cardiac stem cells receptome. Proteomics. 2015;15(7):1332–7.

24. Valiente-Alandi I, Albo-Castellanos C, Herrero D, Sanchez I, Bernad A. Bmi1$^+$ cardiac progenitor cells contribute to myocardial repair following acute injury. Stem Cell Res Therapy. 2016;7(1):1–11.

25. Aghila Rani KG, Kartha CC. Effects of epidermal growth factor on proliferation and migration of cardiosphere-derived cells expanded from adult human heart. Growth Fact. 2010;28(3):157–65.

26. Cheung TH, Rando TA. Molecular regulation of stem cell quiescence. Nat Rev Mol Cell Biol. 2013;14(6):329–40.

27. Groenendyk J, Agellon LB, Michalak M. Coping with endoplasmic reticulum stress in the cardiovascular system. Ann Rev Physiol. 2013;75:49–67.

28. Urbanek K, Torella D, Sheikh F, De Angelis A, Nurzynska D, Silvestri F, et al. Myocardial regeneration by activation of multipotent cardiac stem cells in ischemic heart failure. Proc Natl Acad Sci USA. 2005;102(24):8692–7.

29. Urbanek K, Cesselli D, Rota M, Nascimbene A, De Angelis A, Hosoda T, et al. Stem cell niches in the adult mouse heart. Proc Natl Acad Sci USA. 2006;103(24):9226–31.

30. Bearzi C, Rota M, Hosoda T, Tillmanns J, Nascimbene A, De Angelis A, et al. Human cardiac stem cells. Proc Natl Acad Sci USA. 2007;104(35):14068–73.

31. Bearzi C, Leri A, Monaco FL, Rota M, Gonzalez A, Hosoda T, et al. Identification of a coronary vascular progenitor cell in the human heart. Proc Natl Acad Sci USA. 2009;106(37):15885–90.

32. Yamahara K, Fukushima S, Coppen SR, Felkin LE, Varela-Carver A, Barton PJ, et al. Heterogeneic nature of adult cardiac side population cells. Biochem Biophys Res Commun. 2008;371(4):615–20.

33. Laugwitz K-L, Moretti A, Lam J, Gruber P, Chen Y, Woodard S, et al. Postnatal isl1$^+$ cardioblasts enter fully differentiated cardiomyocyte lineages. Nature. 2005;433(7026):647–53.

34. Messina E, De Angelis L, Frati G, Morrone S, Chimenti S, Fiordaliso F, et al. Isolation and expansion of adult cardiac stem cells from human and murine heart. Circ Res. 2004;95(9):911–21.

35. Kubo H, Jaleel N, Kumarapeli A, Berretta RM, Bratinov G, Shan X, et al. Increased cardiac myocyte progenitors in failing human hearts. Circulation. 2008;118(6):649–57.

36. Flanders KC, Holder MG, Winokur TS. Autoinduction of mRNA and protein expression for transforming growth factor-beta S in cultured cardiac cells. J Mol Cell Cardiol. 1995;27(2):805–12.

37. Behfar A, Zingman LV, Hodgson DM, Rauzier JM, Kane GC, Terzic A, et al. Stem cell differentiation requires a paracrine pathway in the heart. FASEB J: Off Publ Federat Am Soc Exp Biol. 2002;16(12):1558–66.

38. Deb A. Cell–cell interaction in the heart via Wnt/β-catenin pathway after cardiac injury. Cardiovasc Res. 2014;102(2):214–23.

39. Hosoda T, Zheng H, Cabral-da-Silva M, Sanada F, Ide-Iwata N, Ogórek B, et al. Human cardiac stem cell differentiation is regulated by a mircrine mechanism. Circulation. 2011;123(12):1287–96.

40. Gonzalez A, Rota M, Nurzynska D, Misao Y, Tillmanns J, Ojaimi C, et al. Activation of cardiac progenitor cells reverses the failing heart senescent phenotype and prolongs lifespan. Circ Res. 2008;102(5):597–606.

41. Samarel AM. Costameres, focal adhesions, and cardiomyocyte mechanotransduction. Am J Physiol Heart Circ Physiol. 2005;289(6):H2291-301.

42. Urbich C, Aicher A, Heeschen C, Dernbach E, Hofmann WK, Zeiher AM, et al. Soluble factors released by endothelial progenitor cells promote migration of endothelial cells and cardiac resident progenitor cells. J Mol Cell Cardiol. 2005;39(5):733–42.

43. Lushaj EB, Lozonschi L, Barnes M, Anstadt E, Kohmoto T. Mitochondrial DNA deletion mutations in adult mouse cardiac side population cells. Mutation Res/Fund Molecular Mech Mutagenesis. 2012;734(1–2):62–8.

44. Yoon J, Choi S-C, Park C-Y, Shim W-J, Lim D-S. Cardiac side population cells exhibit endothelial differentiation potential. Exp Mol Med. 2007;39(5):653–62.

45. Hsieh PC, Davis ME, Gannon J, MacGillivray C, Lee RT. Controlled delivery of PDGF-BB for myocardial protection using injectable self-assembling peptide nanofibers. J Clin Investig. 2006;116(1):237–48.

46. Vannella KM, Wynn TA. Mechanisms of organ injury and repair by macrophages. Ann Rev Physiol. 2017;79:593–617.

47. De Couto G, Liu W, Tseliou E, Sun B, Makkar N, Kanazawa H, et al. Macrophages mediate cardioprotective cellular postconditioning in acute myocardial infarction. J Clin Investig. 2015;125(8):3147–62.

48. Boukouaci W, Lauden L, Siewiera J, Dam N, Hocine H-R, Khaznadar Z, et al. Natural killer cell crosstalk with allogeneic human cardiac-derived stem/progenitor cells controls persistence. Cardiovasc Res. 2014;104(2):290–302.

49. Zhou Y, Pan P, Yao L, Su M, He P, Niu N, et al. CD117-positive cells of the heart: progenitor cells or mast cells? J Histochem Cytochem. 2010;58(4):309–16.

50. Takeda N, Manabe I. Cellular interplay between cardiomyocytes and nonmyocytes in cardiac remodeling. Int J Inflam. 2011;Article ID 535241. https://doi.org/10.4061/2011/535241.

51. Zhang X, Shen M-R, Xu Z-D, Hu Z, Chen C, Chi Y-L, et al. Cardiomyocyte differentiation induced in cardiac progenitor cells by cardiac fibroblast-conditioned medium. Exp Biol Med. 2014;239(5):628–37.

52. Lepilina A, Coon AN, Kikuchi K, Holdway JE, Roberts RW, Burns CG, et al. A dynamic epicardial injury response supports progenitor cell activity during zebrafish heart regeneration. Cell. 2006;127(3):607–19.

53. Chen TH-P, Chang T-C, Kang J-O, Choudhary B, Makita T, Tran CM, et al. Epicardial induction of fetal cardiomyocyte proliferation via a retinoic acid-inducible trophic factor. Dev Biol. 2002;250(1):198–207.

54. Winter EM, van Oorschot AA, Hogers B, van der Graaf LM, Doevendans PA, Poelmann RE, et al. A new direction for cardiac regeneration therapy: application of synergistically acting epicardium-derived cells and cardiomyocyte progenitor cells. Circul: Heart Failure. 2009;2(6):643–53.

55. Popescu L, Faussone-Pellegrini MS. TELOCYTES–a case of serendipity: the winding way from Interstitial Cells of Cajal (ICC), via Interstitial Cajal-Like Cells (ICLC) to TELOCYTES. J Cell Mol Med. 2010;14(4):729–40.

56. Gherghiceanu M, Popescu L. Cardiomyocyte precursors and telocytes in epicardial stem cell niche: electron microscope images. J Cell Mol Med. 2010;14(4):871–7.

57. Popescu LM, Fertig ET, Gherghiceanu M. Reaching out: junctions between cardiac telocytes and cardiac stem cells in culture. J Cell Mol Med. 2016;20(2):370–80.

58. Manole C, Cismaşiu V, Gherghiceanu M, Popescu L. Experimental acute myocardial infarction: telocytes involvement in neo-angiogenesis. J Cell Mol Med. 2011;15(11):2284–96.

59. Cismasiu V, Radu E, Popescu L. miR-193 expression differentiates telocytes from other stromal cells. J Cell Mol Med. 2011;15(5):1071–4.

60. Bei Y, Wang F, Yang C, Xiao J. Telocytes in regenerative medicine. J Cell Mol Med. 2015;19(7):1441–54.

61. Bani D, Formigli L, Gherghiceanu M, Faussone-Pellegrini MS. Telocytes as supporting cells for myocardial tissue organization in developing and adult heart. J Cell Mol Med. 2010;14(10):2531–8.

62. Kwon C, Qian L, Cheng P, Nigam V, Arnold J, Srivastava D. A regulatory pathway involving Notch1/β-catenin/Isl1 determines cardiac progenitor cell fate. Nat Cell Biol. 2009;11(8):951–7.

63. Boni A, Urbanek K, Nascimbene A, Hosoda T, Zheng H, Delucchi F, et al. Notch1 regulates the fate of cardiac progenitor cells. Proc Natl Acad Sci USA. 2008;105(40):15529–34.

64. Gude NA, Emmanuel G, Wu W, Cottage CT, Fischer K, Quijada P, et al. Activation of Notch-mediated protective signaling in the myocardium. Circ Res. 2008;102(9):1025–35.

65. Collesi C, Zentilin L, Sinagra G, Giacca M. Notch1 signaling stimulates proliferation of immature cardiomyocytes. J Cell Biol. 2008;183(1):117–8.

66. Kratsios P, Catela C, Salimova E, Huth M, Berno V, Rosenthal N, et al. Distinct roles for cell-autonomous Notch signaling in cardiomyocytes of the embryonic and adult heart. Circul Res. 2009;106(3):559–2.
67. Yu B, Song B. Notch 1 signalling inhibits cardiomyocyte apoptosis in ischaemic postconditioning. Heart Lung Circ. 2014;23(2):152–8.
68. Gude N, Sussman M. Notch signaling and cardiac repair. J Mol Cell Cardiol. 2012;52(6):1226–32.
69. Rizzo P, Mele D, Caliceti C, Pannella M, Fortini C, Clementz AG, et al. The role of notch in the cardiovascular system: potential adverse effects of investigational notch inhibitors. Front Oncol. 2015;4:384.
70. Øie E, Sandberg WJ, Ahmed MS, Yndestad A, Lærum OD, Attramadal H, et al. Activation of Notch signaling in cardiomyocytes during post-infarction remodeling. Scand Cardiovasc J. 2010;44(6):359–6.
71. Gude N, Joyo E, Toko H, Quijada P, Villanueva M, Hariharan N, et al. Notch activation enhances lineage commitment and protective signaling in cardiac progenitor cells. Basic Res Cardiol. 2015;110(3):29.
72. Nakayama KH, Hou L, Huang NF. Role of extracellular matrix signaling cues in modulating cell fate commitment for cardiovascular tissue engineering. Adv Healthcare Mater. 2014;3(5):628–1.
73. French KM, Boopathy AV, DeQuach JA, Chingozha L, Lu H, Christman KL, et al. A naturally derived cardiac extracellular matrix enhances cardiac progenitor cell behavior in vitro. Acta Biomater. 2012;8(12):4357–64.
74. Gaetani R, Yin C, Srikumar N, Braden R, Doevendans PA, Sluijter JP, et al. Cardiac-derived extracellular matrix enhances cardiogenic properties of human cardiac progenitor cells. Cell Transplant. 2016;25(9):1653–63.
75. French KM, Maxwell JT, Bhutani S, Ghosh-Choudhary S, Fierro MJ, Johnson TD, et al. Fibronectin and cyclic strain improve cardiac progenitor cell regenerative potential in vitro. Stem Cells Int. 2016. https://doi.org/10.1155/2016/8364382.
76. Konstandin MH, Toko H, Gastelum GM, Quijada P, De La Torre A, Quintana M, et al. Fibronectin is essential for reparative cardiac progenitor cell response after myocardial infarction. Circ Res. 2013;113(2):115–25.
77. Williams C, Budina E, Stoppel WL, Sullivan KE, Emani S, Emani SM, et al. Cardiac extracellular matrix–fibrin hybrid scaffolds with tunable properties for cardiovascular tissue engineering. Acta Biomater. 2015;14:84–95.
78. Young JL, Engler AJ. Hydrogels with time-dependent material properties enhance cardiomyocyte differentiation in vitro. Biomaterials. 2011;32(4):1002–9.
79. Choi M-Y, Kim J-T, Lee W-J, Lee Y, Park KM, Yang Y-I, et al. Engineered extracellular microenvironment with a tunable mechanical property for controlling cell behavior and cardiomyogenic fate of cardiac stem cells. Acta Biomater. 2017;50:234–48.
80. Mohyeldin A, Garzón-Muvdi T, Quiñones-Hinojosa A. Oxygen in stem cell biology: a critical component of the stem cell niche. Cell Stem Cell. 2010;7(2):150–61.
81. Israeli-Rosenberg S, Manso AM, Okada H, Ross RS. Integrins and integrin-associated proteins in the cardiac myocyte. Circ Res. 2014;114(3):572–86.
82. Mauretti A, Bax NA, van Marion MH, Goumans MJ, Sahlgren C, Bouten CV. Cardiomyocyte progenitor cell mechanoresponse unrevealed: strain avoidance and mechanosome development. Integr Biol. 2016;8(9):991–1001.
83. Mauretti A, Spaans S, Bax NA, Sahlgren C, Bouten CV. Cardiac progenitor cells and the interplay with their microenvironment. Stem Cells Int. 2017;20 p.https://doi.org/10.1155/2017/7471582.
84. van Marion MH, Bax NA, van Turnhout MC, Mauretti A, van der Schaft DW, Goumans MJT, et al. Behavior of CMPCs in unidirectional constrained and stress-free 3D hydrogels. J Mol Cell Cardiol. 2015;87:79–91.
85. Keith B, Simon MC. Hypoxia-inducible factors, stem cells, and cancer. Cell. 2007;129(3):465–72.

86. Gustafsson MV, Zheng X, Pereira T, Gradin K, Jin S, Lundkvist J, et al. Hypoxia requires notch signaling to maintain the undifferentiated cell state. Dev Cell. 2005;9(5):17–628.
87. Adams J, Difazio L, Rolandelli R, Lujan J, Hasko G, Csoka B, et al. HIF-1: a key mediator in hypoxia. Acta Physiol Hung. 2009;96(1):19–28.
88. Jürgensen JS, Rosenberger C, Wiesener MS, Warnecke C, Hörstrup JH, Gräfe M, et al. Persistent induction of HIF-1α and -2α in cardiomyocytes and stromal cells of ischemic myocardium. FASEB J. 2004;18(12):1415–7.
89. Staller P, Sulitkova J, Lisztwan J, Moch H, Oakeley EJ, Krek W. Chemokine receptor CXCR4 downregulated by von Hippel-Lindau tumour suppressor pVHL. Nature. 2003;425(6955):307–11.
90. Zernecke A, Schober A, Bot I, von Hundelshausen P, Liehn EA, Möpps B, et al. SDF-1α/CXCR4 axis is instrumental in neointimal hyperplasia and recruitment of smooth muscle progenitor cells. Circ Res. 2005;96(7):784–91.
91. Jujo K, Hamada H, Iwakura A, Thorne T, Sekiguchi H, Clarke T, et al. CXCR4 blockade augments bone marrow progenitor cell recruitment to the neovasculature and reduces mortality after myocardial infarction. Proc Natl Acad Sci USA. 2010;107(24):11008–13.
92. Zheng H, Fu G, Dai T, Huang H. Migration of endothelial progenitor cells mediated by stromal cell-derived factor-1α/CXCR4 via PI3K/Akt/eNOS signal transduction pathway. J Cardiovasc Pharmacol. 2007;50(3):274–80.
93. Tang YL, Zhu W, Cheng M, Chen L, Zhang J, Sun T, et al. Hypoxic preconditioning enhances the benefit of cardiac progenitor cell therapy for treatment of myocardial infarction by inducing CXCR4 expression. Circ Res. 2009;104(10):1209–16.
94. Chen D, Xia Y, Zuo K, Wang Y, Zhang S, Kuang D, et al. Crosstalk between SDF-1/CXCR4 and SDF-1/CXCR7 in cardiac stem cell migration. Sci Rep. 2015;5(1):1–9.
95. van Oorschot AA, Smits AM, Pardali E, Doevendans PA, Goumans MJ. Low oxygen tension positively influences cardiomyocyte progenitor cell function. J Cell Mol Med. 2011;15(12):2723–34.
96. Ceradini DJ, Gurtner GC. Homing to hypoxia: HIF-1 as a mediator of progenitor cell recruitment to injured tissue. Trends Cardiovasc Med. 2005;15(2):57–63.
97. Wagner MJ, Khan M, Mohsin S. Healing the broken heart; the immunomodulatory effects of stem cell therapy. Front Immunol. 2020;11:639. https://doi.org/10.3389/fimmu.2020.00639.
98. Zlatanova I, Pinto C, Silvestre J-S. Immune modulation of cardiac repair and regeneration: the art of mending broken hearts. Front Cardiovasc Med. 2016;3:40. https://doi.org/10.3389/fcvm.2016.00040.
99. Sanganalmath SK, Bolli R. Cell therapy for heart failure: a comprehensive overview of experimental and clinical studies, current challenges, and future directions. Circ Res. 2013;113(6):810–34.
100. Sattler S, Fairchild P, Watt FM, Rosenthal N, Harding SE. The adaptive immune response to cardiac injury—the true roadblock to effective regenerative therapies? NPJ Regen Med. 2017;2(1):1–5.
101. Mohsin S, Houser SR. Cortical bone derived stem cells for cardiac wound healing. Korean Circul J. 2019;49(4):314.
102. Williams AR, Hare JM. Mesenchymal stem cells: biology, pathophysiology, translational findings, and therapeutic implications for cardiac disease. Circ Res. 2011;109(8):923–40.
103. Li T-S, Cheng K, Malliaras K, Smith RR, Zhang Y, Sun B, et al. Direct comparison of different stem cell types and subpopulations reveals superior paracrine potency and myocardial repair efficacy with cardiosphere-derived cells. J Am Coll Cardiol. 2012;59(10):942–53.
104. Shi Y, Su J, Roberts AI, Shou P, Rabson AB, Ren G. How mesenchymal stem cells interact with tissue immune responses. Trends Immunol. 2012;33(3):136–43.
105. Aurora AB, Olson EN. Immune modulation of stem cells and regeneration. Cell Stem Cell. 2014;15(1):14–25.
106. Mooney DJ, Vandenburgh H. Cell delivery mechanisms for tissue repair. Cell Stem Cell. 2008;2(3):205–13.

107. Bernardo ME, Fibbe WE. Mesenchymal stromal cells: sensors and switchers of inflammation. Cell Stem Cell. 2013;13(4):392–402.
108. Landolina M, Gasparini M, Lunati M, Iacopino S, Boriani G, Bonanno C, et al. Long-term complications related to biventricular defibrillator implantation: rate of surgical revisions and impact on survival: insights from the Italian clinicalservice database. Circulation. 2011;123(22):2526–35.
109. Kingsley DM. The TGF-beta superfamily: new members, new receptors, and new genetic tests of function in different organisms. Genes Dev. 1994;8(2):133–46.
110. Cao W, Cao K, Cao J, Wang Y, Shi Y. Mesenchymal stem cells and adaptive immune responses. Immunol Lett. 2015;168(2):147–53.
111. Lin L, Du L. The role of secreted factors in stem cells-mediated immune regulation. Cell Immunol. 2018;326:24–32.
112. Drago D, Basso V, Gaude E, Volpe G, Peruzzotti-Jametti L, Bachi A, et al. Metabolic determinants of the immune modulatory function of neural stem cells. J Neuroinflam. 2016;13(1):1–18.
113. Volpe G, Bernstock JD, Peruzzotti-Jametti L, Pluchino S. Modulation of host immune responses following non-hematopoietic stem cell transplantation: Translational implications in progressive multiple sclerosis. J Neuroimmunol. 2019;331:11–27.
114. Gouadon E, Moore-Morris T, Smit NW, Chatenoud L, Coronel R, Harding SE, et al. Concise review: pluripotent stem cell-derived cardiac cells, a promising cell source for therapy of heart failure: where do we stand? Stem Cells. 2016;34(1):34–43.
115. Dong F, Harvey J, Finan A, Weber K, Agarwal U, Penn MS. Myocardial CXCR4 expression is required for mesenchymal stem cell mediated repair following acute myocardial infarction. Circulation. 2012;126(3):314–24.
116. Hatzistergos KE, Quevedo H, Oskouei BN, Hu Q, Feigenbaum GS, Margitich IS, et al. Bone marrow mesenchymal stem cells stimulate cardiac stem cell proliferation and differentiation. Circ Res. 2010;107(7):913–22.
117. Lee RH, Pulin AA, Seo MJ, Kota DJ, Ylostalo J, Larson BL, et al. Intravenous hMSCs improve myocardial infarction in mice because cells embolized in lung are activated to secrete the anti-inflammatory protein TSG-6. Cell Stem Cell. 2009;5(1):54–63.
118. Loffredo FS, Steinhauser ML, Gannon J, Lee RT. Bone marrow-derived cell therapy stimulates endogenous cardiomyocyte progenitors and promotes cardiac repair. Cell Stem Cell. 2011;8(4):389–98.
119. Tang X-L, Li Q, Rokosh G, Sanganalmath SK, Chen N, Ou Q, et al. Long-term outcome of administration of c-kitPOS cardiac progenitor cells after acute myocardial infarction: transplanted cells do not become cardiomyocytes, but structural and functional improvement and proliferation of endogenous cells persist for at least one year. Circ Res. 2016;118(7):1091–105.
120. Chimenti I, Smith RR, Li TS, Gerstenblith G. Messina E. Giacomello A. Marban E. Relative roles of direct regeneration versus paracrine effects of human cardiosphere-derived cells transplanted into infarcted mice. Circ Res. 2010;106:971–80.
121. Spaan JA, Piek JJ, Hoffman JI, Siebes M. Physiological basis of clinically used coronary hemodynamic indices. Circulation. 2006;113(3):446–55.
122. Weirather J, Hofmann UD, Beyersdorf N, Ramos GC, Vogel B, Frey A, et al. Foxp3+ CD4+ T cells improve healing after myocardial infarction by modulating monocyte/macrophage differentiation. Circ Res. 2014;115(1):55–67.
123. Courties G, Heidt T, Sebas M, Iwamoto Y, Jeon D, Truelove J, et al. In vivo silencing of the transcription factor IRF5 reprograms the macrophage phenotype and improves infarct healing. J Am Coll Cardiol. 2014;63(15):1556–66.
124. Barile L, Milano G, Vassalli G. Beneficial effects of exosomes secreted by cardiac-derived progenitor cells and other cell types in myocardial ischemia. Stem Cell Invest. 2017;4:93. https://doi.org/10.21037/sci.2017.11.06.
125. Ben-Mordechai T, Palevski D, Glucksam-Galnoy Y, Elron-Gross I, Margalit R, Leor J. Targeting macrophage subsets for infarct repair. J Cardiovasc Pharmacol Therapeutics. 2015;20(1):36–51.

126. de Couto G, Gallet R, Cambier L, Jaghatspanyan E, Makkar N, Dawkins JF, et al. Exosomal microRNA transfer into macrophages mediates cellular postconditioning. Circulation. 2017;136(2):200–14.
127. Latronico MV, Condorelli G. The might of microRNA in mitochondria. Circ Res. 2012;110:1540–2.

Part III
Cardiomyocyte Aging and Death

Chapter 12
Cardiomyocyte Senescence

Abstract Aging is well-recognized as an independent risk factor for heart diseases. Aging is associated with defects in metabolism and dysfunction of cardiomyocytes in addition to fibrosis and reduced angiogenesis in the heart. Aging or senescent cardiomyocytes have several features such as hypertrophy, DNA damage, endoplasmic reticulum stress, dysfunction of mitochondria and contractile defects. Another characteristic is a senescence-associated secreting phenotype (SASP), which secretes increased levels of pro-inflammatory factors, growth factors, proteases and exosomes. The senescent cardiomyocytes alter the phenotype of the endothelial cells, fibroblasts, and immune cells of the heart and thus contribute to pathological remodelling. Non-cardiomyocytes regulate the senescence of cardiomyocyte as well.

Keywords Cardiomyocyte · Aging · Senescence · Secretory phenotype · Telomere · Telomerase · Autophagy · Senolysis · Senolytics

Introduction

As age advances, senescent cells accumulate in tissues throughout the body as a result of the Hayflick limit on replication of cells, damage from toxins and suppression of uncontrolled cell proliferation.

Aging is well-recognized as an independent risk factor for heart diseases. With advancing age, there is reduction in the number of cardiomyocytes (CMs). In the young adult, cardiomyocyte loss is compensated by hypertrophy of the surviving CMs, as capacity for cardiomyocyte renewal is limited in adult CMs [1, 2]. Senescent CMs however do not have the capacity to enlarge [1, 3]. Cardiomyocyte loss along with an accumulation of extracellular matrix and reduced angiogenesis in the aged heart is hence associated with gradual deterioration in cardiac function and can result in heart failure [4, 5]. Physiological aging of CMs is primarily driven by intrinsic aging factors. Extrinsic factors (for example smoking) contribute to pathological accelerated aging of the heart.

© The Author(s), under exclusive license to Springer Nature Switzerland AG 2021   187
C. C. Kartha, *Cardiomyocytes in Health and Disease*,
https://doi.org/10.1007/978-3-030-85536-9_12

Cellular Senescence

Aging at the cellular level is termed as 'cellular senescence' [6]. Cellular senescence is defined as an irreversible loss of the potential of mitotic cells to divide, followed by a senescence-associated secretory phenotype (SASP). p53, p21, p16^{Ink4a}, p38MAPK, and γH2AX, telomere attrition and enhanced signals for SA-β-gal are indirect indicators of cellular senescence. They are generally used to detect senescent cells. Genetic changes in senescent cells result in SASP and the altered cells can secrete pro-inflammatory molecules [7]. Senescent cells through these secreted factors can be pathogenic by causing chronic inflammation and tissue remodelling.

Types of Cellular Senescence

There are two types of cellular senescence. Somatic cells eventually enter a phase of irreversible growth arrest or 'replicative senescence' [8]. Replicative cellular senescence, via the p53 or p16^{Ink4a} signalling pathways is initiated when telomere shortening exceeds the physiological range. Telomeres which are necessary for chromosomal stability and DNA replication, replicate incompletely during cell division, leading to telomere attrition.

Premature senescence can result through various external as well as internal stress signals (stress-induced premature senescence). The p53 or p16^{Ink4a} signalling pathways also mediate this type of cellular senescence. In human cells, telomere dysfunction activates either p53 or p16^{Ink4a} signalling [9]. While DNA damage and telomere dysfunction triggers p53 signalling, p16^{Ink4a} signalling is associated with mitogenic and general cellular stress [10–12]. Most of the stressors that induce cellular senescence activate either or both the p53/p21 or p16^{Ink4a} /retinoblastoma protein pathways [13].

Role of Cellular Senescence in Heart Diseases

Pathways involved in cellular senescence and cardiovascular diseases have strong links. p53 levels and apoptosis in the heart are known to be higher in patients with end-stage heart failure [14–16]. Shorter telomere length in circulating leucocytes predict adverse events in patients with coronary artery disease [17, 18]. Insulin-like growth factor-binding protein-7 (IGFBP7), a known inhibitor of cell proliferation is recognized as a senescence secretome and is linked with tissue aging and obesity as well as with poor prognosis in patients with heart failure with preserved ejection fraction (HFpEF) [19].

The Senescent Cardiomyocyte

Cardiomyocytes affected by the aging process, have several physiological and morphological characteristics. The response to loss of CMs in the aged heart is hypertrophy of the remaining CMs to preserve heart function [1]. However, the ability to enlarge is lost in the senescent cardiomyocyte [3, 20]. The senescent cardiomyocyte is also the end stage of an irreversible cell cycle arrest. This is mediated by an upregulation of p16/INK4A and p53 and a reduction in telomere length to approximately 15 kb from 30 kb; uncapped telomeres are also noted [11, 14, 21, 22]. Cardiomyocyte function is also impaired. There are changes in gene expression related to myocyte activation, contraction, and relaxation in senescent CMs. Genes involved in heat shock/stress response, mitochondrial death signalling and function, cytoskeletal organization, survival/growth, and transcription are differentially expressed in young and old CMs [23, 24].

Aging alters the nuclear structure which is important for gene expression and genomic stability. Nuclear changes include downregulation of the intermediate filaments, lamin A and C which are vital for the maintenance of the nuclear shape, deterioration in electrical coupling and cell–cell communication, and downregulation of connexin 43 expression in cardiac gap junctions [25, 26]. There is repression of enzymes and pathways such as PKC, AKT, AP1, which are involved in survival and growth, and upregulation of cell cycle inhibitors such as p21 and p16 [27]. Survivin, an inhibitor of apoptosis is also downregulated in senescent cells [28]. Some metallothioneins which inhibit oxidant activity and cell death are downregulated, favouring development of senescence-associated phenotypes [29, 30]. The balance between pro and anti-death signals is tilted towards death in senescent CMs [27].

Aging significantly reduces the sensitivity of myofilaments to Ca^{2+}. Functional impairment in sarcoplasmic reticulum (SR) or downregulation of SR Ca^{2+}-ATPase and Na^+/Ca^{2+} exchanger may contribute to prolongation of duration of relaxation of CMs [23]. Biomechanical properties of CMs are affected because of decrease in elasticity of the plasma membrane of the aged cardiomyocyte [31]. Cardiomyocyte function is decreased also because of changes in signalling, which affects Ca^{2+} transients, and the deposition of lipofuscin granules secondary to impaired autophagy [32]. Adrenergic and angiotensin-signalling pathways are also altered in aged CMs [33]. The senescent cardiomyocyte having gradually lost its structural and functional integrity either undergoes cell death either by apoptosis or by necrosis.

Several structural and functional alterations are seen in senescent CMs [27]. Expressions of heat shock proteins (HSP70) and anti-oxidative enzymes (hemeoxygenase-1) are reduced, which affect ability of CMs to cope with stress. (ii) There is alteration in function of the mitochondrial respiratory chain, which leads to electron leakage and oxidative damage. (iii) Downregulation of sarcoplasmic reticulum Ca^{2+}-ATPase and change in the expression of cytoskeletal proteins result in increased stiffness and reduced contractility/decelerated relaxation in CMs. (iv) A transcriptional switch occurs in contractile protein isoforms (e.g. fast type V 1 switches to slow type V 3 myosin). (v) There is decrease in the expression of survivin and shift in the Bcl2 rheostat to a pro-apoptotic state, which promote cell death signalling.

Features of Senescent Cardiomyocytes

Salient features of senescent CMs are hypertrophy, DNA damage, endoplasmic reticulum (ER) stress, dysfunction of mitochondria, contractile defects, and senescence-associated secreting phenotype (SASP) (Fig. 12.1) [34]. As adult CMs are terminally differentiated, cell cycle arrest is not a feature of cardiomyocyte senescence.

Hypertrophy is a common feature of senescence in cardiomyocytes [27]. There is increased expression of hypertrophic genes in senescent cardiomyocytes [23, 24]. Cell shortening and re-lengthening are defective in senescent cardiomyocytes. Pacing frequency of these cells is also higher [35]. DNA damage and NAD$^+$ depletion regulate defects in contraction [36].

DNA Damage

DNA damage is induced in senescent CMs by accumulation of mitochondrial reactive oxygen species. A common feature of DNA damage in senescent cells is shortening of telomere [37]. There is evidence from studies in both animals and humans that length-independent telomere damage mediate cardiomyocyte senescence [38, 39].

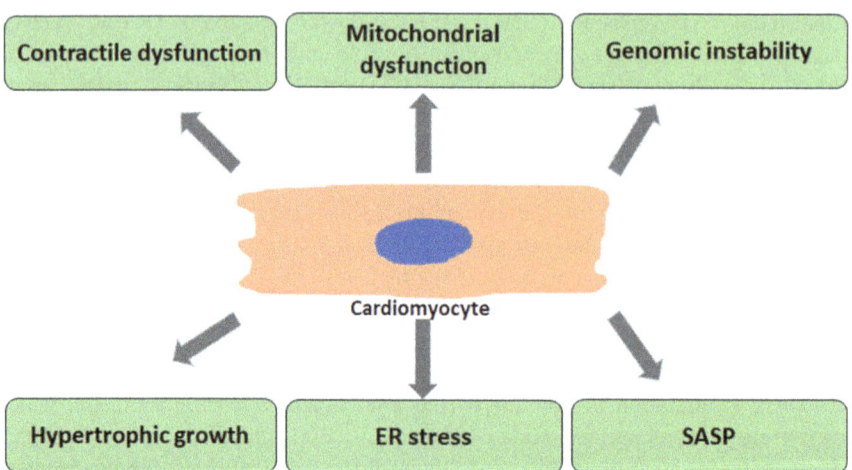

Fig. 12.1 Indicators of cardiomyocyte senescence. ER—endoplasmic reticulum. SASP—senescence-associated secreting phenotype

Endoplasmic Reticulum (ER) Stress

ER stress increases in senescent CMs with defective contractility. ER stress assists apoptosis and hypertrophy of CMs [40, 41]. Cardiomyocyte senescence can be prevented and contractile function improved by reducing ER stress [42, 43].

Mitochondrial Dysfunction

The fission–fusion mechanism in mitochondria is impaired in senescent CMs. This leads to decline in mitochondrial function [44]. P53 contributes to cell-cycle arrest in senescent cells via inhibition of cyclin-dependent kinases. P53 facilitates cellular senescence by inhibiting Parkin-mediated mitophagy as well [45]. Senescence of CMs can be repressed by targeting Drp1/Parkin/PINK1 signalling and thus improving the function of mitochondria [44, 46].

Senescence-Associated Secreting Phenotype (SASP)

Cardiomyocytes of SASP have increased expression of CCN family member 1 (CCN1), interleukins (IL1α, IL1β, and IL6), tumour necrosis factor-alpha (TNFα), monocyte chemoattractant protein-1 (MCP1), endothelin (ET-1), tumour growth factor-beta (TGFβ), and growth and differentiation factor 15 (GDF15) [34, 47]. These cells secrete higher amounts of pro-inflammatory cytokines and chemokines, growth modulators, proteases, angiogenetic factors, matrix metalloproteinases and exosomes [48]. The secreted factors can alter the phenotype of non-myocytes such as endothelial cells, fibroblasts, and immune cells of the heart aiding pathological remodelling of the heart (Fig. 12.2).

Metabolic Switch in Senescent Cardiomyocytes

Aging of the heart is associated with defects in metabolism and metabolic dysfunction of CMs [49, 50]. Metabolic alterations in senescent CMs contribute to decline in cardiac function [51]. While fatty acid oxidation and not glycolysis is the principal source of energy in adult CMs of mammalian hearts, in CMs of aging hearts, glucose utilization is increased for ATP production; the use of fatty acid is reduced. This metabolic pattern is similar to that of foetal and neonatal CMs [52].

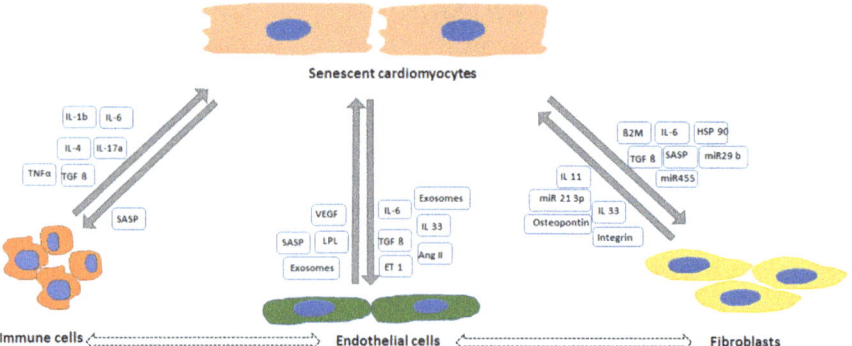

Fig. 12.2 Signalling among cardiomyocytes, fibroblasts, endothelial cells and extracellular matrix during cardiomyocyte senescence. Ang—Angiotensin. β2M—β2 macroglobulin. EC—Endothelial cells. ECM—extracellular matrix. EGFR—Epithelial cell growth factor receptor. ET—Endothelin. HSP—Heat shock protein. IL—Interleukin. LPL—lipoprotein lipase. miR—micro ribonucleic acid. SASP—Senescence-associated secretory phenotype. TGF—Transforming growth factor. TNF—Tissue necrosis factor. VEGF—Vascular endothelial growth factor

Fatty acyl-coenzyme A (CoA) and pyruvate are the major resources for the production of energy (ATP) in mitochondria of CMs. These are metabolites of the pathways of both fatty acid oxidation and glucose oxidation. Carnitine–palmitoyl transferase-1 (CPT1) regulates the entry of long-chain acyl-CoA into mitochondria and pyruvate dehydrogenase (PDH) control oxidation of pyruvate [52]. In the hearts of aging rats, the level of CPT1 is significantly decreased [53]. The deficiency of CPT1 causes lipotoxicity and has been found to exacerbate cardiomyocyte senescence [54]. Peroxisome proliferator-activated receptor α (PPARα) and PGC-1α are also regulators of fatty acid metabolism. The expression of PPARα and PGC-1α decreases with age [55]. In mice prone to accelerated senescence, the decrease is associated with rise in ceramide levels and cardiac hypertrophy [56].

Activation of insulin signalling via insulin growth factor receptor (IGFR), induces SASP and promotes cardiomyocyte senescence [57]. P53 activation aids glycolysis and promotes cardiomyocyte senescence, which can be prevented by inhibition of P53 [58].

Increase in ketone bodies (acetoacetate, β-hydroxybutyrate, and acetone) occurs in the aging heart and in patients with heart failure [59, 60]. Ketone bodies are the source of oxidative ATP production in hearts with age-associated cardiac hypertrophy and failure [60]. β-hydroxybutyrate, the primary ketone body produced by the body during ketosis, protects the cells against hypoxic stress and represses senescence of CMs [61]. Knocking out succinyl-CoA:3-oxoacid CoA transferase (SCOT) promotes mitochondrial stress and cell senescence [62].

Senescence of CMs is modulated by the hexosamine biosynthetic pathway as well. O-linked-GlcNAc transferase (OGT) involved in the pathway catalyzes O-GlcNAcylation (O-GlcNAc) of proteins [63]. CMs are protected from senescence by O-GlcNAcylation of proteins, which aid to reduce calcium overload, open the mitochondrial permeability transition pore, decrease ER stress and alter inflammatory and heat shock responses [64]. Increasing O-GlcNAc levels has been found to have protective effects in several pathologic conditions [65, 66].

AMP-activated protein kinase (AMPK), NAD+ -dependent sirtuins, FOXOs, and mammalian target of rapamycin (mTOR) also regulate senescence of CMs. Activation of AMPK can improve mitochondrial dynamics, reduce ER stress and repress senescence of CMs [67–69]. SIRT1 (Sirtuin1), SIRT2, SIRT3, SIRT6, and SIRT7 can regulate senescence through their effects on metabolism and hypertrophic growth of CMs [68, 70–75]. ROS metabolism in mitochondria is modulated by SIRT3 and SIRT4 [76].

Non-Myocytes and Cardiomyocyte Senescence

Dysfunction of non-myocytes in the aged heart affects cardiomyocyte senescence (Fig. 12.2) [34, 77–79]. Intracellular signalling pathways such as metabolic sensors/regulators and the paracrine effects of the non-myocytes (endothelial cells and fibroblasts as well as immune cells) regulate senescence of CMs.

Effect of Endothelial Cells on Cardiomyocyte Senescence

Metabolism of endothelial cells is also altered during aging of the heart. The dysfunction of endothelial cells facilitates the senescence of CMs. Ang II, NRG1, ET-1, apelin, and proinflammatory factors (e.g., TGFβ and IL6), and eicosanoids secreted by endothelial cells modulate the senescence of CMs [80]. Ang II and ET-1 released by endothelial cells bind to their receptors on CMs and can promote mitochondrial dysfunction, ER stress and hypertrophic growth of CMs.

Loss of endothelin receptor A (ETAR) in CMs modulates autophagy response and rescues aging-associated hypertrophy and contractile dysfunction of CMs [81, 82]. Exosomes or extracellular vehicles produced by endothelial cells in the heart also affect cardiomyocyte functions [83, 84]. Cardiomyocyte senescence is modulated by metabolic changes in endothelial cells as well [85, 86].

Effects of Fibroblasts on Cardiomyocyte Senescence

Cardiac fibroblasts (CFs) do regulate senescence of CMs. The mechanisms involved are paracrine effects and remodelling of extracellular matrix (ECM) [87]. Integrins expressed by fibroblasts are important for ECM homeostasis and for governing aging of CMs [78, 88].

Ang II, cardiotrophin 1, fibroblast growth factor (FGF), IL-6, insulin-like growth factor 1 (IGF1), TGFβ, and TNFα secreted by CFs mediate fibroblast–cardiomyocyte communications [88]. IL-33 from CFs can retard CM senescence induced by hypertrophy and hypoxia [89]; IL-11 cause dysfunction of CMs [90]. The plasma membrane calcium ATPase 4 signalling in CFs mediates hypertrophy of CMs. This is achieved by upregulation of the expression frizzled-related protein 2 (sFRP2) and release of the protein [91].

miR-21-3p derived from fibroblast exosomes is a potent inducer of hypertrophy of CMs in rodents. Its targets are sorbin, SH3 domain-containing protein 2 (SORBS2), PDZ proteins and LIM domain 5 (PDLIM5) [92]. Angiotensin II increases the release of exosomes with osteopontin and epidermal growth factor receptor (EGFR) from fibroblasts. These factors activate the renin–angiotensin system and thus enhance hypertrophy of CMs [93]. Exosomes secreted from human fibroblasts have been found to modulate calcium cycling in human induced pluripotent stem cells; they reduce the duration of calcium transient in the cytoplasm [94].

CFs can acquire a fate similar to osteoblasts and thus regulate cardiomyocyte calcification, which is common in aged hearts [95].

Metabolic alterations within CFs may also contribute to the dysfunction and senescence of CMs [96–98].

Role of Immune Cells in Cardiomyocyte Senescence

Macrophages, T cells, and mast cells in the heart can regulate senescence in CMs by altering inflammatory responses.

Senescence can be induced in CMs by macrophage-specific depletion of connexin 43 [99]. Electrical activity is abnormal in such cells [99]. Activation of NLRP3 inflammasome in cardiac macrophages induces secretion of IL-1β, which advances senescence in CMs [100]. While it is known that macrophage-mediated inflammation can contribute to hypertrophy and senescence of CMs [87], the macrophage subtypes and the specific paracrine factors that are involved are unclear. Whether regulators of metabolism and metabolites in macrophages have any role in senescence of CMs also needs scrutiny.

The paracrine functions of T-regs during injury to the heart are now recognized. In the infarcted hearts. T-reg cells secrete cystatin F, TNF superfamily member 11 (TNFSF11), IL-33, fibrinogen-like protein 2 (FGL-2), matrilin-2, and IGF-2, which stimulate proliferation of CMs and repress senescence of the cells [101].

Mast cells release chymase, which stimulates the production of TGFβ and aggravates hypertrophy and senescence of CMs [102].

Effects of Senescent Cardiomyocytes on Nonmyocytes

Effects on Endothelial Cells

The SASP factors such as CCN family member 1 (CCN1), IL-1α, TNF-α, TGF-β and monocyte chemoattractant protein released by CMs can induce senescence of endothelial cells. TGF-β secreted by CMs can trigger End-MT (endothelial-mesenchymal transformation), a feature of aged hearts [103, 104].

Cardiomyocytes can modulate the function of endothelial cells by also releasing angiogenic factors such as VEGA-A, lipoprotein lipase, HSP20 or miR-320 enriched exosomes [105–111].

Effects on Fibroblasts

Cardiomyocytes synthesize and secrete interleukins, TGF-β, CCN1, β-2 microglobulin (β2M), fibroblast growth factor (FGF), placental growth factor (PGF). CMs regulate the function of CFs via the paracrine effects of these factors [88, 112–117]. microRNA-92a of exosomes derived from CMs has been found to mediate myofibroblast activation [118]. Hsp90 derived from myocytes activates STAT-3 signalling and modulates collagen upregulation in fibroblasts during cardiac hypertrophy [119].

Mechanisms in Cardiomyocyte Aging

Several factors participate in the aging process in CMs [27]. These include oxidative stress, inflammation, telomere damage, dysfunction of mechanisms for cellular protection and repair, alterations in survival and death mechanisms, changes in cell metabolism, and switch in gene expression.

Oxidative Stress

Oxidative stress is a common factor of both intrinsic and extrinsic causes of aging. Oxidative stress results in defective contractile structure and biochemical function of CMs. Intrinsic oxidative stress and damage triggers physiological aging of CMs. The mitochondrial respiratory chain, lysosomes, and enzymes such as NADPH oxidase are the main sources for oxidants.

Reactive oxygen or nitrogen species and/or radicals generated during oxidative stress, can alter the properties of nucleic acids, lipids, sugars, and proteins leading to deterioration of cellular function [120]. Oxidative damage to DNA may affect proteins of the mitochondrial respiratory chain and thus alter respiratory-chain function, cause intrinsic oxidative stress and further cellular damage [121, 122]. Radicals and oxidants produced by respiratory chain function are considered key factors in aging of the cardiomyocyte [123].

There are also other contributors to the intrinsic generation of oxidative stress. These include NADPH oxidase. NADPH oxidase subunits gp91 phox and p47 phox are increased in aged cells [124].

Some of the aging factors reduce defense and repair mechanisms in the cells and thus increase oxidative stress. Oxidant damage can affect the genes involved in DNA and thus decrease the cell's capacity to repair damaged DNA. Expression of genes such as Ku86/Ku70, PARP, and TRF2 are reduced during aging [23, 24, 125].

A deletion of proteins such as HSP 70 and catalase which protect cells against oxidative stress, accelerates aging-induced decrease in cardiac function; a transgenic overexpression of these proteins decreases myocardial dysfunction and damage [126, 127].

Inflammation

The balance between inflammatory and anti-inflammatory factors shifts towards inflammation during aging. Necrosis of cells and increase in autoantigens may promote inflammation. Inflammation can induce oxidative stress and damage; conversely oxidative damage can trigger inflammation. A vicious circle comprising both contributes to cardiomyocyte senescence [27].

Post-Translational Modifications and Metabolism

Post-translational modification of proteins has a key role in aging and heart disease. Advanced glycation end products (AGE) are increased 2.5-fold in aged hearts and contribute to prolongation of cardiomyocyte re-lengthening as seen in the aging heart [124].

Cellular Waste

The accumulation of waste secondary to damage of lysosomal membranes and the release of oxidants and radicals, as well as lysosomal enzymes contributes to oxidative stress and cell death [128]. Giant mitochondria, which accumulate during aging contain mutated DNA sequences and defective respiratory chain components which contribute to oxidative stress and damage [32, 128].

Telomere Damage

Telomere shortening is a major inducer of senescence [129]. Telomeres protect the exposed ends of DNA in chromosomes, from damage [130, 131]. Repeated cell divisions disrupt the shield. This can result in shortening of telomeres, a DNA damage response (DDR) and activation of the senescence programme [130]. While this mechanism may explain age-related degeneration in skin and hematopoietic tissues, which have stem cell pools, it does not explain senescence in post-mitotic cells such as adult CMs. The mechanisms that drive senescence in post-mitotic cells and their role in ageing associated tissue degeneration are a topic of recent investigations [132, 133]

Senescent CMs have permanent DNA damage at telomere regions. As most of the CMs in the adult heart are post-mitotic, replicative senescence-induced telomere shortening is considered unlikely to be the cause of normal physiological ageing of CMs. Stress induced telomere damage can cause telomere dysfunction and induce senescence [134, 135].

Rhys Anderson and colleagues have shown that a senescence-like phenotype is a feature of normal physiological ageing of human and murine CMs [38]. They provide a mechanistic model for senescence in rarely dividing post-mitotic cells. Foci of permanent DNA damage in telomere regions seem to increase in CMs with age. This phenomenon is independent of telomere length, telomerase activity or DNA replication and can be induced by mitochondrial dysfunction. Several senescence markers are induced in CMs on specific induction of length-independent telomere dysfunction. Classical pathways of senescence growth arrest are activated and a distinct SASP with functional effects emerge. Length-independent telomere damage activates the classical senescence-inducing pathways, $p21^{CIP}$ and $p16^{INK4a}$ in CMs.

The shortening of telomeres and telomere attrition contribute to oxidative stress, linked with aging of the cardiomyocyte. In a telomerase knockout mouse line, the lack of enzyme activity and resulting telomere shortening were seen to cause a phenotype characterized by cardiac dilatation and heart failure, which was associated with upregulation of p53 [22].

Imbalance Between Growth and Death

During the aging process, there is alteration in the balance between growth and death as well as in the type of cell death (apoptosis or necrosis) [29, 30, 136, 137]. These changes lead to cell death as well as hypertrophy of surviving CMs. Cytokines of the IL-6 family seems to have a role in cardiomyocyte hypertrophy as well as in the protection of CMs against apoptosis [138]. Necrosis causes leakage of cytosolic contents which trigger an inflammatory immune response and secondary increase in oxidative stress. The degree (low or medium or high) of oxidative stress would determine whether a cell survives, undergoes apoptosis, or dies by necrosis.

Altered Gene Expression

The expression of specific genes is altered directly by induction of mutations and indirectly by posttranslational modification of proteins, lipids, and sugars, resulting in functional alterations. Modified (pro-aging) gene expression is a key determinant of cardiomyocyte senescence and its progression [23, 24].

Changes in Metabolism

Proteins involved in metabolism and IGF signalling also have a key role in cardiomyocyte aging [125].

Senolytics

Several molecules able to remove senescent cells (senolysis) have been identified. These are known as senolytics. Several studies indicate that elimination of senescent cells or senolysis is potentially beneficial for reversal of age-related dysfunction in the heart via promotion of regenerative capacity. Senolytics could hopefully be a therapy for heart diseases in the future [139–142].

Conclusions

Oxidative stress, inflammation, mechanisms for cellular protection, death and repair, telomere integrity, metabolism, post-translational modifications, and altered gene expression play important roles in the aging of cardiomyocytes. Alterations in these factors modulate cardiomyocyte senescence and dysfunction. Aging processes in non-cardiomyocytes in the heart also interfere with cardiomyocyte aging. While physiological aging of the cardiomyocyte is primarily driven by intrinsic aging factors, extrinsic factors contribute to pathologically accelerated aging of the heart.

The core mechanisms underlying cardiomyocyte senescence are not yet clear. The functions of senescent cardiomyocytes during pathological remodelling and regeneration in the heart and how senescent cardiomyocytes contribute to aging of the heart and cardiac failure need further elucidation. An interesting question is whether the mechanisms for cardiomyocyte senescence could be a therapeutic target.

References

1. Anversa P, Palackal T, Sonnenblick EH, Olivetti G, Meggs LG, Capasso JM. Myocyte cell loss and myocyte cellular hyperplasia in the hypertrophied aging XE "Aging" rat heart XE "Heart." Circ Res. 1990;67:871–85.
2. Fleg JL, O'Connor F, Gerstenblith G, Becker LC, Clulow J, Schulman SP, Lakatta EG. Impact of age on the cardiovascular response to dynamic upright exercise in healthy men and women. J Appl Physiol. 1995;78:890–900.
3. Chimenti C, Kajstura J, Torella D, Urbanek K, Heleniak H, Colussi C, Di MF, Nadal-Ginard B, Frustaci A, Leri A, Maseri A, Anversa P. Senescence and death of primitive cells and myocytes lead to premature cardiac aging and heart failure. Circ Res. 2003;93:604–13.
4. Chang KC, Peng YI, Dai SH, Tseng YZ. Age related changes in pumping mechanical behavior of rat ventricle in terms of systolic elastance and resistance. J Gerontol A Biol Sci Med Sci. 2000;55:B440–7.
5. Lakatta EG. Why cardiovascular function may decline with age. Geriatrics. 1987;42:84–7.
6. Shimizu I, Minamino T. Cellular senescence in cardiac diseases. J Cardiol. 2019;74:313–9.
7. Tchkonia T, Zhu Y, van Deursen J, Campisi J, Kirkland JL. Cellular senescence and the senescent secretory phenotype: therapeutic opportunities. J Clin Invest. 2013;123:966–72.
8. Hayflick L, Moorhead PS. The serial cultivation of human diploid cell strains. Exp Cell Res. 1961;25:585–621.
9. Jacobs JJ, de Lange T. Significant role for p16INK4a in p53-independent telomere-directed senescence. Curr Biol. 2004;14:2302–8.
10. Campisi J. Senescent cells, tumor suppression, and organismal aging: good citizens, bad neighbors. Cell. 2005;120:513–22.
11. Beausejour CM, Krtolica A, Galimi F, Narita M, Lowe SW, Yaswen P, et al. Reversal of human cellular senescence: roles of the p53 and p16 pathways. EMBO J. 2003;22:4212–22.
12. Donato AJ, Morgan RG, Walker AE, Lesniewski LA. Cellular and molecular biology of aging endothelial cells. J Mol Cell Cardiol. 2015;89:122–35.
13. van Deursen JM. The role of senescent cells in ageing. Nature. 2014;509:439–46.
14. Song H, Conte JV, Foster AH, McLaughlin JS, Wei CM. Increased p53 protein expression in human failing myocardium. J Heart Lung Transplant. 1999;18:744–9.
15. Predmore JM, Wang P, Davis F, Bartolone S, Westfall MV, Dyke DB, et al. Ubiquitin proteasome dysfunction in human hypertrophic and dilated cardiomyopathies. Circulation. 2010;121:997-U33.
16. Birks EJ, Latif N, Enesa K, Folkvang T, Luong LA, Sarathchandra P, et al. Elevated p53 expression is associated with dysregulation of the ubiquitin-proteasome system in dilated cardiomyopathy. Cardiovasc Res. 2008;79:472–80.
17. Hammadah M, Al Mheid I, Wilmot K, Ramadan R, Abdelhadi N, Alkhoder A, et al. Telomere shortening, regenerative capacity, and cardiovascular outcomes. Circ Res. 2017;120:1130–8.
18. Margaritis M, Sanna F, Lazaros G, Akoumianakis I, Patel S, Antonopoulos A.S. et al. Predictive value of telomere length on outcome following acute myocardial infarction: evidence for contrasting effects of vascular vs. blood oxidative stress. Eur Heart J. 2017; 38:3094–3104.
19. Gandhi PU, Chow SL, Rector TS, Krum H, Gaggin HK, McMurray JJ, et al. Prognostic value of insulin XE "Insulin" -like growth factor-binding protein 7 in patients with heart failure and preserved ejection fraction. J Card Fail. 2017;23:20–8.
20. Urbanek K, Quaini F, Tasca G, Torella D, Castaldo C, Nadal-Ginard B, Leri A, Kajstura J, Quaini E, Anversa P. Intense myocyte formation from cardiac stem cells in human cardiac hypertrophy. Proc Natl Acad Sci USA. 2003;100:10440–5.
21. Baker DJ, Childs BG, Durik M, Wijers ME, Sieben CJ, Zhong J, et al. Naturally occurring p16(Ink4a)-positive cells shorten healthy lifespan. Nature. 2016;530:184–9.
22. Leri A, Franco S, Zacheo A, Barlucchi L, Chimenti S, Limana F, Nadal-Ginard B, Kajstura J, Anversa P, Blasco MA. Ablation of telomerase and telomere loss leads to cardiac dilatation and heart failure associated with p53 upregulation. EMBO J. 2003;22:131–9.

23. Bodyak N, Kang PM, Hiromura M, Sulijoadikusumo I, Horikoshi N, Khrapko K, Usheva A. Gene expression profiling of the aging mouse cardiac myocytes. Nucl Acids Res. 2002;30:3788–94.
24. Park SK, Prolla TA. Gene expression profiling studies of aging in cardiac and skeletal muscles. Cardiovasc Res. 2005;66:205–12.
25. Afilalo J, Sebag IA, Chalifour LE, Rivas D, Akter R, Sharma K, Duque G. Age-related changes in lamin A/C expression in cardiomyocytes. Am J Physiol. 2007;293:H1451–6.
26. Boengler K, Heusch G, Schulz R. Connexin 43 and ischemic preconditioning: effects of age and disease. Exp Gerontol. 2006;41:485–8.
27. Bernhard D, Laufer G. The aging cardiomyocyte: a mini-review. Gerontology. 2008;54:24–31.
28. Abbate A, Scarpa S, Santini D, Palleiro J, Vasaturo F, Miller J, Morales C, Vetrovec GW, Baldi A. Myocardial expression of survivin, an apoptosis inhibitor, in aging and heart failure. An experimental study in the spontaneously hypertensive rat. Int J Cardiol. 2006; 111:371–376.
29. Fang CX, Doser TA, Yang X, Sreejayan N, Ren J. Metallothionein antagonizes aging induced cardiac contractile dysfunction: role of PTP1B, insulin receptor tyrosine phosphorylation and Akt. Aging Cell. 2006;5:177–85.
30. Yang X, Doser TA, Fang CX, Nunn JM, Janardhanan R, Zhu M, Sreejayan N, Quinn MT, Ren J. Metallothionein prolongs survival and antagonizes senescence-associated cardiomyocyte diastolic dysfunction: role of oxidative stress. FASEB J. 2006;20:1024–6.
31. Lieber SC, Aubry N, Pain J, Diaz G, Kim SJ, Vatner SF. Aging increases stiffness of cardiac myocytes measured by atomic force microscopy nanoindentation. Am J Physiol. 2004;287:H645–51.
32. Terman A, Brunk UT. Autophagy XE "Autophagy" in cardiac myocyte homeostasis, aging, and pathology. Cardiovasc Res. 2005;68:355–65.
33. Domenighetti AA, Wang Q, Egger M, Richards SM, Pedrazzini T, Delbridge LM. Angiotensin II-mediated phenotypic cardiomyocyte remodeling leads to age-dependent cardiac dysfunction and failure. Hypertension. 2005;46:426–32.
34. Tang X, Li P-H, Chen H-Z. Cardiomyocyte senescence and cellular communications within myocardial microenvironments. Front Endocrinol. 2020. https://doi.org/10.3389/fendo.2020.00280.
35. Lim CC, Apstein CS, Colucci WS, Liao R. Impaired cell shortening relengthening with increased pacing frequency are intrinsic to the senescent mouse cardiomyocyte. J Mol Cell Cardiol. 2000;32:2075–82.
36. Zhang D, Hu X, Li J, Liu J, Baks-te Bulte L, Wiersma M, et al. DNA damage-induced PARP1 activation confers cardiomyocyte dysfunction through NAD+ depletion in experimental atrial fibrillation. Nat Commun. 2019;10:1307. https://doi.org/10.1038/s41467-019-09014-2.
37. Shay JW, Wright WE. Telomeres and telomerase: three decades of progress. Nature Rev Genet. 2019;20:299–309.
38. Anderson R, Lagnado A, Maggiorani D, Walaszczyk A, Dookun E, Chapman J, et al. Length-independent telomere damage drives post-mitotic cardiomyocyte senescence. EMBO J. 2019;38:e100492. https://doi.org/10.15252/embj.2018100492.
39. Ball AJ, Levine F. Telomere-independent cellular senescence in human fetal cardiomyocytes. Aging Cell. 2005;4:21–30.
40. Xie F, Wu D, Huang SF, Cao JG, Li HN, He L, et al. The endoplasmic reticulum stress-autophagy pathway is involved in apelin-13-induced cardiomyocyte hypertrophy in vitro. Acta Pharmacol Sin. 2017;38:1589–600.
41. Zeng Z, Huang N, Zhang Y, Wang Y, Su Y, Zhang H, et al. CTCF inhibits endoplasmic reticulum stress and apoptosis in cardiomyocytes by upregulating RYR2 via inhibiting S100A1. Life Sci. 2020; 242:117158.
42. Bozi LH, Takano AP, Campos JC, Rolim N, Dourado PM, Voltarelli VA, et al. Endoplasmic reticulum stress impairs cardiomyocyte contractility through JNK-dependent upregulation of BNIP3. Int J Cardiol. 2018;272:194–201.

43. Wiersma M, Meijering RA, Qi XY, Zhang D, Liu T, Hoogstra-Berends F, et al. Endoplasmic reticulum stress is associated with autophagy and cardiomyocyte remodeling in experimental and human atrial fibrillation. J Am Heart Assoc. 2017;6:e006458. https://doi.org/10.1161/JAHA.117.006458.

44. Nishimura A, Shimauchi T, Tanaka T, Shimoda K, Toyama T, Kitajima N, et al. Hypoxia-induced interaction of filamin with Drp1 causes mitochondrial hyperfission–associated myocardial senescence. Sci Signal. 2018; 11:eaat5185. https://doi.org/10.1126/scisignal.aat 5185.

45. Hoshino A, Mita Y, Okawa Y, Ariyoshi M, Iwai-Kanai E, Ueyama T, et al. Cytosolic p53 inhibits Parkin-mediated mitophagy and promotes mitochondrial dysfunction in the mouse heart. Nat Commun. 2013;4:2308.

46. Ren X, Chen L, Xie J, Zhang Z, Dong G, Liang J, et al. Resveratrol ameliorates mitochondrial elongation via Drp1/Parkin/PINK1 signaling in senescent-like cardiomyocytes. Oxid Med Cell Longev. 2017;2017:4175353. https://doi.org/10.1155/2017/4175353.

47. Cui S, Xue L, Yang F, Dai S, Han Z, Liu K, et al. Postinfarction hearts are protected by premature senescent cardiomyocytes via GATA4-dependent CCN1 secretion. J Am Heart Assoc. 2018;7:e009111. https://doi.org/10.1161/JAHA.118.009111.

48. Coppe JP, Desprez PY, Krtolica A, Campisi J. The senescence-associated secretory phenotype: the dark side of tumour suppression. Annu Rev Pathol. 2010;5:99–118.

49. Picca A, Mankowski RT, Burman JL, Donisi L, Kim JS, Marzetti E, et al. Mitochondrial quality control mechanisms as molecular targets in cardiac ageing. Nat Rev Cardiol. 2018;15:543–54.

50. Gude NA, Broughton KM, Firouzi F, Sussman MA. Cardiac ageing: extrinsic and intrinsic factors in cellular renewal and senescence. Nat Rev Cardiol. 2018;15:523–42.

51. Li H, Hastings MH, Rhee J, Trager LE, Roh JD, Rosenzweig A. Targeting age-related pathways in heart failure. Circ Res. 2020;126:533–51.

52. Kolwicz SC Jr, Purohit S, Tian R. Cardiac metabolism and its interactions with contraction, growth, and survival of cardiomyocytes. Circ Res. 2013;113:603–16.

53. Zhang X, Liu C, Liu C, Wang Y, Zhang W, Xing Y. Trimetazidine and L-carnitine prevent heart aging and cardiac metabolic impairment in rats via regulating cardiac metabolic substrates. Exp Gerontol. 2019;119:120–7.

54. Long Q, Liu J, Wang P, Zhou Y, Ding Y, Prasain J, et al. Carnitine Palmitoyltransferase-1b (CPT1b) deficiency aggravates pressure-overload-induced cardiac hypertrophy due to lipotoxicity. Circulation. 2012;126:1705–16.

55. Dillon LM, Rebelo AP, Moraes CT. The role of PGC-1 coactivators in aging skeletal muscle and heart. IUBMB Life. 2012;64:231–41.

56. Rodriguez-Calvo R, Serrano L, Barroso E, Coll T, Palomer X, Camins A, et al. Peroxisome proliferator-activated receptor alpha down-regulation is associated with enhanced ceramide levels in age-associated cardiac hypertrophy. J Gerontol. 2007;62:1326–36.

57. Ock S, Lee WS, Ahn J, Kim HM, Kang H, Kim HS, et al. Deletion of IGF-1 receptors in cardiomyocytes attenuates cardiac aging in male mice. Endocrinology. 2016;157:336–45.

58. Gu J, Wang S, Guo H, Tan Y, Liang Y, Feng A, et al. Inhibition of p53 prevents diabetic cardiomyopathy by preventing early-stage apoptosis and cell senescence, reduced glycolysis, impaired angiogenesis. Cell Death Dis. 2018;9:82.

59. Aubert G, Martin OJ, Horton JL, Lai L, Vega RB, Leone TC, et al. The failing heart relies on ketone bodies as a fuel. Circulation. 2016;133:698–705.

60. Bedi KC Jr, Snyder NW, Brandimarto J, Aziz M, Mesaros C, Worth AJ, et al. Evidence for intramyocardial disruption of lipid metabolism and increased myocardial ketone utilization in advanced human heart failure. Circulation. 2016;133:706–16.

61. Klos M, Morgenstern S, Hicks K, Suresh S, Devaney EJ. The effects of the ketone body β-hydroxybutyrate on isolated rat ventricular myocyte excitation-contraction coupling. Arch Biochem Biophys. 2019;662:143–50.

62. Schugar RC, Moll AR, Andre d'Avignon D, Weinheimer CJ, Kovacs A, Crawford PA. Cardiomyocyte-specific deficiency of ketone body metabolism promotes accelerated pathological remodeling. Mol Metab. 2014;3:754–69.

63. Wells L, Vosseller K, Hart GW. Glycosylation of nucleocytoplasmic proteins: signal transduction and O-GlcNAc. Science. 2001;291:2376–8.
64. Jensen RV, Andreadou I, Hausenloy DJ, Botker HE. The role of O-GlcNAcylation for protection against ischemia-reperfusion injury. Int J Mol Sci. 2019;20:404.
65. Champattanachai V, Marchase RB, Chatham JC. Glucosamine protects neonatal cardiomyocytes from ischemia-reperfusion injury via increased protein O-GlcNAc and increased mitochondrial Bcl-2. Am J Physiol Cell Physiol. 2008;294:C1509–20.
66. Jones SP, Zachara NE, Ngoh GA, Hill BG, Teshima Y, Bhatnagar A, et al. Cardioprotection by N-acetylglucosamine linkage to cellular proteins. Circulation. 2008;117:1172–82.
67. Turdi S, Fan X, Li J, Zhao J, Huff AF, Du M, et al. AMP-activated protein kinase deficiency exacerbates aging-induced myocardial contractile dysfunction. Aging Cell. 2010;9:592–606.
68. Tang X, Chen XF, Wang NY, Wang XM, Liang ST, Zheng W, et al. SIRT2 acts as a cardioprotective deacetylase in pathological cardiac hypertrophy. Circulation. 2017;136:2051–67.
69. Gélinas R, Mailleux F, Dontaine J, Bultot L, Demeulder B, Ginion A, et al. AMPK activation counteracts cardiac hypertrophy by reducing O-GlcNAcylation. Nat Commun. 2018;9:1–17.
70. Tang X, Chen XF, Chen HZ, Liu DP. Mitochondrial Sirtuins in cardiometabolic diseases. Clin Sci. 2017;131:2063–78.
71. Kane AE, Sinclair DA. Sirtuins and NAD+ in the development and treatment of metabolic and cardiovascular diseases. Circ Res. 2018;123:868–85.
72. Vakhrusheva O, Smolka C, Gajawada P, Kostin S, Boettger T, Kubin T, et al. Sirt7 increases stress resistance of cardiomyocytes and prevents apoptosis and inflammatory cardiomyopathy in mice. Circ Res. 2008;102:703–10.
73. Tang X, Ma H, Han L, Zheng W, Lu YB, Chen XF, et al. SIRT1 deacetylates the cardiac transcription factor Nkx2.5 and inhibits its transcriptional activity. Sci Rep. 2016; 6:36576. https://doi.org/10.1038/srep36576.
74. Hsu YJ, Hsu SC, Hsu CP, Chen YH, Chang YL, Sadoshima J, et al. Sirtuin 1 protects the aging heart from contractile dysfunction mediated through the inhibition of endoplasmic reticulum stress-mediated apoptosis in cardiac-specific Sirtuin 1 knockout mouse model. Int J Cardiol. 2017;228:543–52.
75. Sundaresan NR, Vasudevan P, Zhong L, Kim G, Samant S, Parekh V, et al. The sirtuin SIRT6 blocks IGF-Akt signaling and development of cardiac hypertrophy by targeting c-Jun. Nat Med. 2012;18:1643–50.
76. Luo YX, Tang X, An XZ, Xie XM, Chen XF, Zhao X, et al. Sirt4 accelerates Ang II-induced pathological cardiac hypertrophy by inhibiting manganese superoxide dismutase activity. Eur Heart J. 2017;38:1389–98.
77. Saucerman JJ, Tan PM, Buchholz KS, McCulloch AD, Omens JH. Mechanical regulation of gene expression in cardiac myocytes and fibroblasts XE "Fibroblast." Nat Rev Cardiol. 2019;16:361–78.
78. Costantino S, Paneni F, Cosentino F. Ageing, metabolism and cardiovascular disease. J Physiol. 2016;594:2061–73.
79. Hernandez-Segura A, Nehme J, Demaria M. Hallmarks of cellular senescence. Trends Cell Biol. 2018;28:436–53.
80. Colliva A, Braga L, Giacca M, Zacchigna S. Endothelial cell–cardiomyocyte crosstalk in heart development and disease. J Physiol. 2020;598:2923–39.
81. Ceylan-Isik AF, Dong M, Zhang Y, Dong F, Turdi S, Nair S, et al. Cardiomyocyte-specific deletion of endothelin XE "Endothelin" receptor A rescues aging-associated cardiac hypertrophy and contractile dysfunction: role of autophagy. Basic Res Cardiol. 2013;108:335.
82. Ceylan AF, Wang S, Kandadi MR, Chen J, Hua Y, Pei Z, et al. Cardiomyocyte-specific knockout of endothelin receptor a attenuates obesity cardiomyopathy. Biochim Biophys Acta. 2018;1864:3339–52.
83. Hu J, Wang S, Xiong Z, Cheng Z, Yang Z, Lin J, et al. Exosomal Mst1 transfer from cardiac microvascular endothelial cells to cardiomyocytes deteriorates diabetic cardiomyopathy. Biochim Biophys Acta. 2018;1864:3639–49.

84. Akbar N, Digby JE, Cahill TJ, Tavare AN, Corbin AL, Saluja S, et al. Endothelium-derived extracellular vesicles promote splenic monocyte mobilization in myocardial infarction. JCI Insight. 2017;2:e93344. https://doi.org/10.1172/jci.insight.93344.
85. Zhang W, Wang Q, Wu Y, Moriasi C, Liu Z, Dai X, et al. Endothelial Cell–specific liver kinase B1 deletion causes endothelial dysfunction and hypertension in mice in vivo. Circulation. 2014;129:1428–39.
86. Omura J, Satoh K, Kikuchi N, Satoh T, Kurosawa R, Nogi M, et al. Protective roles of endothelial AMP-activated protein kinase against hypoxia-induced pulmonary hypertension in mice. Circ Res. 2016;119:197–209.
87. Kamo T, Akazawa H, Komuro I. Cardiac nonmyocytes in the hub of cardiac hypertrophy. Circ Res. 2015;117:89–98.
88. Civitarese RA, Kapus A, McCulloch CA, Connelly KA. Role of integrins in mediating cardiac fibroblast–cardiomyocyte cross talk: a dynamic relationship in cardiac biology and pathophysiology. Basic Res Cardiol. 2017;112:6. https://doi.org/10.1007/s00395-016-0598-6.
89. Kakkar R, Lee RT. Intramyocardial fibroblast myocyte communication. Circ Res. 2010;106:47–57.
90. Schafer S, Viswanathan S, Widjaja AA, Lim WW, Moreno-Moral A, DeLaughter DM, et al. IL-11 is a crucial determinant of cardiovascular fibrosis. Nature. 2017;552:110–5.
91. Mohamed MA, Abou-Leisa R, Stafford N, Maqsood A, Zi M, Prehar S, et al. The plasma membrane calcium ATPase 4 signalling in cardiac fibroblasts mediates cardiomyocyte hypertrophy. Nat Commun. 2016;7:11074.
92. Bang C, Batkai S, Dangwal S, Gupta SK, Foinquinos A, Holzmann A, et al. Cardiac fibroblast–derived microRNA passenger strand-enriched exosomes mediate cardiomyocyte hypertrophy. J Clin Invest. 2014;124:2136–46.
93. Lyu L, Wang H, Li B, Qin Q, Qi L, Nagarkatti M, et al. A critical role of cardiac fibroblast-derived exosomes in activating renin angiotensin system in cardiomyocytes. J Mol Cell Cardiol. 2015;89:268–79.
94. Wang BX, Couch L, MacLeod KT, Harding SE, Terracciano CM. Extracellular vesicles secreted from human fibroblasts modulate human induced pluripotent stem cell-cardiomyocyte calcium cycling. Circulation. 2017;136:A19928.
95. Pillai CL, Li S, Romay M, Lam L, Lu Y, Huang J, et al. Cardiac fibroblasts adopt osteogenic fates and can be targeted to attenuate pathological heart calcification. Cell Stem Cell. 2017;20:218–32.
96. Dadson K, Chasiotis H, Wannaiampikul S, Tungtrongchitr R, Xu A, Sweeney G, et al. Adiponectin mediated APPL1-AMPK signaling induces cell migration, MMP activation, and collagen remodeling in cardiac fibroblasts. Cell Biochem. 2014;115:785–93.
97. Cieslik KA, Taffet GE, Crawford JR, Trial J, Mejia Osuna P, Entman ML, et al. AICAR-dependent AMPK activation improves scar formation in the aged heart in a murine model of reperfused myocardial infarction. Mol Cell Cardiol. 2013;63:26–36.
98. Vivar R, Humeres C, Muñoz C, Boza P, Bolivar S, Tapia F, et al. FoxO1 mediates TGF-beta1-dependent cardiac myofibroblast differentiation. Biochim Biophys Acta. 2016;1863:128–38.
99. Hulsmans M, Clauss S, Xiao L, Aguirre AD, King KR, Hanley A, et al. Macrophages facilitate electrical conduction in the heart. Cell. 2017;169:510–22.
100. Monnerat G, Alarcon ML, Vasconcellos LR, Hochman-Mendez C, Brasil G, Bassani RA, et al. Macrophage-dependent IL-1beta production induces cardiac arrhythmias in diabetic mice. Nat Commun. 2016;7:13344.
101. Zacchigna S, Martinelli V, Moimas S, Colliva A, Anzini M, Nordio A, et al. Paracrine effect of regulatory T cells promotes cardiomyocyte proliferation during pregnancy and after myocardial infarction. Nat Commun. 2018;9:2432.
102. Li J, Jubair S, Janicki JS. Estrogen inhibits mast cell chymase release to prevent pressure overload-induced adverse cardiac remodeling. Hypertension. 2015;65:328–34.
103. Zeisberg EM, Tarnavski O, Zeisberg M, Dorfman AL, McMullen JR, Gustafsson E, et al. Endothelial-to-mesenchymal transition contributes to cardiac fibrosis. Nat Med. 2007;13:952–61.

104. Song S, Liu L, Yu Y, Zhang R, Li Y, Cao W, et al. Inhibition of BRD4 attenuates transverse aortic constriction- and TGF-β-induced endothelial-mesenchymal transition and cardiac fibrosis. J Mol Cell Cardiol. 2019;127:83–96.

105. Wang Y, Zhang D, Chiu AP, Wan A, Neumaier K, Vlodavsky I, et al. Endothelial heparanase regulates heart metabolism by stimulating lipoprotein lipase secretion from cardiomyocytes. Arterioscler Thromb Vasc Biol. 2013;33:894–902.

106. Zhang D, Wan A, Chiu AP, Wang Y, Wang F, Neumaier K, et al. Hyperglycemia-induced secretion of endothelial heparanase stimulates a vascular endothelial growth factor autocrine network in cardiomyocytes that promotes recruitment of lipoprotein lipase. Arterioscler Thromb Vasc Biol. 2013;33:2830–8.

107. Dallinga-Thie GM, Franssen R, Mooij HL, Visser ME, Hassing HC, Peelman F, et al. The metabolism of triglyceride-rich lipoproteins revisited: new players, new insight. Atherosclerosis. 2010;211:1–8.

108. Wang F, Wang Y, Kim MS, Puthanveetil P, Ghosh S, Luciani DS, et al. Glucose-induced endothelial heparanase secretion requires cortical and stress actin reorganization. Cardiovasc Res. 2010;87:127–36.

109. Chiu AP, Wan A, Lal N, Zhang D, Wang F, Vlodavsky I, et al. Cardiomyocyte VEGF regulates endothelial cell GPIHBP1 to relocate lipoprotein lipase to the coronary lumen during diabetes mellitus. Arterioscler Thromb Vasc Biol. 2016;36:145–55.

110. Zhang X, Wang X, Zhu H, Kranias EG, Tang Y, Peng T, et al. Hsp20 functions as a novel cardiokine in promoting angiogenesis via activation of VEGFR2. PLoS One. 2012;7:e32765. https://doi.org/10.1371/journal.pone.0032765.

111. Wang X, Huang W, Liu G, Cai W, Millard RW, Wang Y, et al. Cardiomyocytes mediate anti-angiogenesis in type 2 diabetic rats through the exosomal transfer of miR-320 into endothelial cells. J Mol Cell Cardiol. 2014;74:139–50.

112. Feng T, Meng J, Kou S, Jiang Z, Huang X, Lu Z, et al. CCN1-induced cellular senescence promotes heart regeneration. Circulation. 2019;139:2495–8.

113. Fujiu K, Nagai R. Contributions of cardiomyocyte-cardiac fibroblast-immune cell interactions in heart failure development. Basic Res Cardiol. 2013;108:357.

114. Major JL, McKinsey TA. Putting the heat on cardiac fibrosis, Hsp20 regulates myocyte-to-fibroblast crosstalk. JACC Basic Transl Sci. 2019;4:200–3.

115. Yuan J, Liu H, Gao W, Zhang L, Ye Y, Yuan L, et al. MicroRNA-378 suppresses myocardial fibrosis through a paracrine mechanism at the early stage of cardiac hypertrophy following mechanical stress. Theranostics. 2018;8:2565–82.

116. Accornero F, Berlo JHV, Benard MJ, Lorenz JN, Carmeliet P, Molkentin JD. Placental growth factor regulates cardiac adaptation and hypertrophy through a paracrine mechanism. Circ Res. 2011;109:272–80.

117. Frangogiannis NG. The functional pluralism of fibroblasts in the infarcted myocardium. Circ Res. 2016;119:1049–51.

118. Wang X, Morelli MB, Matarese A, Sardu C, Santulli G. Cardiomyocyte-derived exosomal microRNA-92a mediates post-ischemic myofibroblast activation both in vitro and ex vivo. ESC Heart Fail. 2020.https://doi.org/10.1002/ehf2.12584.

119. Datta R, Bansal T, Rana S, Datta K, Datta Chaudhuri R, Chawla-Sarkar M, et al. Myocyte-Derived Hsp90 modulates collagen upregulation via biphasic activation of STAT-3 in fibroblasts during cardiac hypertrophy. Mol Cell Biol. 2017;37:e00611–6. https://doi.org/10.1128/MCB.00611-16.

120. Bokov A, Chaudhuri A, Richardson A. The role of oxidative damage and stress in aging. Mech Age Dev. 2004;125:811–26.

121. Di LF, Bernardi P. Mitochondrial function and myocardial aging: a critical analysis of the role of permeability transition. Cardiovasc Res. 2005;66:222–32.

122. Thompson LV. Oxidative stress, mitochondria and mtDNA-mutator mice. Exp Gerontol. 2006;41:1220–2.

123. Judge S, Leeuwenburgh C. Cardiac mitochondrial bioenergetics, oxidative stress, and aging. Am J Physiol. 2007;292:C1983–92.

124. Li SY, Du M, Dolence EK, Fang CX, Mayer GE, Ceylan-Isik AF, LaCour KH, Yang X, Wilbert CJ, Sreejayan N, Ren J. Aging induces cardiac diastolic dysfunction, oxidative stress, accumulation of advanced glycation endproducts and protein modification. Aging Cell. 2005;4:57–64.
125. Torella D, Rota M, Nurzynska D, Musso E, Monsen A, Shiraishi I, Zias E, Walsh K, Rosenzweig A, Sussman MA, Urbanek K, Nadal-Ginard B, Kajstura J, Anversa P, Leri A. Cardiac stem cell and myocyte aging, heart failure, and insulin-like growth factor-1 overexpression. Circ Res. 2004;94:514–24.
126. Kim YK, Suarez J, Hu Y, McDonough PM, Boer C, Dix DJ, Dillmann WH. Deletion of the inducible 70-kDa heat shock protein genes in mice impairs cardiac contractile function and calcium handling associated with hypertrophy. Circulation. 2006;113:2589–97.
127. Ren J, Li Q, Wu S, Li SY, Babcock SA. Cardiac overexpression of antioxidant catalase attenuates aging-induced cardiomyocyte relaxation dysfunction. Mech Age Dev. 2007;128:276–85.
128. Terman A, Gustafsson B, Brunk UT. The lysosomal-mitochondrial axis theory of postmitotic aging and cell death. Chem Biol Interact. 2006;163:29–37.
129. Bodnar AG, Ouellette M, Frolkis M, Holt SE, Chiu CP, Morin GB, Harley CB, Shay JW, Lichtsteiner S, Wright WE. Extension of life-span by introduction of telomerase into normal human cells. Science. 1998;279:349–52.
130. Griffith JD, Comeau L, Rosenfield S, Stansel RM, Bianchi A, Moss H, de Lange T. Mammalian telomeres end in a large duplex loop. Cell. 1999;97:503–14.
131. d'Adda di Fagagna F, Reaper PM, Clay-Farrace L, Fiegler H, Carr P, von Zglinicki T, Saretzki G, Carter NP, Jackson SP. A DNA damage checkpoint response in telomere-initiated senescence. Nature. 2003; 426:194–198.
132. Anderson R, Richardson GD, Passos JF. Mechanisms driving the ageing heart. Exp Gerontol. 2018;109:5–15.
133. Sapieha P, Mallette FA. Cellular senescence in postmitotic cells: beyond growth arrest. Trends Cell Biol. 2018;28:595–607.
134. Fumagalli M, Rossiello F, Clerici M, Barozzi S, Cittaro D, Kaplunov JM, Bucci G, Dobreva M, Matti V, Beausejour CM, Herbig U, Longhese MP, d'Adda di Fagagna F. Telomeric DNA damage is irreparable and causes persistent DNA-damage-response activation. Nat Cell Biol. 2012; 14:355–365.
135. Hewitt G, Jurk D, Marques FDM, Correia-Melo C, Hardy T, Gackowska A, Anderson R, Taschuk M, Mann J, Passos JF. Telomeres are favoured targets of a persistent DNA damage response in ageing and stressinduced senescence. Nat Commun. 2012;3:708.
136. Purcell NH, Wilkins BJ, York A, Saba-El-Leil MK, Meloche S, Robbins J, Molkentin JD. Genetic inhibition of cardiac ERK1/2 promotes stress-induced apoptosis and heart failure but has no effect on hypertrophy in vivo. Proc Natl Acad Sci USA. 2007;104:14074–9.
137. Tsujita Y, Muraski J, Shiraishi I, Kato T, Kajstura J, Anversa P, Sussman MA. Nuclear targeting of Akt antagonizes aspects of cardiomyocyte hypertrophy. Proc Natl Acad Sci USA. 2006;103:11946–51.
138. Ancey C, Menet E, Corbi P, Fredj S, Garcia M, Rucker-Martin C, Bescond J, Morel F, Wijdenes J, Lecron JC, Potreau D. Human cardiomyocyte hypertrophy induced in vitro by gp130 stimulation. Cardiovasc Res. 2003;59:78–85.
139. Lewis-McDougall FC, Ruchaya PJ, Domenjo-Vila E, Shin Teoh T, Prata L, Cottle B.J, et al. Aged-senescent cells contribute to impaired heart regeneration. Aging Cell. 2019; 18:e12931.
140. Walaszczyk A, Dookun E, Redgrave R, Tual-Chalot S, Victorelli S, Spyridopoulos I, et al. Pharmacological clearance of senescent cells improves survival and recovery in aged mice following acute myocardial infarction. Aging Cell. 2019; 18:e12945.
141. Kirkland JL, Tchkonia T. Cellular senescence: a translational perspective. Ebiomedicine. 2017; 21:21–28.
142. Kirkland JL, Tchkonia T, Zhu Y, Niedernhofer LJ, Robbins PD. The clinical potential of senolytic drugs. J Am Geriatr Soc. 2017;65:2297–301.

Chapter 13
Mechanisms of Cardiomyocyte Death

Abstract Several forms of cell death have been recognized. They include apoptosis, autophagy, necrosis, necroptosis, pyroptosis and ferroptosis. Apoptosis, necrosis and autophagic cell death are seen in cardiomyocytes during progression of heart disease. Either progressive or acute cell death is a feature of myocardial infarction, ischemia/reperfusion, heart failure as well as drug-induced cardiac toxicity. Several proteins are involved in cardiomyocyte death signalling. Pharmacological and genetic inhibition of autophagy, apoptosis or necrosis can improve cardiac function in pathological states.

Keywords Cardiomyocytes · Cardiotoxicity · Cell death · Apoptosis · Autophagy · Necrosis · Necroptosis · Pyroptosis · Ferroptosis · Death signalling

Introduction

Several types of cell death are known. These include apoptosis, autophagy and necrosis, necroptosis, pyroptosis and ferroptosis. Autophagic cell death, apoptosis and necrosis, are seen during progression of heart disease [1]. Either progressive or acute cell death is a hallmark of myocardial infarction, ischemia/reperfusion and heart failure. and drug-induced drug toxicity [1]. Several drugs also induce death of cardiomyocytes through various mechanisms [2].

Types of Cell Death

Apoptosis

Apoptosis is the most widely studied form of cell death. Cell shrinkage, increase in cytoplasmic density, reduced mitochondrial membrane potential (MMP) and changes in permeability are the features of apoptosis. Apoptotic bodies that are formed are taken up and degraded by neighbouring cells.

© The Author(s), under exclusive license to Springer Nature Switzerland AG 2021
C. C. Kartha, *Cardiomyocytes in Health and Disease*,
https://doi.org/10.1007/978-3-030-85536-9_13

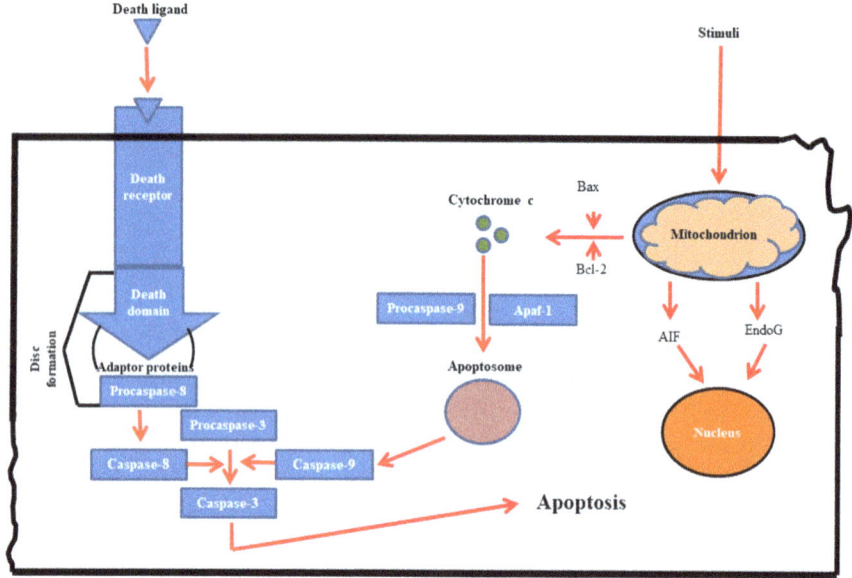

Fig. 13.1 Intrinsic, extrinsic and terminal signalling pathways for apoptosis. AIF-Apoptosis-inducing factor. Endo G-Endonuclease G

Apoptosis has intrinsic and extrinsic mechanisms (Fig. 13.1). Intrinsic apoptosis is caused by DNA damage, failure of mitosis, increased oxidative stress, endoplasmic reticulum (ER) stress and loss of growth factor signalling [3–6]. Intrinsic cell apoptosis is regulated primarily by pro-apoptotic members of the B cell lymphoma-2 (Bcl-2) family (Bax, Bak, and BH3-only protein) which act on the mitochondria. Bcl-2 induces translocation of Bax/Bak into the mitochondria and permeabilization of the mitochondrial membrane, resulting in the release of cytochrome C into the cytoplasm, formation of the apoptosome and initiation of the caspases cascade (Fig. 13.2) [7].

Extrinsic apoptosis is triggered by Fas cell surface death receptor (Fas) and tumour necrosis factor receptor (TNF) super family members (TNFR1, TNFRSF10A, and TNFRSF10B), located in the plasma membrane coupled with their specific homologous ligands. Death-inducing signalling complex (DISC) composed of death ligands and their receptors on the cell membrane link death receptor signalling with caspases cascade pathway (Fig. 13.1). Caspase enzyme cleaves the pro-apoptotic proteins caspase-8 and caspase10 to trigger apoptosis [8, 9].

Both extrinsic and the intrinsic pathways of apoptosis are present in cardiomyocytes (CMs) [1]. CMs express both Fas and TNF receptor 1. The extrinsic apoptotic pathway can be triggered by Fas ligand, tumour necrosis factor-α (TNF-α), or TRAIL (TNF-related apoptosis inducing ligand) [1]. The cells release TRAIL as well [10]. Cardiomyocyte-specific overexpression of TNF-α in transgenic mice causes dilated

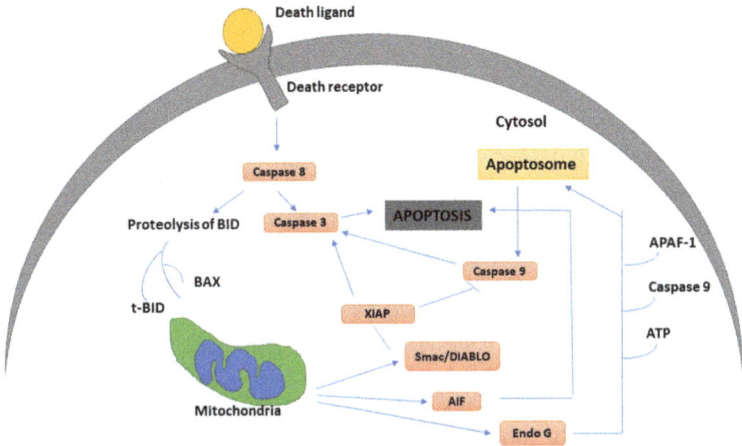

Fig. 13.2 Pathways for apoptosis in cardiomyocytes. AIF—Apoptosis inducing factor. APAF—Apoptosis protease activating factor. ATP—Adenosine triphosphate. BAX—BCL-2 associated X protein. BID—BH3 interacting domain death agonist. DIABLO—Direct IAP (inhibitor of apoptosis protein) binding protein. Endo G—Endonuclease G. SMAC—Second mitochondria-derived activator of caspases. XIAP—X-linked inhibitor of apoptosis protein

cardiomyopathy and heart failure, indicating the damaging effect of activation of the death receptor pathway by TNF-α [11].

CMs have strategies to regulate the intrinsic apoptotic pathway [1, 12]. They express various members of the BCL2 family, which include antiapoptotic and pro-apoptotic Bcl-2 proteins [13]. They also express low levels of Apaf1 (apoptosis protease activating factor 1), which controls caspase activation [14].

Mitochondria in Apoptosis of Cardiomyocytes

The mitochondrion is the primary organelle involved in the intrinsic apoptotic pathway (Fig. 13.2). Activation of the mitochondrial apoptotic pathway leading to activation of executioner caspase is important in cardiac injury [1, 15, 16]. Levels of all caspases in CMs decrease with age [17]. In the heart which lacks caspase activation, apoptosis can be induced through caspase-independent pathways [18]. Apoptosis repressor with caspase recruitment domain (ARC) can inhibit both the intrinsic and extrinsic pathways. ARC inhibits the extrinsic pathway by interacting with caspase 8 and components of the death-inducing signalling complex, such as FADD (Fas associated via death domain). Intrinsic pathway is inhibited by blocking Bax activation and translocation into mitochondria [19].

Apoptosis inducing factor (AIF) which is anchored to the mitochondrial inner membrane is released from mitochondria during ischemia/reperfusion and has been implicated in cardiomyocyte death induced by oxidative stress [20]. AIF is cleaved and released into the cytosol upon induction of apoptosis. AIF then translocates to the nucleus and mediates large-scale DNA fragmentation [21].

Endonuclease G (Endo G), localized to the intermembrane space of mitochondria also participates in cardiomyocyte apoptosis mediated by I/R [17]. Endo G translocates to the nucleus, where it cleaves DNA.

The intrinsic pathway can be activated by the extrinsic apoptotic pathway through caspase 8-dependent cleavage of Bcl2-interacting protein (BID) (Fig. 13.2). The C-terminal fragment of the resulting truncated BID (t-BID) translocates to the mitochondrial outer membrane, where it is thought to activate the intrinsic pathway [1].

There is continuous counterbalanced events of fusion and fission in mitochondria [22]. DRP-1 (dynamin related protein-1) and FIS1 (mitochondrial fission 1 protein) regulate mitochondrial fission [22]. Mitofusins 1/2 (MFN1/2) and the GTPase OPA1 (optic atrophy gene 1) regulate mitochondrial fusion [22]. Apoptotic stimuli induce fragmentation of the mitochondrial network in CMs [23]. Mitochondrial membrane damage is mediated by mechanisms that involve DRP-1, MFN2, and the pro-apoptotic protein Bax [22–24].

Autophagy

The term 'autophagy' refers to a cellular catabolic process in which intracellular substances are degraded by itself through lysosome-based mechanisms [25–27]. Autophagy is a pro-survival mechanism for maintaining metabolic homeostasis in cells. It helps to destroy unwanted or damaged cellular components [28]. During the autophagy process, some long-lived or misfolded proteins as well as damaged organelles are delivered into lysosomes and degraded into nutrients such as amino acids for recycling and new use [29, 30]. More than 30 kinds of autophagy-related genes (Atgs) and proteins involved in the process of autophagy have been discovered till date [31, 32].

Autophagy has been mainly classified into macroautophagy, microautophagy, and chaperone-mediated autophagy (Fig. 13.3) [33]. There are other selective forms of autophagy. These are mitophagy, pexophagy, ribophagy, xenophagy, and secretory autophagy [34–38].

Macroautophagy, is the most common form of autophagy. In this process, substances are sequestered within cytosolic vesicles composed of a lipid bilayer membrane, termed phagophore. Phagophore matures into autophagosomes and the cargo is later transferred to lysosomes for degradation [39]. Macroautophagy occurs in two major steps [25, 39–42]. In the first step, the Unc-51-like kinase 1 (ULK1), Atg13, and Atg101 combine to form a complex. The Atg complex triggers the assembly of Beclin-1, Atg14, VPS15 (vacuolar protein sorting 15) and VPS34. The resulting Class III phosphatidylinositol 3-hydroxy kinase (PI3K) complex (Class III PI3K complex) induces the formation of the phagophore. Membrane expansion and fusion with Atg5-Atg12-Atg16L1 and light chain 3 (LC3) results in the formation of autophagosomes. In the second step, autophagosomes integrate with lysosomes to form autolysosomes. They are the functional units of autophagy for degradation and recycling of cytoplasmic constituents.

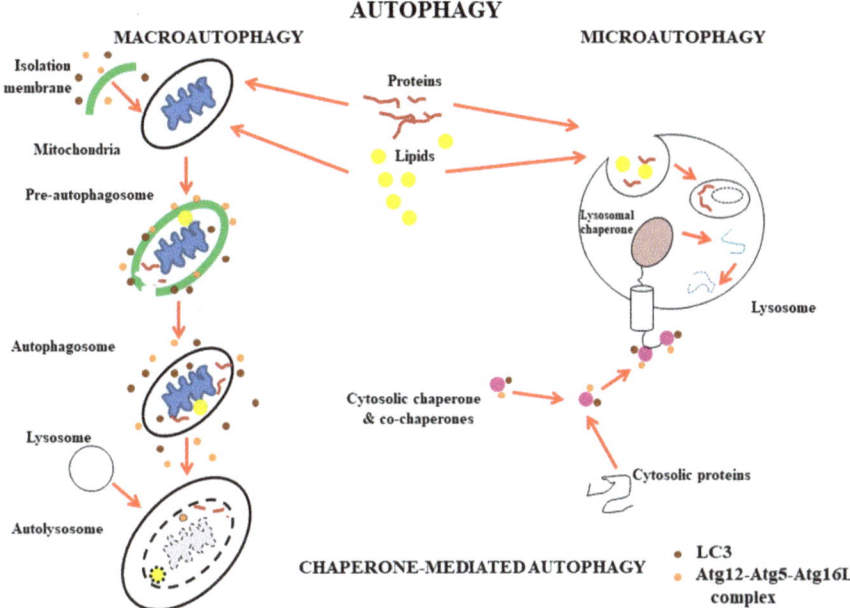

Fig. 13.3 Mechanisms of different types of autophagy. Atg—autophagy related genes. LC3—Microtubule associated protein 1A/1B—Light chain 3

Microautophagy involves engulfment of cytoplasmic constituents through invagination of the lysosomal/vacuolar membranes [43]. Chaperone-mediated autophagy, which depends on chaperones permits selective degradation of soluble cytosolic proteins in lysosomes [44].

Two classical signalling pathways regulate autophagy [41, 45]. The Class I PI3K-mammalian target of rapamycin (mTOR) signalling pathway is an inhibitory pathway. It is triggered by a nutrient enriched milieu and activates mTOR and the mTOR complex (mTORC1) via protein kinase B (Akt) pathway. Formation of the Atg1 complex is thus inhibited [46, 47]. The second signalling pathway is induced by AMP-activated protein kinase (AMPK), which senses stress. It promotes autophagy by inactivating mTORC1 or phosphorylating serine residues of ULK1 and thus activating the ULK1 kinase complex [48, 49].

Autophagy has also a protective role in organisms. For example, autophagy has been shown to inhibit apoptosis of cells [50, 51]. Autophagy can as well suppress inflammatory and immune reactions in several diseases [52–62]. Over induction of autophagy may however lead to 'autophagic cell death' [39, 60].

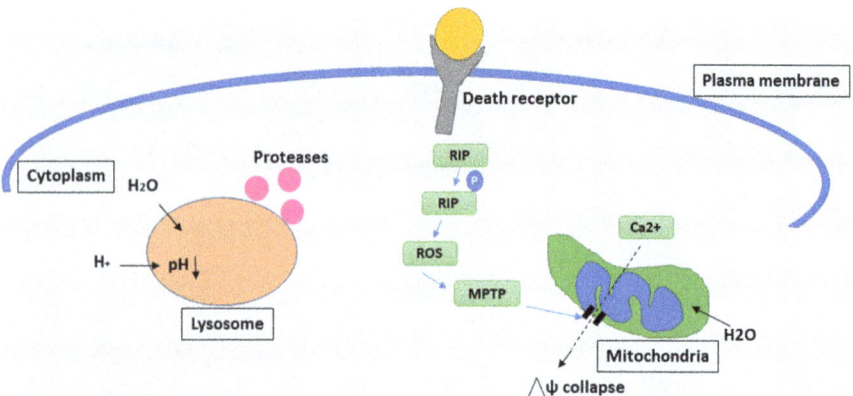

Fig. 13.4 Pathways for necrosis in cardiomyocytes. RIP—Receptor interacting protein. ROS—reactive oxygen species. MPTP—Mitochondrial permeability transition pore. Ψ—Membrane potential

Necrosis

Necrosis is a form of passive and irreversible cell death and is always associated with diseases. Necrosis is characterized by swelling of cell and organelles, damage to the cell membrane, loss of ATP and disintegration of organelles. Release of cellular contents initiate an inflammatory response resulting in pathological changes in tissues [1]. Necrosis is chiefly caused by physical or chemical injury to the cell and is considered as an unregulated form of cell death [63]. A number of mechanisms that may initiate and execute necrosis have been discovered. These involve death receptors, reactive oxygen species (ROS), Ca^{2++}, and mitochondrial permeability transition pore (MPTP) (Fig. 13.4) [63, 64].

Recent evidences suggest that a proportion of necrosis may be governed by regulated signalling events. This type of necrosis is known as programmed necrosis, caspase-independent cell death and necroptosis [65].

Necroptosis

Specific transduction mechanisms regulate necroptosis. Morphological features of necroptosis are similar to necrosis and pryoptosis and ferroptosis, the other regulated forms of cell death.

Death receptor TNFR1 is involved in necroptosis [66]. TNFR1 activation stimulates RIPK1 (receptor interacting protein kinase 1)) which recruits RIPK3 and result in the formation of necrosomes [67]. Sequential activation of RIPK3/MLKL (mixed lineage kinase domain like pseudokinase) is also important in necroptosis signalling

[68]. Phosphorylated MLKL can damage plasma membrane and organelles. Inflammatory factors thus released elicit an immune response [69]. Caspase-8 can inactivate RIPK1 and RIPK3 and play an inhibitory role in necroptosis [70, 71].

Pyroptosis

Pyroptosis is an inflammatory and regulated form of cell death and occurs in defense of pathogens such as virus, bacteria and fungi [72, 73]. Activation of caspase-1, caspase-3, caspase-4 and caspase-11 is necessary for triggering pyroptosis. Caspases cleaves GSDMD (Gasdermin D) or GSDME to damage the cell membrane and release interleukin-1β (IL-1β) and IL-18, inducing pyroptosis [74, 75].

Ferroptosis

Ferroptosis is characterized by the intracellular increase of iron and lipid ROS, which can deplete anti-oxidases and damage mitochondria, and thus cause cell death [76]. Inhibition of Toll like receptor 4 (TLR4) and triphosphopyridine nucleotide oxidase 4 (NOX4) can significantly reduce ferroptosis [54]. Glutathione peroxidase 4 (GPx4) suppresses lipid peroxidation and thus can prevent ferroptosis induced by erastin and RSL3 (RAS selective lethal) [77].

Cardiac Myocyte Death Mechanisms in Heart Diseases

Ischemia and Ischemia/Reperfusion (I/R)

Most of the in vivo and in vitro studies have revealed that ischemia or hypoxia induces autophagy [78–80]. The two pathways responsible for ischemia or hypoxia-induced autophagy involve either BNIP38 or AMPK [81]. If ischemia is prolonged, the autophagic response becomes dysfunctional. During reperfusion, autophagy is further upregulated despite restoration of oxygen and nutrients [78, 82]. The mechanisms for sustained activation of autophagy during reperfusion is different from those in ischemia. Oxidative stress, mitochondrial damage/BNIP3, endoplasmic reticulum stress, and calcium overload are possibly more important in sustaining autophagy during reperfusion [83].

While autophagy has a protective role during mild-to-moderate ischemia, upregulation of autophagy can be either beneficial or detrimental during I/R [78, 82]. Clearance of autophagosome is impaired during I/R. Ischemia induces a decline in the levels of LAMP2, important for autophagosome–lysosome fusion. Upregulation of Beclin 1 during reperfusion impairs autophagosome processing further, leading to increased production of ROS, mitochondrial permeabilization, and death of CMs [84].

Myocardial Infarction

Permanent coronary occlusion induces maximum cardiac myocyte apoptosis at 4.5 h, whereas necrosis peaks at 24 h [85]. Reperfusion accelerates the timing of apoptosis [86]. FAS, and not TNFR, is the key contributor to the activation of the extrinsic apoptotic pathway during MI [87–89]. The intrinsic apoptosis pathway also has a major role in MI [90–92].

There is no direct evidence for any specific role of autophagy in myocardial infarction (MI). [93]. Autophagy could be important in the peri-infarct zone where cardiomyocytes have sublethal injury. Autophagy may also contribute to postinfarction remodelling. There are studies which indicate that agents which can modulate autophagy enhancing pathways are protective [94, 95].

During MI, activation of anaerobic glycolysis results in accumulation of H^+ and acidosis [1]. This leads to elevated levels of intracellular Na^{++} and finally in increased levels of Ca^{2++} in the cytoplasm. The mitochondrial Ca^{2++} uniporter transports Ca^{2++} into mitochondria, increasing the levels of Ca^{2++} in the mitochondria. Ca^{2++}-dependent dehydrogenase is activated leading to decline in NADH and electron flux through the electron transport chain, increase in ROS, decrease in ATP levels and loss of mitochondrial membrane potential. Oxygen and ATP-generating capabilities are restored upon reperfusion. ATP levels and mitochondrial membrane potential are also recovered. These alterations cause prolonged opening of MPTP and mitochondrial swelling finally leading to necrosis of cells [1].

Cardiac Failure

About 0.12–0.70% apoptotic cells can be found in the hearts of patients with severe heart failure [96]. This is significantly higher than the reported prevalence of 0.01–0.001% in the normal human myocardium. Even low levels of apoptosis can have significant effects given the limited capacity of adult cardiomyocytes to proliferate. An apoptotic rate of 0.1% has been estimated to result in a 37% loss of cardiomyocyte during one year [97]. Downregulation of XIAP and cIAP1/2 in cardiomyocytes is considered as the basis for increased apoptosis of cardiomyocytes in the failing heart [98–100].

In patients with heart failure (HF) secondary to dilated cardiomyopathy, cardiomyocyte death with features of autophagy has been observed at a rate of 0.03% [15]. Studies in mice after genetic manipulation of key autophagic pathways reveal that autophagy facilitates development of pathological hypertrophy [101, 102]. Indirect pharmaceutical manipulations of autophagy however indicate that autophagy can have an anti-hypertrophic role [103–105]. A concept of an optimal, 'adaptive zone' of autophagic activation has been advanced to explain the conflicting observations [12].

Though early studies suggested that apoptosis is the key factor for cell death in end-stage heart failure, later findings indicate that necrosis contributes many times more to disease pathogenesis than apoptosis [106]. Ca^{2+} stress, along with sustained activation of adrenergic receptors, triggers necrotic cell death [107]. MPTP opening is also involved [108, 109].

Cardiotoxic Drugs and Cardiomyocyte Death

Anticancer Drug—Doxorubicin (DOX)

Increased production of ROS is known to be the typical mechanism for DOX induced death of CMs. ROS production may be caused by the chemical interaction of DOX and NADPH. The generated ROS further causes DNA damage [110, 111]. In addition, DOX produces changes in intracellular iron which also contributes to ROS generation [112, 113]. Intracellular iron also has a role in in the degradation of hypoxia-inducible factors (HIF) [114]. Topoisomerase 2 beta (Top2β) also seems to be involved in ROS formation by DOX [115]. Top2β interaction with DOX leads to DNA double-strand breaks and transcriptome changes in cardiomyocytes [116]. The DOX-Top2β alliance may inhibit the transcription of peroxisome proliferator-activated receptor gamma coactivator-1 (PGC1) including PGC1α and PGC1β, involved in mitochondrial biogenesis [117, 118].

Doxorubicin-induced oxidative stress can activate death reporters, which in turn can stimulate caspase cascades [119]. The matricellular protein CCN1 triggered by DOX reacts with integrin α6β1 to activate p38 mitogen-activated protein kinase (p38-MAPK), which aids the release of second mitochondrial activator of caspase (SMAC) and high-temperature requirement protein A2 (HtrA2). They synergize with Fas L to induce apoptosis in cardiomyocytes [120].

TNFR1 is also involved in DOX-associated death of CMs. DOX alters the level of TNFα resulting in an increase in TNFR1 expression and reduction in TNFR2 expression, followed by the activation of caspase-8 and suppression of IκBα [121]. Another suggested mechanism is alteration of mitochondrial oxido-reduction balance in the cells caused by oxidative stress induced by DOX. DOX promotes ROS overproduction within the mitochondria and induces mitochondrial dysfunction [122]. DOX also down regulates antioxidant enzymes such as copper, manganese and zinc superoxide dismutases (SODs), glutathione peroxidase (GSH-Px) and catalase [123]. The imbalance between oxidation and antioxidation intensifies mitochondrial damage.

Several studies indicate that different DOX dosages induce apoptosis through varying pathways. A high concentration of DOX tend to promote ROS accumulation, while a lower concentration is more likely to inhibit the expression of haem oxygenase

1 (HO-1). Down regulation of HO-1 activates caspase-3 and thus induce cardiomyocyte apoptosis [124]. A high concentration of DOX also tends to cause DNA damage, PARP-1 dissociation and severe apoptosis; a low concentration activates the p53-related mitochondrial apoptosis pathway [125]. A dose-dependent increase in p53 expression caused by DOX inhibits transcription of type 1 insulin-like growth factor receptor (IGF-1R) and induces transcription of IGF binding protein-3 (IGFBP-3), resulting in resistance to IGF-1 and apoptosis [126]. Sirtuin 1 (SIRT1)-mediated deacetylation of p53 may be involved in the regulation of p53 by DOX [127].

Though autophagy is initially activated to thwart the toxic effects of DOX, overwhelming oxidative stress inhibits the degradation of lysosomes and leads to autophagic cell death. Increase in ROS induced by DOX increases the ratio of LC3II/LC3I and Beclin 1 level [128]. DOX also up-regulates the levels of pro-autophagy factors such as p53, p38-MAPK, and JNK-MAPK, and down-regulates the expression of p85, the catalytic subunit of phosphoinosmde-3-kinase (PI3K) as well as Akt phosphorylation [129, 130].

Increase in ubiquitinated proteins secondary to perturbations in protein degradation in cardiomyocytes cause the accumulation of autophagy flux and autophagosomes [131]. DOX suppresses acidification of lysosomes and degradation of autolysosomes, which stop the autophagic flux and increases cellular damage [132]. DOX also upregulates histone deacetylase 6 (HDAC6) leading to mitochondrial dysfunction and disturbance in autophagy flux [133]. DOX can suppress the expression of transcription factor EB (TFEB) and impair lysosomal cathepsin B, followed by inhibition of lysosomal autophagy, increase in the levels of ROS and caspase-3 cleavage [134].

DOX modulates autophagy-related factors as well and cause autophagic cell death. DOX increases the expression of high mobility group box 1 (HMGB1), which has an important role in autophagy mechanisms. HMGB1 silencing attenuates autophagy and can reverse the damage to CMs [135]. The transcription factor GATA4 is inhibited in DOX-treated CMs. GATA4 is a silencer of autophagy. GATA4 induces the expression of Bcl2, which can interact with Beclin 1 and silence autophagy [136].

A high dosage or prolonged exposure to DOX initiate cardiomyocyte necrosis. The rate of necrosis in CMs increases with the accumulation of ROS and peroxyl nitrite [111, 137]. During prolonged exposure of CMs to DOX, initial apoptosis develops into necrosis because of damage to DNA [138]. The delayed autophagy in CMs results in severe necrosis secondary to apoptosis [131, 139]. DOX initiates necrotic cell death also by modulating necrosis-related factors within the cell. DOX induces degradation of titin [140]. In addition, DOX activates BH3-only protein BNIP3 and damage the respiratory chain complex IV subunit 1 (COX1) and uncoupling protein 3 (UCP3), which affects respiratory efficiency and finally, leads to necrotic cell death [141].

DOX can activate RIPK3 which binds with phosphorylated-CaMKII, leading to the opening of mitochondrial permeability transition pore (MTPT), and necroptosis [142].

DOX can cause pyroptosis by increasing secretion of IL-1β and IL-18, and activating TLR4, NLRP3 inflammasome and caspases [143]. Dox-induced pyroptosis is mediated by activation of DRP1/NOX signalling as well as increase in the expression of NLRP3, mediated via recruitment of IGF2BP1 by upregulated lncRNA [144, 145]. Embryonic stem cells-derived exosomes and heat-shock protein 22 have been shown to inhibit TLR4/NLRP3/caspase1 signalling and reverse DOX-induced pyroptosis in CMs [146, 147]. DOX-induced pyroptosis in CMs can be reduced by ROS suppression which inhibits sirtuin 1/NLRP3 signalling pathway [148, 149].

DOX-induced accumulation of ROS and lipid peroxidation can cause ferroptosis of CMs [150]. Activation of TLR 4 and NOX 4 also promote DOX-induced ferroptosis [151].

Antidiabetic Drugs—Glitazones

Rosiglitazone a known PPARγ agonist can cause cardiomyocyte toxicity independent of PPARγ, via mitochondrial dysfunction induced by oxidative stress. Rosiglitazone can increase the levels of NAPHD and iNOS, and also exhaust antioxidant enzymes such as SOD and glutathione reductase, leading to apoptosis [152]. Pioglitazone upregulates the levels of sphingomyelinase and ceramidase, a mediator of apoptosis of CMs [153]. Pioglitazone can induce apoptosis in a VEGFR-2 dependent manner, by activating Bax and phosphorylated p53 as well as by suppressing phosphorylated Akt and mTOR [154].

Antiviral Drug—Zidovudine

Zidovudine, a nucleoside reverse transcriptase inhibitor, widely used for human immunodeficiency virus type 1 (HIV-1) infection can cause accumulation of ROS and peroxynitrite, leading to single-strand DNA breaks and mitochondrial energy depletion [155]. Zidovudine promotes the transport of protein kinase C δ (PKCδ) from cytoplasm to membrane, enhancing the activation of NADPH oxidases [156]. Increase in mitochondrial ROS brought about by zidovudine activates caspase-3/7 and results in apoptosis [157]. Fas/Fas L is also involved in apoptosis of cardiomyocytes actuated by zidovudine [158]. Zidovudine inhibits maturation of autophagosome and decreases autophagic flux as well, leading to accumulation of ROS [159]. Zidovudine-induced necrosis of CMs is considered to involve activation of PARP [157].

Conclusion

Either progressive or acute cell death is a feature of myocardial infarction, ischemia/reperfusion, heart failure as well as drug-induced cardiac toxicity. Six types of cardiomyocyte death are recognized. They are apoptosis, autophagy, necrosis, necroptosis, pyroptosis and ferroptosis. Mechanisms for different forms of cell death and the related signalling pathways have been much elucidated. They regulate loss of cells in ischemic and failing hearts. Caspase-8 seems to be the molecular switch that regulates intrinsic and extrinsic apoptosis, necroptosis and pyroptosis. It is important to unravel the contribution of each form of cell death in various cardiac diseases and drug-induced cardiotoxicity to prevent cardiomyocyte death in pathological states.

References

1. Whelan RS, Kaplinskiy V, Kitsis RN. Cell death in the pathogenesis of heart disease: mechanisms and significance. Annu Rev Physiol. 2010;72:19–44.
2. Wanjun MA, Weil S, Zhang B, Li W. Molecular mechanisms of cardiomyocyte death in drug-induced cardiotoxicity. Front Cell Dev Biol. 2020; 8:434. https://doi.org/10.3389/fcell.2020.00434.
3. Brumatti G, Salmanidis M, Ekert PG. Crossing paths: interactions between the cell death machinery and growth factor XE "Growth factors" survival signals. Cell Mol Life Sci. 2010;67:1619–30.
4. Wu CC, Bratton SB. Regulation of the intrinsic apoptosis pathway by reactive oxygen species. Antioxid Redox Signal. 2013;19:546–58.
5. Czabotar PE, Lessene G, Strasser A, Adams JM. Control of apoptosis by the BCL-2 protein family: implications for physiology and therapy. Nat Rev Mol Cell Biol. 2014;15:49–63.
6. Roos WP, Thomas AD, Kaina B. DNA damage and the balance between survival and death in cancer biology. Nat Rev Cancer. 2016;16:20–33.
7. Hutt KJ. The role of BH3-only proteins in apoptosis within the ovary. Reproduction. 2015;149:R81–9.
8. Barnhart BC, Alappat EC, Peter ME. The CD95 type I/type II model. Semin Immunol. 2003;15:185–93.
9. Yang JK. Death effecter domain for the assembly of death-inducing signaling complex. Apoptosis. 2015;20:235–9.
10. Liao X, Wang X, Gu Y, Chen Q, Chen LY. Involvement of death receptor signaling in mechanical stretch-induced cardiomyocyte apoptosis. Life Sci. 2005;77:160–74.
11. Kubota T, McTiernan CF, Frye CS, Slawson SE, Lemster BH, Koretsky AP, et al. Dilated cardiomyopathy in transgenic mice with cardiac-specific overexpression of tumor necrosis factor-a. Circ Res. 1997;81:627–35.
12. Lee Y, Gustafsson AB. Role of apoptosis in cardiovascular disease. Apoptosis. 2009;14:536–48.
13. Condorelli G, Morisco C, Stassi G, Notte A, Farina F, Sgaramella G, et al. Increased cardiomyocyte apoptosis and changes in proapoptotic and antiapoptotic genes bax and bcl-2 during left ventricular adaptations to chronic pressure overload in the rat. Circulation. 1999;99:3071–8.
14. Potts MB, Vaughn AE, McDonough H, Patterson C, Deshmukh M. Reduced Apaf-1 levels in cardiomyocytes engage strict regulation of apoptosis by endogenous XIAP. J Cell Biol. 2005;171:925–30.
15. Kostin S, Pool L, Elsasser A, Hein S, Drexler HC, Arnon E, et al. Myocytes die by multiple mechanisms in failing human hearts. Circ Res. 2003;92:715–24.
16. Holly TA, Drincic A, Byun Y, Nakamura S, Harris K, Klocke FJ, et al. Caspase inhibition reduces myocyte cell death induced by myocardial ischemia and reperfusion in vivo. J Mol Cell Cardiol. 1999;31:1709–15.
17. Bahi N, Zhang J, Llovera M, Ballester M, Comella JX, Sanchis D. Switch from caspase dependent to caspase-independent death during heart development: essential role of endonuclease G in ischemia-induced DNA processing of differentiated cardiomyocytes. J Biol Chem. 2006;281:22943–52.

18. Bae S, Siu PM, Choudhury S, Ke Q, Choi JH, Koh YY, et al. Delayed activation of caspase independent apoptosis during heart failure in transgenic mice overexpressing caspase inhibitor Crm A. Am J Physiol. 2010;299:H1374–81.

19. Gustafsson AB, Tsai JG, Logue SE, Crow MT, Gottlieb RA. Apoptosis repressor with caspase recruitment domain protects against cell death by interfering with Bax activation. J Biol Chem. 2004;279:21233–8.

20. Chen M, Zsengeller Z, Xiao CY, Szabo C. Mitochondrial-to-nuclear translocation of apoptosis-inducing factor in cardiac myocytes during oxidant stress: potential role of poly (ADP-ribose) polymerase-1. Cardiovasc Res. 2004;63:682–8.

21. Daugas E, Susin SA, Zamzami N, Ferri KF, Irinopoulou T, Larochette N, et al. Mitochondri-onuclear translocation of AIF in apoptosis and necrosis. FASEB J. 2000;14:729–39.

22. Parra V, Verdejo H, Del CA, Pennanen C, Kuzmicic J, Iglewski M, et al. The complex interplay between mitochondrial dynamics and cardiac metabolism. J Bioenerg Biomembr. 2011;43:47–51.

23. Parra V, Eisner V, Chiong M, Criollo A, Moraga F, Garcia A, et al. Changes in mitochon-drial dynamics during ceramide-induced cardiomyocyte early apoptosis. Cardiovasc Res. 2008;77:387–97.

24. Kuzmicic J, del Campo A, Lopez-Crisosto C, Morales PE, Pennanen C, Bravo-Sagua R, et al. Mitochondrial dynamics: a potential new therapeutic target for heart failure. Rev Esp Cardiol. 2011;64:916–23.

25. Klionsky DJ, Abdelmohsen K, Abe A, Abedin MJ, Abeliovich H, Acevedo Arozena A, et al. Guidelines for the use and interpretation of assays for monitoring autophagy (3rd edition). Autophagy. 2016; 12: 1–222.

26. Ktistakis NT. In praise of M. Anselmier who first used the term "autophagie" in 1859. Autophagy. 2017; 13:2015–2017.

27. Hewitt G, Korolchuk VI. Repair, reuse, recycle: the expanding role of autophagy in genome maintenance. Trends Cell Biol. 2017;27:340–51.

28. Yang S, Liu J, Qu C, Sun J, Zhang BQ, Sun YR, et al. Potassium channels and autophagy. Sheng Li Xue Bao. 2017; 69:509–514.

29. Boya P, Codogno P, Rodriguez-Muela N. Autophagy in stem cells: repair, remodelling and metabolic reprogramming. Development. 2018; 145:pii: dev146506. https://doi.org/10.1242/dev.146506.

30. Li W, Li S, Li Y, Lin X, Hu Y, Meng T, et al. Immunofluorescence staining protocols for major autophagy proteins including LC3, P62, and ULK1 in mammalian cells in response to normoxia and hypoxia. Methods Mol Biol. 2018;1854:175–85. 10.1007/7651_2018_124.

31. Diaz M, Garcia C, Sebastiani G, De Zegher F, Lopez-Bermejo A, Ibanez L. Placental and cord blood methylation of genes involved in energy homeostasis: association with fetal growth and neonatal body composition. Diabetes. 2017; 66:779–784.

32. Wildenberg ME, Koelink PJ, Diederen K, Te Velde AA, Wolfkamp SC, Nuij VJ, et al. The ATG16L1 risk allele associated with Crohn's disease results in a Rac1-dependent defect in dendritic cell migration that is corrected by thiopurines. Mucosal Immunol. 2017;10:352–60.

33. Zhang X, Evans TD, Jeong SJ, Razani B. Classical and alternative roles for autophagy in lipid metabolism. Curr Opin Lipidol. 2018;29:203–11.

34. Ponpuak M, Mandell MA, Kimura T, Chauhan S, Cleyrat C, Deretic V. Secretory autophagy. Curr Opin Cell Biol. 2015; 35:106–116.

35. Mao K, Klionsky DJ. Xenophagy: a battlefield between host and microbe, and a possible avenue for cancer treatment. Autophagy. 2017;13:223–4.

36. An H, Harper JW. Systematic analysis of ribophagy in human cells reveals bystander flux during selective autophagy. Nat Cell Biol. 2018;20:135–43.

37. Broda M, Millar AH, Van Aken O. Mitophagy: a mechanism for plant growth and survival. Trends Plant Sci. 2018;23:434–50.

38. Tsuchiya M, Ogawa H, Koujin T, Mori C, Osakada H, Kobayashi S, et al. p62/SQSTM1 promotes rapid ubiquitin conjugation to target proteins after endosome rupture during xenophagy. FEBS Open Biol. 2018;8:470–80.

39. Wang P, Shao BZ, Deng Z, Chen S, Yue Z, Miao CY. Autophagy in ischemic stroke. Prog Neurobiol. 2018;163–164:98–117.
40. Bento CF, Renna M, Ghislat G, Puri C, Ashkenazi A, Vicinanza M, et al. Mammalian autophagy: how does it work? Annu Rev Biochem. 2016;85:685–713.
41. Shao BZ, Han BZ, Zeng YX, Su DF, Liu C. The roles of macrophage autophagy in atherosclerosis. Acta Pharmacol Sin. 2016;37:150–6.
42. Zachari M, Ganley IG. The mammalian ULK1 complex and autophagy initiation. Essays Biochem. 2017;61:585–96.
43. Kalachev AV, Yurchenko OV. Microautophagy in nutritive phagocytes of sea urchins. Protoplasma. 2017;254:609–14.
44. Alfaro IE, Albornoz A, Molina A, Moreno J, Cordero K, Criollo A, et al. Chaperone mediated autophagy in the crosstalk of neurodegenerative diseases and metabolic disorders. Front Endocrinol. 2018;9:778. https://doi.org/10.3389/fendo.2018.00778.
45. Inoki K, Kim J, Guan KL. AMPK and mTOR in cellular energy homeostasis and drug targets. Annu Rev Pharmacol Toxicol. 2012;52:381–400.
46. Kaur A, Sharma S. Mammalian target of rapamycin (mTOR) as a potential therapeutic target in various diseases. Inflammo Pharmacology. 2017;25:293–312.
47. Perez-Alvarez MJ, Villa Gonzalez M, Benito-Cuesta I, Wandosell FG. Role of mTORC1 controlling proteostasis after brain ischemia. Front Neurosci. 2018;12:60. https://doi.org/10.3389/fnins.2018.00060.
48. Dodson M, Darley-Usmar V, Zhang J. Cellular metabolic and autophagic pathways: traffic control by redox signaling. Free Radic Biol Med. 2013;63:207–21.
49. Zhao X, Luo G, Cheng Y, Yu W, Chen R, Xiao B, et al. Compound C induces protective autophagy in human cholangiocarcinoma cells via Akt/ mTOR-independent pathway. J Cell Biochem. 2018;119:5538–50.
50. Pott J, Kabat AM, Maloy KJ. Intestinal epithelial cell autophagy is required to protect against TNF-induced apoptosis during chronic colitis in mice. Cell Host Microbe. 2018;23:191–202.
51. Wang B, Nie J, Wu L, Hu Y, Wen Z, Dong L, et al. AMPKalpha2 protects against the development of heart failure by enhancing mitophagy via PINK1 phosphorylation. Circ Res. 2018;122:712–29.
52. Burger E, Araujo A, Lopez-Yglesias A, Rajala MW, Geng L, Levine B, et al. Loss of Paneth cell autophagy causes acute susceptibility to Toxoplasma gondii-mediated inflammation. Cell Host Microbe. 2018;23:177–90.
53. Pankratz F, Hohnloser C, Bemtgen X, Jaenich C, Kreuzaler S, Hoefer I, et al. MicroRNA-100 suppresses chronic vascular inflammation by stimulation of endothelial autophagy. Circ Res. 2018;122:417–32.
54. Chen Z, Li C, Qian YH, Fu Y, Feng ZM. Enhancement of autophagy flux by isopsoralen ameliorates interleukin XE "Interleukin"-1beta-stimulated apoptosis in rat chondrocytes. J Asian Nat Prod Res. 2019;8:1–14.
55. Gogiraju R, Hubert A, Fahrer J, Straub BK, Brandt M, Wenzel P, et al. Endothelial leptin receptor deletion promotes cardiac autophagy and angiogenesis following pressure overload by suppressing Akt/mTOR signaling. Circ Heart Fail. 2019;12:e005622. https://doi.org/10.1161/CIRCHEARTFAILURE.118.005622.
56. Crino PB. The mTOR signalling cascade: paving new roads to cure neurological disease. Nat Rev Neurol. 2016;12:379–92.
57. Cosin-Roger J, Simmen S, Melhem H, Atrott K, Frey-Wagner I, Hausmann M, et al. Hypoxia ameliorates intestinal inflammation through NLRP3/ mTOR downregulation and autophagy activation. Nat Commun. 2017;8:98. https://doi.org/10.1038/s41467-017-00213-3.
58. Schwerd T, Pandey S, Yang HT, Bagola K, Jameson E, Jung J, et al. Impaired antibacterial autophagy links granulomatous intestinal inflammation in Niemann-Pick disease type C1 and XIAP deficiency with NOD2 variants in Crohn's disease. Gut. 2017;66:1060–73.
59. Zhao Z, Zhang L, Guo XD, Cao LL, Xue TF, Zhao XJ, et al. Rosiglitazone exerts an anti-depressive effect in unpredictable chronic mild-stress-induced depressive mice by maintaining essential neuron autophagy and inhibiting excessive astrocytic apoptosis. Front Mol Neurosci. 2017;10:293. https://doi.org/10.3389/fnmol.2017.00293.

60. Lambelet M, Terra LF, Fukaya M, Meyerovich K, Labriola L, Cardozo AK, et al. Dysfunctional autophagy following exposure to pro-inflammatory cytokines contributes to pancreatic beta-cell apoptosis. Cell Death Dis. 2018;9:96. https://doi.org/10.1038/s41419-017-0121-5.

61. Liu X, Deng Y, Xu Y, Jin W, Li H. MicroRNA-223 protects neonatal rat cardiomyocytes and H9c2 cells from hypoxia-induced apoptosis and excessive autophagy via the Akt/mTOR pathway by targeting PARP-1. J Mol Cel Cardiol. 2018;118:133–46.

62. Sciarretta S, Forte M, Frati G, Sadoshima J. New insights into the role of mTOR signaling in the cardiovascular system. Circ Res. 2018;122:489–505.

63. Vanlangenakker N, Vanden Berghe T, Krysko DV, Festjens N, Vandenabeele P. Molecular mechanisms and pathophysiology of necrotic cell death. Curr Mol Med. 2008;8:207–20.

64. Kroemer G, Galluzzi L, Brenner C. Mitochondrial membrane permeabilization in cell death. Physiol Rev. 2007;87:99–163.

65. Henriquez M, Armisen R, Stutzin A, Quest AF. Cell death by necrosis, a regulated way to go. Curr Mol Med. 2008;8:187–206.

66. Kaiser WJ, Sridharan H, Huang C, Mandal P, Upton JW, Gough PJ, et al. Toll-like receptor 3-mediated necrosis via TRIF, RIP3, and MLKL. J BiolChem. 2013;288:31268–79.

67. Grootjans S, Vanden Berghe T, Vandenabeele P. Initiation and execution mechanisms of necroptosis: an overview. Cell Death Differ. 2017;24:1184–95.

68. Song BW, Wang L. Necroptosis: a programmed cell necrosis. Sheng Li Ke Xue Jin Zhan. 2013;44:281–6.

69. Galluzzi L, Vitale I, Aaronson S, Abrams JM, Adam D, Agostinis P, et al. Molecular mechanisms of cell death: recommendations of the nomenclature committee on cell death 2018. Cell Death Differ. 2018;25:486–541.

70. Belmonte F, Das S, Sysa-Shah P, Sivakumaran V, Stanley B, Guo X, et al. ErbB2 overexpression upregulates antioxidant enzymes, reduces basal levels of reactive oxygen species, and protects against doxorubicin cardiotoxicity. Am J Physiol Heart Circ Physiol. 2015;309:H1271–80.

71. Tummers B, Green DR. Caspase-8: regulating life and death. Immunol Rev. 2017;277:76–89.

72. Cookson BT, Brennan MA. Pro-inflammatory programmed cell death. Trends Microbiol. 2001;9:113–4.

73. Jorgensen I, Miao EA. Pyroptotic cell death defends against intracellular pathogens. Immunol Rev. 2015;265:130–42.

74. Shi J, Zhao Y, Wang K, Shi X, Wang Y, Huang H, et al. Cleavage of GSDMD by inflammatory caspases determines pyroptotic cell death. Nature. 2015;526:660–5.

75. Man SM, Karki R, Kanneganti TD. Molecular mechanisms and functions of pyroptosis, inflammatory caspases and inflammasomes in infectious diseases. Immunol Rev. 2017;277:61–75.

76. Sumneang N, Siri-Angkul N, Kumfu S, Chattipakorn SC, Chattipakorn N. The effects of iron overload on mitochondrial function, mitochondrial dynamics, and ferroptosis in cardiomyocytes. Arch Biochem Biophys. 2020;680:108241. https://doi.org/10.1016/j.abb.2019.108241.

77. Imai H, Matsuoka M, Kumagai T, Sakamoto T, Koumura T. Lipid peroxidation-dependent cell death regulated by GPx4 and ferroptosis. Curr Top Microbiol Immunol. 2017;403:143–70.

78. Matsui Y, Takagi H, Qu X, Abdellatif M, Sakoda H, Asano T, et al. Distinct roles of autophagy in the heart during ischemia and reperfusion: roles of AMP-activated protein kinase and Beclin 1 in mediating autophagy. Circ Res. 2007;100:914–22.

79. Yan L, Vatner DE, Kim SJ, Ge H, Masurekar M, Massover WH, et al. Autophagy in chronically ischemic myocardium. Proc Natl Acad Sci USA. 2005;102:13807–12.

80. French CJ, Taatjes DJ, Sobel BE. Autophagy in myocardium of murine hearts subjected to ischemia followed by reperfusion. Histochem Cell Biol. 2010;134:519–26.

81. Russell RR 3rd, Li J, Coven DL, Pypaert M, Zechner C, Palmeri M, et al. AMP-activated protein kinase mediates ischemic glucose uptake and prevents postischemic cardiac dysfunction, apoptosis, and injury. J Clin Invest. 2004;114:495–503.

82. Hamacher-Brady A, Brady NR, Gottlieb RA. Enhancing macroautophagy protects against ischemia/reperfusion injury in cardiac myocytes. J Biol Chem. 2006;281:29776–87.
83. Gustafsson AB, Gottlieb RA. Autophagy in ischemic heart disease. Circ Res. 2009;104:150–8.
84. Chiong M, Wang ZV, Pedrozo Z, Cao DJ, Troncoso R, Ibacache M, Criollo A, Nemchenko A, Hill JA, Lavandero S. Cardiomyocyte death: mechanisms and translational implications. Cell Death Dis. 2011;2:e244. https://doi.org/10.1038/cddis.2011.130.
85. Kajstura J, Cheng W, Reiss K, Clark WA, Sonnenblick EH, Krajewski S, et al. Apoptotic and necrotic myocyte cell deaths are independent contributing variables of infarct size in rats. Lab Invest. 1996;74:86–107.
86. Fliss H, Gattinger D. Apoptosis in ischemic and reperfused rat myocardium. Circ Res. 1996;79:949–56.
87. Nakamura T, Ueda Y, Juan Y, Katsuda S, Takahashi H, Koh E. Fas-mediated apoptosis in adriamycin-induced cardiomyopathy in rats: in vivo study. Circulation. 2000;102:572–8.
88. Lee P, Sata M, Lefer DJ, Factor SM, Walsh K, Kitsis RN. Fas pathway is a critical mediator of cardiac myocyte death and MI during ischemia-reperfusion in vivo. Am J Physiol. 2003;284:H456–63.
89. Kurrelmeyer KM, Michael LH, Baumgarten G, Taffet GE, Peschon JJ, Sivasubramanian N, et al. Endogenous tumor necrosis factor protects the adult cardiac myocyte against ischemic-induced apoptosis in a murine model of acute myocardial infarction. Proc Natl Acad Sci USA. 2000;97:5456–61.
90. Chen Z, Chua CC, Ho YS, Hamdy RC, Chua BH. Overexpression of Bcl-2 attenuates apoptosis and protects against myocardial I/R injury in transgenic mice. Am J Physiol. 2001;280:H2313–20.
91. Hochhauser E, Cheporko Y, Yasovich N, Pinchas L, Offen D, Barhum Y, et al. Bax deficiency reduces infarct size and improves long-term function after myocardial infarction. Cell Biochem Biophys. 2007;47:11–20.
92. Toth A, Jeffers JR, Nickson P, Min JY, Morgan JP, Zambetti GP, et al. Targeted deletion of Puma attenuates cardiomyocyte death and improves cardiac function during ischemiareperfusion. Am J Physiol. 2006;291:H52–60.
93. Wu D, Zhang K, Hu P. The role of autophagy in acute myocardial infarction. Front Pharmacol. 2019;10:551. https://doi.org/10.3389/fphar.2019.00551.
94. Buss SJ, Muenz S, Riffel JH, Malekar P, Hagenmueller M, Weiss CS, et al. Beneficial effects of mammalian target of rapamycin inhibition on left ventricular remodeling after myocardial infarction. J Am Coll Cardiol. 2009;54:2435–46.
95. McCormick J, Suleman N, Scarabelli TM, Knight RA, Latchman DS, Stephanou A. STAT1 deficiency in the heart protects against myocardial infarction by enhancing autophagy. J Cell Mol Med. 2012;16:386–93.
96. van Empel VP, Bertrand AT, Hofstra L, Crijns HJ, Doevendans PA, De Windt LJ. Myocyte apoptosis in heart failure. Cardiovasc Res. 2005;67:21–9.
97. Mani K. Programmed cell death in cardiac myocytes: strategies to maximize post-ischemic salvage. Heart Fail Rev. 2008;13:193–209.
98. Scheubel RJ, Bartling B, Simm A, Silber RE, Drogaris K, Darmer D, et al. Apoptotic pathway activation from mitochondria and death receptors without caspase-3 cleavage in failing human myocardium: fragile balance of myocyte survival? J Am Coll Cardiol. 2002;39:481–8.
99. Chua CC, Gao J, Ho YS, Xiong Y, Xu X, Chen Z, et al. Overexpression of IAP-2 attenuates apoptosis and protects against myocardial ischemia/reperfusion injury in transgenic mice. Biochim Biophys Acta. 2007;1773:577–83.
100. Bhuiyan MS, Fukunaga K. Inhibition of HtrA2/Omi ameliorates heart dysfunction following ischemia/reperfusion injury in rat heart in vivo. Eur J Pharmacol. 2007;557:168–77.
101. Cao DJ, Wang ZV, Battiprolu PK, Jiang N, Morales CR, Kong Y, et al. Histone deacetylase (HDAC) inhibitors attenuate cardiac hypertrophy by suppressing autophagy. Proc Natl Acad Sci USA. 2011;108:4123–8.
102. Zhu H, Tannous P, Johnstone JL, Kong Y, Shelton JM, Richardson JA, et al. Cardiac autophagy is a maladaptive response to hemodynamic stress. J Clin Invest. 2007;117:1782–93.

103. Shende P, Plaisance I, Morandi C, Pellieux C, Berthonneche C, Zorzato F, et al. Cardiac raptor ablation impairs adaptive hypertrophy, alters metabolic gene expression, and causes heart failure in mice. Circulation. 2011;123:1073–82.

104. Shioi T, McMullen JR, Tarnavski O, Converso K, Sherwood MC, Manning WJ, et al. Rapamycin attenuates load-induced cardiac hypertrophy in mice. Circulation. 2003;107:1664–70.

105. Zhang D, Contu R, Latronico MV, Zhang J, Rizzi R, Catalucci D, et al. mTORC1 regulates cardiac function and myocyte survival through 4E-BP1 inhibition in mice. J Clin Invest. 2010;120:556–63.

106. Guerra S, Leri A, Wang X, Finato N, Di Loreto C, Beltrami CA, et al. Myocyte death in the failing human heart is gender dependent. Circ Res. 1999;85:856–66.

107. Nakayama H, Chen X, Baines CP, Klevitsky R, Zhang X, Zhang H, et al. Ca^{2+}- and mitochondrial-dependent cardiomyocyte necrosis as a primary mediator of heart failure. J Clin Invest. 2007;117:2431–44.

108. Baines CP, Kaiser RA, Purcell NH, Blair NS, Osinska H, Hambleton MA, et al. Loss of cyclophilin D reveals a critical role for mitochondrial permeability transition in cell death. Nature. 2005;434:658–62.

109. Nakagawa T, Shimizu S, Watanabe T, Yamaguchi O, Otsu K, Yamagata H, et al. Cyclophilin D-dependent mitochondrial permeability transition regulates some necrotic but not apoptotic cell death. Nature. 2005;434:652–8.

110. Kim Y, Ma AG, Kitta K, Fitch SN, Ikeda T, Ihara Y, et al. Anthracycline-induced suppression of GATA-4 transcription factor: implication in the regulation of cardiac myocyte apoptosis. Mol Pharmacol. 2003;63:368–77.

111. Mukhopadhyay P, Rajesh M, Batkai S, Kashiwaya Y, Hasko G, Liaudet L, et al. Role of superoxide, nitric oxide, and peroxynitrite in doxorubicininduced cell death in vivo and in vitro. Am J Physiol Heart Circ Physiol. 2009;296:H1466–83.

112. Malisza KL, Hasinoff BB. Production of hydroxyl radical by iron (III)-anthraquinone complexes through self-reduction and through reductive activation by the xanthine oxidase/hypoxanthine system. Arch Biochem Biophys. 1995;321:51–60.

113. Childs AC, Phaneuf SL, Dirks AJ, Phillips T, Leeuwenburgh C. Doxorubicin treatment in vivo causes cytochrome C release and cardiomyocyte apoptosis, as well as increased mitochondrial efficiency, superoxide dismutase activity, and Bcl-2: Bax ratio. Cancer Res. 2002;62:4592–8.

114. Peyssonnaux C, Nizet V, Johnson RS. Role of the hypoxia inducible factors HIF in iron metabolism. Cell Cycle. 2008;7:28–32.

115. Vejpongsa P, Yeh ET. Topoisomerase 2beta: a promising molecular target for primary prevention of anthracycline-induced cardiotoxicity. Clin Pharmacol Ther. 2014;95:45–52.

116. Zhang S, Liu X, Bawa-Khalfe T, Lu LS, Lyu YL, Liu LF, et al. Identification of the molecular basis of doxorubicin-induced cardiotoxicity. Nat Med. 2012;18:1639–42.

117. Finkel T. Cell biology: a clean energy programme. Nature. 2006;444:151–2.

118. Finck BN, Kelly DP. Peroxisome proliferator-activated receptor gamma coactivator-1 (PGC-1) regulatory cascade in cardiac physiology and disease. Circulation. 2007;115:2540–8.

119. Zhao L, Zhang B. Doxorubicin induces cardiotoxicity through upregulation of death receptors mediated apoptosis in cardiomyocytes. Sci Rep. 2017;7:44735.

120. Hsu PL, Mo FE. Matricellular protein CCN1 mediates doxorubicin induced cardiomyopathy in mice. Oncotarget. 2016;7:36698–710.

121. Chiosi E, Spina A, Sorrentino A, Romano M, Sorvillo L, Senatore G, et al. Change in TNF-alpha receptor expression is a relevant event in doxorubicin-induced H9c2 cardiomyocyte cell death. J Interferon Cytokine Res. 2007;27:589–97.

122. Giulivi C, Boveris A, Cadenas E. Hydroxyl radical generation during mitochondrial electron transfer and the formation of 8- hydroxy desoxy guanosine in mitochondrial DNA. Arch Biochem Biophys. 1995;316:909–16.

123. Costa VM, Carvalho F, Duarte JA, Bastos M, de L, Remiao F. The heart as a target for xenobiotic toxicity: the cardiac susceptibility to oxidative stress. Chem Res Toxicol. 2013; 26:1285–1311.

124. Bernuzzi F, Recalcati S, Alberghini A, Cairo G. Reactive oxygen species-independent apoptosis in doxorubicin-treated H9c2 cardiomyocytes: role for heme oxygenase-1 down-modulation. Chem Biol Interact. 2009;177:12–20.

125. Cunha-Oliveira T, Ferreira LL, Coelho AR, Deus CM, Oliveira PJ. 2018 Doxorubicin triggers bioenergetic failure and p53 activation in mouse stem cell-derived cardiomyocytes. Toxicol Appl Pharmacol. 2009;348:1–13.

126. Fabbi P, Spallarossa P, Garibaldi S, Barisione C, Mura M, Altieri P, et al. Doxorubicin impairs the insulin-like growth factor-1 system and causes insulin-like growth factor-1 resistance in cardiomyocytes. PLoS One. 2015;10:e0124643. https://doi.org/10.1371/journal.pone.012 4643.

127. Zhang C, Feng Y, Qu S, Wei X, Zhu H, Luo Q, et al. Resveratrol attenuates doxorubicin-induced cardiomyocyte apoptosis in mice through SIRT1-mediated deacetylation of p53. Cardiovasc Res. 2011;90:538–45.

128. Zhang YY, Meng C, Zhang XM, Yuan CH, Wen M, Chen Z, et al. Ophiopogonin D attenuates doxorubicin-induced autophagic cell death by relieving mitochondrial damage in vitro and in vivo. J Pharmacol Exp Ther. 2015;352:166–74.

129. Ludke A, Akolkar G, Ayyappan P Sharma, AK, Singal PK. Time course of changes in oxidative stress and stress-induced proteins in cardiomyocytes exposed to doxorubicin and prevention by vitamin C. PLoS One. 2017; 12:e0179452. https://doi.org/10.1371/journal.pone.0179452

130. Yu W, Sun H, Zha W, Cui W, Xu L, Min Q, et al. Apigenin attenuates adriamycin-induced cardiomyocyte apoptosis via the PI3K/AKT/mTOR pathway. Evid Based Complement Alternat Med. 2017:2590676.

131. Dimitrakis P, Romay-Ogando MI, Timolati F, Suter TM, Zuppinger C. Effects of doxorubicin cancer therapy on autophagy and the ubiquitinproteasome system in long-term cultured adult rat cardiomyocytes. Cell Tissue Res. 2012;350:361–72.

132. Li DL, Wang ZV, Ding G, Tan W, Luo X, Criollo A, et al. Doxorubicin blocks cardiomyocyte autophagic flux by inhibiting lysosome acidification. Circulation. 2016;133:1668–87.

133. Song R, Yang Y, Lei H, Wang G, Huang Y, Xue W, et al. HDAC6 inhibition protects cardiomyocytes against doxorubicin-induced acute damage by improving alpha-tubulin acetylation. J Mol Cell Cardiol. 2018;124:58–69.

134. Bartlett JJ, Trivedi PC, Yeung P, Kienesberger PC, Pulinilkunnil T. Doxorubicin impairs cardiomyocyte viability by suppressing transcription factor EB expression and disrupting autophagy. Biochem J. 2016;473:3769–89.

135. Luo P, Zhu Y, Chen M, Yan H, Yang B, Yang X, et al. HMGB1 contributes to adriamycin-induced cardiotoxicity via up-regulating autophagy. Toxicol Lett. 2018;292:115–22.

136. Kobayashi S, Volden P, Timm D, Mao K, Xu X, Liang Q. Transcription factor GATA4 inhibits doxorubicin-induced autophagy and cardiomyocyte death. J Biol Chem. 2010;285:793–804.

137. Fulbright JM, Egas-Bejar DE, Huh WW, Chandra J. Analysis of redox and apoptotic effects of anthracyclines to delineate a cardioprotective strategy. Cancer Chemother Pharmacol. 2015;76:1297–307.

138. Rharass T, Gbankoto A, Canal C, Kursunluoglu G, Bijoux A, Panakova D, et al. Oxidative stress does not play a primary role in the toxicity induced with clinical doses of doxorubicin in myocardial H9c2 cells. Mol Cell Biochem. 2016;413:199–215.

139. Li S, Wang W, Niu T, Wang H, Li B, Shao L, et al. Nrf2 deficiency exaggerates doxorubicin-induced cardiotoxicity and cardiac dysfunction. Oxid Med Cell Longev. 2014:748524.

140. Lim CC, Zuppinger C, Guo X, Kuster GM, Helmes M, Eppenberger HM, et al. Anthracyclines induce calpain-dependent titin proteolysis and necrosis in cardiomyocytes. J Biol Chem. 2004;279:8290–9.

141. Dhingra R, Margulets V, Chowdhury SR, Thliveris J, Jassal D, Fernyhough P, et al. Bnip3 mediates doxorubicin-induced cardiac myocyte necrosis and mortality through changes in mitochondrial signaling. Proc Natl Acad Sci USA. 2014;111:E5537–44.

142. Zhang T, Zhang Y, Cui M, Jin L, Wang Y, Lv F, et al. CaMKII is a RIP3 substrate mediating ischemia- and oxidative stress-induced myocardial necroptosis. Nat Med. 2016;22:175–82.

143. Singla DK, Johnson TA, Tavakoli DZ. Exosome treatment enhances anti-inflammatory M2 macrophages and reduces inflammationinduced pyroptosis in doxorubicin-induced cardiomyopathy. Cells. 2019;8:1224. https://doi.org/10.3390/cells8101224.

144. Zeng C, Duan F, Hu J, Luo B, Huang B, Lou X, et al. NLRP3 inflammasome-mediated pyroptosis contributes to the pathogenesis of non-ischemic dilated cardiomyopathy. Redox Biol. 2020; 101523.https://doi.org/10.1016/j.redox.2020.101523

145. Meng L, Lin H, Zhang J, Lin N, Sun Z, Gao F, et al. Doxorubicin induces cardiomyocyte pyroptosis via the TINCR-mediated posttranscriptional stabilization of NLR family pyrin domain containing 3. J Mol Cell Cardiol. 2019;136:15–26.

146. Tavakoli R, Dargani Z, Singla DK. Embryonic stem cell-derived exosomes inhibit doxorubicin-induced TLR4-NLRP3-mediated cell deathpyroptosis. Am J Physiol Heart Circ Physiol. 2019;317:H460–71.

147. Lan Y, Wang Y, Huang K, Zeng Q. Heat shock protein 22 attenuates doxorubicin-induced cardiotoxicity via regulating inflammation and apoptosis. Front Pharmacol. 2020;11:257. https://doi.org/10.3389/fphar.2020.00257.

148. Sun Z, Lu W, Lin N, Lin H, Zhang J, Ni T, et al. Dihydromyricetin alleviates doxorubicin-induced cardiotoxicity by inhibiting NLRP3 inflammasome through activation of SIRT1. Biochem Pharmacol. 2020;175:113888. https://doi.org/10.1016/j.bcp.2020.113888.

149. Zhai J, Tao L, Zhang S, Gao H, Zhang Y, Sun J, et al. Calycosin ameliorates doxorubicin-induced cardiotoxicity by suppressing oxidative stress and inflammation via the sirtuin 1-NOD-like receptor protein 3 pathway. Phytother Res. 2020;34:649–59.

150. Koleini N, Nickel BE, Edel AL, Fandrich RR, Ravandi A, Kardami E. Oxidized phospholipids in Doxorubicin-induced cardiotoxicity. Chem Biol Interact. 2019;303:35–9. https://doi.org/10.1016/j.cbi.2019.01.032.

151. Chen X, Xu S, Zhao C, Liu B. Role of TLR4/NADPH oxidase 4 pathway in promoting cell death through autophagy and ferroptosis during heart failure. Biochem Biophys Res Commun. 2019;516:37–43.

152. Mishra P, Singh SV, Verma AK, Srivastava P, Sultana S, Rath SK. Rosiglitazone induces cardiotoxicity by accelerated apoptosis. Cardiovasc Toxicol. 2014;14:99–119.

153. Baranowski M, Blachnio A, Zabielski P, Gorski J. Pioglitazone induces de novo ceramide synthesis in the rat heart. Prostaglandins Other Lipid Mediat. 2007;83:99–111.

154. Zhong W, Jin W, Xu S, Wu Y, Luo S, Liang M, et al. Pioglitazone induces cardiomyocyte apoptosis and inhibits cardiomyocyte hypertrophy Via VEGFR-2 signaling pathway. Arq Bras Cardiol. 2018;111:162–9.

155. Szabados E, Fischer GM, Toth K, Csete B, Nemeti B, Trombitas K, et al. Role of reactive oxygen species and poly-ADP-ribose polymerase in the development of AZT-induced cardiomyopathy in rat. Free Radic Biol Med. 1999;26:309–17.

156. Papparella I, Ceolotto G, Berto L, Cavalli M, Bova S, Cargnelli G, et al. Vitamin C prevents zidovudine-induced NAD(P)H oxidase activation and hypertension in the rat. Cardiovasc Res. 2007;73:432–8.

157. Gao RY, Mukhopadhyay P, Mohanraj R, Wang H, Horváth B, Yin S. Resveratrol attenuates azidothymidine-induced cardiotoxicity by decreasing mitochondrial reactive oxygen species generation in human cardiomyocytes. Mol Med Rep. 2011;4:151–5.

158. Purevjav E, Nelson DP, Varela JJ, Jimenez S, Kearney DL, Sanchez XV, et al. Myocardial fas ligand expression increases susceptibility to AZT-induced cardiomyopathy. Cardiovasc Toxicol. 2007;7:255–63.

159. Lin H, Stankov MV, Hegermann J, Budida R, Panayotova-Dimitrova D, Schmidt RE, et al. Zidovudine-mediated autophagy inhibition enhances mitochondrial toxicity in muscle cells. Antimicrob Agents Chemother. 2019;63:e01443-e1518. https://doi.org/10.1128/AAC.014 43-18.

Chapter 14
Cardiomyocyte Response to Ischemic Injury

Abstract Ischemic injury to the heart as may occur in myocardial infarction and ischemia/reperfusion injury is commonly accompanied by death of cardiomyocytes (CMs). During acute myocardial infarction (AMI), functioning CMs are lost through necrosis, apoptosis and autophagy, which occur concurrently as cells struggle to survive ischemic injury. Hypoxia secondary to arterial occlusion leads to necrosis of CMs adjacent to the occluded artery. Consequent inflammation causes release of pro-apoptotic cytokines resulting in apoptosis of ischemic CMs. Apoptotic cell death, an important contributor to myocardial damage in patients with AMI has a major role not only in the loss of CMs after infarction but also in the left ventricular remodelling that ensues. Survival of CMs in hypoxic conditions would be determined by autophagic flux in these cells. The balance between upregulation of autophagy and downregulation of apoptosis and inflammation in the peri-infarct area dictate maintenance of functional capacity of CMs. An understanding of the cellular pathways of death and survival of CMs is requisite for developing appropriate strategies to salvage CMs and preserve cardiac function during AMI.

Keywords Cardiomyocytes · Ischemia · Myocardial infarction · Ischemia · Reperfusion · Necrosis · Apoptosis · Autophagy · Myocardial inflammation

Introduction

Cell death, either acute or progressive, is a feature of several cardiac diseases, including myocardial infarction (MI), and ischemia/reperfusion (I/R) injury [1]. All three types of cell death, necrosis, apoptosis, autophagic cell death and inflammation occur concurrently during progression of ischemic injury to the heart as cardiomyocytes (CMs) cope to survive ischemic injury (Fig. 14.1) [1]. Signalling pathways common for necrosis, apoptosis, autophagy and inflammation during MI are depicted in Fig. 14.2. When the ischemic injury is irreversible, rise in cytosolic Ca^{2+} induced by the inhibition of Na^+, K^+–ATPase and damage to mitochondria, activates several proteases, disrupts anchoring cytoskeletal proteins and gradually increase permeability of cell membrane. All these lead to death of CMs [2–6].

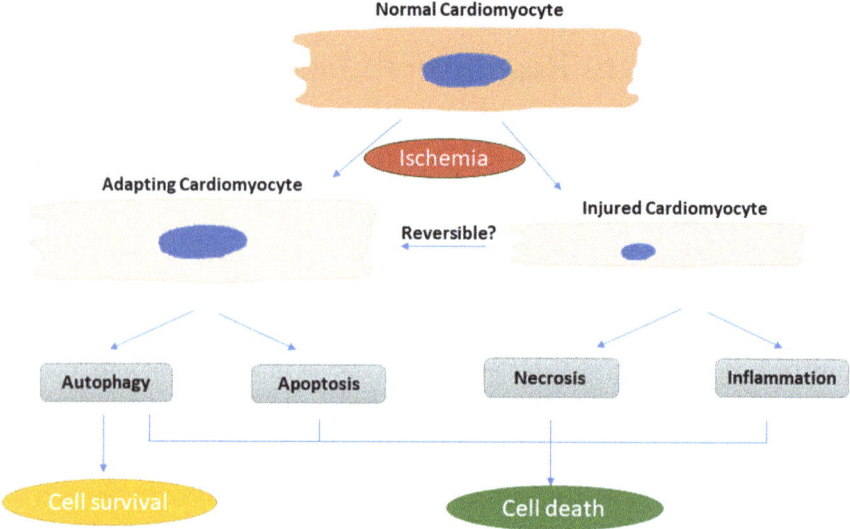

Fig. 14.1 Pathways of cardiomyocyte death during acute ischemic injury

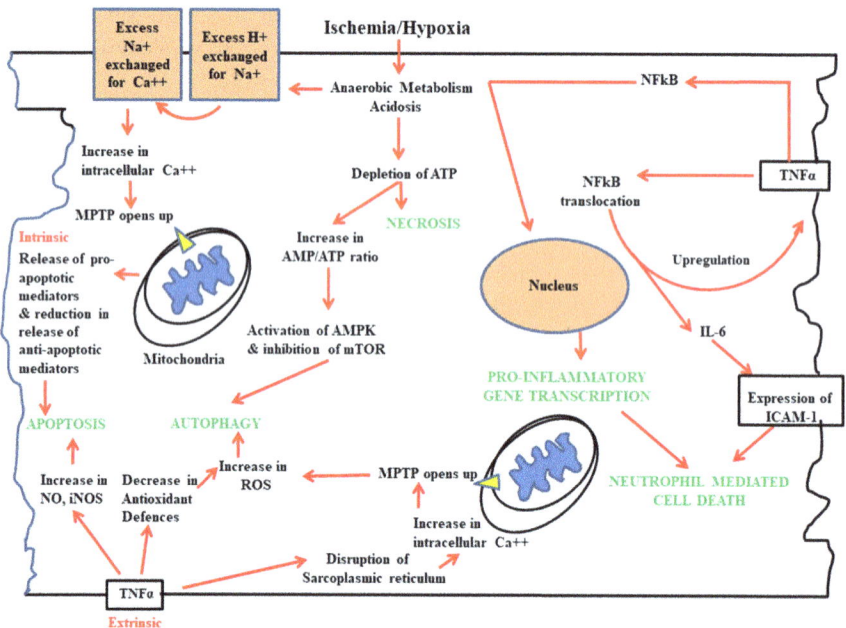

Fig. 14.2 Signalling pathways which are common to necrosis, apoptosis, autophagy and inflammation during acute ischemic injury to cardiomyocytes

Necrosis in Ischemic Injury

During the first 24 h of acute myocardial infarction (AMI), the loss of CMs is because of ischemic necrosis, which begins as early as 2 h after the initial ischemic insult, peaks around 24 h and subsides 72 h post-MI [7–10]. Gap junctions diffuse the factors that promote cellular necrosis [11].

Hypoxic cells under ischemia switch to anaerobic metabolism for survival [1]. Activation of anaerobic glycolysis results in the accumulation of H^+ in CMs. Ion pumps on the sarcolemma respond to remove excess H^+ in exchange of Na^{++}. Intracellular Na^{++} increases. As the Na^{++}/Ca^{++} exchanger in reverse mode has less capacity to remove intracellular Ca^{++}, cytoplasmic Ca^{++} levels rise. The mitochondrial Ca^{++} uniporter transports Ca^{++} into mitochondria, inducing activation of Ca^{++}-dependent dehydrogenase, reduction in NADH and electron flux through the electron transport chain, elevation of reactive oxygen species (ROS) and decrease in the levels of ATP. The rise in Ca^{++} in the mitochondrial matrix finally attains a plateau under hypoxia. Reperfusion restores oxygen and ATP-generation as well as mitochondrial membrane potential. These alterations regenerate the ion gradient necessary for further entry of Ca^{++} into mitochondria, resulting in sustained opening of mitochondrial permeability transition pores (MPTP) and swelling of mitochondria eventually leading to cellular necrosis [1]. MPTP opening seems to be the link between cardiomyocyte necrosis and I/R injury [12, 13].

Blocking late Na + channels can prevent these changes and decrease necrosis [14, 15]. Cellular constituents released from ruptured mitochondria increase apoptosis in addition to cell necrosis. Both apoptosis and necrotic cell death during ischemia/reperfusion (I/R) are decreased when MPTP is inhibited [16].

Hyperoxic preconditioning of CMs has been found to prevent MPTP opening and cytochrome c release during I/R [17]. Calcium/calmodulin-dependent protein kinase II (CaMKII) is also a modulator of necrotic cell death. Its inhibition decreases necrotic cell death in CMs exposed to I/R [18]. Cyclophilin D is another regulator of necrosis mediated by MPTP [12, 13].

Apoptosis in Ischemic Injury

Apoptosis peaks around 4.5 h after an ischemic insult [9, 19]. Apoptosis is an important cause for the death of CMs in the peri-infarct region [20–22]. Apoptosis determines infarct size, extent of cardiac remodelling and development of heart failure after AMI [23, 24]. Apoptosis is mediated by both the extrinsic and the intrinsic pathway; both pathways converge on the mitochondria. CMs are however less susceptible to apoptosis than nonmyocyte cells within the heart as they have low levels of Apaf1 and caspases [25].

Extrinsic Pathway

Activation of the extrinsic pathway occurs through the interaction between Fas-ligand and tumour necrosis factor alpha (TNF-α). Levels of TNF-α and its receptors are elevated in patients with AMI and they can predict infarct size, left ventricular dysfunction and prognosis [26–33]. High mobility group box 1 (HMGB1), a stress protein released by CMs during ischemic injury enhances the apoptotic effects of TNF-α [34]. There is evidence from animal experiments that apoptosis-stimulating fragment (Fas) and not TNFR, is the major mechanism for stimulating the extrinsic apoptotic pathway during MI [35–38].

In a rodent model of MI, it was seen that TNFR1 (TNF related apoptosis inducing ligand) mediates the cardiotoxic effect and that the cardioprotective effect is through TNFR2. Compared to TNFR1-knockout mice, mice lacking TNFR2 had poor survival, severe ventricular dysfunction, and exacerbated hypertrophy of CMs and interstitial fibrosis in the non-infarct regions [39].

Levels of TNF-related apoptosis stimulating ligand (TRAIL) are decreased in patients with AMI [40–42]. There are evidences for TRAIL to be a potential marker of severity of coronary artery disease and also a predictor of prognosis in patients after AMI [40–42]. Low levels of TRAIL are associated with worse prognosis in patients with AMI while higher levels of TRAIL levels appear to be protective.

(TRAIL) is released soon after the onset of reperfusion in ischemic hearts [43]. TRAIL-R1 and TRAIL-R2 appear to mediate pro-survival and proliferation signals [44]. Soluble recombinant TRAIL has protective effects in experimental animals [45, 46].

Intrinsic Pathway

Animal studies indicate that the intrinsic apoptosis pathway also has a role in MI [47, 48]. Compared to the extrinsic pathway, the intrinsic pathway has a more significant role in regulating death of CMs during ischemia [49]. Pro-apoptotic Bcl-2-associated-X protein (BAX) and BH3-only proteins such as BNIP3 aid to intensify the apoptotic cascade that leads to activation of intracellular caspases [50]. Bcl-2 is down-regulated during ischemia, thus speeding death of CMs [51]. Ischemic-preconditioning is seen to augment Bcl-2 expression via the JAK-STAT signalling pathway and thus could have a cardioprotective role during early ischemic period [52]. Suppression of BAX and phosphorylation of p38 mitogen-activated protein kinase (p38 MAPK) along with increased Akt-Bcl-2 signalling decrease apoptosis in the ischemic myocardium [53].

Other Regulators of Apoptosis in Ischemic Heart

There are also several other regulators of apoptosis in the ischemic myocardium. Reduction in p53 or deletion of p53 decreases apoptosis in CMs during ischemia [54, 55]. Rapamycin activation of mammalian target of rapamycin (mTOR) also results in decreased apoptosis [56]. Upregulation of nicotinamide phosphoribosyl transferase (Nampt), which salvages NAD + in the cell is associated with downregulation of apoptosis [57].

Caspases-3 and -7 are also relevant in ischemic injury.[58]. Infarct size is reduced when caspase activation is blocked in animal models of MI [59–61]. The levels of cleaved caspase-3 p17 peptide was found to be nearly four times higher in the acute phase of AMI in patients undergoing percutaneous angioplasty [62]. In patients with MI, downregulation of caspase-3 is associated with reduction in the size of the infarct and decrease in the apoptotic index of cardiomyocytes [63].

Apoptosis in Ischemia/Reperfusion Injury

Reperfusion seem to quicken the timing of apoptosis [64]. Cardiac specific overexpression of Bcl-2, an inhibitor of apoptosis is found to significantly reduce infarct size after I/R injury and correlates with decrease in apoptosis of CMs [65, 66].

(Fas) is a key mediator of cardiomyocyte apoptosis during ischemia/reperfusion (I/R) injury in animals [37]. Experimental studies have revealed the role of TNF-α in mediating apoptosis of CMs during I/R. TNF-α has been shown in vitro to induce apoptosis in cardiomyocytes [67]. Low dose of TNF-α improves cardiac function while high dose of TNF-α aggravates ischemia and reperfusion injury [68].

Overexpression of cardiac specific caspase-3, the most important caspase of the terminal apoptotic pathway in transgenic mice has been found to result in a more extensive infarct after I/R injury [69].

Endonuclease-G (EndoG) is involved in DNA degradation during apoptosis in CMs subjected to ischemia [70]. Infarct size in models of I/R can be reduced by inhibiting mitochondrial fission via mitofusion proteins [71].

Autophagy in the Ischemic Heart

Autophagy is a mechanism that removes protein aggregates within cells and organelles for degradation to sustain energy homeostasis [72, 73]. During ischemia, this mechanism aids cell survival by promoting removal of intracellular waste and recycling substrates for ATP generation [74, 75].

Activation of autophagy mechanisms are vital for the maintenance of normal functions of CMs in the initial period of ischemic injury [76, 77]. This is a pro-survival mechanism to protect CMs from ischemic or ischemia/reperfusion injury and to replenish energy [73, 78–80]. Excessive activation of autophagy may however have adverse effects [81, 82]. The autophagic response can become dysfunctional if ischemia is prolonged.

Most studies, though not all, have revealed that ischemia/hypoxia induces autophagy both in vivo and in vitro [73, 78, 83]. The two pathways responsible for ischemia/hypoxia-induced autophagy involve either BNIP38 or AMPK [84]. Ischemia reduces oxygen supply to the myocardium, leading to decline in ATP (adenosine tri phosphate). Increase in AMP causes activation of AMP activated protein kinase (AMPK), which along with inhibition of mTOR activates autophagy. During ischemia, reactive oxygen species (ROS) activates serine and cysteine proteases which causes decrease in the levels of LAMP2, critical for autophagosome–lysosome fusion and thus impairs autophagosome processing.

Bcl-2-associated athanogene (BAG-1) is found to increase during ischemia. Cardioprotection through enhanced autophagy can be abolished by inhibition of autophagy through si-RNA of BAG-1 [85]. BAG-1 can also induce protective autophagy by linking heat shock proteins, (HSP) Hsc70/Hsp70 with the proteasome [85]. HSP 20 is involved in the blockade of autophagy during myocardial ischemia [86].

Autophagy in Response to Reperfusion

During reperfusion, autophagy is further upregulated even despite restoration of oxygen and nutrients [73, 87]. Upregulation of autophagy can be either advantageous or deleterious during reperfusion Fig. 14.3) [73, 87]. The mechanisms for sustained activation of autophagy during reperfusion is different from those in ischemia.

Autophagy during reperfusion is activated mainly through beclin-1 [88]. Upregulation of Beclin 1 during reperfusion further impairs autophagosome processing which is decreased by ischemia and also lead to increased production of ROS, mitochondrial permeabilization and death of CMs [89, 90]. Upregulation of p62 (a microtubule associated protein-1 light chain-3 (LC3)-binding protein) involved in protein aggregation and the ratio of LC3-II/I, are indicators of autophagy during I/R [91, 92].

BNIP3 has also been found to activate autophagy during I/R [74, 93]. Oxidative stress, mitochondrial damage/BNIP3, endoplasmic reticulum stress, and calcium overload are possibly more important in sustaining autophagy during reperfusion [94–96]. While autophagy has a protective role during mild-to-moderate ischemia, upregulation of autophagy can be either beneficial or detrimental during I/R [73, 87].

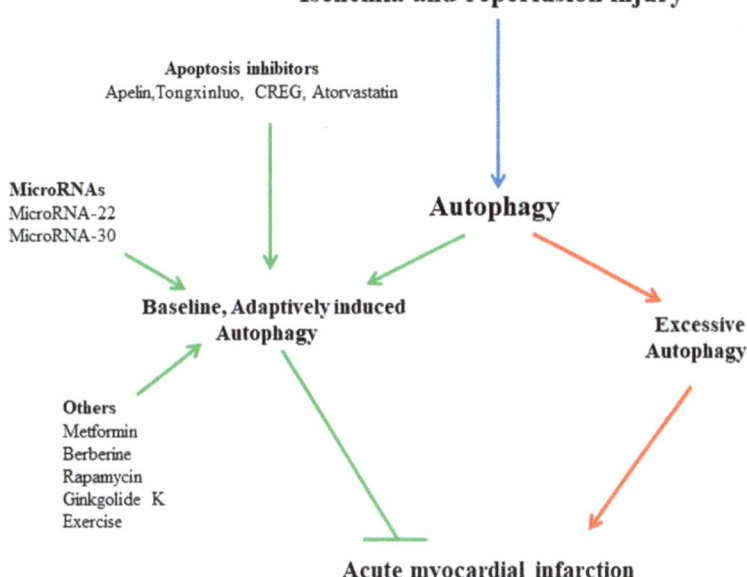

Fig. 14.3 Ischemia induced autophagy types in acute myocardial infarction and agents which can maintain baseline autophagy

Effects of Autophagy During Ischemia/Reperfusion

Several studies have provided evidences for the role of autophagy in protecting CMs from I/R injury [97–101]. 'Adaptive' induction of autophagy alleviates damage to CMs under I/R injury (Fig. 14.3). There is a positive association between pharmacological upregulation of autophagy and increased resistance to I/R injury to the myocardium in swine [75]. Several agents and pathways which maintains baseline autophagy or adaptively induced autophagy have been found to alleviate acute MI (Fig. 14.3). Quality control of mitophagy preserves mitochondrial homoeostasis and thus protects CMs from I/R injury [102].

Autophagy could have a detrimental effect as well during I/R. I/R induced autophagy could lead to the induction of apoptosis, necrosis, and inflammatory reaction, all of which could affect the viability of CMs [103].

Autophagy in Acute Myocardial Infarction

In patients with AMI, upregulation of autophagy has been observed [104, 105]. Upregulation of autophagy can protect CMs against ischemia and decrease adverse remodelling after MI [106]. Mitophagy, a specific autophagy mechanism for the

maintenance of mitochondrial homeostasis, is considered to have a key cardioprotective role in ischemic injury [107]. The 'adaptive' induction of autophagy aids
to reduce the formation of aggresomes and the consequent deleterious effects in
the heart [72]. In rats, autophagy has been found to decrease the infarct size after
ligation of the coronary artery [108]. Autophagy inhibitor, bafilomycin A1, significantly increased the infarct size in animals with AMI suggesting that autophagy has
a cardioprotective effect during AMI [109].

The upregulation of autophagy mechanism in myocardial infarcts is found to be
via the AMPK-mTOR signalling pathway in CMs [110, 111]. Genetic inhibition
of AMPK signalling pathway can result in dysfunction of autophagy mechanisms
and increase in infarct size [112]. Autophagy inhibitor, 3-MA also causes adverse
cardiac remodelling after infarction, by activating nuclear factor-κB (NF-κB) [113].
NF-κB-induced autophagy aggravates injury to CMs in AMI [114, 115].

During ischemia, hypoxia is a key factor for the death of CMs; hypoxia induces
both apoptosis and an excessive autophagy process in CMs [116]. Hypoxia-induced
autophagy can be suppressed in both animals with MI and hypoxic H9C2 cells,
using exosome-transported miRNA-93-5p, which targets Atg7, a vital autophagy-
related gene and toll-like receptor 4 [117]. MicroRNA-223 also protects neonatal
rat cardiomyocytes and H9C2 cells from hypoxia-induced apoptosis and excessive
autophagy through the Akt/mTOR pathway; its target is PARP-1 [118].

The phosphatase and tensin homolog deleted on chromosome 10 (PTEN)-PI3K-
Akt signalling pathway has also been found to participate in the induction of
autophagy in cardiomyocytes subjected to hypoxia in vitro [119].

Effects of Autophagy in Myocardial Infarction

Severe ischemia, can induce an exaggerated autophagy response which can promote
death of CMs and worsen heart function [120–122].

Autophagy during repair after ischemic injury aids to maintain transforming
growth factor receptor I levels [123]. Loss of Sirt7 during AMI induces a down
regulation of transforming growth factor-β receptor I; autophagy inhibitor can block
this effect [123].

Autophagy may also contribute to the postinfarction remodelling process. Activation of AMPK (adenosine monophosphate kinase) blunts development of heart
failure (HF) induced by myocardial infarction (MI). mTOR inhibition has been found
to reduce remodelling and improve heart function after MI [124]. In an ex vivo model
of MI, deficiency of STAT1 was seen to enhance autophagy and thus protective [125].

As a basal level of autophagy is required to maintain cardiac structure and function, a decline of autophagy as may happen in diseases associated with metabolic
derangements such as diabetes, can impair host response to hypoxic-ischemic injury
and cause a damaging effect on CMs [126–128].

Inflammatory Mediators from Ischemic Cardiomyocytes

The injured CMs secrete pro-inflammatory mediators, which activate vascular endothelial cells, neutrophils and macrophages. Ischemia, hypoxia and necrosis initiate as well as augment this response [129]. Hypoxia-driven mechanisms activate proinflammatory transcription factors and inflammatory cells during acute MI. Increased levels of hypoxia inducible factor-1α (HIF-1α) result in the expression of several pro-inflammatory proteins [130–134]. There is also upregulation of JNK and p38 MAPK [135]. Hypoxic activation of PKCα and AGEs/RAGE/PKCβII/c-Jun pathways induce expression of early growth response-1 (EGR-1) which is implicated in the regulation of TNF-α gene in endothelial cells [136–138]. Hypoxia also results in production of ROS through multiple mechanisms [139–142]. ROS induced damage is aggravated from down regulation of antioxidant defense mechanisms induced by TNF-α [143].

TNF-α signalling can regulate inflammatory response genes in addition to causing cell damage and apoptosis. TNF-α increases levels of nitric oxide (NO) in CMs via activation of inducible nitric oxide synthase (iNOS) [144]. NO upregulates p53, BAX, and leads to release of cytochrome c in mitochondria and thus in apoptosis [145]. Rise in the levels of NO and ROS together with activation of hypoxia-induced PKC dependent signalling result in the upregulation of TNF-α, IL-6, and IL-10 via nuclear translocation of NF-κB [146–148]. TNF-α induced caspase-8 causes leak in ryanodine receptor-2 channel resulting in an increase in intracellular calcium and dysfunction of CMs [149]. Interleukins, cytokines and ROS also cause cell injury. Inflammation, matrix degradation, and apoptosis are significantly less in TNF-α-KO mice during MI [150].

Activation of interleukin and TNF-α is prominent in the peri-infarct zone [151]. Inhibition of adenosine kinase by hypoxia induced dysfunction of purine salvage pathway can increase the levels of adenosine in CMs and inosine extracellularly and activate mast cells [152–154]. Mast cell activation and degranulation of TNF-α, histamine, tryptase, and chymase may aggravate injury to CMs [152, 155–159]. Inhibition of mast cell degranulation has been found to reduce oxidative injury of CMs and size of infarct [160, 161].

Constitutively expressed TLR-4 is activated on CMs and expression of IL-6 is increased by the rise in levels of DAMPs (Death associated molecular pattern) [162, 163]. IL-6 induces ICAM1 expression in CMs and targets them for damage by neutrophils [164].

Several signalling mediators of mechanisms which are common to necrosis, apoptosis and autophagy regulate the inflammatory response in AMI and the balance among them determines the proportion of CMs that survive and die post-MI [88]. Ischemia induced inflammation may dampen necrosis, apoptosis and autophagy.

Conclusions

The initial ischemic injury to cardiomyocytes (CMs) initiates the cellular mechanisms of necrosis, apoptosis, autophagy and inflammation. Hypoxia causes necrosis of CMs and induces an inflammatory response. Release of inflammatory pro-apoptotic cytokines and activation of the intrinsic apoptotic pathway results in the apoptosis of ischemic CMs. CMs can concurrently elicit autophagy mechanisms to survive the hypoxic environment and prevent cell death. The fate of ischemic CMs is dependent on the balance among these mechanisms, appropriate autophagic flux and reduction of apoptosis and inflammation in the peri-infarct area. Necrosis, apoptosis, autophagy, and inflammation occur concurrently as cells struggle to survive ischemic injury. While necrotic cell death may not be reversible, apoptosis, autophagy, and inflammation are amenable to manipulation to permit injured CMs to survive during myocardial infarction. An in-depth understanding of the factors that determine the balance between cellular survival and death of CMs in response to myocardial ischemia may aid the development of strategies to promote survival of CMs during the early phases of ischemic injury and thus enhance cardiac function after myocardial infarction.

References

1. Whelan RS, Kaplinskiy V, Kitsis RN. Cell death in the pathogenesis of heart disease: mechanisms and significance. Ann Rev Physiol. 2010;72:19–44.
2. Jennings RB, Ganote CE. Structural changes in myocardium during acute ischemia. Circ Res. 1974;35(Suppl 3):156–72.
3. Jennings RB. Steenbergen CJr, Reimer KA. Myocardial ischemia and reperfusion Monogr Pathol. 1995;37:47–80.
4. Stanley WC. Cardiac energetics during ischaemia and the rationale for metabolic interventions. Cor Art Dis. 2001;12(Suppl 1):S3–7.
5. Buja LM. Myocardial ischemia and reperfusion injury. Cardiovasc Pathol. 2005;14:170–5.
6. Burke AP, Virmani R. Pathophysiology of acute myocardial infarction. Med Clin North Am. 2007;91:553–72.
7. Tavernarakis N. Cardiomyocyte necrosis: alternative mechanisms, effective interventions. Biochim Biophys Acta. 2007;1773:480–2.
8. Golstein P, Kroemer G. Cell death by necrosis: towards a molecular definition. Trends Biochem Sci. 2007;32:37–43.
9. Kajstura J, Cheng W, Reiss K, Clark WA, Sonnenblick EH, et al. Apoptotic and necrotic myocyte cell deaths are independent contributing variables of infarct size in rats. Lab Invest. 1996;74:86–107.
10. Huang S, Chen HH, Yuan H, Dai G, Schuhle DT, et al. Molecular MRI of acute necrosis with a novel DNA-binding gadolinium chelate: kinetics of cell death and clearance in infarcted myocardium. Circ Cardiovasc Imaging. 2011;4:729–37.
11. Shintani-Ishida K, Unuma K, Yoshida K. Ischemia enhances translocation of connexin43 and gap junction intercellular communication, thereby propagating contraction band necrosis after reperfusion. Circ J. 2009;73:1661–8.
12. Baines CP, Kaiser RA, Purcell NH, Blair NS, Osinska H, Hambleton MA, et al. Loss of cyclophilin D reveals a critical role for mitochondrial permeability transition in cell death. Nature. 2005;434:658–62.

13. Nakagawa T, Shimizu S, Watanabe T, Yamaguchi O, Otsu K, Yamagata H, et al. Cyclophilin D-dependent mitochondrial permeability transition regulates some necrotic but not apoptotic cell death. Nature. 2005;434:652–8.

14. Yamada K, Matsui K, Ogawa S, Yamamoto S, Mori M, et al. Reduction of myocardial infarct size by SM-198110, a novel Na+/H+ exchange inhibitor, in rabbits. Naunyn Schmiedebergs Arch Pharmacol. 2005;371:408–19.

15. Hale SL, Leeka JA, Kloner RA. Improved left ventricular function and reduced necrosis after myocardial ischemia/reperfusion in rabbits treated with ranolazine, an inhibitor of the late sodium channel. J Pharmacol Exp Ther. 2006;318:418–23.

16. Bognar Z, Kalai T, Palfi A, Hanto K, Bognar B, et al. A novel SOD-mimetic permeability transition inhibitor agent protects ischemic heart by inhibiting both apoptotic and necrotic cell death. Free Radic Biol Med. 2006;41:835–48.

17. Petrosillo G, Di Venosa N, Moro N, Colantuono G, Paradies V, et al. In vivo hyperoxic preconditioning protects against rat-heart ischemia/reperfusion injury by inhibiting mitochondrial permeability transition pore opening and cytochrome c release. Free Radic Biol Med. 2011;50:477–83.

18. Vila-Petroff M, Salas MA, Said M, Valverde CA, Sapia L, et al. CaMKII inhibition protects against necrosis and apoptosis XE "Apoptosis" in irreversible ischemia/reperfusion injury. Cardiovasc Res. 2007;73:689–98.

19. Anversa P, Cheng W, Liu Y, Leri A, Redaelli G, et al. Apoptosis and myocardial infarction. Basic Res Cardiol. 1998;93:8–12.

20. Abbate A, Bussani R, Biondi-Zoccai GG, Santini D, Petrolini A, et al. Infarct-related artery occlusion, tissue markers of ischaemia, and increased apoptosis in the peri-infarct viable myocardium. Eur Heart J. 2005;26:2039–45.

21. Saraste A, Pulkki K, Kallajoki M, et al. Apoptosis in human acute myocardial infarction. Circulation. 1997;95:320–3.

22. Olivetti G, Quaini F, Sala R, et al. Acute myocardial infarction in humans is associated with activation of programmed myocyte cell death in the surviving portion of the heart. J Mol Cell Cardiol. 1996;28:2005–16.

23. Abbate A, Biondi-Zoccai GG, Bussani R, Dobrina A, Camilot D, Feroce F, Rossiello R, Baldi F, Silvestri F, Biasucci LM, Baldi A. Increased myocardial apoptosis in patients with unfavorable left ventricular remodeling and early symptomatic post-infarction heart failure. J Am Coll Cardiol. 2003;41(5):753–60.

24. Baldi A, Abbate A, Bussani R, Patti G, Melf R, Angelini A, Dobrina A, Rossiello R, Silvestri F, Baldi F, Di Sciascio G. Apoptosis and post-infarction left ventricular remodeling. J Mol Cell Cardiol. 2002;34(2):165–74.

25. Teringova E, Tousek P. Apoptosis in ischemic heart disease. J Transl Med. 2017;15:87. https://doi.org/10.1186/s12967-017-1191-y.

26. Nilsson L, Szymanowski A, Swahn E, et al. Soluble TNF receptors are associated with infarct size and ventricular dysfunction in ST-elevation myocardial infarction. PLoS ONE. 2013;8(2):e55477.

27. Kehmeier ES, Lepper W, Kropp M, Heiss C, Hendgen-Cotta U, Balzer J, Neizel M, Meyer C, Merx MW, Verde PE, Ohmann C, Heusch G, Kelm M, Rassaf T. TNF-α, myocardial perfusion and function in patients with ST-segment elevation myocardial infarction and primary percutaneous coronary intervention. Clin Res Cardiol. 2012;101(10):815–27.

28. Mielczarek-Palacz A, Sikora J, Kondera-Anasz Z, Smycz M. Changes in concentrations of tumor necrosis factor TNF and its soluble receptors type 1 (sTNF-r1) and type 2 (sTNF-R2) in serum of patients with ST-segment elevation myocardial infarction. Wiad Lek. 2011;64(2):71–4.

29. Valgimigli M, Ceconi C, Malagutti P, Merli E, Soukhomovskaia O, Francolini G, Cicchitelli G, Olivares A, Parrinello G, Percoco G, Guardigli G, Mele D, Pirani R, Ferrari R. Tumor necrosis factor-alpha receptor 1 is a major predictor of mortality and new-onset heart XE "Heart" failure in patients with acute myocardial infarction: the cytokine-activation and long-term prognosis in myocardial infarction (C-ALPHA) study. Circulation. 2005;111(7):863–70.

30. Ueland T, Kjekshus J, Frøland SS, Omland T, Squire IB, Gullestad L, Dickstein K, Aukrust P. Plasma levels of soluble tumor necrosis factor receptor type I during the acute phase following complicated myocardial infarction predicts survival in high-risk patients. J Am Coll Cardiol. 2005;46(11):2018–21.
31. Gonzálvez M, Ruiz-Ros JA, Pérez-Paredes M, Lozano ML, García-Almagro FJ, Martínez-Corbalán F, Giménez DM, Carrillo A, Carnero A, Cubero T, Gonzálvez JJ, Ureña I, Vicente V. Prognostic value of tumor necrosis factoralpha in patients with ST-segment elevation acute myocardial infarction. Rev Esp Cardiol. 2007;60(12):1233–41.
32. Kaya EB, Ozer N, Deveci OS, Kepez A, Tülümen E, Aksöyek S, Atalar E, Ovünç K, Ozmen F, Ozkutlu H. The early predictors of ventricular remodeling after myocardial infarction: the role of tumor necrosis factor-alpha. Anadolu Kardiyol Derg. 2009;9(2):84–90.
33. Lin XM, Zhang ZY, Wang LF, Zhang L, Liu Y, Liu XL, Yang XC, Cui L, Zhang L. Attenuation of tumor necrosis factor-alpha elevation and improved heart XE "Heart" function by postconditioning for 60 seconds in patients with acute myocardial infarction. Chin Med J. 2010;123(14):1833–9.
34. Xu H, Yao Y, Su Z, Yang Y, Kao R, et al. Endogenous HMGB1 contributes to ischemia-reperfusion-induced myocardial apoptosis by potentiating the effect of TNF-α/JNK. Am J Physiol Heart Circ Physiol. 2011;300:H913–21.
35. Tekin D, Xi L, Kukreja RC. Genetic deletion of fas receptors or Fas ligands does not reduce infarct size after acute global ischemia-reperfusion in isolated mouse heart. Cell Biochem Biophys. 2006;44:111–7.
36. Nakamura T, Ueda Y, Juan Y, Katsuda S, Takahashi H, Koh E. Fas-mediated apoptosis in adriamycin-induced cardiomyopathy in rats: in vivo study. Circulation. 2000;102:572–8.
37. Lee P, Sata M, Lefer DJ, Factor SM, Walsh K, Kitsis RN. Fas pathway is a critical mediator of cardiac myocyte death and MI during ischemia-reperfusion in vivo. Am J Physiol. 2003;284:H456–63.
38. Kurrelmeyer KM, Michael LH, Baumgarten G, Taffet GE, Peschon JJ, Sivasubramanian N, et al. Endogenous tumor necrosis factor protects the adult cardiac myocyte against ischemic-induced apoptosis in a murine model of acute myocardial infarction. Proc Natl Acad Sci USA. 2000;97:5456–61.
39. Monden Y, Kubota T, Inoue T, Tsutsumi T, Kawano S, Ide T, Tsutsui H, Sunagawa K. Tumor necrosis factor-alpha is toxic via receptor 1 and protective via receptor 2 in a murine model of myocardial infarction. Am J Physiol Heart Circ Physiol. 2007;293(1):H743–53.
40. Osmancik P, Teringova E, Tousek P, et al. Prognostic value of TNF-related apoptosis inducing ligand (TRAIL) in acute coronary syndrome patients. PLoS ONE. 2013;8(2):e53860.
41. Mori K, Ikari Y, Jono S, Shioi A, Ishimura E, Emoto M, Inaba M, Hara K, Nishizawa Y. Association of serum TRAIL level with coronary artery disease. Thromb Res. 2010;125(4):322–5.
42. Secchiero P, Corallini F, Ceconi C, Parrinello G, Volpato S, Ferrari R, Zauli G. Potential prognostic signifcance of decreased serum levels of TRAIL after acute myocardial infarction. PLoS ONE. 2009;4(2):e4442.
43. Jeremias I, Kupatt C, Martin-Villalba A, Habazettl H, Schenkel J, Boekstegers P, Debatin KM. Involvement of CD95/Apo1/Fas in cell death after myocardial ischemia. Circulation. 2000;102(8):915–20.
44. LeBlanc HN, Ashkenazi A. Apo2L/TRAIL and its death and decoy receptors. Cell Death Difer. 2003;10:66–75.
45. Tofoli B, Bernardi S, Candido R, Zacchigna S, Fabris B, Secchiero P. TRAIL shows potential cardioprotective activity. Investig New Drugs. 2012;30(3):1257–60.
46. Secchiero P, Candido R, Corallini F, Zacchigna S, Tofoli B, Rimondi E, Fabris B, Giacca M, Zauli G. Systemic tumor necrosis factor-related apoptosisinducing ligand delivery shows antiatherosclerotic activity in apolipoprotein E-null diabetic mice. Circulation. 2006;114(14):1522–30.
47. Hochhauser E, Cheporko Y, Yasovich N, Pinchas L, Offen D, Barhum Y, et al. Bax deficiency reduces infarct size and improves long-term function after myocardial infarction. Cell Biochem Biophys. 2007;47:11–20.

48. Toth A, Jeffers JR, Nickson P, Min JY, Morgan JP, Zambetti GP, et al. Targeted deletion of Puma attenuates cardiomyocyte death and improves cardiac function during is chemiare perfusion. Am J Physiol. 2006;291:H52–60.

49. Gomez L, Chavanis N, Argaud L, Chalabreysse L, Gateau-Roesch O, et al. Fas-independent mitochondrial damage triggers cardiomyocyte death after ischemia-reperfusion. Am J Physiol Heart Circ Physiol. 2005;289:H2153-2158.

50. Kubasiak LA, Hernandez OM, Bishopric NH, Webster KA. Hypoxia and acidosis activate cardiac myocyte death through the Bcl-2 family protein BNIP3. Proc Natl Acad Sci USA. 2002;99:12825–30.

51. Hattori R, Hernandez TE, Zhu L, Maulik N, Otani H, et al. An essential role of the antioxidant gene Bcl-2 in myocardial adaptation to ischemia: an insight with antisense Bcl-2 therapy. Antioxid Redox Signal. 2001;3:403–13.

52. You L, Li L, Xu Q, Ren J, Zhang F. Postconditioning reduces infarct size and cardiac myocyte apoptosis via the opioid receptor and JAK-STAT signaling pathway. Mol Biol Rep. 2011;38:437–43.

53. Kato K, Yin H, Agata J, Yoshida H, Chao L, et al. Adrenomedullin gene delivery attenuates myocardial infarction and apoptosis after ischemia and reperfusion. Am J Physiol Heart Circ Physiol. 2003;285:H1506-1514.

54. Naito AT, Okada S, Minamino T, Iwanaga K, Liu ML, et al. Promotion of CHIP-mediated p53 degradation protects the heart from ischemic injury. Circ Res. 2010;106:1692–702.

55. Matsusaka H, Ide T, Matsushima S, Ikeuchi M, Kubota T, et al. Targeted deletion of p53 prevents cardiac rupture after myocardial infarction in mice. Cardiovasc Res. 2006;70:457–65.

56. Khan S, Salloum F, Das A, Xi L, Vetrovec GW, et al. Rapamycin confers preconditioning-like protection against ischemia-reperfusion injury in isolated mouse heart and cardiomyocytes. J Mol Cell Cardiol. 2006;41:256–64.

57. Hsu CP, Oka S, Shao D, Hariharan N, Sadoshima J. Nicotinamide phosphoribosyltransferase regulates cell survival through NAD+ synthesis in cardiac myocytes. Circ Res. 2009;105:481–91.

58. Chapman JG, Magee WP, Stukenbrok HA, Beckius GE, Milici AJ, et al. A novel nonpeptidic caspase-3/7 inhibitor, (S)-(+)-5-[1-(2- methoxymethylpyrrolidinyl)sulfonyl]isatin reduces myocardial ischemic injury. Eur J Pharmacol. 2002;456:59–68.

59. Kim SJ, Kuklov A, Crystal GJ. In vivo gene delivery of XIAP protects against myocardial apoptosis and infarction following ischemia/reperfusion in conscious rabbits. Life Sci. 2011;88:572–7.

60. Holly TA, Drincic A, Byun Y, Nakamura S, Harris K, et al. Caspase inhibition reduces myocyte cell death induced by myocardial ischemia and reperfusion in vivo. J Mol Cell Cardiol. 1999;31:1709–15.

61. Chandrashekhar Y, Sen S, Anway R, Shuros A, Anand I. Long-term caspase inhibition ameliorates apoptosis, reduces myocardial troponin-I cleavage, protects left ventricular function, and attenuates remodeling in rats with myocardial infarction. J Am Coll Cardiol. 2004;43:295–301.

62. Agosto M, Azrin M, Singh K, Jafe AS, Liang BT. Serum caspase-3 p17 fragment is elevated in patients with ST-segment elevation myocardial infarction: a novel observation. J Am Coll Cardiol. 2011;57(2):220–1.

63. Liu Q. Lentivirus mediated interference of caspase-3 expression ameliorates the heart function on rats with acute myocardial infarction. Eur Rev Med Pharmacol Sci. 2014;18(13):1852–8.

64. Fliss H, Gattinger D. Apoptosis in ischemic and reperfused rat myocardium. Circ Res. 1996;79:949–56.

65. Brocheriou V, Hagege AA, Oubenaissa A, et al. Cardiac functional improvement by a human Bcl-2 transgene in a mouse model of ischemia/reperfusion injury. J Gene Med. 2000;2:326–33.

66. Chen Z, Chua CC, Ho YS, Hamdy RC, Chua BH. Overexpression of Bcl-2 attenuates apoptosis and protects against myocardial I/R injury in transgenic mice. Am J Physiol Heart Circ Physiol. 2001;280:H2313–20.

67. Krown KA, Page MT, Nguyen C, Zechner D, Gutierrez V, Comstock KL, Glembotski CC, Quintana PJ, Sabbadini RA. Tumor necrosis factor α-induced apoptosis in cardiac myocytes. Involvement of the sphingolipid signaling cascade in cardiac cell death. J Clin Invest. 1996;98(12):2854–65.
68. Asgeri M, Pourafkari L, Kundra A, Javadzadegan H, Negargar S, Nader ND. Dual effects of tumor necrosis factor alpha on myocardial injury following prolonged hypoperfusion of the heart. Immunol Invest. 2015;44(1):23–35.
69. Condorelli G, Roncarati R, Ross J Jr, et al. Heart-targeted overexpression of caspase3 in mice increases infarct size and depresses cardiac function. Proc Natl Acad Sci USA. 2001;98:9977–82.
70. Zhang J, Ye J, Altafaj A, Cardona M, Bahi N, et al. EndoG links Bnip3- induced mitochondrial damage and caspase-independent DNA fragmentation in ischemic cardiomyocytes. PLoS One. 2011;6:e17998.
71. Ong SB, Subrayan S, Lim SY, Yellon DM, Davidson SM, et al. Inhibiting mitochondrial fission protects the heart against ischemia/reperfusion injury. Circulation. 2010;121:2012–22.
72. Tannous P, Zhu H, Nemchenko A, Berry JM, Johnstone JL, et al. Intracellular protein aggregation is a proximal trigger of cardiomyocyte autophagy. Circulation. 2008;117:3070–8.
73. Matsui Y, Takagi H, Qu X, Abdellatif M, Sakoda H, et al. Distinct roles of autophagy in the heart during ischemia and reperfusion: roles of AMP-activated protein kinase and Beclin 1 in mediating autophagy. Circ Res. 2007;100:914–22.
74. Dong Y, Undyala VV, Gottlieb RA, Mentzer RM Jr, Przyklenk K. Autophagy: definition, molecular machinery, and potential role in myocardial ischemia-reperfusion injury. J Cardiovasc Pharmacol Ther. 2010;15:220–30.
75. Przyklenk K, Undyala VV, Wider J, Sala-Mercado JA, Gottlieb RA, et al. Acute induction of autophagy as a novel strategy for cardioprotection: getting to the heart of the matter. Autophagy. 2011;7:432–4.
76. Yan L, Sadoshima J, Vatner DE, Vatner SF. Autophagy in ischemic preconditioning and hibernating myocardium. Autophagy. 2009;5:709–12.
77. Zhang QY, Jin HF, Chen S, Chen QH, Tang CS, Du JB, et al. Hydrogen sulfide regulating myocardial structure and function by targeting cardiomyocyte autophagy. Chin Med J. 2018;131:839–44.
78. Yan L, Vatner DE, Kim SJ, Ge H, Masurekar M, Massover WH, et al. Autophagy in chronically ischemic myocardium. Proc Natl Acad Sci USA. 2005;102:13807–12.
79. Sciarretta S, Yee D, Shenoy V, Nagarajan N, Sadoshima J. The importance of autophagy in cardioprotection. High Blood Press & Cardiovasc Prev. 2014;21:21–8.
80. Wang L, Li Y, Ning N, Wang J, Yan Z, Zhang S, et al. Decreased autophagy induced by beta1-adrenoceptor autoantibodies contributes to cardiomyocyte apoptosis. Cell Death Dis. 2018;9:406. https://doi.org/10.1038/s41419-018-0445-9.
81. Ma H, Guo R, Yu L, Zhang Y, Ren J. Aldehyde dehydrogenase 2 (ALDH2) rescues myocardial ischaemia/reperfusion injury: role of autophagy paradox and toxic aldehyde. Eur Heart J. 2011;32:1025–38.
82. Bai YD, Yang YR, Mu XP, Lin G, Wang YP, Jin S, et al. Hydrogen sulfide alleviates acute myocardial ischemia injury by modulating autophagy and inflammation response under oxidative stress. Oxid Med Cell Longev. 2018:3402809. https://doi.org/10.1155/2018/340 2809.
83. French CJ, Taatjes DJ, Sobel BE. Autophagy in myocardium of murine hearts subjected to ischemia followed by reperfusion. Histochem Cell Biol. 2010;134:519–26.
84. Russell RR 3rd, Li J, Coven DL, Pypaert M, Zechner C, Palmeri M, et al. AMP-activated protein kinase mediates ischemic glucose uptake and prevents postischemic cardiac dysfunction, apoptosis, and injury. J Clin Invest. 2004;114:495–503.
85. Gurusamy N, Lekli I, Gorbunov NV, Gherghiceanu M, Popescu LM, et al. Cardioprotection by adaptation to ischaemia augments autophagy in association with BAG-1 protein. J Cell Mol Med. 2009;13:373–87.

86. Qian J, Ren X, Wang X, Zhang P, Jones WK, et al. Blockade of Hsp20 phosphorylation exacerbates cardiac ischemia/reperfusion injury by suppressed autophagy and increased cell death. Circ Res. 2009;105:1223–31.

87. Hamacher-Brady A, Brady NR, Gottlieb RA. Enhancing macroautophagy protects against ischemia/reperfusion injury in cardiac myocytes. J Biol Chem. 2006;281:29776–87.

88. O'Neal WT, Griffin WF, Kent SD, Virag JAI. Cellular pathways of death and survival in acute myocardial infarction. J Clin Exp Cardiolog. 2012;S6:003. https://doi.org/10.4172/2155-9880.S6-003.

89. Chiong M, Wang ZV, Pedrozo Z, Cao DJ, Troncoso R, Ibacache M, Criollo A, Nemchenko A, Hill JA and Lavandero S. Cardiomyocyte death: mechanisms and translational implications. Cell Death Disease. 2011;2:e244. https://doi.org/10.1038/cddis.2011.130.

90. Hariharan N, Zhai P, Sadoshima J. Oxidative stress stimulates autophagic flux during ischemia/reperfusion. Antioxid Redox Signal. 2011;14:2179–90.

91. Komatsu M, Waguri S, Koike M, Sou YS, Ueno T, et al. Homeostatic levels of p62 control cytoplasmic inclusion body formation in autophagy deficient mice. Cell. 2007;131:1149–63.

92. Xiao J, Zhu X, He B, Zhang Y, Kang B, et al. MiR-204 regulates cardiomyocyte autophagy induced by ischemia-reperfusion through LC3-II. J Biomed Sci. 2011;18:35.

93. Hamacher-Brady A, Brady NR, Logue SE, Sayen MR, Jinno M, et al. Response to myocardial ischemia/reperfusion injury involves Bnip3 and autophagy. Cell Death Differ. 2007;14:146–57.

94. Gustafsson AB, Gottlieb RA. Autophagy in ischemic heart disease. Circ Res. 2009;104:150–8.

95. Brady NR, Hamacher-Brady A, Yuan H, Gottlieb RA. The autophagic response to nutrient deprivation in the hl-1 cardiac myocyte is modulated by Bcl-2 and sarco/endoplasmic reticulum calcium stores. FEBS J. 2007;274:3184–97.

96. Scherz-Shouval R, Shvets E, Fass E, Shorer H, Gil L, et al. Reactive oxygen species are essential for autophagy and specifically regulate the activity of Atg4. EMBO J. 2007;26:1749–60.

97. Sala-Mercado JA, Wider J, Undyala VV, Jahania S, Yoo W, Mentzer RM Jr, et al. Profound cardioprotection with chloramphenicol succinate in the swine model of myocardial ischemia-reperfusion injury. Circulation. 2010;122:S179–84.

98. Sengupta A, Molkentin JD, Paik JH, Depinho RA, Yutzey KE. FoxO transcription factors promote cardiomyocyte survival upon induction of oxidative stress. J Biol Chem. 2011;286:7468–78. https://doi.org/10.1074/jbc.M110.179242.

99. Du Sablon A, Parks J, Whitehurst K, Estes H, Chase R, Vlahos E, et al. EphrinA1-Fc attenuates myocardial ischemia/reperfusion injury in mice. PLoS One. 2017;12:e0189307. https://doi.org/10.1371/journal.pone.0189307.

100. Song H, Yan C, Tian X, Zhu N, Li Y, Liu D, et al. CREG protects from myocardial ischemia/reperfusion injury by regulating myocardial autophagy and apoptosis. Biochim Biophys Acta. 2017;1863:1893–903.

101. Fu H, Li X, Tan J. NIPAAm-MMA nanoparticle-encapsulated visnagin ameliorates myocardial ischemia/reperfusion injury through the promotion of autophagy and the inhibition of apoptosis. Oncol Lett. 2018;15:4827–36.

102. Siasos G, Tsigkou V, Kosmopoulos M, Theodosiadis D, Simantiris S, Tagkou NM, et al. Mitochondria and cardiovascular diseases-from pathophysiology to treatment. Ann Transl Med. 2018;6:256. https://doi.org/10.21037/atm.2018.06.21.

103. Qian J, Ren X, Wang X, Zhang P, Jones WK, Molkentin JD, et al. Blockade of Hsp20 phosphorylation exacerbates cardiac ischemia/ reperfusion injury by suppressed autophagy and increased cell death. Circ Res. 2009;105:1223–31.

104. Bullon P, Cano-Garcia FJ, Alcocer-Gomez E, Varela-Lopez A, Roman-Malo L, Ruiz-Salmeron RJ, et al. Could NLRP3-inflammasome be a cardiovascular risk biomarker in acute myocardial infarction patients? Antioxid Redox Signal. 2017;27:269–75.

105. Demircan G, Kaplan O, Ozdas SB. Role of autophagy in the progress of coronary total occlusion. Bratisl Lek Listy. 2018;119:103.

106. Wu D, Zhang K, Hu P. The role of autophagy in acute myocardial infarction. Front Pharmacol. 2019;10:551. https://doi.org/10.3389/fphar.2019.00551).
107. Tahrir FG, Langford D, Amini S, Mohseni Ahooyi T, Khalili K. Mitochondrial quality control in cardiac cells: mechanisms and role in cardiac cell injury and disease. J Cell Physiol. 2019;234:8122–33.
108. Aisa Z, Liao GC, Shen XL, Chen J, Li L, Jiang SB. Effect of autophagy on myocardial infarction and its mechanism. Eur Rev Med Pharmacol Sci. 2017;21:3705–13.
109. Kanamori H, Takemur G, Goto K, Maruyama R, Ono K, Nagao K, et al. Autophagy limits acute myocardial infarction induced by permanent coronary artery occlusion. Am J Physiol Heart Circ Physiol. 2011;300:H2261–71.
110. Li Q, Dong QT, Yang YJ, Tian XQ, Jin C, Huang PS, et al. AMPK-mediated cardio protection of atorvastatin relates to the reduction of apoptosis and activation of autophagy in infarcted rat hearts. Am J Transl Res. 2016;8:4160–71.
111. Foglio E, Puddighinu G, Germani A, Russo MA, Limana F. HMGB1 inhibits apoptosis following MI and induces autophagy via mTORC1 inhibition. J Cell Physiol. 2017;232:1135–43.
112. Sciarretta S, Zhai P, Shao D, Maejima Y, Robbins J, Volpe M, et al. Rheb is a critical regulator of autophagy during myocardial ischemia: pathophysiological implications in obesity and metabolic syndrome. Circulation. 2012;125:1134–46.
113. Wu X, He L, Chen F, He X, Cai Y, Zhang G, et al. Impaired autophagy contributes to adverse cardiac remodeling in acute myocardial infarction. PLoS One. 2014; 9:e112891. https://doi.org/10.1371/journal.pone.0112891.
114. Zeng M, Wei X, Wu Z, Li W, Li B, Zhen Y, et al. NF-kappaB-mediated induction of autophagy in cardiac ischemia/reperfusion injury. Biochem Biophys Res Commun. 2013;436:180–5.
115. Zeng M, Wei X, Wu Z, Li W, Zheng Y, Li B, et al. Simulated ischemia/reperfusion-induced p65-Beclin 1-dependent autophagic cell death in human umbilical vein endothelial cells. Sci Rep. 2016;6:37448. https://doi.org/10.1038/srep37448.
116. Zhang Z, Yang M, Wang Y, Wang L, Jin Z, Ding L, et al. Autophagy regulates the apoptosis of bone marrow-derived mesenchymal stem cells under hypoxic condition via AMP-activated protein kinase/ mammalian target of rapamycin pathway. Cell Biol Int. 2016;40:671–85.
117. Liu J, Jiang M, Deng S, Lu J, Huang H, Zhang Y, et al. miR-93-5p-containing exosomes treatment attenuates acute myocardial infarction-induced myocardial damage. Mol Ther Nucleic Acids. 2018;11:103–15.
118. Liu X, Deng Y, Xu Y, Jin W, Li H. MicroRNA-223 protects neonatal rat cardiomyocytes and H9c2 cells from hypoxia-induced apoptosis and excessive autophagy via the Akt/mTOR pathway by targeting PARP-1. J Mol Cell Cardiol. 2018;118:133–46.
119. Zhang Z, Li H, Chen S, Li Y, Cui Z, Ma J. Knockdown of MicroRNA-122 protects H9c2 cardiomyocytes from hypoxia-induced apoptosis and promotes autophagy. Med Sci Monit. 2017;23:4284–90.
120. Li J, Zhang D, Wiersma M, Brundel B. Role of autophagy in proteostasis: friend and foe in cardiac diseases. Cell. 2018;7:E279. https://doi.org/10.3390/cells7120279.
121. Liu CY, Zhang YH, Li RB, Zhou LY, An T, Zhang RC, et al. LncRNA CAIF inhibits autophagy and attenuates myocardial infarction by blocking p53-mediated myocardin transcription. Nat Commun. 2018;9:29. https://doi.org/10.1038/s41467-017-02280-y.
122. Xiao C, Wang K, Xu Y, Hu H, Zhang N, Wang Y, et al. Transplanted mesenchymal stem cells reduce autophagic flux in infarcted hearts via the exosomal transfer of mir-125b. Circ Res. 2018;123:564–78.
123. Araki S, Izumiya Y, Rokutanda T, Ianni A, Hanatani S, Kimura Y, et al. Sirt7 Contributes to myocardial tissue repair by maintaining transforming growth factor-beta signaling pathway. Circulation. 2015;132:1081–93.
124. Buss SJ, Muenz S, Riffel JH, Malekar P, Hagenmueller M, Weiss CS, et al. Beneficial effects of mammalian target of rapamycin inhibition on left ventricular remodeling after myocardial infarction. J Am Coll Cardiol. 2009;54:2435–46.

125. McCormick J, Suleman N, Scarabelli TM, Knight RA, Latchman DS, Stephanou A. STAT1 deficiency in the heart protects against myocardial infarction by enhancing autophagy. J Cell Mol Med. 2012;16:386–93.
126. Ham PB, Raju R. Mitochondrial function in hypoxic ischemic injury and influence of aging. Prog Neurobiol. 2017;157:92–116.
127. Baranyai T, Nagy CT, Koncsos G, Onodi Z, Karolyi-Szabo M, Makkos A, et al. Acute hyperglycemia abolishes cardioprotection by remote ischemic percondition. Cardiovasc Diabetol. 2015;14:151. https://doi.org/0.1186/s12933-015-0313-1.
128. Sciarretta S, Boppana VS, Umapathi M, Frati G, Sadoshima J. Boosting autophagy in the diabetic heart: a translational perspective. Cardiovasc Diagn Ther. 2015;5:394–402. https://doi.org/10.3978/j.issn.2223-3652.2015.07.02.
129. Harris BH, Gelfand JA. The immune response to trauma. Semin Pediatr Surg. 1995;4:77–82.
130. Salceda S, Caro J. Hypoxia-inducible factor 1alpha (HIF-1alpha) protein is rapidly degraded by the ubiquitin-proteasome system under normoxic conditions. Its stabilization by hypoxia depends on redox-induced changes. J Biol Chem. 1997;272:22642–22647.
131. Lee SH, Wolf PL, Escudero R, Deutsch R, Jamieson SW, et al. Early expression of angiogenesis factors in acute myocardial ischemia and infarction. N Engl J Med. 2000;342:626–33.
132. Roberts WG, Palade GE. Increased microvascular permeability and endothelial fenestration induced by vascular endothelial growth factor. J Cell Sci. 1995;108:2369–79.
133. Jeong HJ, Moon PD, Kim SJ, Seo JU, Kang TH, et al. Activation of hypoxia-inducible factor-1 regulates human histidine decarboxylase expression. Cell Mol Life Sci. 2009;66:1309–19.
134. Dong F, Khalil M, Kiedrowski M, O'Connor C, Petrovic E, et al. Critical role for leukocyte hypoxia inducible factor-1alpha expression in post-myocardial infarction left ventricular remodeling. Circ Res. 2010;106:601–10.
135. Sun HY, Wang NP, Halkos M, Kerendi F, Kin H, et al. Postconditioning attenuates cardiomyocyte apoptosis via inhibition of JNK and p38 mitogenactivated protein kinase signaling pathways. Apoptosis. 2006;11:1583–93.
136. Chang JS, Wendt T, Qu W, Kong L, Zou YS, et al. Oxygen deprivation triggers upregulation of early growth response-1 by the receptor for advanced glycation end products. Circ Res. 2008;102:905–13.
137. Lo LW, Cheng JJ, Chiu JJ, Wung BS, Liu YC, et al. Endothelial exposure to hypoxia induces Egr-1 expression involving PKCalpha-mediated Ras/Raf-1/ERK1/2 pathway. J Cell Physiol. 2001;188:304–12.
138. Kramer B, Meichle A, Hensel G, Charnay P, Kronke M. Characterization of an Krox-24/Egr-1-responsive element in the human tumor necrosis factor promoter. Biochim Biophys Acta. 1994;1219:413–21.
139. Such L, Alberola A, Gil F, Bendala E, Vina J, et al. Effect of glutathione on canine myocardial ischaemia without reperfusion. J Pharm Pharmacol. 1993;45:298–302.
140. Waypa GB, Guzy R, Mungai PT, Mack MM, Marks JD, et al. Increases in mitochondrial reactive oxygen species trigger hypoxia-induced calcium responses in pulmonary artery smooth muscle cells. Circ Res. 2006;99:970–8.
141. Duilio C, Ambrosio G, Kuppusamy P, DiPaula A, Becker LC, et al. Neutrophils are primary source of O2 radicals during reperfusion after prolonged myocardial ischemia. Am J Physiol Heart Circ Physiol. 2001;280:H2649-2657.
142. Scarabelli T, Stephanou A, Rayment N, Pasini E, Comini L, et al. Apoptosis of endothelial cells precedes myocyte cell apoptosis in ischemia/reperfusion injury. Circulation. 2001;104:253–6.
143. Kaur K, Sharma AK, Dhingra S, Singal PK. Interplay of TNF-alpha and IL-10 in regulating oxidative stress in isolated adult cardiac myocytes. J Mol Cell Cardiol. 2006;41:1023–30.
144. Song W, Lu X, Feng Q. Tumor necrosis factor-alpha induces apoptosis via inducible nitric oxide synthase in neonatal mouse cardiomyocytes. Cardiovasc Res. 2000;45:595–602.
145. Brune B, von Knethen A, Sandau KB. Nitric oxide (NO): an effector of apoptosis. Cell Death Differ. 1999;6:969–75.
146. Xuan YT, Tang XL, Banerjee S, Takano H, Li RC, et al. Nuclear factorkappaB plays an essential role in the late phase of ischemic preconditioning in conscious rabbits. Circ Res. 1999;84:1095–109.

147. Zingarelli B, Hake PW, Yang Z, O'Connor M, Denenberg A, et al. Absence of inducible nitric oxide synthase modulates early reperfusion-induced NF-kappaB and AP-1 activation and enhances myocardial damage. FASEB J. 2002;16:327–42.
148. Onai Y, Suzuki J, Kakuta T, Maejima Y, Haraguchi G, et al. Inhibition of IkappaB phosphorylation in cardiomyocytes attenuates myocardial ischemia/ reperfusion injury. Cardiovasc Res. 2004;63:51–9.
149. Fauconnier J, Meli AC, Thireau J, Roberge S, Shan J, et al. Ryanodine receptor leak mediated by caspase-8 activation leads to left ventricular injury after myocardial ischemia-reperfusion. Proc Natl Acad Sci USA. 2011;108:13258–63.
150. Sun M, Dawood F, Wen WH, Chen M, Dixon I, et al. Excessive tumor necrosis factor activation after infarction contributes to susceptibility of myocardial rupture and left ventricular dysfunction. Circulation. 2004;110:3221–8.
151. LaFramboise WA, Bombach KL, Dhir RJ, Muha N, Cullen RF, et al. Molecular dynamics of the compensatory response to myocardial infarct. J Mol Cell Cardiol. 2005;38:103–17.
152. Tilley SL, Wagoner VA, Salvatore CA, Jacobson MA, Koller BH. Adenosine and inosine increase cutaneous vasopermeability by activating A(3) receptors on mast cells. J Clin Invest. 2000;105:361–7.
153. Bowditch J, Brown AK, Dow JW. Accumulation and salvage of adenosine and inosine by isolated mature cardiac myocytes. Biochim Biophys Acta. 1985;844:119–28.
154. Decking UK, Schlieper G, Kroll K, Schrader J. Hypoxia-induced inhibition of adenosine kinase potentiates cardiac adenosine release. Circ Res. 1997;81:154–64.
155. Annecke T, Fischer J, Hartmann H, Tschoep J, Rehm M, et al. Shedding of the coronary endothelial glycocalyx: effects of hypoxia/reoxygenation vs ischaemia/reperfusion. Br J Anaesth. 2011;107:679–86.
156. Frangogiannis NG, Lindsey ML, Michael LH, Youker KA, Bressler RB, et al. Resident cardiac mast cells degranulate and release preformed TNFalpha, initiating the cytokine cascade in experimental canine myocardial ischemia/reperfusion. Circulation. 1998;98:699–710.
157. Asako H, Kurose I, Wolf R, DeFrees S, Zheng ZL, et al. Role of H1 receptors and P-selectin in histamine-induced leukocyte rolling and adhesion in postcapillary venules. J Clin Invest. 1994;93:1508–15.
158. Wei CC, Hase N, Inoue Y, Bradley EW, Yahiro E, et al. Mast cell chymase limits the cardiac efficacy of Ang I-converting enzyme inhibitor therapy in rodents. J Clin Invest. 2010;120:1229–39.
159. Kinoshita M, Okada M, Hara M, Furukawa Y, Matsumori A. Mast cell tryptase in mast cell granules enhances MCP-1 and interleukin XE "Interleukin" -8 production in human endothelial cells. Arterioscler Thromb Vasc Biol. 2005;25:1858–63.
160. Rork TH, Wallace KL, Kennedy DP, Marshall MA, Lankford AR, et al. Adenosine A2A receptor activation reduces infarct size in the isolated, perfused mouse heart by inhibiting resident cardiac mast cell degranulation. Am J Physiol Heart Circ Physiol. 2008;295:H1825-1833.
161. Nistri S, Cinci L, Perna AM, Masini E, Bani D. Mast cell inhibition and reduced ventricular arrhythmias in a swine model of acute myocardial infarction upon therapeutic administration of relaxin. Inflamm Res. 2008;57:S7-8.
162. Boyd JH, Mathur S, Wang Y, Bateman RM, Walley KR. Toll-like receptor stimulation in cardiomyoctes decreases contractility and initiates an NF-kappaB dependent inflammatory response. Cardiovasc Res. 2006;72:384–93.
163. Piccinini AM, Midwood KS. DAMPening inflammation by modulating TLR signalling. Mediators Inflamm. 2010;ID672395. https://doi.org/10.1155/2010/672395.
164. Kukielka GL, Smith CW, Manning AM, Youker KA, Michael LH, et al. Induction of interleukin-6 synthesis in the myocardium. Potential role in postreperfusion inflammatory injury. Circulation. 1995;92:1866–1875.

Chapter 15
Cardiomyocytes in Heart Failure

Abstract The cause for heart failure (HF) is the death or dysfunction of a significant number of cardiomyocytes (CMs), which can result from several cardiovascular diseases. The distinct processes that characterize cardiomyocyte remodelling in failing hearts are hypertrophy of surviving CMs, metabolic abnormalities, mitochondrial dysfunction, defective autophagy and death of CMs. The neuro-hormonal axis is activated in HF and in turn induces insulin resistance and an increased use of non-carbohydrate substrates for energy production. Oxidation of fat is increased in patients with HF. Adenosine triphosphate, phosphocreatine and creatine kinase levels in the heart are reduced leading to decreased efficiency of mechanical work. Molecular mechanisms that lead to the progressive loss of CMs include oxidative stress, metabolic switch, alterations of proteins involved in calcium transport and inflammation. Both gradual and acute cell death via apoptosis, necrosis, and autophagy are hallmarks of HF. Given that the pathophysiology of HF involves death or dysfunction of CMs, understanding their mechanisms is imperative for identifying strategies for the treatment of HF and reducing related complications.

Keyword Heart failure · Cardiomyocyte remodelling · Cardiomyocyte hypertrophy · Apoptosis · Autophagy · Necrosis · Mitochondrial dysfunction · Cardiomyocyte metabolism · Calcium dysregulation

Introduction

Heart failure (HF) is a chronic heart disease, which is one of the leading causes of morbidity and mortality worldwide [1]. In HF, the heart is unable to maintain the blood flow required to meet metabolic requirements of the body [2]. HF results from a variety of pathologic insults that cause structural damage to the heart. The progression of HF is associated with molecular and cellular changes in cardiomyocytes (CMs), together known as cardiomyocyte remodelling [3]. Several factors contribute to the development and progression of cardiomyocyte remodelling that leads to cardiac dysfunction [4].

© The Author(s), under exclusive license to Springer Nature Switzerland AG 2021   245
C. C. Kartha, *Cardiomyocytes in Health and Disease*,
https://doi.org/10.1007/978-3-030-85536-9_15

Cardiomyocyte Remodelling

Remodelling of CMs may occur in many pathological conditions which result in pressure and volume overload, a decrease in contractility and or an increase in wall stress, neuroendocrine activation and because of genetic defects [3, 5, 6].

Initially, remodelling of CMs is adaptive and the structural changes have a compensatory effect for the maintenance of normal heart function [7, 8]. Prolonged stress result in maladaptive changes that lead to continued and irreversible cardiac dysfunction [9]. Death of CMs is a key event that contributes to the development of cardiac dysfunction. Cardiomyocyte loss in HF occurs through necrosis, necroptosis, apoptosis, or autophagy [10].

Many molecular pathways intersect in the pathway that leads to the progressive loss of CMs. These include oxidative stress, changes in energy metabolism, alterations of proteins involved in calcium transport, inflammation and neurohormonal activation [11–13].

Dysregulation of physiological mechanisms, such as the those which tightly regulates calcium influx and uptake, is a common feature of several pathophysiological cellular alterations in cardiac remodelling. In failing CMs, calcium uptake mediated by sarco/endoplasmic reticulum Ca^{2+}–ATPase (SERCA)-2a is impaired and there is unregulated calcium efflux through ryanodine receptors (RyRs) [12]. Calcium dysregulation affects excitation–contraction coupling (ECC), hypertrophic growth, energy metabolism, mitochondrial function as well as cell survival [10].

Adaptive Cardiomyocyte Hypertrophy

Remodelling of cardiomyocytes starts initially as a compensatory adaptive response to mechanical and physiological stress and eventually evolves into maladaptive remodelling, triggering the transition to heart failure [14].

Cardiomyocyte hypertrophy (CH) in response to hemodynamic overload on cardiac walls activates complex biological responses. Stromal Interaction Molecule (STIM)-1, is one of the molecular initiators of the hypertrophic response. STIM-1 can also prevent as well as reverse hypertrophy. STIM-1 in association with calcium release-activated calcium channel protein (ORAI)-1/3 mediates increased Ca^{2+} influxes in response to hypertrophic stress. Activation of the mammalian target of rapamycin complex (mTORC)-2 follows. Silencing of STIM-1 prevents mTORC-2 phosphorylation of Akt kinase, suppress GSK-3β activity, and thus inhibit hypertrophic responses [15].

During maladaptive remodelling, sarcomere is added end-to-end and gradually force production decreases resulting in contractile dysfunction [14]. Several cytokines and growth factors participate in the remodelling in response to hemodynamic overload. Defects in the activity of the sarcoplasmic reticulum (SR)

Ca^{2+} pump, sarcolemmal Ca^{2+} ATPase, and the Na^+ /Ca^{2+} exchanger may result initially in diastolic dysfunction.

Effect of Mechanical Stress

Mechanical wall stress activates mechanosensitive ion channels and integrins and the signals are transduced through the Akt pathway. Short-term activation of the Akt pathway induces hypertrophy and long-term activation leads to HF [16]. Mechanical stretch assists the release of angiotensin (AT)-II and endothelin-1, which together activates G-protein coupled receptor (GPCR) signalling through Gαq subunits. Activation of phospholipase C (PLC), a downstream effector of Gαq, leads to hypertrophy through the PI3K/Akt pathway [17].

Epigenetic Mechanisms

Epigenetics has a key role in the downstream network activated by hypertrophic stimuli. Class II histone deacetylases, HDAC4 and HDAC5, normally interfere with DNA binding of pro-hypertrophic transcription factors, such as NFAT, myocyte enhancement factor (MEF), and GATA-4. Pro-hypertrophic transcriptional activity can be dampened by the cytoplasmic translocation of Class II HDACs, caused by oxidation or phosphorylation of specific residues, by Ca^{2+} /calmodulin-dependent protein kinase (CaMK)-II, GPCR kinase (GRK)-5, PKC (protein kinase C), and PKD [18, 19].

Other Factors

Endothelial cells play a role in paracrine pro-hypertrophic signalling during pressure overload [20]. Upregulation of signalling molecules such as brain natriuretic peptide (BNP) and atrial natriuretic peptide (ANP), and structural proteins such as β-myosin heavy chain (β-myHC) are also involved in the hypertrophic process [14].

Oxidative stress levels are increased during hemodynamic overload [21]. Peptidylprolyl cis–trans isomerase NIMA-interacting (PIN)-1 an important player in several physiological processes has been found to be upregulated during pressure overload [22].

Inflammation and Metabolism

ATP and phosphocreatine levels are decreased in the remodelling heart because of chronic inflammation. This reduction impairs carbohydrate metabolism and fatty acid oxidation in the mitochondria [23]. Energy demand is met through anaerobic metabolism. Further, inflammatory cytokines impede SERCA2a activity [24].

PGC-1α, modulates the expression of several genes involved in mitochondrial biogenesis, β-oxidation, glucose oxidative metabolism and the electron transport chain [25]. PGC-1α expression is decreased in hypertrophy [26, 27] and heart failure [28]. PGC-1α is important for regulating the inflammatory response as well [29–31].

Mitochondrial Dysfunction

Dysfunction of mitochondria is well known as an important cause for the progression of HF. Mitochondrial dysfunction associated with HF is characterized by several factors such as alterations in mitochondrial dynamics (fission, fusion, autophagy), membrane potential and ion homeostasis, switch in substrate metabolism, and increase in reactive oxygen species (ROS) and other free radicals (nitric oxide, hydroxyl). All of them contribute to adverse outcomes in HF [32].

Iron imbalance and oxidative stress are also major factors for the evolution of HF. Mitochondrial ATP–binding cassette (ABC) transporters regulate iron metabolism and maintenance of redox status in cells. Deficiency of mitochondrial ABC transporters has been linked to impaired mitochondrial electron transport chain complex activity, iron overload, and increased levels of ROS, all of which can result in dysfunction of mitochondria [33, 34].

Mitochondrial ROS (mt-ROS), generated mainly from complexes I and III of the electron transport chain (ETC) act at low levels as intracellular messengers during remodelling of CMs. At high levels, they cause damage to mitochondrial DNA (mt-DNA) and proteins. Energy production is impaired because of the effect on transcription of mitochondrial genes coding for components of the ETC [35]. ERK1/2 activation seems to mediate mt-ROS-induced remodelling [36].

Sirtuin 4 overexpression is also associated with an increase of mt-ROS and simultaneous decrease in manganese superoxide dismutase (MnSOD) [37]. During progression of cardiac remodelling, there is significant downregulation of genes such as PGC-1α and PGC-1β, p38-MAPK, and mitochondrial transcription factor A (TFAM), all of which are involved in mitochondrial biogenesis [38, 39].

Deranged Cellular Metabolism in Heart Failure

Oxidative phosphorylation in the mitochondria generates more than 95% of ATP produced in the heart and about 70 – 90% of the ATP produced is via oxidation of fatty acids [40, 41]. The rest is from oxidation of glucose, lactate, small amounts of ketone bodies and amino acids. In the normal healthy heart, the Randle cycle regulates the glucose and fatty acid metabolism [40]. Increase in FFA oxidation in the heart decreases oxidation of glucose; FFA oxidation is inhibited when there is rise in oxidation of glucose. In terms of ATP produced per O2 molecules consumed, FFA oxidation is a less efficient source of energy than glucose oxidation. Increase in FFA oxidation is also associated with significant decrease in cardiac efficiency [42]. During progressive HF, there is an imbalance between the demand for oxygen by CMs and supply and availability of metabolic substrates. Glucose uptake is decreased in CMs and conversion to lactate is increased in CMs. Pyruvate is mostly transformed into lactate, and the rise in lactate production results in cell acidosis. The free fatty acid (FFA) pathway also slows down and hence less ATP is produced. These metabolic changes lead to impaired cell homeostasis, structural changes in membrane and finally, cell death [42, 43].

Effects of the Neuro-Hormonal Activation on Metabolism of the Failing Heart

Neuro-hormonal activation is an important contributor for both mechanical and metabolic inefficiency of CMs in HF. Increase in peripheral lipolysis mediated by adrenergic activation leads to increase in the availability of FFA. Adrenergic activation may also induce insulin resistance and may thus contribute to increase in the plasma levels of FFA [44].

Another key regulator of cardiac energy metabolism and function is angiotensin II [45]. Angiotensin II via production of ROS, damages mitochondria in CMs [36] and affects mitochondrial oxidative phosphorylation and FFA oxidation [46, 47]. Angiotensin II regulates glucose oxidation as well [45, 48]. Angiotensin II can reduce the availability of ATP by decreasing oxidative metabolism and thus decreasing production of ATP [49].

Cardiomyocyte Death in Heart Failure

All three types of cell death, autophagic cell death, apoptosis, and necrosis are seen during progression of HF [50, 51]. Progressive loss of CMs in HF result in cardiac dysfunction.

Necrosis

Though earlier studies indicated apoptosis as an important contributor to the death of CMs in end-stage HF, later studies indicate that necrosis is more prominent in the failing human heart and contribute significantly more than apoptosis to HF pathogenesis [52].

Necrotic death is triggered by sustained Ca^{++} stress and activation of adrenergic receptors. An increase in activity of L-type calcium channel promotes progressive cell death; an adrenergic receptor agonist would augment this effect [53]. There is also evidence to indicate the involvement of opening of the mitochondrial permeability transition pore (MPTP) in necrosis of CMs [54, 55].

Apoptosis

In the normal adult human myocardium apoptosis is rare. In hearts of patients with severe HF, 0.12–0.70% apoptotic cells have been identified [56]. Even at low levels, apoptosis causes significant effects [57, 58]. As the adult CMs do not have significant proliferative capacity, a low, but increased rate of apoptosis can have an important role in the pathogenesis of HF.

Several cytokines initiate cardiomyocyte apoptosis in the failing overloaded heart. The cytokine effect is via an increase in ROS levels and GPCR signalling [35].

Autophagy

In patients with HF because of dilated cardiomyopathy, death of CMs with characteristics of autophagy has been observed at a rate of 0.03% [59]. In comparison, the rate of apoptotic cell death was 0.002% [59]. The specific role of autophagy in the pathogenesis of HF-related CM remodelling is however not explicit. Genetic manipulations of the key autophagic mechanisms in mice have suggested that autophagy aids the development of pathological hypertrophy [60, 61]. Pharmaceutical manipulations of upstream pathways of autophagy have indicated that autophagy may have an anti-hypertrophic role as well [62–64].

Autophagy is certainly required for the process of cardiomyocyte hypertrophy (CH). Moderate pressure overload induces and also activates autophagy. Hypertrophy in response to pressure overload is intensified on overexpression of Beclin 1 [61]. Hypertrophic response to hypertrophic agonists can be reduced by knockdown of ATG5 and Beclin 1 [60].

Several signalling pathways modulate the autophagic flux. mTOR signalling is a major inhibitor of autophagy [65]. Inhibition of mTOR, an upstream repressor of autophagy, lessens CH [63, 66]. Inactivation of cardiac mTOR also results in increase in apoptosis and rapid progression to HF [64]. Though this may not have an important role in the development of HF, there is evidence for a specific role of mTOR in preserving cardiac function [62].

While enormous activation of autophagy may lead to cell death, physiologic activation protects cells from apoptotic death [67]. Extensive crosstalk between autophagy and apoptosis exists in the adult heart. Interaction between Beclin1 and Bcl-2 family members determine the essence of these two phenomena [67].

Cardiomyocytes in Heart Failure with Preserved Ejection Fraction

Nearly one half of patients with HF have HF with preserved ejection fraction (HFpEF) [68]. The functional cardiomyocyte phenotype that typifies HFpEF is not clearly known. There are no reports of studies on Ca^{2+} handling and contractile responsiveness of intact CMs derived from either patients or animals with HFpEF. The trigger for HFpEF is not systolic loading and hence the cardiomyocyte phenotype and associated pathophysiology in HFpEF are likely to be distinctly different from that of HF with reduced ejection fraction (HFrEF) [69].

The HFpEF phenotype is heterogenous [68, 70, 71]. Cardiomyocyte hypertrophy is prominent and is an important prognostic indicator [72, 73]. In a selectively inbred rat strain with hypertrophic heart, validated as a model for HFpEF, the progression to diastolic failure with preservation of systolic performance was found to be associated with hypercontractile CMs and increased availability of systolic activator Ca^{2+} and increased L-type Ca^{2+} channel current [69]. The mechanism of remodelling of CMs in this model of HFpEF is thus in contrast with the suppressed Ca^{2+} cycling state that typifies HFrEF. While upregulation of Na^+/Ca^{2+} exchanger has been noted as important in transition to failure in HFrEF [74], there seems to be no functional difference in the density of Na^+/Ca^{2+} urrent between CMs from rats with hypertrophic heart and rats with normal heart [69].

Conclusions

Heart failure (HF) is characterized by cardiac remodelling characterized by death of cardiomyocytes (CMs), hypertrophy of surviving CMs, metabolic abnormalities, mitochondrial dysfunction and defective autophagy. Molecular mechanisms that lead to the progressive loss of CMs include oxidative stress, metabolic switch, alterations of proteins involved in calcium transport and inflammation. Metabolic switch to an increased use of non-carbohydrate substrates for energy production in HF, results in metabolic inefficiency in CMs. Oxidation of fat is increased in patients with HF. Adenosine triphosphate, phosphocreatine and creatine kinase levels in the heart are reduced leading to decreased contractile efficiency. Neurohormonal activation partly contributes to the metabolic changes. In HF, there is both gradual and acute death of CMs via apoptosis, necrosis, and autophagy. Given that the pathophysiology of HF involves death or dysfunction of CMs, understanding their mechanisms is imperative for identifying strategies for the treatment of HF and reducing related complications.

References

1. Braunwald E. The war against heart failure: The Lancet lecture. The Lancet. 2015;385:812–24.
2. Goldberg R. In the clinic. Heart failure. Annals of Internal Medicine. 2010; 152: ITC61–15; quiz ITC616.
3. Cohn JN, Ferrari R, Sharpe N. Cardiac remodeling— concepts and clinical implications: a consensus paper from an international forum on cardiac remodeling. Behalf of an international forum on cardiac remodeling. Journal of the American College of Cardiology. 2000; 35:569–582.
4. Schirone L, Forte M, Palmerio S, Yee D, Nocella C, Angelini F, Pagano F, Schiavon S, Bordin A, Carrizzo A, Vecchione C, Valenti V, Chimenti I, De Falco E, Sciarretta S, Frati G. A review of the molecular mechanisms underlying the development and progression of cardiac remodeling. Oxidative Medicine and Cellular Longevity. 2017; Article ID 3920195, 16 pages. https://doi.org/10.1155/2017/3920195.
5. Nian M, Lee P, Khaper N, Liu P. Inflammatory cytokines and post myocardial infarction remodelling. Circ Res. 2004;94:1543–53.
6. Swynghedauw B. Molecular mechanisms of myocardial remodelling. Physiol Rev. 1999;79:215–62.
7. Dorn GW 2nd. The fuzzy logic of physiological cardiac hypertrophy. Hypertension. 2007;49:962–70.
8. Opie LH, Commerford PJ, Gersh BJ, Pfeffer MA. Controversies in ventricular remodelling. Lancet. 2006;367:356–67.
9. Hill JA, Olson EN. Cardiac plasticity. N Engl J Med. 2008;358:1370–80.
10. Burchfield JS, Xie M, Hill JA. Pathological ventricular remodeling: mechanisms: part 1 of 2. Circulation. 2013;128:388–400.
11. Frangogiannis NG. Regulation of the inflammatory response in cardiac repair. Circ Res. 2012;110:159–73.
12. Lehnart SE, Maier LS, Hasenfuss G. Abnormalities of calcium metabolism and myocardial contractility depression in the failing heart. Heart Fail Rev. 2009;14:213–24.
13. Sciarretta S, Paneni F, Palano F, et al. Role of the renin-angiotensin-aldosterone system and inflammatory processes in the development and progression of diastolic dysfunction. Clinical Science (London). 2009;116:467–77.
14. Harvey PA, Leinwand LA. The cell biology of disease: cellular mechanisms of cardiomyopathy. J Cell Biol. 2011;194:355–65.
15. Benard L, Oh JG, Cacheux M, et al. Cardiac Stim1 silencing impairs adaptive hypertrophy and promotes heart failure through inactivation of mTORC2/Akt signaling. Circulation. 2016;133:1458–71.
16. Matsui T, Li L, Wu JC, et al. Phenotypic spectrum caused by transgenic overexpression of activated Akt in the heart. J Biol Chem. 2002;277:22896–901.
17. Sadoshima J, Izumo S. Mechanical stretch rapidly activates multiple signal transduction pathways in cardiac myocytes: potential involvement of an autocrine/paracrine mechanism. EMBO J. 1993;12:1681–92.
18. Backs J, Olson EN. Control of cardiac growth by histone acetylation/deacetylation. Circ Res. 2006;98:15–24.
19. Dassanayaka S, Jones SP. Recent developments in heart failure. Circ Res. 2015;117:e58–63.
20. Appari M, Breitbart A, Brandes F, et al. C1q-TNF-related protein-9 promotes cardiac hypertrophy and failure. Circ Res. 2017;120:66–77.
21. Matsushima S, Kuroda J, Ago T, et al. Increased oxidative stress in the nucleus caused by Nox4 mediates oxidation of HDAC4 and cardiac hypertrophy. Circ Res. 2013;112:651–63.
22. Toko H, Konstandin MH, Doroudgar S, et al. Regulation of cardiac hypertrophic signaling by prolyl isomerase Pin1. Circ Res. 2013;112:1244–52.
23. Neubauer S. The failing heart—an engine out of fuel. N Engl J Med. 2007;356:1140–51.

24. Frati G, Schirone L, Chimenti I, et al. An overview of the inflammatory signalling mechanisms in the myocardium underlying the development of diabetic cardiomyopathy. Cardiovasc Res. 2017;113:378–88.

25. Puigserver P, Spiegelman BM. Peroxisome proliferator-activated receptor-gamma coactivator 1 alpha (PGC-1 alpha): transcriptional coactivator and metabolic regulator. Endocr Rev. 2003;24:78–90.

26. Arany Z, He H, Lin J, et al. Transcriptional coactivator PGC1 alpha controls the energy state and contractile function of cardiac muscle. Cell Metab. 2005;1:259–71.

27. Chen Y, Wang Y, Chen J, et al. Roles of transcriptional corepressor RIP140 and coactivator PGC-1alpha in energy state of chronically infarcted rat hearts and mitochondrial function of cardiomyocytes. Mol Cell Endocrinol. 2012;362:11–8.

28. Sano M, Wang SC, Shirai M, et al. Activation of cardiac Cdk9 represses PGC-1 and confers a predisposition to heart failure. EMBO J. 2004;23:3559–69.

29. Palomer X, Alvarez-Guardia D, Rodríguez-Calvo R, et al. TNF-alpha reduces PGC-1alpha expression through NF-kappa B and p38 MAPK leading to increased glucose oxidation in a human cardiac cell model. Cardiovasc Res. 2009;81:703–12.

30. Alvarez-Guardia D, Palomer X, Coll T, et al. The p65 subunit of NF-kappa B binds to PGC-1alpha, linking inflammation and metabolic disturbances in cardiac cells. Cardiovasc Res. 2010;87:449–58.

31. Meng F, Liu L, Chin PC, D'Mello SR. Akt is a downstream target of NF-kappa B. J Biol Chem. 2002;277:29674–80.

32. O'Rourke B. Metabolism: beyond the power of mitochondria. Nat Rev Cardiol. 2016;13:386–7.

33. Kumar V, Santhosh Kumar TR, Kartha CC. Mitochondrial membrane transporters and metabolic switch in heart failure. Heart Fail Rev. 2019;24:255–67.

34. Kumar V, Kumar AA, Sanawar R, Jaleel A, Santhosh Kumar TR, Kartha CC. Chronic pressure overload results in deficiency of mitochondrial membrane transporter ABCB7 which contributes to iron overload, mitochondrial dysfunction, metabolic shift and worsens cardiac function Scientific Reports. 2019; 9:13170. https://doi.org/10.1038/s41598-019-49666-0

35. Burgoyne JR, Mongue-Din H, Eaton P, Shah AM. Redox signaling in cardiac physiology and pathology. Circ Res. 2012;111:1091–106.

36. Dai DF, Johnson SC, Villarin JJ, et al. Mitochondrial oxidative stress mediates angiotensin II-induced cardiac hypertrophy and Galphaq overexpression-induced heart failure. Circ Res. 2011;108:837–46.

37. Luo YX, Tang X, An XZ, et al. SIRT4 accelerates Ang II-induced pathological cardiac hypertrophy by inhibiting manganese superoxide dismutase activity. Eur Heart J. 2017;38:1389–98.

38. Arany Z, Novikov M, Chin S, Ma Y, Rosenzweig A, Spiegelman BM. Transverse aortic constriction leads to accelerated heart failure in mice lacking PPAR-gamma coactivator 1alpha. Proceedings of the National Academy of Sciences USA. 2006;103:10086–91.

39. Riehle C, Wende AR, Zaha VG, et al. PGC-1beta deficiency accelerates the transition to heart failure in pressure overload hypertrophy. Circ Res. 2011;109:783–93.

40. Doenst T, Dung Nguyen T, Dale AE. Cardiac metabolism in heart failure - Implications beyond ATP production. Circ Res. 2013;113:709–24.

41. Randle PJ, Garland PB, Hales CN, Newsholme EA. The glucose fatty-acid cycle its role in insulin sensitivity and the metabolic disturbances of diabetes mellitus. Lancet. 1963;281:785–9.

42. Lopaschuk GD, Ussher JR, Folmes CD, et al. Myocardial fatty acid metabolism in health and disease. Physiol Rev. 2010;90:207–58.

43. Fragasso G. Deranged cardiac metabolism and the pathogenesis of heart failure. Card Fail Rev. 2016;2(1):8–13.

44. Paolisso G, De Riu S, Marrazzo G, et al. Insulin resistance and hyperinsulinemia in patients with chronic heart failure. Metabolism. 1991;40:972–7.

45. Mori J, Basu R, McLean BA, et al. Agonist-induced hypertrophy and diastolic dysfunction are associated with selective reduction in glucose oxidation: a metabolic contribution to heart failure with normal ejection fraction. Circ Heart Fail. 2012;5:493–503.

46. Pellieux C, Aasum E, Larsen TS, et al. Overexpression of angiotensinogen in the myocardium induces downregulation of the fatty acid oxidation pathway. J Mol Cell Cardiol. 2006;41:459–66.
47. Pellieux C, Montessuit C, Papageorgiou I, Lerch R. Angiotensin II downregulates the fatty acid oxidation pathway in adult rat cardiomyocytes via release of tumour necrosis factor-alpha. Cardiovasc Res. 2009;82:341–50.
48. Mori J, Alrob OA, Wagg CS, et al. Ang II causes insulin resistance and induces cardiac metabolic switch and inefficiency: a critical role of PDK4. Am J Physiol Heart Circ Physiol. 2013;304:H1103–13.
49. Fillmore N, Mori J, Lopaschuk GD. Mitochondrial fatty acid oxidation alterations in heart failure, ischaemic heart disease and diabetic cardiomyopathy. Br J Pharmacol. 2014;171:2080–90.
50. Whelan RS, Kaplinskiy V, Kitsis RN. Cell death in the pathogenesis of heart disease: mechanisms and significance. Annu Rev Physiol. 2010;72:19–44.
51. Chiong M, Wang ZV, Pedrozo Z, Cao DJ, Troncoso R, Ibacache M, Criollo A, Nemchenko A, Hill JA, Lavandero S.Cardiomyocyte death: mechanisms and translational implications. Cell Death and Disease. 2011;2:e244. https://doi.org/10.1038/cddis.2011.130.
52. Guerra S, Leri A, Wang X, Finato N, Di Loreto C, Beltrami CA, et al. Myocyte death in the failing human heart is gender dependent. Circ Res. 1999;85:856–66.
53. Nakayama H, Chen X, Baines CP, Klevitsky R, Zhang X, Zhang H, et al. Ca2+- and mitochondrial-dependent cardiomyocyte necrosis as a primary mediator of heart failure. J Clin Invest. 2007;117:2431–44.
54. Baines CP, Kaiser RA, Purcell NH, Blair NS, Osinska H, Hambleton MA, et al. Loss of cyclophilin D reveals a critical role for mitochondrial permeability transition in cell death. Nature. 2005;434:658–62.
55. Nakagawa T, Shimizu S, Watanabe T, Yamaguchi O, Otsu K, Yamagata H, et al. Cyclophilin D-dependent mitochondrial permeability transition regulates some necrotic but not apoptotic cell death. Nature. 2005;434:652–8.
56. van Empel VP, Bertrand AT, Hofstra L, Crijns HJ, Doevendans PA, De Windt LJ. Myocyte apoptosis in heart failure. Cardiovasc Res. 2005;67:21–9.
57. Mani K. Programmed cell death in cardiac myocytes: strategies to maximize post-ischemic salvage. Heart Fail Rev. 2008;13:193–209.
58. Wencker D, Chandra M, Nguyen K, Miao W, Garantziotis S, Factor SM, et al. A mechanistic role for cardiac myocyte apoptosis in heart failure. J Clin Invest. 2003;111:1497–504.
59. Kostin S, Pool L, Elsasser A, Hein S, Drexler HC, Arnon E, et al. Myocytes die by multiple mechanisms in failing human hearts. Circ Res. 2003;92:715–24.
60. Cao DJ, Wang ZV, Battiprolu PK, Jiang N, Morales CR, Kong Y, et al. Histone deacetylase (HDAC) inhibitors attenuate cardiac hypertrophy by suppressing autophagy. Proc Natl Acad Sci USA. 2011;108:4123–8.
61. Zhu H, Tannous P, Johnstone JL, Kong Y, Shelton JM, Richardson JA, et al. Cardiac autophagy is a maladaptive response to hemodynamic stress. J Clin Invest. 2007;117:1782–93.
62. Shende P, Plaisance I, Morandi C, Pellieux C, Berthonneche C, Zorzato F, et al. Cardiac raptor ablation impairs adaptive hypertrophy, alters metabolic gene expression, and causes heart failure in mice. Circulation. 2011;123:1073–82.
63. Shioi T, McMullen JR, Tarnavski O, Converso K, Sherwood MC, Manning WJ, et al. Rapamycin attenuates load-induced cardiac hypertrophy in mice. Circulation. 2003;107:1664–70.
64. Zhang D, Contu R, Latronico MV, Zhang J, Rizzi R, Catalucci D, et al. mTORC1 regulates cardiac function and myocyte survival through 4E-BP1 inhibition in mice. J Clin Invest. 2010;120:556–63.
65. Sciarretta S, Volpe M, Sadoshima J. Mammalian target of rapamycin signaling in cardiac physiology and disease. Circ Res. 2014;114:549–64.
66. Kushwaha SS, Raichlin E, Sheinin Y, Kremers WK, Chandrasekaran K, Brunn GJ, et al. Sirolimus affects cardiomyocytes to reduce left ventricular mass in heart transplant recipients. Eur Heart J. 2008;29:2742–50.

67. Biala AK, Kirshenbaum LA. The interplay between cell death signaling pathways in the heart. Trends Cardiovasc Med. 2014;24:325–31.

68. Borlaug BA. The pathophysiology of heart failure with preserved ejection fraction. Nat Rev Cardiol. 2014;11:507–15.

69. Curl CL, Danes VR, Bell JR, Raaijmakers AJA, Wendy TK, Chandramouli C, Harding TW, Porrello ER, Erickson JR, Charchar FJ, Kompa AR, Edgley AJ, Crossman DJ, Soeller C, Mellor KM, Kalman JM, Harrap SB, Delbridge L. Cardiomyocyte functional etiology in heart failure with preserved ejection fraction is distinctive—A new preclinical model. J Am Heart Assoc. 2018; 7: e007451.

70. de Simone G, Gottdiener JS, Chinali M, Maurer MS. Left ventricular mass predicts heart failure not related to previous myocardial infarction: the Cardiovascular Health Study. Eur Heart J. 2008;29:741–7.

71. Shah SJ, Kitzman DW, Borlaug BA, van Heerebeek L, Zile MR, Kass DA, Paulus WJ. Phenotype-specific treatment of heart failure with preserved ejection fraction: a multiorgan roadmap. Circulation. 2016;134:73–90.

72. Shah SJ, Katz DH, Selveraj S, Burke MA, Yancy CW, Gheorghiade M, Bonow RO, Huang CC, Deo RC. Phenomapping for novel classification of heart failure with preserved ejection fraction. Circulation. 2015;131:269–79.

73. Kelly JP, Mentz RJ, Mebazaa A, Voors AA, Butler J, Roessig L, Fiuzat M, Zannad F, Pitt B, O'Connor CM, Lam CS. Patient selection in heart failure with preserved ejection fraction clinical trials. J Am Coll Cardiol. 2015;65:1668–82.

74. Rodriguez JS, Velez Rueda JO, Salas M, Becerra R, Di Carlo MN, Said M, Vittone L, Rinaldi G, Portiansky EL, Mundi~na-Weilenmann C, Palomeque J, Mattiazzi A. Increased Na^+/Ca^{2+} exchanger expression/activity constitutes a point of inflection in the progression to heart failure of hypertensive rats PLoS One. 2014; 9: e96400. https://doi.org/10.1371/journal.pone.0096400.

Part IV
Cardiomyocyte Regeneration

Chapter 16
Endogenous Mechanisms for Cardiomyocyte Regeneration

Abstract Cardiomyocyte proliferation is vital for development of the heart. Mammalian cardiomyocytes (CMs) can vigorously proliferate during foetal and neonatal development. In the adult heart, the proliferative capacity of CMs is limited. There is however wide acceptance that CMs proliferate to some degree after birth. Dissecting the cellular and molecular mechanisms that promote proliferation of CMs throughout life, deciphering why proliferative capacity normally diminishes in adult mammalian hearts and finding ways to augment the capacity for regeneration of adult CMs are topics of growing interest. Several endogenous mechanisms for regeneration and proliferation of CMs have been identified. These include the regulators of cell cycle, signal transduction pathways, epigenetic mechanisms and microRNAs.

Keyword Cardiomyocyte differentiation · Cardiomyocyte proliferation · Cardiomyocyte regeneration · Cell cycle · Signalling pathways · Epigenetics · Micro RNA

Introduction

After birth, cardiomyocytes (CMs) transit from proliferative to hypertrophic growth. There is however evidence now that new CMs continue to be generated in humans until the heart attains adult size [1, 2]. Proliferation of CMs is highest in infants and drops to extremely low levels in adults. A latent regenerative potential may exist in the adult heart. Currently, there is considerable interest in deciphering, why proliferative capacity normally diminishes in adult mammalian hearts and in finding ways to augment the capacity for regeneration of adult CMs. Several endogenous mechanisms for regeneration and proliferation of CMs have been identified.

© The Author(s), under exclusive license to Springer Nature Switzerland AG 2021
C. C. Kartha, *Cardiomyocytes in Health and Disease*,
https://doi.org/10.1007/978-3-030-85536-9_16

Cardiomyocyte-Intrinsic Programs for Regeneration in Amphibian Heart

In response to injury, CMs in adult amphibia and zebrafish can synthesize DNA and undergo karyokinesis [3, 4]. Proliferative CMs may originate from either undifferentiated progenitor cells or existing CMs [5–9]. In mice, epicardial cells can be induced to transdifferentiate after injury [10]. The evidence so far suggests that to facilitate division, CMs dedifferentiate to some extent. In regenerating CMs of zebrafish, dissociated sarcomeres and transiently decoupled electrical conductance have been observed [6, 7]. After injury, regulatory sequences of the transcription factor gene Gata4 is activated in zebrafish CMs indicating that during regeneration, programs of embryonic development are reactivated [7]. Gata4 activity is essential for regeneration of CMs [11]. In the regenerating heart of zebrafish, there is also increase in the transcript levels of genes of Gata5, Nkx2.5, Hand2, Tbx5, and Tbx20, which are known to be involved in the development of the embryonic heart [5, 12]. Several other transcriptional programs are also induced during heart regeneration [13, 14]. For example, Jak1/Stat3 downstream mediators are induced upon injury [13]. Transcriptional regulation by NF-κB and telomerase activity have also been implicated [15–17].

Regeneration of Cardiomyocytes in Mammalian Heart

Dynamics of proliferation and differentiation of CMs in the adult heart and the potential for regeneration of CMs after cardiac injury are topics of intense interest for several years now [18]. Several endogenous mechanisms for regeneration and proliferation of CMs have been identified. These include the regulators of cell cycle, signal transduction pathways, epigenetic mechanisms and microRNAs.

Cell Cycle Regulators

Cell cycle regulators such as cyclins, cyclin-dependent kinases, and proto-oncogenes are highly expressed in the foetal and neonatal heart. Their expression is down-regulated in the adult heart. Cyclin-dependent kinase inhibitors which are negative regulators of the cell cycle are up-regulated in adult CMs. Overexpression of cyclin D2, a positive regulator of the G1/S transition can stimulate DNA synthesis and proliferation in mammalian CMs [19, 20]. Cyclin A2, promoter of the G1/S and G2/M transitions, has also been implicated in the proliferation of CMs [21–23].

The E2F family of transcription factors can induce DNA synthesis in adult CMs [24, 25]. Adenoviral transfection of E2F2 expression in mouse CMs have been seen to result in proliferation of the cells [26]. This effect is associated with induction of cyclin A and E. The expression of cyclin-dependent kinase inhibitors such as p21 is however not affected [26].

Signal Transduction Pathways

Several endogenous ligands and their receptors which modulate proliferation and regeneration of CMs have been identified. Some of them induce proliferation of CMs in vivo as well. Target molecules identified include fibroblast growth factors (FGF-1, FGF-2), insulin growth factor (IGF), epidermal growth factor (EGF), neuregulin1 (NRG1), interferon-1β (IF-1β) and erythropoietin (EPO) [27, 28].

The Notch signalling pathway is known to regulate genes important for proliferation, differentiation and apoptosis. The pathway has a key role in specification of cell fate as well as patterning of tissue in the embryo and is thus important for the developing heart. Notch activation is essential to sustain proliferation of CMs at birth [29]. Proliferative potential of neonatal rat CMs has been found to be prolonged when the cells are transfected with the intracellular domain of Notch. Activation of the Notch pathway is however ineffective in adult CMs [30].

The role of neuregulin-1 (NRG1) in inducing proliferation of CMs has been much investigated [31–35]. NRG1 and its co-receptors ErbB2 and ErbB4 (estrogen receptors B2 and B4) are essential for the development of heart in mice [36–38]. NRG1-induced proliferation of CMs decreases after birth and is linked to reduction in ErbB2 expression. When constitutively active ErbB2 is inducted, proliferation is seen in neonatal, juvenile, and adult CMs. Transient expression of ErbB2 improves cardiac performance after ischemic injury and is accompanied by cardiomyocyte dedifferentiation and proliferation followed by redifferentiation [39]. Pro-survival pathways are activated in ErbB2-over-expressing transgenic mice [40].

NRG1 signalling stimulates DNA synthesis, karyokinesis, and cytokinesis in zebrafish heart and in cultured CMs [41, 42]. In zebrafish, NRG1 gene is necessary for heart regeneration; myocardial injury activates the gene [43]. Recombinant NRG1 induces the proliferation of mononucleated mouse CMs as well as proliferation of CMs in myocardial explants from human infants [31, 32].

A possible upstream regulator of NRG1 is innervation of the myocardium [44]. Studies indicate that nerve function is important for heart regeneration in both zebrafish and mouse [45].

Signal transduction cascades seem to integrate several extracellular factors to regulate proliferation of CMs. Many downstream effectors of NRG1/ErbB such as PI3K/Akt, MAPK/ErK, and FAK have been discovered [46]. NRG-1β signalling was found to activate the ErbB2-ERK-SIRT1 signalling pathway in a model of radiation-induced heart disease [47]. Other mediators of ErbB signalling pathways include the Rac-PAK intracellular kinases. The Rac1-PAK2 pathway is necessary for regeneration of the heart in zebrafish [48]. In mice hearts, transient ErbB2 overexpression activates the ERK, AKT, and GSK3β/β-catenin pathways and promotes the dedifferentiation and proliferation of CMs induced by the activation of the NRG1/ErbB2 signalling pathway [39].

Follistatin-like 1 (Fstl1) and insulin-like-growth factor2 (IGF2), produced by the epicardium are two extracellular regulators of cell cycle entry in CMs [49–51]. IGF2 is the primary mitogenic factor produced by embryonic epicardial cells of mouse heart [50]. Inhibition of this pathway prevents proliferation of CMs [50]. IGF2 is initially induced by EPO and later regulated by oxygen and glucose [51].

Inflammatory and immune-mediated signalling pathways may also modulate proliferation of CMs. Oncostatin M (OSM), a proinflammatory cytokine was found to be protective in a mouse model of myocardial infarction [52]. Macrophages are considered important for regeneration of CMs in neonatal and adult mice [53, 54].

Negative regulators of proliferation of CMs have also been discovered. Activated mitogen-activated protein (MAP) kinase (p38) phosphorylates downstream signalling molecules important for differentiation and hypertrophy of CMs [55]. Overexpression of p38 inhibits proliferation of rat foetal CMs and inhibition of p38 promotes proliferation of CMs [41].

Hippo signalling pathway is another signal transduction pathway which regulates proliferation of CMs [56, 57]. Hippo signalling seems to negatively regulate a subset of Wnt target genes and thus restrict proliferation of CMs. The terminal effector is the transcriptional co-activator Yes-associated protein (YAP). Enlarged hearts resulting from alterations in the Hippo pathway are attributed to cardiomyocyte proliferation regulated by YAP [58].

Epigenetic Regulation

Temporal and spatial regulation of gene expression during development of the embryonic heart is regulated by specific epigenetic modifications of DNA and involves chromatin remodelling processes [59, 60]. Chromatin remodelling is also possibly involved in the transcriptional reprograming of gene expression during the transition of neonatal CMs with high replication rates to quiescent adult cells.

Brg1 (Brahma-related gene 1), a chromatic remodelling protein promotes proliferation of embryonic CMs. Brg1 is inactivated in adults. Brg1is reactivated after stress [61]. Bromodomain proteins facilitate transcription activation by recruiting coregulatory complexes.

LEN, a regeneration-specific enhancer element has been identified to regulate leptin b in the injured heart [62–64]. LEN mediates the upregulation of leptin b after injury [62]. LEN can also upregulate NRG1 in transgenic zebrafish lines [62].

Down-regulation and inhibition of TERT (telomerase reverse transcriptase) which result in loss of telomerase activity can inhibit proliferation of cells [65, 66]. TERT and telomerase activity are down-regulated in the heart of adult mouse [67]. hTERT expression synergistically increases both S-phase entry of neonatal CMs and proliferation induced by knockdown of cyclin-dependent kinase inhibitors (CKI) [68].

Reactive oxygen species (ROS) is another potential regulator of proliferation of CMs [69].

Micro RNAs

miRNAs which modulate proliferation of neonatal and adult CMs have been identified [57, 70–74]. Proliferation is mediated by the partial dedifferentiation of adult CMs. Homer1, involved in the regulation of calcium signalling in CMs, Hopx, a regulator suppressing proliferation of embryonic CMs and Clic5, a chloride intracellular ion channel, are considered potential downstream mediators of has-miR-590 and hasmiR-199a in increasing cardiomyocyte proliferation. miRNA-204 also promotes proliferation of both adult and neonatal CMs in vitro.

miRNA-204 increases the expression of several cell cycle regulators and directly targets Jarid2, involved in the regulation of gene expression during development. Knock-down of Jarid2 promotes proliferation of CMs [75].

Conclusions

Cell cycle regulators, signal transduction pathways, epigenetic mechanisms and microRNAs modulate regeneration and proliferation of CMs in the mammalian heart. Advance in knowledge in this domain could pave way for identifying strategies to regenerate CMs after cardiac injury. A key target could be chromatin. Ways to modulate its regions involved in regeneration and proliferation of CMs are being identified.

References

1. Bergmann O, Zdunek S, Felker A, Salehpour M, Alkass K, Bernard S, et al. Dynamics of cell generation and turnover in the human heart. Cell. 2015;161(7):1566–75.
2. Mollova M, Bersell K, Walsh S, Savla J, Das LT, Park SY, et al. Cardiomyocyte proliferation contributes to heart growth in young humans. Proc Natl Acad Sci USA. 2013;110(4):1446–51.
3. Bettencourt-Dias M, Mittnacht S, Brockes JP. Heterogeneous proliferative potential in regenerative adult newt cardiomyocytes. J Cell Sci. 2003;116:4001–9.
4. Poss KD, Wilson LG, Keating MT. Heart regeneration in zebrafish. Science. 2002;298:2188–90.
5. Lepilina A, Coon AN, Kikuchi K, Holdway JE, Roberts RW, Burns CG, Poss KD. A dynamic epicardial injury response supports progenitor cell activity during zebrafish heart regeneration. Cell. 2006;127:607–19.
6. Jopling C, Sleep E, Raya M, Martı M, Raya A, Belmonte JCI. Zebrafish heart regeneration occurs by cardiomyocyte dedifferentiation and proliferation. Nature. 2010;464:606–9.
7. Kikuchi K, Holdway JE, Werdich AA, Anderson RM, Fang Y, Egnaczyk GF, Evans T, Macrae CA, Stainier DY, Poss KD. Primary contribution to zebrafish heart regeneration by gata4+ cardiomyocytes. Nature. 2010;464:601–5.

8. Wang J, Panáková D, Kikuchi K, Holdway JE, Gemberling M, Burris JS, Singh SP, Dickson AL, Lin Y-F, Sabeh MK, et al The regenerative capacity of zebrafish reverses cardiac failure caused by genetic cardiomyocyte depletion. Development 2011; 138:3421–30.

9. Porrello ER, Mahmoud AI, Simpson E, Hill JA, Richardson JA, Olson EN, Sadek HA. Transient regenerative potential of the neonatal mouse heart. Science. 2011;331:1078–80.

10. Smart N, Bollini S, Dubé KN, Vieira JM, Zhou B, Davidson S, Yellon D, Riegler J, Price AN, Lythgoe MF, et al De novo cardiomyocytes from within the activated adult heart after injury. Nature 2011; 474:640–644.

11. Gupta V, Gemberling M, Karra R, Rosenfeld GE, Evans T, Poss KD. An injury-responsive gata4 program shapes the zebrafish cardiac ventricle. Curr Biol. 2013;23:1221–7.

12. Kikuchi K, Holdway JE, Major RJ, Blum N, Dahn RD, Begemann G, Poss KD. Retinoic acid production by endocardium and epicardium is an injury response essential for zebrafish heart regeneration. Dev Cell. 2011;20:397–404.

13. Fang Y, Gupta V, Karra R, Holdway JE, Kikuchi K, Poss KD. Translational profiling of cardiomyocytes identifies an early Jak1/Stat3 injury response required for zebrafish heart regeneration. Proc Natl Acad Sci USA. 2013;110:13416–21.

14. Lien CL, Schebesta M, Makino S, Weber GJ, Keating MT. Gene expression analysis of zebrafish heart regeneration. PLoS Biol. 2006; 4: e260.

15. Karra R, Knecht AK, Kikuchi K, Poss KD. Myocardial NF-κB activation is essential for zebrafish heart regeneration. Proc Natl Acad Sci USA. 2015;112:13255–60.

16. Bär C, de Jesus BB, Serrano R, Tejera A, Ayuso E, Jimenez V, Formentini I, Bobadilla M, Mizrahi J, de Martino A, et al. Telomerase expression confers cardioprotection in the adult mouse heart after acute myocardial infarction. Nature Commun. 2014;5:5863.

17. Bednarek D, González-Rosa JM, Guzmán-Martınez G, Gutiérrez-Gutiérrez Ó, Aguado T, Sánchez-Ferrer C, Marques IJ, Galardi-Castilla M, de Diego I, Gómez MJ, et al Telomerase is essential for zebrafish heart regeneration. Cell Rep. 2015; 12:1691–1703.

18. Yester JW, Kühn B. Mechanisms of cardiomyocyte proliferation and differentiation in development and regeneration. Curr Cardiol Rep. 2017;19(2):13. https://doi.org/10.1007/s11886-017-0826-1.

19. Busk PK, Bartkova J, Strom CC, Wulf-Andersen L, Hinrichsen R, Christoffersen TE, et al. Involvement of cyclin D activity in left ventricle hypertrophy in vivo and in vitro. Cardiovasc Res. 2002;56(1):64–75.

20. Pasumarthi KB, Nakajima H, Nakajima HO, Soonpaa MH, Field LJ. Targeted expression of cyclin D2 results in cardiomyocyte DNA synthesis and infarct regression in transgenic mice. Circ Res. 2005;96(1):110–8.

21. Chaudhry HW, Dashoush NH, Tang H, Zhang L, Wang X, Wu EX, et al. Cyclin A2 mediates cardiomyocyte mitosis in the postmitotic myocardium. J Biol Chem. 2004;279(34):35858–66.

22. Cheng RK, Asai T, Tang H, Dashoush NH, Kara RJ, Costa KD, et al. Cyclin A2 induces cardiac regeneration after myocardial infarction and prevents heart failure. Circ Res. 2007;100(12):1741–8.

23. Shapiro SD, Ranjan AK, Kawase Y, Cheng RK, Kara RJ, Bhattacharya R, et al Cyclin A2 induces cardiac regeneration after myocardial infarction through cytokinesis of adult cardiomyocytes. Sci Transl Med. 2014; 6(224):224ra27.

24. Kirshenbaum LA, Abdellatif M, Chakraborty S, Schneider MD. Human E2F–1 reactivates cell cycle progression in ventricular myocytes and represses cardiac gene transcription. Dev Biol. 1996;179(2):402–11.

25. Liu Y, Kitsis RN. Induction of DNA synthesis and apoptosis in cardiac myocytes by E1A oncoprotein. J Cell Biol. 1996;133(2):325–34.

26. Ebelt H, Zhang Y, Kampke A, Xu J, Schlitt A, Buerke M, et al. E2F2 expression induces proliferation of terminally differentiated cardiomyocytes in vivo. Cardiovasc Res. 2008;80(2):219–26.

27. Pasumarthi KB, Field LJ. Cardiomyocyte cell cycle regulation. Circ Res. 2002;90(10):1044–54.

28. Senyo SE, Lee RT, Kuhn B. Cardiac regeneration based on mechanisms of cardiomyocyte proliferation and differentiation. Stem Cell Res. 2014; 13(3 Pt B):532–41.

29. Collesi C, Zentilin L, Sinagra G, Giacca M. Notch1 signaling stimulates proliferation of immature cardiomyocytes. J Cell Biol. 2008;183(1):117–28.
30. Felician G, Collesi C, Lusic M, Martinelli V, Ferro MD, Zentilin L, et al. Epigenetic modification at Notch responsive promoters blunts efficacy of inducing notch pathway reactivation after myocardial infarction. Circ Res. 2014;115(7):636–49.
31. Bersell K, Arab S, Haring B, Kuhn B. Neuregulin1/ErbB4 signaling induces cardiomyocyte proliferation and repair of heart injury. Cell. 2009;138(2):257–70.
32. Polizzotti BD, Ganapathy B, Walsh S, Choudhury S, Ammanamanchi N, Bennett DG, et al Neuregulin stimulation of cardiomyocyte regeneration in mice and human myocardium reveals a therapeutic window. Sci Transl Med 2015; 7(281):281ra45.
33. Ganapathy B, Nandhagopal N, Polizzotti BD, Bennett D, Asan A, Wu Y, et al Neuregulin-1 administration protocols sufficient for stimulating cardiac regeneration in young mice do not induce somatic, organ, or neoplastic growth. PLoS One. 2016; 11(5):e0155456.
34. Wadugu B, Kuhn B. The role of neuregulin/ErbB2/ErbB4 signaling in the heart with special focus on effects on cardiomyocyte proliferation. Am J Physiol Heart Circ Physiol. 2012;302(11):H2139–47.
35. Parodi EM, Kuhn B. Signalling between microvascular endothelium and cardiomyocytes through neuregulin. Cardiovasc Res. 2014;102(2):194–204.
36. Lee KF, Simon H, Chen H, Bates B, Hung MC, Hauser C. Requirement for neuregulin receptor erbB2 in neural and cardiac development. Nature. 1995;378(6555):394–8.
37. Gassmann M, Casagranda F, Orioli D, Simon H, Lai C, Klein R, et al. Aberrant neural and cardiac development in mice lacking the ErbB4 neuregulin receptor. Nature. 1995;378(6555):390–4.
38. Meyer D, Birchmeier C. Multiple essential functions of neuregulin in development. Nature. 1995;378(6555):386–90.
39. D'Uva G, Aharonov A, Lauriola M, Kain D, Yahalom-Ronen Y, Carvalho S, et al. ERBB2 triggers mammalian heart regeneration by promoting cardiomyocyte dedifferentiation and proliferation. Nat Cell Biol. 2015;17(5):627–38.
40. Belmonte F, Das S, Sysa-Shah P, Sivakumaran V, Stanley B, Guo X, et al. ErbB2 overexpression upregulates antioxidant enzymes, reduces basal levels of reactive oxygen species, and protects against doxorubicin cardiotoxicity. Am J Physiol Heart Circ Physiol. 2015;309(8):H1271–80.
41. Engel FB, Schebesta M, Duong MT, Lu G, Ren S, Madwed JB, et al. p38 MAP kinase inhibition enables proliferation of adult mammalian cardiomyocytes. Genes Dev. 2005;19(10):1175–87.
42. Gemberling M, Karra R, Dickson AL, Poss KD. Nrg1 is an injury-induced cardiomyocyte mitogen for the endogenous heart regeneration program in zebrafish. Elife. 2015; 4:e05871.
43. Gemberling M, Bailey TJ, Hyde DR, Poss KD. The zebrafish as a model for complex tissue regeneration. Trends Genet. 2013;29(11):611–20.
44. Lockhart ST, Turrigiano GG, Birren SJ. Nerve growth factor modulates synaptic transmission between sympathetic neurons and cardiac myocytes. J Neurosci. 1997;17(24):9573–82.
45. Mahmoud AI, O'Meara CC, Gemberling M, Zhao L, Bryant DM, Zheng R, et al. Nerves regulate cardiomyocyte proliferation and heart regeneration. Dev Cell. 2015;34(4):387–99.
46. Pentassuglia L, Sawyer DB. The role of Neuregulin-1beta/ErbB signaling in the heart. Exp Cell Res. 2009;315(4):627–37.
47. Gu A, Jie Y, Sun L, Zhao SEM, You Q. RhNRG-1beta protects the myocardium against irradiation-induced damage via the ErbB2-ERK-SIRT1 signaling pathway. PLoS One 2015; 10(9):e0137337.
48. Peng X, He Q, Li G, Ma J, Zhong TP. Rac1-PAK2 pathway is essential for zebrafish heart regeneration. Biochem Biophys Res Commun. 2016;472(4):637–42.
49. Wei K, Serpooshan V, Hurtado C, Diez-Cunado M, Zhao M, Maruyama S, et al. Epicardial FSTL1 reconstitution regenerates the adult mammalian heart. Nature. 2015;525(7570):479–85.
50. Li P, Cavallero S, Gu Y, Chen TH, Hughes J, Hassan AB, et al. IGF signaling directs ventricular cardiomyocyte proliferation during embryonic heart development. Development. 2011;138(9):1795–805.

51. Shen H, Cavallero S, Estrada KD, Sandovici I, Kumar SR, Makita T, et al. Extracardiac control of embryonic cardiomyocyte proliferation and ventricular wall expansion. Cardiovasc Res. 2015;105(3):271–8.
52. Kubin T, Poling J, Kostin S, Gajawada P, Hein S, Rees W, et al. Oncostatin M is a major mediator of cardiomyocyte dedifferentiation and remodeling. Cell Stem Cell. 2011;9(5):420–32.
53. Aurora AB, Porrello ER, Tan W, Mahmoud AI, Hill JA, Bassel-Duby R, et al. Macrophages are required for neonatal heart regeneration. J Clin Invest. 2014;124(3):1382–92.
54. Lavine KJ, Epelman S, Uchida K, Weber KJ, Nichols CG, Schilling JD, et al. Distinct macrophage lineages contribute to disparate patterns of cardiac recovery and remodeling in the neonatal and adult heart. Proc Natl Acad Sci USA. 2014;111(45):16029–34.
55. Liang Q, Molkentin JD. Redefining the roles of p38 and JNK signaling in cardiac hypertrophy: dichotomy between cultured myocytes and animal models. J Mol Cell Cardiol. 2003;35(12):1385–94.
56. Heallen T, Zhang M, Wang J, Bonilla-Claudio M, Klysik E, Johnson RL, et al. Hippo pathway inhibits Wnt signaling to restrain cardiomyocyte proliferation and heart size. Science. 2011;332(6028):458–61.
57. Tian Y, Liu Y, Wang T, Zhou N, Kong J, Chen L, et al. A microRNA-Hippo pathway that promotes cardiomyocyte proliferation and cardiac regeneration in mice. Sci Transl Med. 2015;7(279):279ra38.
58. von Gise A, Lin Z, Schlegelmilch K, Honor LB, Pan GM, Buck JN, et al. YAP1, the nuclear target of Hippo signaling, stimulates heart growth through cardiomyocyte proliferation but not hypertrophy. Proc Natl Acad Sci USA. 2012;109(7):2394–9.
59. Wamstad JA, Alexander JM, Truty RM, Shrikumar A, Li F, Eilertson KE, et al. Dynamic and coordinated epigenetic regulation of developmental transitions in the cardiac lineage. Cell. 2012;151(1):206–20.
60. Paige SL, Thomas S, Stoick-Cooper CL, Wang H, Maves L, Sandstrom R, et al. A temporal chromatin signature in human embryonic stem cells identifies regulators of cardiac development. Cell. 2012;151(1):221–32.
61. Hang CT, Yang J, Han P, Cheng HL, Shang C, Ashley E, et al. Chromatin regulation by Brg1 underlies heart muscle development and disease. Nature. 2010;466(7302):62–7.
62. Kang J, Hu J, Karra R, Dickson AL, Tornini VA, Nachtrab G, et al. Modulation of tissue repair by regeneration enhancer elements. Nature. 2016;532(7598):201–6.
63. Zhang Y, Proenca R, Maffei M, Barone M, Leopold L, Friedman JM. Positional cloning of the mouse obese gene and its human homologue. Nature. 1994;372(6505):425–32.
64. Fang Y, Gupta V, Karra R, Holdway JE, Kikuchi K, Poss KD. Translational profiling of cardiomyocytes identifies an early Jak1/Stat3 injury response required for zebrafish heart regeneration. Proc Natl Acad Sci USA. 2013;110(33):13416–21.
65. Bodnar AG, Ouellette M, Frolkis M, Holt SE, Chiu CP, Morin GB, et al. Extension of life-span by introduction of telomerase into normal human cells. Science. 1998;279(5349):349–52.
66. Hahn WC, Stewart SA, Brooks MW, York SG, Eaton E, Kurachi A, et al. Inhibition of telomerase limits the growth of human cancer cells. Nat Med. 1999;5(10):1164–70.
67. Oh H, Taffet GE, Youker KA, Entman ML, Overbeek PA, Michael LH, et al. Telomerase reverse transcriptase promotes cardiac muscle cell proliferation, hypertrophy, and survival. Proc Natl Acad Sci USA. 2001;98(18):10308–13.
68. Di Stefano V, Giacca M, Capogrossi MC, Crescenzi M, Martelli F. Knockdown of cyclin-dependent kinase inhibitors induces cardiomyocyte re-entry in the cell cycle. J Biol Chem. 2011;286(10):8644–54.
69. Puente BN, Kimura W, Muralidhar SA, Moon J, Amatruda JF, Phelps KL, et al. The oxygen-rich postnatal environment induces cardiomyocyte cell-cycle arrest through DNA damage response. Cell. 2014;157(3):565–79.
70. Eulalio A, Mano M, Dal Ferro M, Zentilin L, Sinagra G, Zacchigna S, et al. Functional screening identifies miRNAs inducing cardiac regeneration. Nature. 2012;492(7429):376–81.
71. Liu N, Bezprozvannaya S, Williams AH, Qi X, Richardson JA, Bassel-Duby R, et al. microRNA-133a regulates cardiomyocyte proliferation and suppresses smooth muscle gene expression in the heart. Genes Dev. 2008;22(23):3242–54.

72. Zhao Y, Samal E, Srivastava D. Serum response factor regulates a muscle-specific microRNA that targets Hand2 during cardiogenesis. Nature. 2005;436(7048):214–20.
73. Zhao Y, Ransom JF, Li A, Vedantham V, von Drehle M, Muth AN, et al. Dysregulation of cardiogenesis, cardiac conduction, and cell cycle in mice lacking miRNA-1-2. Cell. 2007;129(2):303–17.
74. Porrello ER, Johnson BA, Aurora AB, Simpson E, Nam YJ, Matkovich SJ, et al. MiR-15 family regulates postnatal mitotic arrest of cardiomyocytes. Circ Res. 2011;109(6):670–9.
75. Liang D, Li J, Wu Y, Zhen L, Li C, Qi M, et al. miRNA-204 drives cardiomyocyte proliferation via targeting Jarid2. Int J Cardiol. 2015;201:38–48.

Chapter 17
Mechanisms to Induce Cardiomyocyte Proliferation

Abstract Endogenous mechanisms for renewal of adult cardiomyocytes (CMs) are insufficient for repair of extensive injury to the myocardium as adult CMs have limited capacity for proliferation. Several recent studies suggest that latent regenerative potential in the adult heart can be exploited and that in the injured heart, regeneration can be induced by modulating the endogenous molecular signals that stimulate proliferation of CMs. These signals include growth factors, intrinsic signalling pathways, microRNAs, and cell cycle regulators. Adult CMs can be induced to re-enter the cell cycle by regulating cell-intrinsic and extrinsic signalling pathways. Appropriate growth molecules can be delivered by injection into the myocardium or using epicardial patches or systemic administration.

Keywords Cardiomyocyte regeneration · Cardiac repair · Cell cycle regulators · Growth factors · Signalling pathways · microRNAs

Introduction

The adult human heart following injury is more prone for scarring than regeneration of significant quantity of functional CMs. Endogenous mechanisms for renewal of adult cardiomyocytes (CMs) are insufficient for repair of extensive injury to the myocardium as adult CMs have limited capacity for proliferation. The adult heart however does have latent regenerative potential and these may be exploited for repair of the injured heart [1, 2]. Three main strategies are currently investigated for regeneration of CMs and heart repair in mammals. They are: (i) transplantation of stem and progenitor cells, (ii) transdifferentiation of resident cardiac fibroblasts and (iii) reactivation of endogenous mechanisms for regeneration [3]. Among them, targeting endogenous mechanisms for heart regeneration through stimulation of cardiomyocyte self-renewal, is an exciting approach to improve cardiac function in patients with heart failure [1]. Studies during the past one decade have explored diverse ways to unlock the cell-cycle arrest of CMs in adult hearts and thus stimulate the proliferation of pre-existing CMs [4].

Activation of Endogenous Mechanisms

Studies in zebrafish and mouse have provided much insights into the endogenous regeneration process in the heart [5]. They indicate the possibility of stimulating endogenous mechanisms for heart repair after myocardial damage and heart failure. The endogenous molecular signals that stimulate proliferation of CMs include growth factors, intrinsic signalling pathways, microRNAs and cell cycle regulators. In the rodent heart, proliferation of CMs can be increased by stimulation of signals and pathways, such as IGF (insulin growth factor), microRNAs and cyclin proteins (Figs. 17.1 and 17.2). The evidences are however not adequate for translation into clinical application.

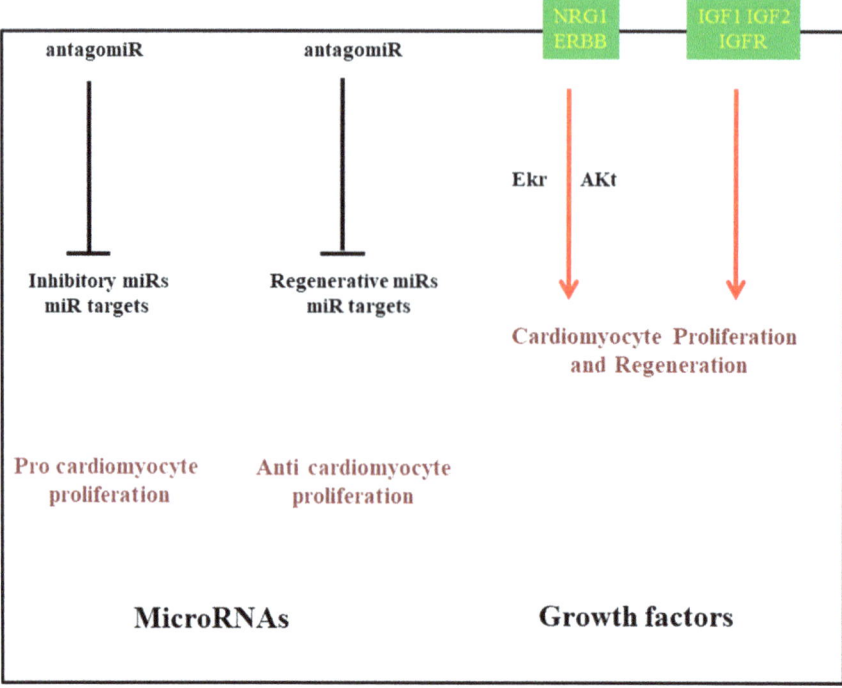

Fig. 17.1 Growth factors, cell cycle regulators and intrinsic signalling pathways which promote proliferation of cardiomyocytes. ERK—Extracellular signal regulated kinase. ERBB—Epidermal growth factor receptor. IGF—Insulin growth factor. IGFR—insulin growth factor receptor. miR—microRNA. NRG—Neuregulin

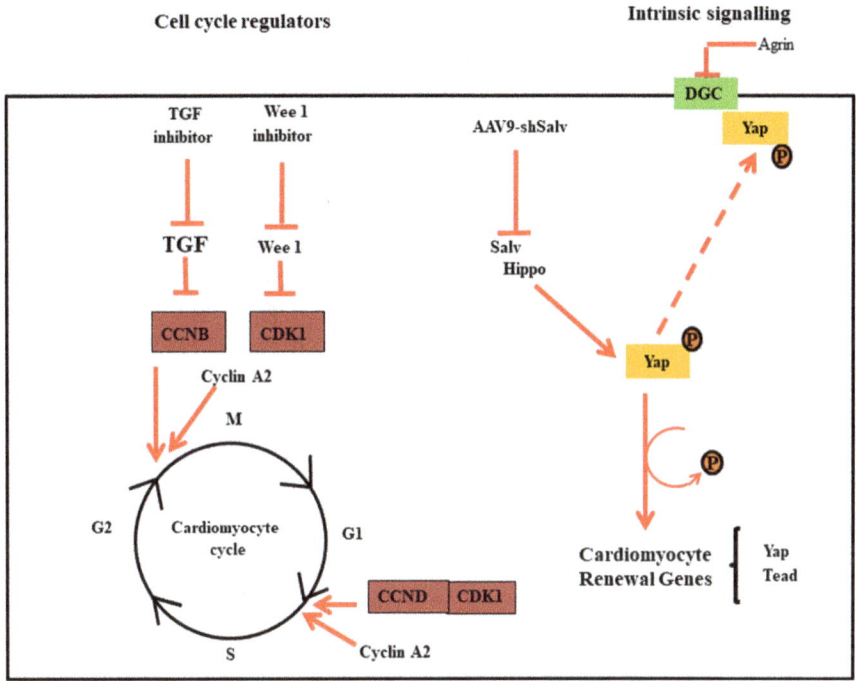

Fig. 17.2 Intrinsic signalling pathways and cell cycle regulators which promote proliferation of cardiomyocytes. CCNB—Cyclin B. CDK—Cyclin D kinase. DGC—Dystrophin glycoprotein complex. Salv—Salvador. TGF—Transforming growth factor. Yap—Yes associated protein

Growth Factors

Several classes of ligand/cell surface receptor complexes can induce re-entry of CMs into the cell cycle and aid regeneration (Fig. 17.1).

The NRG1 (neuregulin 1) ligand which signals through ERBB2-4 (Erb-B2 receptor tyrosine kinase 2–4) tyrosine kinase receptors, is known to control development [6]. NRG1/ERBB2/ERBB4 signalling stimulates proliferation of CMs in the injured heart [7, 8]. The effect may be more effective during the early period of life [9]. Enhanced proliferation, dedifferentiation, and hypertrophy of CMs are seen in neonate and adult mice with ERBB2 overexpression [10]. After ischemic injury, transient induction of ERBB2 expression results in regeneration of CMs and significant improvement in cardiac function [10]. Thus, increasing NRG1/ERBB activity is a potential strategy to induce regeneration of CMs. All the upstream regulators and downstream targets of NRG1 are yet to be identified. The ERK (extracellular signal-regulated kinase) and AKT (alpha serine/threonine-protein kinase) pathways are indispensable downstream mediators of ERBB2-induced dedifferentiation and proliferation of CMs [10].

After myocardial infarction (MI) when rats are treated with a combination of FGF1 (fibroblast growth factor) and a p38 MAPK (mitogen-activated protein kinase) inhibitor (p38i) there is increased mitoses in CMs [11]. p38i treatment alone causes proliferation but does not rescue heart function [11]. In patients with acute MI, losmapimod, a p38i did not however improve the outcome [12].

Epicardial growth factors, including FST1 (follistatin-like 1) and IGF2, have also been found to promote cell cycle re-entry of CMs [13]. The effect of FST1 is only when it is derived from the epicardium [13]. In infarcted pig hearts, proliferation of CMs was observed when IGF1 and HGF (hepatocyte growth factor) together were continuously administered [14]. In pigs with ischemic injury to the heart, injection of growth hormone–releasing hormone, which is known to activate IGF1 signalling decreases scarring after 2 weeks [15].

Periostin, an extracellular matrix component also stimulates proliferation of CMs. IN rats, delivery of recombinant periostin after MI promotes cell cycle re-entry of differentiated CMs, reduce scarring and improve cardiac function [16]. Recombinant periostin stimulated regeneration of CMs, and improved heart function in swine as well [17].

Intrinsic Signalling Mechanisms

Several intrinsic signalling pathways in the cardiomyocyte that regulate cardiac regeneration have been discovered. These pathways are important for cardiac development, cardiomyocyte maturation, and postnatal growth.

Transcription Factors

Transcription factor Meis1 (Meis homeobox 1) is a key regulator of proliferation of CMs in the postnatal period [18]. Deletion of Meis1 in CMs after the neonatal regenerative phase extends the time period for normal cell cycles and regeneration in CMs [18]. In adult hearts, Meis1 deletion re-activates mitosis in CMs whereas Meis1 overexpression inhibits cardiomyocyte regeneration in the neonatal heart [18]. Transcriptional activation targets of Meis1 are the CDK inhibitors p15, p16, and p21 [18].

Transcription factor GATA4 (GATA-binding protein 4) is necessary for cardiac regeneration. GATA4 regulates FGF16 [19]. The observation that overexpression of FGF16 in GATA4-ablated hearts partially rescues the injury-induced defects, suggests that paracrine factors can be used as targets for cardiac repair [19]. Overexpression of Tbx20 (T-box 20) activates the YAP (Yes-associated protein), BMP (bone morphogenetic protein), and AKT proliferative pathways [20]. Tbx20 represses antiproliferation genes p21 and Meis1 [20]. Tbx20 overexpression in adult CMs decreases the infarct size and improves cardiac function after MI [20].

The Hippo Signaling Pathway

The Hippo signaling pathway has important roles in heart development, cardiomyocyte homeostasis and regeneration [21]. During development, Hippo signalling restrains proliferation of CMs and aid to preserve appropriate tissue pattern and heart size [22]. Cardiac-specific deletion of the core Hippo component Salvador during embryogenesis results in increased proliferation of CMs and cardiomegaly. A similar phenotype results when Mst (mammalian sterile 20-like kinase/hippo) and Lats kinases (large tumor suppressor kinase) are knocked out [22]. Mst and Lats kinases are components of the Mst/Hpo and Lats/Wts which regulates phosphorylation of the downstream effector Yap/ Yki (yes-associated protein/yorkie), a transcriptional coactivator that promotes expression of survival and pro-growth genes [23].

Hippo signalling is now known to repress several cellular mechanisms that are vital for heart repair (Fig. 17.2). There is evidence that Hippo inhibits regeneration of CMs in the adult heart [24–28]. In rodents, direct delivery of a small molecule Hippo pathway inhibitor (short hairpin Salvador RNA) via an adeno-associated virus 9 into the heart muscle during ischemic injury or even weeks after injury, improved heart repair and function after MI [26].

Proliferation of CMs and reduction in infarct size have been observed after Yap overexpression in the infarcted heart of mice [25, 29]. Studies indicate that Yap regulates proliferation of CMs by interacting with the dystrophin-glycoprotein complex (DGC) and the extracellular protein agrin at the plasma membrane [30, 31]. Agrin stimulates proliferation of CMs by assisting DGC disassembly, disrupting the Yap-DGC interaction and Yap translocation [31].

After MI, in Hippo-deficient hearts, Paired-like homeodomain-2 (Pitx2), a transcription factor is upregulated [32]. Pitx2 and Yap may coregulate the expression of antioxidant genes and thus protect the heart from oxidative stress after injury and aid regeneration [32]. Park2, the quality control stress response gene in mitochondria is also upregulated in Hippo-deficient hearts and facilitates regeneration [26]. These studies indicate that genetic programs for heart repair are present in failing hearts. Hippo signalling is upregulated in the hearts of patients with heart failure, as revealed by study of tissue samples from them [26]. Inhibition of the Hippo pathway can possibly mobilize endogenous CMs and reverse the loss of cardiac function in patients with heart failure.

MicroRNAs (miRs)

microRNAs regulate cardiomyocyte division during normal aging or in response to injury. In mouse and pigs, miR-590, miR-199a, and miR-99/100 stimulate proliferation of CMs and improve cardiac function after MI [33–35]. miR590 and miR199a are known to aid re-entry of CMs into the cell cycle and promote proliferation of

CMs in neonatal and adult mice [33]. Simultaneous delivery of these miRs after MI assists regeneration of CMs and rescues heart function [33].

miRs of the miR15 family inhibit proliferation of neonatal CMs [36]. Inhibition of miR15 promotes proliferation of CMs in adult hearts and improves function of the infarcted heart [36]. In mice and pigs, anti-miR oligonucleotides have been found to reduce infarct size after ischemia–reperfusion [37].

miR-17–92, an oncogene is necessary for proliferation of CMs [38]. Transgenic miR-17–92 overexpression in mice hearts increased proliferation of CMs in the border zone of ischemia, decreased scarring and improved heart function [38].

A positive regulatory role for miR-31a-5p in proliferation of CMs has been suggested. This microRNA is upregulated in neonatal rat CMs in the early phase of regeneration [39]. Markers of cardiomyocyte proliferation were reduced in neonatal rats after injection of miR-31a-5p antagomiRs [39].

miR302-367 is required for proliferation of CMs during development. After its overexpression, dedifferentiation of CMs and reduction in fibrosis after MI have been found in adult mice [40]. miR302-367 may possibly regulate renewal and regeneration of CMs through Hippo inhibition. miR302-367 targets the Mst and Lats kinases of the Hippo pathway [40].

Thus, miRs can either positively or negatively regulate proliferation of CMs. Anti-miR antagomir molecules are considered useful for cardiac repair (Fig. 17.1).

Cell Cycle Regulators

One among the several strategies attempted to stimulate proliferation of pre-existing CMs in the adult heart is to overexpress the cell cycle genes for stimulation of mitotic division [41].

Expression of cell cycle regulators such as cyclins and cyclin-dependent kinases is largely silenced in the adult heart [42]. Overexpression studies in mice have found that cyclin A2 and cyclin D2 promote proliferation of mature CMs in adult hearts. In transgenic mice, constitutive cardiac expression of cyclin A2 causes increase in proliferation of CMs [43]. In cyclin A2 transgenic mice after ischemic injury, cell cycle re-entry has been observed in CMs of the infracted myocardium and adjacent zone [44]. Studies in rats and pigs reveal that viral delivery of cyclin A2 protects CMs against ischemic injury after infarction [45, 46]. Transgenic overexpression of cyclin D1 and D2 also promotes cell cycle activity in CMs of infarcted hearts [47, 48].

In another study, combinatorial overexpression of cell cycle regulators CDK1/CCNB/CDK4/CCND (cyclin dependent kinase 1/cyclin B1/cyclin-dependent kinase 4/cyclin D1) was found to stimulate proliferation of postmitotic CMs [49]. Cell cycle re-entry of CMs is enhanced when CDK1/CCNB is substituted with Wee1 (mitosis inhibitor protein kinase) and TGF (transforming growth factor)-β inhibitor molecules [49]. These findings suggest that cell cycle reactivation may be a useful strategy for regeneration of CMs (Fig. 17.2).

Other Factors

Transient fibroblast senescence has been reported to induce proliferation of neonatal CMs. Senescence-associated secretory phenotype factors and p53-mediated pathways are involved in this effect [50, 51]. Physiologic conditions such as exercise, pregnancy, and hypoxia can also cause proliferation of CMs through multiple mechanisms [52–56]. Their role in stimulating heart regeneration after injury needs further scrutiny.

Conclusions

The current knowledge on mechanisms for inducing cardiomyocyte proliferation in the diseased heart indicates the possibility of stimulating endogenous mechanisms for heart repair after ischemic myocardial damage and in failing hearts. Several pathways and regulatory nodes, which could be targeted and synergized to attain better functional outcome after cardiac injury have been identified. Discovery of small molecules that target them would provide avenues for inducing proliferation of CMs and cardiac repair in the diseased heart.

References

1. Foglia MJ, Poss KD. Building and re-building the heart by cardiomyocyte proliferation. Development. 2016;143:729–40.
2. Heallen TR, Kadow ZA, Kim JH, Wang J, Martin JF. Stimulating cardiogenesis as a treatment for heart failure. Circ Res. 2019;124:1647–57.
3. Lin Z, Pu WT. Strategies for cardiac regeneration and repair. Sci Transl Med. 2014; 6(239):239rv1.
4. He L, Nguyen NB, Ardehali R, Zhou B. Heart regeneration by endogenous stem cells and cardiomyocyte proliferation: controversy, fallacy, and progress. Circulation. 2020;142:275–91.
5. Cahill TJ, Choudhury RP, Riley PR. Heart regeneration and repair after myocardial infarction: translational opportunities for novel therapeutics. Nat Rev Drug Discov. 2017;16:699–717.
6. Citri A, Yarden Y. EGF-ERBB signalling: towards the systems level. Nat Rev Mol Cell Biol. 2006;7:505–16.
7. Bersell K, Arab S, Haring B, Kühn B. Neuregulin1/ErbB4 signaling induces cardiomyocyte proliferation and repair of heart injury. Cell. 2009;138:257–70.
8. Wadugu B, Kühn B. The role of neuregulin/ErbB2/ErbB4 signaling in the heart with special focus on effects on cardiomyocyte proliferation. Am J Physiol Heart Circ Physiol. 2012;302:H2139–47.
9. Polizzotti BD, Ganapathy B, Walsh S, Choudhury S, Ammanamanchi N, Bennett DG, dos Remedios CG, Haubner BJ, Penninger JM, Kühn B. Neuregulin stimulation of cardiomyocyte regeneration in mice and human myocardium reveals a therapeutic window. Sci Transl Med. 2015; 7:281ra45.
10. D'Uva G, Aharonov A, Lauriola M, et al. ERBB2 triggers mammalian heart regeneration by promoting cardiomyocyte dedifferentiation and proliferation. Nat Cell Biol. 2015;17:627–38.

11. Engel FB, Hsieh PC, Lee RT, Keating MT. FGF1/p38 MAP kinase inhibitor therapy induces cardiomyocyte mitosis, reduces scarring, and rescues function after myocardial infarction. Proc Natl Acad Sci USA. 2006;103:15546–51.
12. O'Donoghue ML, Glaser R, Cavender MA, et al. LATITUDE-TIMI 60 investigators. effect of losmapimod on cardiovascular outcomes in patients hospitalized with acute myocardial infarction: a randomized clinical trial. JAMA. 2016; 315:1591–1599.
13. Wei K, Serpooshan V, Hurtado C, et al. Epicardial FSTL1 reconstitution regenerates the adult mammalian heart. Nature. 2015;525:479–85.
14. Koudstaal S, Bastings MM, Feyen DA, Waring CD, van Slochteren FJ, Dankers PY, Torella D, Sluijter JP, Nadal-Ginard B, Doevendans PA, Ellison GM, Chamuleau SA. Sustained delivery of insulin XE "Insulin" -like growth factor-1/hepatocyte growth factor XE "Growth factors" stimulates endogenous cardiac repair in the chronic infarcted pig heart. J Cardiovasc Transl Res. 2014;7:232–41.
15. Bagno LL, Kanashiro-Takeuchi RM, Suncion VY, et al. Growth hormone releasing hormone agonists reduce myocardial infarct scar in swine with subacute ischemic cardiomyopathy. J Am Heart Assoc. 2015; 4:e001464.
16. Kühn B, del Monte F, Hajjar RJ, Chang YS, Lebeche D, Arab S, Keating MT. Periostin induces proliferation of differentiated cardiomyocytes and promotes cardiac repair. Nat Med. 2007;13:962–9.
17. Ladage D, Yaniz-Galende E, Rapti K, Ishikawa K, Tilemann L, Shapiro S, Takewa Y, Muller-Ehmsen J, Schwarz M, Garcia MJ, Sanz J, Hajjar RJ, Kawase Y. Stimulating myocardial regeneration with periostin peptide in large mammals improves function post-myocardial infarction but increases myocardial fibrosis. PLoS One. 2013; 8:e59656.
18. Mahmoud AI, Kocabas F, Muralidhar SA, Kimura W, Koura AS, Thet S, Porrello ER, Sadek HA. Meis1 regulates postnatal cardiomyocyte cell cycle XE "Cell cycle" arrest. Nature. 2013;497:249–53.
19. Yu W, Huang X, Tian X, Zhang H, He L, Wang Y, Nie Y, Hu S, Lin Z, Zhou B, Pu W, Lui KO, Zhou B. GATA4 regulates Fgf16 to promote heart repair after injury. Development. 2016;143:936–49.
20. Xiang FL, Guo M, Yutzey KE. Overexpression of Tbx20 in adult cardiomyocytes promotes proliferation and improves cardiac function after myocardial infarction. Circulation. 2016;133:1081–92.
21. Dong J, Feldmann G, Huang J, Wu S, Zhang N, Comerford SA, Gayyed MF, Anders RA, Maitra A, Pan D. Elucidation of a universal size-control mechanism in Drosophila and mammals. Cell. 2007;130:1120–33.
22. Heallen T, Zhang M, Wang J, Bonilla-Claudio M, Klysik E, Johnson RL, Martin JF. Hippo pathway inhibits Wnt signaling to restrain cardiomyocyte proliferation and heart size. Science. 2011;332:458–61.
23. Halder G, Johnson RL. Hippo signaling: growth control and beyond. Development. 2011;138:9–22.
24. Heallen T, Morikawa Y, Leach J, Tao G, Willerson JT, Johnson RL, Martin JF. Hippo signaling impedes adult heart regeneration. Development. 2013;140:4683–90.
25. Xin M, Kim Y, Sutherland LB, Murakami M, Qi X, McAnally J, Porrello ER, Mahmoud AI, Tan W, Shelton JM, Richardson JA, Sadek HA, Bassel-Duby R, Olson EN. Hippo pathway effector Yap promotes cardiac regeneration. Proc Natl Acad Sci USA. 2013;110:13839–44.
26. Leach JP, Heallen T, Zhang M, Rahmani M, Morikawa Y, Hill MC, Segura A, Willerson JT, Martin JF. Hippo pathway deficiency reverses systolic heart failure after infarction. Nature. 2017;550:260–4.
27. Wang J, Liu S, Heallen T, Martin JF. The hippo pathway in the heart: pivotal roles in development, disease, and regeneration. Nat Rev Cardiol. 2018;5:672–84.
28. Xiao Y, Leach J, Wang J, Martin JF. Hippo/Yap signaling in cardiac development and regeneration. Curr Treat Options Cardiovasc Med. 2016;18:38.
29. Lin Z, von Gise A, Zhou P, Gu F, Ma Q, Jiang J, Yau AL, Buck JN, Gouin KA, van Gorp PR, Zhou B, Chen J, Seidman JG, Wang DZ, Pu WT. Cardiac-specific YAP activation improves cardiac function and survival in an experimental murine MI model. Circ Res. 2014;115:354–63.

30. Morikawa Y, Heallen T, Leach J, Xiao Y, Martin JF. Dystrophin glycoprotein complex sequesters Yap to inhibit cardiomyocyte proliferation. Nature. 2017;547:227–31.
31. Bassat E, Mutlak YE, Genzelinakh A, et al. The extracellular matrix protein agrin promotes heart regeneration in mice. Nature. 2017;547:179–84.
32. Tao G, Kahr PC, Morikawa Y, Zhang M, Rahmani M, Heallen TR, Li L, Sun Z, Olson EN, Amendt BA, Martin JF. Pitx2 promotes heart repair by activating the antioxidant response after cardiac injury. Nature. 2016;534:119–23.
33. Eulalio A, Mano M, Dal Ferro M, Zentilin L, Sinagra G, Zacchigna S, Giacca M. Functional screening identifies miRNAs inducing cardiac regeneration. Nature. 2012;492:376–81.
34. Gabisonia K, Prosdocimo G, Aquaro GD, Carlucci L, Zentilin L, Secco I, Ali H, Braga L, Gorgodze N, Bernini F, et al. MicroRNA therapy stimulates uncontrolled cardiac repair after myocardial infarction in pigs. Nature. 2019;569:418–22.
35. Aguirre A, Montserrat N, Zacchigna S, Nivet E, Hishida T, Krause MN, Kurian L, Ocampo A, Vazquez-Ferrer E, Rodriguez-Esteban C, et al. In vivo activation of a conserved microRNA program induces mammalian heart regeneration. Cell Stem Cell. 2014;15:589–604.
36. Porrello ER, Mahmoud AI, Simpson E, Johnson BA, Grinsfelder D, Canseco D, Mammen PP, Rothermel BA, Olson EN, Sadek HA. Regulation of neonatal and adult mammalian heart regeneration by the miR-15 family. Proc Natl Acad Sci USA. 2013;110:187–92.
37. Hullinger TG, Montgomery RL, Seto AG, Dickinson BA, Semus HM, Lynch JM, Dalby CM, Robinson K, Stack C, Latimer PA, Hare JM, Olson EN, van Rooij E. Inhibition of miR-15 protects against cardiac ischemic injury. Circ Res. 2012;110:71–81.
38. Chen J, Huang ZP, Seok HY, Ding J, Kataoka M, Zhang Z, Hu X, Wang G, Lin Z, Wang S, Pu WT, Liao R, Wang DZ. mir-17-92 cluster is required for and sufficient to induce cardiomyocyte proliferation in postnatal and adult hearts. Circ Res. 2013;112:1557–66.
39. Xiao J, Liu H, Cretoiu D, Toader DO, Suciu N, Shi J, Shen S, Bei Y, Sluijter JP, Das S, Kong X, Li X. miR-31a-5p promotes postnatal cardiomyocyte proliferation by targeting RhoBTB1. Exp Mol Med. 2017; 49:e386.
40. Tian Y, Liu Y, Wang T, et al. A microRNA-Hippo pathway that promotes cardiomyocyte proliferation and cardiac regeneration in mice. Sci Transl Med. 2015; 7:279ra38.
41. Pasumarthi KB, Field LJ. Cardiomyocyte cell cycle regulation. Circ Res. 2002;90:1044–54.
42. Ahuja P, Sdek P, MacLellan WR. Cardiac myocyte cell cycle control in development, disease, and regeneration. Physiol Rev. 2007;87:521–44.
43. Chaudhry HW, Dashoush NH, Tang H, Zhang L, Wang X, Wu EX, Wolgemuth DJ. Cyclin A2 mediates cardiomyocyte mitosis in the postmitotic myocardium. J Biol Chem. 2004;279:35858–66.
44. Cheng RK, Asai T, Tang H, Dashoush NH, Kara RJ, Costa KD, Naka Y, Wu EX, Wolgemuth DJ, Chaudhry HW. Cyclin A2 induces cardiac regeneration after myocardial infarction and prevents heart failure. Circ Res. 2007;100:1741–8.
45. Shapiro SD, Ranjan AK, Kawase Y, Cheng RK, Kara RJ, Bhattacharya R, Guzman-Martinez G, Sanz J, Garcia MJ, Chaudhry HW. Cyclin A2 induces cardiac regeneration after myocardial infarction through cytokinesis of adult cardiomyocytes. Sci Transl Med. 2014; 6:224ra27.
46. Woo YJ, Panlilio CM, Cheng RK, Liao GP, Atluri P, Hsu VM, Cohen JE, Chaudhry HW. Therapeutic delivery of cyclin A2 induces myocardial regeneration and enhances cardiac function in ischemic heart failure. Circulation. 2006;114:I206–13.
47. Pasumarthi KB, Nakajima H, Nakajima HO, Soonpaa MH, Field LJ. Targeted expression of cyclin D2 results in cardiomyocyte DNA synthesis and infarct regression in transgenic mice. Circ Res. 2005;96:110–8.
48. Soonpaa MH, Koh GY, Pajak L, Jing S, Wang H, Franklin MT, Kim KK, Field LJ. Cyclin D1 overexpression promotes cardiomyocyte DNA synthesis and multinucleation in transgenic mice. J Clin Invest. 1997;99:2644–54.
49. Mohamed TMA, Ang YS, Radzinsky E, Zhou P, Huang Y, Elfenbein A, Foley A, Magnitsky S, Srivastava D. Regulation of cell cycle to stimulate adult cardiomyocyte proliferation and cardiac regeneration. Cell. 2018;173(104):e12-116.e12.

50. Feng T, Meng J, Kou S, Jiang Z, Huang X, Lu Z, Zhao H, Lau LF, Zhou B, Zhang H. CCN1-induced cellular senescence promotes heart regeneration. Circulation. 2019;139:2495–8.
51. Sarig R, Rimmer R, Bassat E, Zhang L, Umansky KB, Lendengolts D, Perlmoter G, Yaniv K, Tzahor E. Transient p53-mediated regenerative senescence in the injured heart. Circulation. 2019;139:2491–4.
52. Boström P, Mann N, Wu J, Quintero PA, Plovie ER, Panáková D, Gupta RK, Xiao C, MacRae CA, Rosenzweig A, et al. C/EBPβ controls exercise-induced cardiac growth and protects against pathological cardiac remodeling. Cell. 2010;143:1072–83.
53. Vujic A, Lerchenmüller C, Wu TD, Guillermier C, Rabolli CP, Gonzalez E, Senyo SE, Liu X, Guerquin-Kern JL, Steinhauser ML, et al Exercise induces new cardiomyocyte generation in the adult mammalian heart. Nature Commun. 2018; 9:1659.
54. Zacchigna S, Martinelli V, Moimas S, Colliva A, Anzini M, Nordio A, Costa A, Rehman M, Vodret S, Pierro C, et al. Paracrine effect of regulatory T cells promotes cardiomyocyte proliferation during pregnancy and after myocardial infarction. Nature Commun. 2018;9:2432.
55. Kimura W, Xiao F, Canseco DC, Muralidhar S, Thet S, Zhang HM, Abderrahman Y, Chen R, Garcia JA, Shelton JM, et al. Hypoxia fate mapping identifies cycling cardiomyocytes in the adult heart. Nature. 2015;523:226–30.
56. Nakada Y, Canseco DC, Thet S, Abdisalaam S, Asaithamby A, Santos CX, Shah AM, Zhang H, Faber JE, Kinter MT, et al. Hypoxia induces heart regeneration in adult mice. Nature. 2017;541:222–7.

Chapter 18
Cell Sources of Cardiomyocytes for Heart Repair

Abstract Several studies have aimed at developing a cell transplantation–based strategy for repairing extensively damaged heart. Cardiomyocytes can be grown in vitro thanks to the insights gained on the cellular and molecular mechanisms underlying development of the heart. Cell-based and paracrine factor-based strategies to generate new cardiomyocytes in vitro utilize peptides, recombinant proteins, plasmid, adenoviral, adeno-associated viruses, retroviruses, miRNA and RNAi. All these strategies have natural advantages as well as limitations. Many sources of cells have been investigated for their potential use in regenerating the injured myocardium. They include induced pluripotent stem cells, cardiac fibroblasts, endogenous cardiac progenitor cells, cardiosphere-derived cells, side population cells, bone marrow cells, embryonic stem cells and mesenchymal stem cells.

Keywords Cardiomyocyte · Cardiomyogenesis · Lineage conversion · Pluripotent stem cells · Cardiac stem cells · Cardiac progenitor cells · Direct reprogramming

Introduction

Preserving the number and function of cardiomyocytes (CMs) is considered as a possible strategy for both prevention and treatment of ischemic heart disease, which is associated with considerable loss of CMs. Transplantation of CMs generated in vitro is one among the various approaches investigated for increasing CMs in the dysfunctional adult heart [1].

Mechanisms of Cardiomyogenesis

The initial step in producing functional and mature CMs in vitro is to understand the complex mechanisms underlying the formation of the various subtypes of CMs specified by their location and function in vivo. Several developmental signalling pathways, cardiac-specific transcription factors and transcriptional regulators interact and regulate development of the heart (see Chap. 2 in this book).

© The Author(s), under exclusive license to Springer Nature Switzerland AG 2021   279
C. C. Kartha, *Cardiomyocytes in Health and Disease*,
https://doi.org/10.1007/978-3-030-85536-9_18

Three families of extracellular signalling molecules primarily regulate definite populations of cardiac progenitor cells. These molecules are the wingless integrated (Wnt), fibroblast growth factor (FGF) and transforming growth factor-beta (TGFβ) superfamily ligands such as Wnt3a, bone morphogenetic protein 4 (BMP4), Nodal and activin A [2]. These ligands are expressed in gradients depending on the spatiotemporal context. Signalling pathways and downstream transcriptional events, finally result in the expression of cardiac-specific factors in a specified group of mesodermal cells [2].

The major sources of CMs in the heart are the first heart field (FHF) and second heart field (SHF) [3]. Proepicardium is a minor source. After lineage specification, paracrine signals from the epicardium and endocardium have a key role in coordinating the proliferation and terminal differentiation in newly formed CMs to finally produce sufficient numbers of fully functional cells.

The long noncoding RNA Braveheart (Bvht) mediates the epigenetic changes that facilitate specification of the cardiac lineage. (Bry) is a direct target of canonical Wnt signalling [4]. In the primitive streak, Bvht induces the expression of transcription factor Mesp1 [5–7]. Mesp1 acts upstream of key cardiac-specific transcription factors, such as GATA binding protein 4 (Gata4), Isl1, myocyte enhancer factor 2c (Mef2c) and NK2 homeobox 5 (Nkx2.5) [8–10]. These factors are necessary for cardiogenesis [11]. During the initiation of differentiation of CMs, several transcription factors are expressed. These include the zinc finger factors Gata4 and Gata5, the homeodomain factor Nkx2.5, T-box factors such as Tbx5, the MADS-box factor Mef2c, and the basic helix-loop-helix transcription factors, Hand1 and Hand2. Differentiation of CMs is also associated with chromatin remodelling [12].

Generation of Cardiomyocytes in Vitro

Both pluripotent stem cells (PSCs) and somatic cells have been induced for a cardiomyocyte fate in vitro [13–17]. Growing of cells in vitro offers the advantage of the availability of a large number of cells for manipulation before transplantation into the heart.

Directed differentiation approaches for conversion of PSCs into CMs in vitro follow the steps of embryonic development and utilize the signalling pathways and growth factors that initiate commitment to the cardiomyogenic lineage (Fig. 18.1). Direct lineage conversion of other somatic cells into CMs consider the specific gene regulatory network that define the cardiomyocyte phenotype and necessitates the delivery of transcription factors.

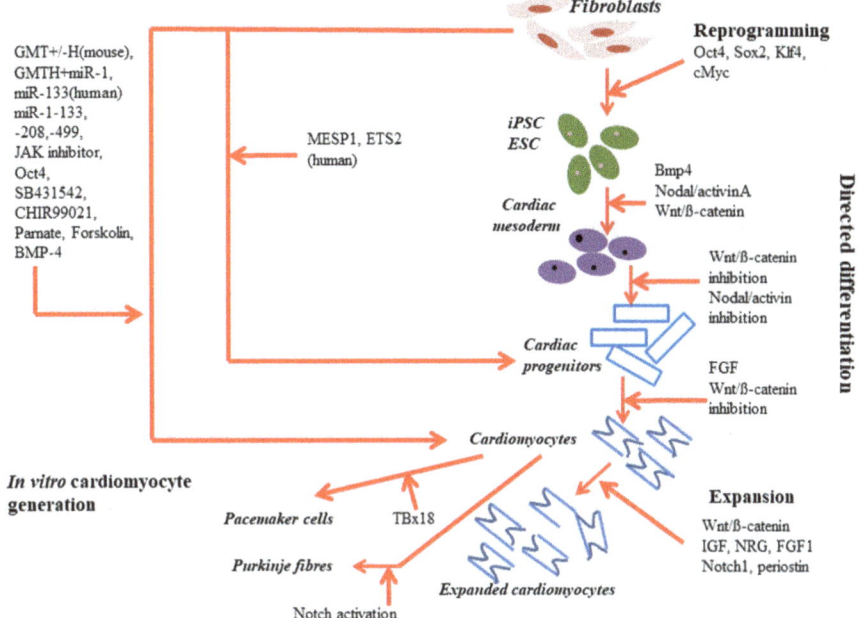

Fig. 18.1 A summary of various approaches for in vitro generation of cardiomyocytes

Directed Differentiation of Pluripotent Stem Cells into Cardiomyocytes

The Wnt, Activin/Nodal/TGFβ, BMP and FGF signalling pathways, which coordinate cardiogenesis in vivo modulate also the differentiation of pluripotent stem cells (PSCs) into cardiovascular progenitor cells and CMs in vitro (Fig. 18.1). During cardiac differentiation in vitro, these pathways have to be activated and inhibited repetitively as the cells in culture traverse the different stages of development [2, 18–22]. Endogenous canonical Wnt signalling is necessary for differentiation of human embryonic stem cells (ESCs) into CMs [36]. Temporal modulation of canonical Wnt signaling was found adequate to generate a supply of 80–98% CMs from several human ESC lines [21, 23].

The differentiation efficiency can be improved by the use of defined factors as well as timing the supplementation and control of the dose of molecules, which specifically activate or inhibit precise intracellular signalling pathways [2, 18–20, 24, 25].

Several approaches have been tried to obtain a relatively unadulterated population of CMs. These strategies consist of enrichment by cell sorting using an antibody against signal-regulatory protein α (SIRPA) or cell purification using mitochondria-specific fluorescent dyes [26]. Or taking advantage of the biochemical differences in metabolism between CMs and non-CMs [26–28].

By manipulating retinoic acid signalling, cardiomyogenic differentiation can be directed towards an atrial versus ventricular cardiomyocyte subtype [29–31]. A conduction system-like phenotype can be generated through activation of the Notch signalling pathway (Fig. 18.1) [32].

Direct Lineage Conversion of Somatic Cells into Cardiomyocytes

Fully differentiated fibroblasts can be reprogrammed into pluripotent cells by forced expression of the pluripotency transcription factors octamer-binding transcription factor 4 (Oct4), SRY (sex determining region Y)-box 2 (Sox2), Kruppel-like factor 4 (Klf4) and myelocytomatosis oncogene (Myc), which are collectively known as OSKM factors (Fig. 18.1). This observation has prompted attempts to directly convert fully differentiated cells to those of another cell lineage, without passing through the pluripotent state. Partial dedifferentiation of fibroblasts and subsequent cardiomyocyte differentiation sans a pluripotent stage has been achieved [33].

Islas et al., succeeded in transdifferentiation of fibroblasts through a cardiac progenitor stage instead of a pluripotent stage by the induced expression of the transcription factors Mesp1 and Ets2 (Fig. 18.1) [34]. Wang et al., accomplished the same by the induced expression of Oct4 and a blend of small molecules [35].

Direct lineage conversion from fibroblasts to cardiomyocyte-like cells without passing through a stem cell-like stage has also been achieved via the forced expression of essential cardiac lineage transcription factors [13, 17, 36–38]. The risk of tumour formation is reduced by circumventing pluripotency.

The forced expression of miR-1, -133, -208 and -499 is also adequate for direct conversion of mouse fibroblasts into CMs, both in vitro and in vivo. Inhibition of Janus kinase 1 (JAK1) can augment the conversion (Fig. 18.1) [39].

Direct lineage conversion to functional cardiac-like myocytes has also been achieved using fibroblasts from human adult heart and dermis. Reprogramming was done by using a combination of factors, such as GATA4, HAND2, TBX5, MYOCD, as well as miR-1 and miR-133 [37]. A cardiomyocyte-like phenotype could be induced in human fibroblasts derived from embryonic stem cells, foetal heart and neonatal skin by the expression of GMT plus estrogen-related receptor gamma (ESRRG), MESP1, MYOCD and zinc finger protein, FOG family member 2 (ZFPM2) [36].

A sinoatrial-like phenotype has been induced by forced expression of Tbx18 or Tbx3. Activation of Notch signalling in CMs generated Purkinje-like cells [32, 40, 41].

The mechanistic aspects of directed reprogramming to cardiac lineages are yet to be entirely delineated. An important problem is that the CMs generated through reprogramming are not entirely mature cells and that they are also heterogenous [42].

Other Sources of Cells for Regeneration of Heart

Several cell sources have been identified for their potential use in regenerating the injured myocardium. They include endogenous cardiac progenitor cells, side population cells, cardiosphere derived cells, bone marrow cells and embryonic stem cells [43–46].

Resident c-kit+ Cardiac Stem Cells

Beltrami et al. reported the presence of multipotent resident stem cells (c-Kit+ CSCs) in the adult heart and that they can undergo clonal expansion in vitro [47]. They also found that these cells when transplanted into the periphery of myocardial infarcts, differentiate into new CMs [47]. In patients with chronic ischemic heart disease, cardiac function improved significantly after transplantation of c-Kit+ CSCs harvested from atrial appendages [48]. Subsequently, two independent studies did not find differentiation of transplanted c-Kit+ CSCs into CMs [49, 50]. van Berlo et al. observed that the new CMs were a result of cell fusion [51]. Several studies have noted that c-Kit+ CSCs do not aid regeneration of CMs in the injured heart [52–55]. Whether the adult mammalian heart does harbour CSCs with potential for cardiomyocyte regeneration is under debate [56].

SCA-1+ Cardiac Progenitor Cells

Sca-1 (stem cell antigen–1) is another marker for cardiomyocyte stem cells. Sca-1+ stem cells isolated from the adult heart can differentiate into CMs in vitro [57, 58]. These cells are multipotent [58]. Transplantation of clonally expanded Sca1+ cells improves cardiac function after myocardial infarction. Both cardiomyocyte differentiation as well as paracrine mechanisms mediate the effect [59].

Resident Sca-1+ cells in the adult heart are heterogenous [60–62]. The potential for differentiation into CMs is determined by the nature of the sub population [63, 64]. Whether endogenous Sca-1+ cells have the capacity for differentiation into CMs in vivo is unclear. Though Uchida et al. reported that Sca-1+ cells contributed to CMs in the aging heart and in hearts damaged by pressure overload [65], several later studies however concluded that Sca-1+ cells do not produce new CMs [66–70].

Cardiosphere-Derived Stem Cells

Cardiospheres formed from subcultures of adult atrial or ventricular heart tissues were considered to have the capacity for long-term self-renewal and differentiation into CMs after transplantation into injured hearts. When cardiosphere-derived cells (CDCs) were cocultured with neonatal rat CMs, the CDCs differentiated into CMs. In mice, rats, and pigs, CDCs injected into myocardial infarcts generated CMs and vascular endothelial cells, resulting in improved cardiac function [71–75]. The paracrine effects of CDCs could have contributed to the beneficial effects [76]. Several studies have however questioned the utility of CDCs in cardiac repair [77–79]. There is an opinion that the beneficial effects of CDCs are mediated by exosomes from CDCs [80].

ABCG2+ Side Population Cells

ABCG2+ side population cells (ATP binding cassette subfamily G member 2) of the adult heart differentiate into CMs in vitro [81]. These cells differentiated into CMs, endothelial cells, and smooth muscle cells, when injected into rats with cardiac injury [82]. Genetic lineage tracing studies have revealed that ABCG2+ cells are able to differentiate into multiple cardiac cell lineages only during the embryonic stages. This ability is lost in adulthood [83, 84]. These cells can fuse with pre-existing CMs and may cause proliferation of CMs by stimulating their entry into cell cycle. ABCG2+ cells by themselves may not differentiate into CMs [85]. Genetic fate mapping studies indicate that the adult endogenous ABCG2+ side population do not have the characteristics of CSCs and lack the capacity for myogenic regeneration in vivo.

BMI1+ Cardiomyocyte Stem Cells

An additional endogenous cardiac progenitor contributing to homeostasis is Bmi1+ cells [86, 87]. Bmi1+ cells can be induced to differentiate into CMs by coculturing with neonatal rat CMs. One study reported that after myocardial infarction, Bmi1+ cells contribute to de novo generation of CMs [87]. This finding is not supported by studies on DNA synthesis in CMs [43, 44, 88].

Caveat on Cardiac Stem Cells

CSCs such as c-Kit, Sca-1, and side populations may not be legitimate stem cells for CMs. An enigma that remains is whether the adult heart has resident CSCs and whether they generate even a single cardiomyocyte in the adult heart. Fate mapping, lineage tracing and labelling studies seem to indicate that the adult heart lacks an endogenous stem cell for regeneration of CMs [89–92].

Bone Marrow Cells

Early studies reported that c-Kit+ hematopoietic cells from the bone marrow of both mice and humans differentiate into CMs and express sarcomere markers [93, 94]. Later investigators did not however find evidence for the differentiation of transplanted c-Kit+ BMCs into CMs [95, 96]. It is possible that these cells may stimulate neovascularization through a paracrine effect and modulate inflammatory mechanisms and a few of the cells may fuse with CMs [97–100].

Embryonic Stem Cells

Studies have demonstrated that human embryonic stem cells and CMs derived from pluripotent stem cells when transplanted into large animals, survive and improve heart function after cardiac injury [101, 102].

Human Induced Pluripotent Stem Cells

Cardiac regeneration using human-induced pluripotent stem cells (hiPSCs) has attracted increased attention in recent times [103, 104]. In 2006, Yamanaka et al. introduced the genes Oct3/4, Sox2, Klf4, and c-Myc (referred to as Yamanaka factors) into somatic cells in mice and successfully developed induced pluripotent stem cells (iPSCs) [103]. These cells were similar to embryonic stem cells (ESCs) in their characteristics. A year later, they developed human-iPSCs (hiPSCs) [104]. hiPSCs are considered promising for use in regenerative medicine as they do not have the drawbacks of ESCs. hiPSC-derived CMs are possibly excellent for use in patients with severe heart failure [105].

Several protocols are currently available for generation of human CMs from pluripotent stem cells [106]. Human-PSCs (hPSCs) can be induced to differentiate into CMs at various locations within the heart. These cells have electrical activity similar to human CMs [107]. Three- and two-dimensional culture techniques are available for cardiac differentiation. The three-dimensional culture method needs the use of recombinant proteins, and hence expensive [108]. In 2-dimensional culture techniques, low-molecular-weight compounds such as CHIR99021 and inhibitors of Wnt, such as IWR-1 and IWP-2 are used. This method is considered cost-effective [109]. Despite these advancements, generating PSC-derived CMs with adult-like phenotypes in vitro remains a challenge [110].

Many concerns with respect to the preparation of clinical-grade hiPSCs, large-scale generation of hiPSCs and CMs, prevention of tumours and an effective transplantation strategy have been dealt with satisfactorily. Contamination with residual undifferentiated stem cells and noncardiomyocytes remains a matter of concern. Hopefully, advances in technologies would resolve the issue.

Another promising novel approach for heart repair is through cardiomyocyte secreted vesicles [111, 112].

Conclusions

Several cell-based and paracrine factor-based strategies utilizing peptides, recombinant proteins, plasmid, adenoviral, adeno-associated viruses, retroviruses, miRNA and RNAi have been developed to generate new cardiomyocytes in vitro. Pluripotent cells, endogenous progenitor cells and somatic cells have also been identified for their potential use in regenerating the injured myocardium. Despite natural advantages, they all have limitations as well. Further investigations are hence warranted to identify clinically useful approaches.

References

1. Später D, Hansson EM, Zangi L, Chien KR. How to make a cardiomyocyte Development. 2014;141:4418–31.
2. Noseda M, Peterkin T, Simões FC, Patient R, Schneider MD. Cardiopoietic factors: extracellular signals for cardiac lineage commitment. Circ Res. 2011;108:129–52.
3. Brade T, Pane LS, Moretti A, Chien KR, Laugwitz KL. Embryonic heart progenitors and cardiogenesis. Cold Spring Harb Perspect Med. 2013;3: a013847.
4. Yamaguchi TP, Takada S, Yoshikawa Y, Wu N, McMahon AP. T (Brachyury) is a direct target of Wnt3a during paraxial mesoderm specification. Genes Dev. 1999;13:3185–90.
5. Costello I, Pimeisl IM, Dräger S, Bikoff EK, Robertson EJ, Arnold SJ. The T-box transcription factor Eomesodermin acts upstream of Mesp1 to specify cardiac mesoderm during mouse gastrulation. Nat Cell Biol. 2011;13:1084–91.
6. Klattenhoff CA, Scheuermann JC, Surface LE, Bradley RK, Fields PA, Steinhauser ML, Ding H, Butty VL, Torrey L, Haas S, et al. Braveheart, a long noncoding RNA required for cardiovascular lineage commitment. Cell. 2013;152:570–83.
7. Chan SSK, Shi X, Toyama A, Arpke RW, Dandapat A, Iacovino M, Kang J, Le G, Hagen HR, Garry DJ, et al. Mesp1 patterns mesoderm into cardiac, hematopoietic, or skeletal myogenic progenitors in a context-dependent manner. Cell Stem Cell. 2013;12:587–601.

8. Bondue A, Lapouge G, Paulissen C, Semeraro C, Iacovino M, Kyba M, Blanpain C. Mesp1 acts as a master regulator of multipotent cardiovascular progenitor specification. Cell Stem Cell. 2008;3:69–84.
9. Kitajima S, Takagi A, Inoue T, Saga Y. MesP1 and MesP2 are essential for the development of cardiac mesoderm. Development. 2000;127:3215–26.
10. Saga Y, Miyagawa-Tomita S, Takagi A, Kitajima S, Miyazaki JI, Inoue T. MesP1 is expressed in the heart precursor cells and required for the formation of a single heart tube. Development. 1999;126:3437–47.
11. Vincent SD, Buckingham ME. How to make a heart: the origin and regulation of cardiac progenitor cells. Curr Top Dev Biol. 2010;90:1–41.
12. Takeuchi JK, Bruneau BG. Directed transdifferentiation of mouse mesoderm to heart tissue by defined factors. Nature. 2009;459:708–11.
13. Ieda M, Fu JD, Delgado-Olguin P, Vedantham V, Hayashi Y, Bruneau BG, Srivastava D. Direct reprogramming of fibroblasts into functional cardiomyocytes by defined factors. Cell. 2010;142:375–86.
14. Kehat I, Kenyagin-Karsenti D, Snir M, Segev H, Amit M, Gepstein A, Livne E, Binah O, Itskovitz-Eldor J, Gepstein L. Human embryonic stem cells can differentiate into myocytes with structural and functional properties of cardiomyocytes. J Clin Invest. 2001;108:407–14.
15. Maltsev VA, Wobus AM, Rohwedel J, Bader M, Hescheler J. Cardiomyocytes differentiated in vitro from embryonic stem cells developmentally express cardiac-specific genes and ionic currents. Circ Res. 1994;75:233–44.
16. Metzger JM, Lin WI, Johnston RA, Westfall MV, Samuelson LC. Myosin heavy chain expression in contracting myocytes isolated during embryonic stem cell cardiogenesis. Circ Res. 1995;76:710–9.
17. Song K, Nam YJ, Luo X, Qi X, Tan W, Huang GN, Acharya A, Smith CL, Tallquist MD, Neilson EG, et al. Heart repair by reprogramming non-myocytes with cardiac transcription factors. Nature. 2012;485:599–604.
18. Yang L, Soonpaa MH, Adler ED, Roepke TK, Kattman SJ, Kennedy M, Henckaerts E, Bonham K, Abbott GW, Linden RM, et al. Human cardiovascular progenitor cells develop from a KDR+ embryonic-stem cell-derived population. Nature. 2008;453:524–8.
19. Kattman SJ, Witty AD, Gagliardi M, Dubois NC, Niapour M, Hotta A, Ellis J, Keller G. Stage-specific optimization of activin/nodal and BMP XE "bone morphogenic protein (BMP)" signaling promotes cardiac differentiation of mouse and human pluripotent stem cell lines. Cell Stem Cell. 2011;8:228–40.
20. Laflamme MA, Chen KY, Naumova AV, Muskheli V, Fugate JA, Dupras SK, Reinecke H, Xu C, Hassanipour M, Police S, et al. Cardiomyocytes derived from human embryonic stem cells in pro-survival factors enhance function of infarcted rat hearts. Nat Biotechnol. 2007;25:1015–24.
21. Lian X, Hsiao C, Wilson G, Zhu K, Hazeltine LB, Azarin SM, Raval KK, Zhang J, Kamp TJ, Palecek SP. Cozzarelli Prize Winner: robust cardiomyocyte differentiation from human pluripotent stem cells via temporal modulation of canonical Wnt signaling. Proc Natl Acad Sci USA. 2012;109:E1848–57.
22. Paige SL, Osugi T, Afanasiev OK, Pabon L, Reinecke H, Murry CE. Endogenous Wnt/beta-catenin signaling is required for cardiac differentiation in human embryonic stem cells. PLoS ONE. 2010;5:e11134.
23. Lian X, Zhang J, Azarin SM, Zhu K, Hazeltine LB, Bao X, Hsiao C, Kamp TJ, Palecek SP. Directed cardiomyocyte differentiation from human pluripotent stem cells by modulating Wnt/β-catenin signaling under fully defined conditions. Nat Protoc. 2013;8:162–75.
24. Cai W, Albini S, Wei K, Willems E, Guzzo RM, Tsuda M, Giordani L, Spiering S, Kurian L, Yeo GW, et al. Coordinate Nodal and BMP inhibition directs Baf60c-dependent cardiomyocyte commitment. Genes Dev. 2013;27:2332–44.
25. Willems E, Cabral-Teixeira J, Schade D, Cai W, Reeves P, Bushway PJ, Lanier M, Walsh C, Kirchhausen T, Izpisua Belmonte JC, et al. Small molecule-mediated TGF-β type II receptor degradation promotes cardiomyogenesis in embryonic stem cells. Cell Stem Cell. 2012;11:242–52.

26. Dubois NC, Craft AM, Sharma P, Elliott DA, Stanley EG, Elefanty AG, Gramolini A, Keller G. SIRPA is a specific cell-surface marker for isolating cardiomyocytes derived from human pluripotent stem cells. Nat Biotechnol. 2011;29:1011–8.

27. Hattori F, Chen H, Yamashita H, Tohyama S, Satoh Y-S, Yuasa S, Li W, Yamakawa H, Tanaka T, Onitsuka T, et al. Nongenetic method for purifying stem cell-derived cardiomyocytes. Nat Methods. 2010;7:61–6.

28. Tohyama S, Hattori F, Sano M, Hishiki T, Nagahata Y, Matsuura T, Hashimoto H, Suzuki T, Yamashita H, Satoh Y, et al. Distinct metabolic flow enables large-scale purification of mouse and human pluripotent stem cell-derived cardiomyocytes. Cell Stem Cell. 2013;12:127–37.

29. Jiang J, Han P, Zhang Q, Zhao J, Ma Y. Cardiac differentiation of human pluripotent stem cells. J Cell Mol Med. 2012;16:1663–8.

30. Weng Z, Kong C-W, Ren L, Karakikes I, Geng L, He J, Chow MZY, Mok CF, Keung W, Chow H, et al. A simple, cost-effective but highly efficient system for deriving ventricular cardiomyocytes from human pluripotent stem cells. Stem Cells Dev. 2014;23:1704–16.

31. Zhang Q, Jiang J, Han P, Yuan Q, Zhang J, Zhang X, Xu Y, Cao H, Meng Q, Chen L, et al. Direct differentiation of atrial and ventricular myocytes from human embryonic stem cells by alternating retinoid signals. Cell Res. 2011;21:579–87.

32. Rentschler S, Yen AH, Lu J, Petrenko NB, Lu MM, Manderfield LJ, Patel VV, Fishman GI, Epstein JA. Myocardial Notch signaling reprograms cardiomyocytes to a conduction-like phenotype. Circulation. 2012;126:1058–66.

33. Efe JA, Hilcove S, Kim J, Zhou H, Ouyang K, Wang G, Chen J, Ding S. Conversion of mouse fibroblasts into cardiomyocytes using a direct reprogramming strategy. Nat Cell Biol. 2011;13:215–22.

34. Islas JF, Liu Y, Weng K-C, Robertson MJ, Zhang S, Prejusa A, Harger J, Tikhomirova D, Chopra M, Iyer D, et al. Transcription factors ETS2 and MESP1 transdifferentiate human dermal fibroblasts into cardiac progenitors. Proc Natl Acad Sci USA. 2012;109:13016–21.

35. Wang H, Cao N, Spencer CI, Nie B, Ma T, Xu T, Zhang Y, Wang X, Srivastava D, Ding S. Small molecules enable cardiac reprogramming of mouse fibroblasts with a single factor. Cell Rep. 2014;6:951–60.

36. Fu J-D, Stone NR, Liu L, Spencer CI, Qian L, Hayashi Y, Delgado Olguin P, Ding S, Bruneau BG, Srivastava D. Direct reprogramming of human fibroblasts toward a cardiomyocyte-like state. Stem Cell Rep. 2013;235–47.

37. Nam Y-J, Song K, Luo X, Daniel E, Lambeth K, West K, Hill JA, DiMaio JM, Baker LA, Bassel-Duby R, et al. Reprogramming of human fibroblasts toward a cardiac fate. Proc Natl Acad Sci USA. 2013;110:5588–93.

38. Qian L, Huang Y, Spencer CI, Foley A, Vedantham V, Liu L, Conway SJ, Fu J-d, Srivastava D. In vivo reprogramming of murine cardiac fibroblasts into induced cardiomyocytes. Nature. 2012; 485:593–98.

39. Jayawardena TM, Egemnazarov B, Finch EA, Zhang L, Payne JA, Pandya K, Zhang Z, Rosenberg P, Mirotsou M, Dzau VJ. MicroRNA-mediated in vitro and in vivo direct reprogramming of cardiac fibroblasts to cardiomyocytes. Circ Res. 2012;110:1465–73.

40. Bakker ML, Boink GJJ, Boukens BJ, Verkerk AO, van den Boogaard M, den Haan AD, Hoogaars WMH, Buermans HP, de Bakker JMT, Seppen J, et al. T-box transcription factor TBX3 reprogrammes mature cardiac myocytes into pacemaker cells. Cardiovasc Res. 2012;94:443–9.

41. Kapoor N, Liang W, Marbán E, Cho HC. Direct conversion of quiescent cardiomyocytes to pacemaker cells by expression of Tbx18. Nat Biotechnol. 2013;31:54–62.

42. Chen JX, Krane M, Deutsch M-A, Wang L, Rav-Acha M, Gregoire S, Engels MC, Rajarajan K, Karra R, Abel ED, et al. Inefficient reprogramming of fibroblasts into cardiomyocytes using Gata4, Mef2c, and Tbx5. Circ Res. 2012;111:50–5.

43. Eschenhagen T, Bolli R, Braun T, Field LJ, Fleischmann BK, Frisén J, Giacca M, Hare JM, Houser S, Lee RT, et al. Cardiomyocyte regeneration: a consensus statement. Circulation. 2017;136:680–6.

44. Cahill TJ, Choudhury RP, Riley PR. Heart regeneration and repair after myocardial infarction: translational opportunities for novel therapeutics. Nat Rev Drug Discov. 2017;16:699–717.
45. Telukuntla KS, Suncion VY, Schulman IH, Hare JM. The advancing field of cell-based therapy: insights and lessons from clinical trials. J Am Heart Assoc. 2013;2:e000338. doi: https://doi.org/10.1161/JAHA.113.000338.
46. Tang JN, Cores J, Huang K, Cui XL, Luo L, Zhang JY, Li TS, Qian L, Cheng K. Concise review: is cardiac cell therapy dead? embarrassing trial outcomes and new directions for the future. Stem Cells Transl Med. 2018;7:354–9.
47. Beltrami AP, Barlucchi L, Torella D, Baker M, Limana F, Chimenti S, Kasahara H, Rota M, Musso E, Urbanek K, et al. Adult cardiac stem cells are multipotent and support myocardial regeneration. Cell. 2003;114:763–76.
48. Bolli R, Chugh AR, D'Amario D, Loughran JH, Stoddard MF, Ikram S, Beache GM, Wagner SG, Leri A, Hosoda T, et al. Cardiac stem cells in patients with ischaemic cardiomyopathy (SCIPIO): initial results of a randomised phase 1 trial. Lancet. 2011;378:1847–57.
49. Zaruba MM, Soonpaa M, Reuter S, Field LJ. Cardiomyogenic potential of C-kit (+)-expressing cells derived from neonatal and adult mouse hearts. Circulation. 2010;121:1992–2000.
50. Pouly J, Bruneval P, Mandet C, Proksch S, Peyrard S, Amrein C, Bousseaux V, Guillemain R, Deloche A, Fabiani JN, et al. Cardiac stem cells in the real world. J Thorac Cardiovasc Surg. 2008;135:673–8.
51. van Berlo JH, Kanisicak O, Maillet M, Vagnozzi RJ, Karch J, Lin SC, Middleton RC, Marbán E, Molkentin JD. c-kit+ cells minimally contribute cardiomyocytes to the heart. Nature. 2014;509:337–41.
52. Sultana N, Zhang L, Yan J, Chen J, Cai W, Razzaque S, Jeong D, Sheng W, Bu L, Xu M, et al. Resident c-kit(+) cells in the heart are not cardiac stem cells. Nat Commun. 2015;6:8701.
53. Liu Q, Yang R, Huang X, Zhang H, He L, Zhang L, Tian X, Nie Y, Hu S, Yan Y, et al. Genetic lineage tracing identifies in situ Kit-expressing cardiomyocytes. Cell Res. 2016;26:119–30.
54. He L, Li Y, Li Y, Pu W, Huang X, Tian X, Wang Y, Zhang H, Liu Q, Zhang L, et al. Enhancing the precision of genetic lineage tracing using dual recombinases. Nat Med. 2017;23:1488–98.
55. He L, Han M, Zhang Z, Li Y, Huang X, Liu X, Pu W, Zhao H, Wang QD, Nie Y, et al. Reassessment of c-kit+ cells for cardiomyocyte contribution in adult heart. Circulation. 2019;140:164–6.
56. Cai CL, Molkentin JD. The elusive progenitor cell in cardiac regeneration: slip slidin' away. Circ Res. 2017;120:400–6.
57. Oh H, Bradfute SB, Gallardo TD, Nakamura T, Gaussin V, Mishina Y, Pocius J, Michael LH, Behringer RR, Garry DJ, et al. Cardiac progenitor cells from adult myocardium: homing, differentiation, and fusion after infarction. Proc Natl Acad Sci USA. 2003;100:12313–8.
58. Matsuura K, Nagai T, Nishigaki N, Oyama T, Nishi J, Wada H, Sano M, Toko H, Akazawa H, Sato T, et al. Adult cardiac Sca-1-positive cells differentiate into beating cardiomyocytes. J Biol Chem. 2004;279:11384–91.
59. Matsuura K, Honda A, Nagai T, Fukushima N, Iwanaga K, Tokunaga M, Shimizu T, Okano T, Kasanuki H, Hagiwara N, et al. Transplantation of cardiac progenitor cells ameliorates cardiac dysfunction after myocardial infarction in mice. J Clin Invest. 2009;119:2204–17.
60. Pfister O, Mouquet F, Jain M, Summer R, Helmes M, Fine A, Colucci WS, Liao R. CD31- but not CD31+ cardiac side population cells exhibit functional cardiomyogenic differentiation. Circ Res. 2005;97:52–61.
61. Noseda M, Harada M, McSweeney S, Leja T, Belian E, Stuckey DJ, Abreu Paiva MS, Habib J, Macaulay I, de Smith AJ, et al. PDGFRα demarcates the cardiogenic clonogenic Sca1+ stem/progenitor cell in adult murine myocardium. Nat Commun. 2015;6:6930. https://doi.org/10.1038/ncomms7930.
62. Takamiya M, Haider KH, Ashraf M. Identification and characterization of a novel multipotent sub-population of Sca-1+ cardiac progenitor cells for myocardial regeneration. PLoS One. 2011;6:e25265. https://doi.org/10.1371/journal.pone.0025265
63. Wang X, Hu Q, Nakamura Y, Lee J, Zhang G, From AH, Zhang J. The role of the Sca-1+/CD31− cardiac progenitor cell population in postinfarction left ventricular remodeling. Stem Cells. 2006;24:1779–88.

64. Liang SX, Tan TY, Gaudry L, Chong B. Differentiation and migration of Sca1+/CD31−cardiac side population cells in a murine myocardial ischemic model. Int J Cardiol. 2010;138:40–9.
65. Uchida S, De Gaspari P, Kostin S, Jenniches K, Kilic A, Izumiya Y, Shiojima I, Grosse Kreymborg K, Renz H, Walsh K, et al. Sca1-derived cells are a source of myocardial renewal in the murine adult heart. Stem Cell Reports. 2013;1:397–410.
66. Vagnozzi RJ, Sargent MA, Lin SJ, Palpant NJ, Murry CE, Molkentin JD. Genetic lineage tracing of Sca-1+ cells reveals endothelial but not myogenic contribution to the murine heart. Circulation. 2018;138:2931–9.
67. Zhang L, Sultana N, Yan J, Yang F, Chen F, Chepurko E, Yang FC, Du Q, Zangi L, Xu M, et al. Cardiac Sca-1+ cells are not intrinsic stem cells for myocardial development, renewal, and repair. Circulation. 2018;138:2919–30.
68. Neidig LE, Weinberger F, Palpant NJ, Mignone J, Martinson AM, Sorensen DW, Bender I, Nemoto N, Reinecke H, Pabon L, et al. Evidence for minimal cardiogenic potential of stem cell antigen 1-positive cells in the adult mouse heart. Circulation. 2018;138:2960–2.
69. Tang J, Li Y, Huang X, He L, Zhang L, Wang H, Yu W, Pu W, Tian X, Nie Y, et al. Fate mapping of Sca1+ cardiac progenitor cells in the adult mouse heart. Circulation. 2018;138:2967–9.
70. Soonpaa MH, Lafontant PJ, Reuter S, Scherschel JA, Srour EF, Zaruba MM, Rubart-von der Lohe M, Field LJ. Absence of cardiomyocyte differentiation following transplantation of adult cardiac-resident Sca-1+ cells into infarcted mouse hearts. Circulation. 2018;138:2963–66.
71. Smith RR, Barile L, Cho HC, Leppo MK, Hare JM, Messina E, Giacomello A, Abraham MR, Marbán E. Regenerative potential of cardiosphere-derived cells expanded from percutaneous endomyocardial biopsy specimens. Circulation. 2007;115:896–908.
72. Malliaras K, Li TS, Luthringer D, Terrovitis J, Cheng K, Chakravarty T, Galang G, Zhang Y, Schoenhoff F, Van Eyk J, et al. Safety and efficacy of allogeneic cell therapy in infarcted rats transplanted with mismatched cardiosphere-derived cells. Circulation. 2012;125:100–12.
73. Tseliou E, Pollan S, Malliaras K, Terrovitis J, Sun B, Galang G, Marbán L, Luthringer D, Marbán E. Allogeneic cardiospheres safely boost cardiac function and attenuate adverse remodeling after myocardial infarction in immunologically mismatched rat strains. J Am Coll Cardiol. 2013;61:1108–19.
74. Malliaras K, Smith RR, Kanazawa H, Yee K, Seinfeld J, Tseliou E, Dawkins JF, Kreke M, Cheng K, Luthringer D, et al. Validation of contrast-enhanced magnetic resonance imaging to monitor regenerative efficacy after cell therapy in a porcine model of convalescent myocardial infarction. Circulation. 2013;128:2764–75.
75. Grigorian-Shamagian L, Liu W, Fereydooni S, Middleton RC, Valle J, Cho JH, Marbán E. Cardiac and systemic rejuvenation after cardiospherederived cell therapy in senescent rats. Eur Heart J. 2017;38:2957–67.
76. Chimenti I, Smith RR, Li TS, Gerstenblith G, Messina E, Giacomello A, Marbán E. Relative roles of direct regeneration versus paracrine effects of human cardiosphere-derived cells transplanted into infarcted mice. Circ Res. 2010;106:971–80.
77. Redgrave RE, Tual-Chalot S, Davison BJ, Singh E, Hall D, Amirrasouli MM, Gilchrist D, Medvinsky A, Arthur HM. Cardiosphere-derived cells require endoglin for paracrine-mediated angiogenesis. Stem Cell Reports. 2017;8:1287–98.
78. Zhao ZA, Han X, Lei W, Li J, Yang Z, Wu J, Yao M, Lu XA, He L, Chen Y, et al. Lack of cardiac improvement after cardiosphere-derived cell transplantation in aging mouse hearts. Circ Res. 2018;123:266–87.
79. Kasai-Brunswick TH, Costa AR, Barbosa RA, Farjun B, Mesquita FC, Silva Dos Santos D, Ramos IP, Suhett G, Brasil GV, Cunha ST, et al. Cardiosphere-derived cells do not improve cardiac function in rats with cardiac failure. Stem Cell Res Ther. 2017;8:36. https://doi.org/10.1186/s13287-017-0481-x.
80. Ibrahim AG, Cheng K, Marbán E. Exosomes as critical agents of cardiac regeneration triggered by cell therapy. Stem Cell Reports. 2014;2:606–19.
81. Martin CM, Meeson AP, Robertson SM, Hawke TJ, Richardson JA, Bates S, Goetsch SC, Gallardo TD, Garry DJ. Persistent expression of the ATP-binding cassette transporter, Abcg2, identifies cardiac SP cells in the developing and adult heart. Dev Biol. 2004;265:262–75.

82. Oyama T, Nagai T, Wada H, Naito AT, Matsuura K, Iwanaga K, Takahashi T, Goto M, Mikami Y, Yasuda N, et al. Cardiac side population cells have a potential to migrate and differentiate into cardiomyocytes in vitro and in vivo. J Cell Biol. 2007;176:329–41.

83. Maher TJ, Ren Y, Li Q, Braunlin E, Garry MG, Sorrentino BP, Martin CM. ATP-binding cassette transporter Abcg2 lineage contributes to the cardiac vasculature after oxidative stress. Am J Physiol Heart Circ Physiol. 2014;306:H1610–8.

84. Doyle MJ, Maher TJ, Li Q, Garry MG, Sorrentino BP, Martin CM. Abcg2- labeled cells contribute to different cell populations in the embryonic and adult heart. Stem Cells Dev. 2016;25:277–84.

85. Yellamilli A, Ren Y, McElmurry RT, Lambert JP, Gross P, Mohsin S, Houser SR, Elrod JW, Tolar J, Garry DJ, et al. Abcg2-expressing side population cells contribute to cardiomyocyte renewal through fusion. FASEB J. 2020;34:5642–57.

86. Valiente-Alandi I, Albo-Castellanos C, Herrero D, Arza E, Garcia-Gomez M, Segovia JC, Capecchi M, Bernad A. Cardiac Bmi1(+) cells contribute to myocardial renewal in the murine adult heart. Stem Cell Res Ther. 2015;6:205. https://doi.org/10.1186/s13287-015-0196-9.

87. Valiente-Alandi I, Albo-Castellanos C, Herrero D, Sanchez I, Bernad A. Bmi1 (+) cardiac progenitor cells contribute to myocardial repair following acute injury. Stem Cell Res Ther. 2016;7:100. https://doi.org/10.1186/s13287-016-0355-7.

88. Vagnozzi RJ, Molkentin JD, Houser SR. New myocyte formation in the adult heart: endogenous sources and therapeutic implications. Circ Res. 2018;123:159–76.

89. Li Y, He L, Huang X, Issa Bhaloo S, Zhao H, Zhang S, Pu W, Tian X, Li Y, Liu Q, et al. Genetic lineage tracing of non-myocyte population by dual recombinases. Circulation. 2018;138:793–805.

90. Li Y, Lv Z, He L, Huang X, Zhang S, Zhao H, Pu W, Li Y, Yu W, Zhang L, et al. Genetic tracing identifies early segregation of the cardiomyocyte and nonmyocyte lineages. Circ Res. 2019;125:343–55.

91. Kretzschmar K, Post Y, Bannier-Hélaouët M, Mattiotti A, Drost J, Basak O, Li VSW, van den Born M, Gunst QD, Versteeg D, et al. Profiling proliferative cells and their progeny in damaged murine hearts. Proc Natl Acad Sci USA. 2018;115:E12245–54.

92. Maliken BD, Molkentin JD. Undeniable evidence that the adult mammalian heart lacks an endogenous regenerative stem cell. Circulation. 2018;138:806–8.

93. Orlic D, Kajstura J, Chimenti S, Jakoniuk I, Anderson SM, Li B, Pickel J, McKay R, Nadal-Ginard B, Bodine DM, et al. Bone marrow cells regenerate infarcted myocardium. Nature. 2001;410:701–5.

94. Quaini F, Urbanek K, Beltrami AP, Finato N, Beltrami CA, Nadal-Ginard B, Kajstura J, Leri A, Anversa P. Chimerism of the transplanted heart. N Engl J Med. 2002;346:5–15.

95. Murry CE, Soonpaa MH, Reinecke H, Nakajima H, Nakajima HO, Rubart M, Pasumarthi KB, Virag JI, Bartelmez SH, Poppa V, et al. Haematopoietic stem cells do not transdifferentiate into cardiac myocytes in myocardial infarcts. Nature. 2004;428:664–8.

96. Balsam LB, Wagers AJ, Christensen JL, Kofidis T, Weissman IL, Robbins RC. Haematopoietic stem cells adopt mature haematopoietic fates in ischaemic myocardium. Nature. 2004;428:668–73.

97. Nygren JM, Jovinge S, Breitbach M, Säwén P, Röll W, Hescheler J, Taneera J, Fleischmann BK, Jacobsen SE. Bone marrow-derived hematopoietic cells generate cardiomyocytes at a low frequency through cell fusion, but not transdifferentiation. Nat Med. 2004;10:494–501.

98. Alvarez-Dolado M, Pardal R, Garcia-Verdugo JM, Fike JR, Lee HO, Pfeffer K, Lois C, Morrison SJ, Alvarez-Buylla A. Fusion of bone-marrowderived cells with Purkinje neurons, cardiomyocytes and hepatocytes. Nature. 2003;425:968–73.

99. Fazel S, Cimini M, Chen L, Li S, Angoulvant D, Fedak P, Verma S, Weisel RD, Keating A, Li RK. Cardioprotective c-kit+ cells are from the bone marrow and regulate the myocardial balance of angiogenic cytokines. J Clin Invest. 2006;116:1865–77.

100. Vagnozzi RJ, Maillet M, Sargent MA, Khalil H, Johansen AKZ, Schwanekamp JA, York AJ, Huang V, Nahrendorf M, Sadayappan S, et al. An acute immune response underlies the benefit of cardiac stem cell therapy. Nature. 2020;577:405–9.

101. Chong JJ, Yang X, Don CW, Minami E, Liu YW, Weyers JJ, Mahoney WM, Van Biber B, Cook SM, Palpant NJ, et al. Human embryonic-stem-cell derived cardiomyocytes regenerate non-human primate hearts. Nature. 2014;510:273–7.
102. Shiba Y, Fernandes S, Zhu WZ, Filice D, Muskheli V, Kim J, Palpant NJ, Gantz J, Moyes KW, Reinecke H, et al. Human ES-cell-derived cardiomyocytes electrically couple and suppress arrhythmias in injured hearts. Nature. 2012;489:322–5.
103. Takahashi K, Yamanaka S. Induction of pluripotent stem cells from mouse embryonic and adult fibroblast cultures by defined factors. Cell. 2006;126(4):663–76.
104. Takahashi K, Tanabe K, Ohnuki M, Narita M, Ichisaka T, Tomoda K, et al. Induction of pluripotent stem cells from adult human fibroblasts by defined factors. Cell. 2007;131(5):861–72.
105. Yuasa S, Fukuda K. Recent advances in cardiovascular regenerative medicine: the induced pluripotent stem cell era. Expert Rev Cardiovasc Ther. 2008;6(6):803–10.
106. Kishino Y, Fujita J, Tohyama S, Okada M, Tanosaki S, Someya S, Fukuda K. Toward the realization of cardiac regenerative medicine using pluripotent stem cells. Inflammation and Regeneration. 2020;40: https://doi.org/10.1186/s41232-019-0110-4.
107. Zhang J, Wilson GF, Soerens AG, Koonce CH, Yu J, Palecek SP, et al. Functional cardiomyocytes derived from human induced pluripotent stem cells. Circ Res. 2009;104:e30-41.
108. Hemmi N, Tohyama S, Nakajima K, Kanazawa H, Suzuki T, Hattori F, et al. A massive suspension culture system with metabolic purification for human pluripotent stem cell-derived cardiomyocytes. Stem Cells Transl Med. 2014;3:1473–83.
109. Tohyama S, Fujita J, Fujita C, Yamaguchi M, Kanaami S, Ohno R, et al. Efficient large-scale 2D culture system for human induced pluripotent stem cells and differentiated cardiomyocytes. Stem Cell Rep. 2017;9:1406–14.
110. Karbassi E, Fenix A, Marchiano S, Muraoka N, Nakamura K, Yang X, Murry CE. Cardiomyocyte maturation: advances in knowledge and implications for regenerative medicine. Nat Rev Cardiol. 2020;17(6):341–59.
111. Heallen TR, Martin JF. Heart repair via cardiomyocyte-secreted vesicles. Nat Biomed Eng. 2018;2:271–2.
112. Liu B. Cardiac recovery via extended cell-free delivery of extracellular vesicles secreted by cardiomyocytes derived from induced pluripotent stem cells. Nat Biomed Eng. 2018;2:293–303.

Chapter 19
Reprograming Fibroblasts for Cardiomyocytes and Progenitors

Abstract The search for cell-based strategies to repair the injured heart after ischemic injury has led to the discovery of methods for genetic and pharmacologic reprogramming of fibroblasts into cardiomyocytes. Several studies have demonstrated the feasibility of transdifferentiating fibroblasts into induced cardiomyocytes (iCMs) or induced cardiac progenitor cells (iCPCs) in vitro as well as transdifferentiating fibroblasts into iCMs in vivo. Key transcription factors, cardiogenic genes, small molecules, cytokines, noncoding RNAs, as well as epigenetic modifiers are employed for reprogramming somatic cells. Cardiac reprogramming is currently recognized as a promising option for regenerative therapy in heart diseases.

Keywords Cardiac regeneration · Cardiac reprogramming · Fibroblasts · Induced cardiomyocytes · Cardiac progenitor cells · Transcription factors · Non-coding RNA · Epigenetic modifiers

Introduction

Current treatment strategies aimed at generation of functioning cardiomyocytes (CMs) within the injured heart focus on three approaches: (1) induction of proliferation of endogenous CMs (2) transplantation of cardiovascular progenitor cells (CPCs) or CMs generated from the differentiation of pluripotent stem cells and (3) direct reprogramming of somatic cells to expandable CPCs or CMs [1–4]. Overexpression of specific transcriptional factors (TFs) or using small molecules is a promising strategy for reprogramming of fibroblasts to CMs [5–8]. An overview of the various approaches for in vitro conversion of fibroblasts into induced CMs or induced CPCs is schematically shown in Fig. 19.1.

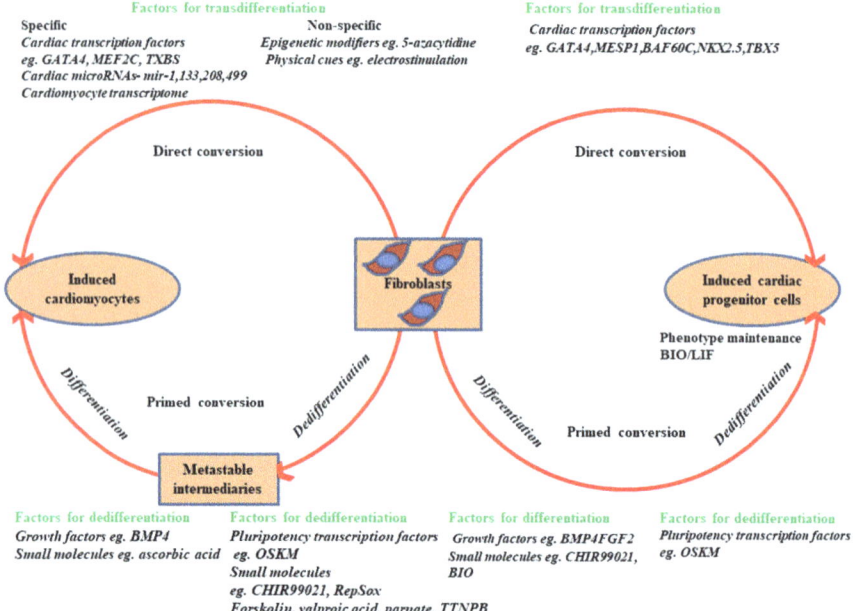

Fig. 19.1 An overview of the approaches for in vitro direct and primed conversion of fibroblasts into inducible cardiomyocytes (iCMs) and inducible cardiac progenitor cells (iCPCs). Various factors for differentiation or dedifferentiation or transdifferentiation are listed

Evolution of Cardiac Reprogramming Strategy

Nearly six decades ago, John B. Gurdon observed that transcriptional reprogramming can be induced in terminally differentiated amphibian somatic cells [9]. Later, reprogramming toward pluripotency was demonstrated also in mature mammalian cells [10–12]. Takahashi and Yamanaka generated iPSCs from mouse fibroblasts by expression of 4 exogenous transcription factors: Oct4 (octamer binding transcription factor 4), Sox2 (SRY-box 2), Klf4 (Kruppel like factor 4) and Myc (MYC proto-oncogene) (OCT4, SOX2, KLF4, and MYC—OSKM) [13]. They generated iPSCs from human fibroblasts as well. Their studies inspired further investigations on nuclear reprogramming [14].

Direct conversion of one terminally differentiated somatic cell into another one, sans an intervening dedifferentiated stage has also been achieved [15]. Mouse fibroblasts could be altered into skeletal myoblasts by enforced expression of the transcription factor-myogenic differentiation 1 (MYOD1) [16]. Attempts to convert fibroblasts to CMs by treating fibroblasts with TGFβ and by other methods were however unsuccessful [17–19].

van Tuyn et al. in 2007, inspired by the emerging iPSC technology, attempted in vitro conversion of fibroblasts from scar in the human heart into iCMs, through lentiviral delivery of a single cardiac TF—myocardin (MYOCD) [20]. They did not achieve complete phenotype conversion. Two years later, it was discovered that a combination of GATA4 (GATA binding protein 4), TBX5 (T-box 5), and BAF60C (SMARCD3) was necessary to induce formation of CMs in the non-cardiac mesoderm of mouse embryos [21].

Ieda et al. successfully accomplished direct conversion of cardiac fibroblasts from healthy mice into beating iCMs in vitro, through retroviral delivery of GATA4, MEF2C (myocyte enhancer factor 2C), and TBX5 – GMT [22]. iCM could be generated in a mouse model of myocardial infarction as well, using GMT or GMTH (GATA4, MEF2C, TBX5, and HAND2—heart and neural crest derivatives expressed 2) [5, 6].

TBX5 is known to coordinate the formation of the primary heart field interacting with other TFs such as GATA4 [23]. GATA4 modifies chromatin structure and permits binding of other TFs such as NKX2–5 to their targets, activating the transcriptional program in CMs [24] MEF2c is a MADS box transcription enhancer factor which interacts with cardiac TFs. MEF2c has a key role in the formation of the secondary heart field [25].

To increase the reprogramming efficiency, addition of other cardiac TFs to GATA4, MEF2c and TBX5 (GMT) have been attempted. bHLH TF HAND2 was one among the first TFs added to GMT. Mouse embryonic fibroblasts treated with the combination (GMHT) resulted in induced CMs (iCMs) with low levels of expression of the sarcomeric proteins [26]. A 15-fold increase in reprogramming efficiency could be achieved when the transcriptional activity of the mix (GMHT) was improved by fusing the transactivation domain of MyoD to MEF2c and overexpressed in mouse fibroblasts [27]. When NKX2–5 was overexpressed in addition to GMHT, the efficiency of cardiac reprogramming in mouse fibroblasts increased to more than 50-fold when compared to the use of only GMT. The resulting iCMs had spontaneous beating, expressed markers of mature CMs and also had strong calcium oscillations [28].

S. Protze and colleagues screened triplet combinations of 10 key cardiac TFs in mouse fibroblasts and found that TBX5, MEF2c, and MyocD, would induce a superior cardiac phenotype of CMs [29]. The effect of GATA4, TBX5, and MEF2C on reprogramming and expression of cardiac sarcomeric genes is enhanced by transcription factors MYOCD, SRF, Mesp1 and SMARCD3 [30]. Approaches used in murine cells have however not largely successful in reprogramming human somatic cells.

There has been some success in reprogramming human cells to iCMs using only TFs. A combination of the E26 transformation specific (ETS) TF family member ETS2 and MESP1 proteins was found to induce reprogramming of human dermal fibroblasts to cardiac progenitors [31]. GMT with MESP1 and MYOCD induced expression of multiple cardiac-specific proteins and increased a variety of cardiac genes in fibroblasts derived from human heart and dermis [32]. Formation of sarcomeres, calcium transients and action potentials were also improved in fibroblasts from human foetal heart and neonatal skin, on expression of GMT along with ESSRG (a transcriptional activator), MESP1, MYOCD, and ZFPM2 (a modulator of GATA proteins) [33].

Generation of Induced Cardiomyocytes

Several different cocktails for cardiac reprogramming have been identified through systematic screening of candidate factors based on several specific criteria and also via in silico approaches [26, 28–30, 34–47].

Cardiac reprogramming success is assessed by measuring promoter activity, protein expression, sarcomere assembly, calcium transients, and contractile activity. Current reprogramming protocols generate diverse iCM populations which include ventricular, atrial, and pacemaker-like cells [5, 22, 26, 27, 29, 30, 32, 35, 37, 40, 48]. Further studies are necessary to direct the specification of iCM subtype.

Direct Conversion with Nonspecific Stimuli

Nonspecific stimuli such as the DNA methyltransferase inhibitor 5-aza, dexamethasone in combination with fibroblast growth factor (FGF), HIF1α (hypoxia-inducible factor 1 subunit α), transforming growth factor-β (TGF-β), cardiomyocyte maintenance medium, various extracellular matrix coatings and electrical triggers have been investigated in vitro for direct conversion into iCM [17, 18, 21, 49–52]. These nonspecific stimuli induced only partial conversion in rodent, rabbit, or human fibroblasts. Contractile iCMs with calcium flux and action potentials were never obtained.

Direct Conversion with Cardiomyocyte-Specific Stimuli

CM-specific stimuli such as combinations of TFs, miRNAs, adult ventricular CM transcriptome and extracts of CMs have been used for direct conversion into iCMs (Fig. 19.1) [53–57]. Some of them produced contractile iCMs. The success rates were lower than that with primed (indirect) cardiac conversion.

Determinants of Efficiency of Direct Conversion Methods

Various technological issues related to direct cardiac reprogramming strategies have been extensively reviewed recently [8].

The feasibility and efficiency of generating functional iCMs by direct conversion is found to be dependent on the fibroblast source, the factors used as well as the method of their delivery [8]. Fibroblasts from large animals and humans were found more resistant to conversion; those from infarct scar are more resistant than healthy fibroblasts. Extra cardiac fibroblasts are refractory to conversion than cardiac fibroblasts.

Viral as well as nonviral vehicles have been tried for direct conversion using cardiac TFs. Efficiency of induction and degree of maturation is varied and dependant on the efficiency of gene transfer and differing profiles of transgene expression [22, 35, 58–61].

Molecular Mechanisms of Direct Conversion into Induced Cardiomyocytes

During the transition from fibroblasts into iCMs, the fibroblast epigenome, transcriptome, proteome, and metabolome are successively reprogrammed. Liu et al. discovered that in mouse fibroblasts which are reprogrammed with GMT, subsequent to epigenetic alterations and changes in gene expression, there is aggregation of active histone marks at CM gene loci, which aids early expression of CM genes, associated with decrease in repressive histone marks at fibroblast loci and silencing of fibroblast genes [62].

Studies using single cell–based global transcriptome profiling have revealed four stages of reprogramming: starting fibroblasts, induced fibroblasts, pre-iCMs, and iCMs [63]. Progression along those stages strongly correlates with GMT expression but is inversely related to cell proliferation. Based on population-based global proteomic studies, Sauls et al. observed temporal changes in the expression of extra cellular matrix (ECM), chromatin, translation, and nucleic acid binding proteins during reprogramming [64]. The key determinants and the specific steps during fibroblast-to-iCM conversion are still unclear.

Indirect Conversion into Induced Cardiomyocytes

Indirect conversion of fibroblasts into iCMs is also known as primed conversion. Efe et al. were the first to record primed cardiac conversion in mouse fibroblasts [65]. The multistage protocol consists of an initial short exposure to pluripotency factors to erase fibroblast identity. This is followed by generation of a dedifferentiated metastable intermediate. Subsequently, differentiation into CMs is initiated by exposure to cardiogenic signalling molecules (Fig. 19.1). Three studies, whose primary goal was to generate iPSCs, surprisingly found beating cells; these studies later switched focus to optimize the yield of iCMs [45, 66, 67]. The iCMs produced by primed conversion, expressed cardiac genes, had ultrastructural features and epigenetic signatures similar to CMs, expressed sarcomeric proteins, had calcium transients and action potentials. They also responded to β-adrenergic and muscarinic receptor agonists. ECM identity seems to have a significant role in primed conversion [68–70].

Wang et al. substituted 3 of the 4 pluripotent TFs with small molecules and showed that, with the SCPF cocktail (SB431542, CHIR99021, parnate, forskolin), only OCT4 was necessary to produce contractile iCMs [46]. When OCT4 was excluded, mouse embryonic fibroblasts were turned into skeletal, and not cardiac muscle-like cells. Park et al. and Fu et al. established the chemical means for primed cardiac conversion of mouse fibroblasts by substituting viral OSKM expression with a chemical cocktail of several small-molecules [45, 66]. The beating clusters of CMs however did not improve with these chemical cocktails.

Ghazizadeh et al. succeeded in programming fibroblasts from human foetal and adult dermis into iCMs with OSKM transactivator of transcription (TAT) proteins and further treatment with chemicals such as CHIR99021, IWP2, SB431542, and purmorphamine [71]. Cao et al. used a cocktail of small-molecules such as CHIR99021, A8301, BIX01294, AS8351, SC1, Y27632, OAC2, SU16F, and JNJ10198409 (9c) and converted dermal fibroblasts from human neonates (HNDFs) into clusters of aMHC-GFP+ beating iCMs. Among these cells, 97% were contractile, 98% expressed multiple cardiac proteins and 100% were of ventricular subtype [72].

Transient pluripotency during primed cardiac conversion is a concern with respect to safety in clinical use [73, 74].

In Vivo Generation of Induced Cardiomyocytes

Given the large pool of proliferating fibroblasts in the hearts during repair following injury, these cells have been identified as targets for in vivo reprogramming through gene transfer. Qian et al. were the pioneers in these efforts. They showed that direct cardiac conversion is possible by retroviral GMT delivery into peri-infarct regions in mouse hearts [6]. Though cardiac performance was improved, an adverse outcome was arrhythmias in the animals which received a viral vector. Song et al. generated contractile iCMs by injecting GMTH ReVs into infarcted mouse hearts [5]. Jayawardena et al. used lentiviral delivery of miR-1, miR-133a, miR-208a, and miR-499 to convert scar fibroblasts into iCMs [54, 75]. Huang et al. achieved in vivo reprogramming using a chemical cocktail of CRFVPT and rolipram (CRFVPTM) [76].

Strategies explored for increasing the efficiency of in vivo reprogramming include increasing the target pool of fibroblasts [54], improving scar vascularization [77], ensuring the efficient transfer of multiple genes [78], considering the ratios of expression levels of factors [79], manipulation of intracellular signalling pathways using small molecules [34] and use of nonintegrating vectors [59, 60].

Generation of Induced Cardiac Progenitor Cells

Low efficiencies of iCM conversion and mitotic quiescence of CMs have prompted the efforts to generate proliferative multipotent inducible cardiac progenitor cells (iCPCs). Not only contractile CMs, but also endothelial and smooth muscle cells can be derived from iCPCs.

The enforced expression of cardiac TFs can accelerate the differentiation of pluripotent stem cells (PSCs) into cardiovascular progenitor cells (CPCs) or CMs [80–83] as well as differentiation of CPCs into CMs [84, 85].

Direct Reprogramming to Cardiac Progenitor Cells

Direct reprogramming has been employed to generate expandable cardiac progenitor cells (CPCs) as well. CPCs are produced and expanded in vitro before transplantation into the damaged heart (Fig. 19.1).

Lalit et al. have produced iCPCs with expansion potential and stable tripotent CPCs from mouse adult cardiac fibroblasts [86]. They generated CPCs from fibroblasts using Mesp1, Tbx5, Gata4, Nkx2–5, and Baf60c, a chromatin remodelling protein [86, 87]. BIO, a Wnt activator, and LIF, a JAK/STAT activator were used for expansion and maintenance of CPCs [86, 87].

Islas et al. converted human neonatal dermal fibroblasts (HNDFs) into NKX2.5-tdTomato+KDR+iCPCs in vitro by lentiviral overexpression of MESP1 and ETS2 (ETS proto-oncogene 2, transcription factor) as well by repeated transduction with corresponding TAT proteins and subsequent exposure to activin A and BMP2 [31]. The iCPC phenotype was, however unipotent and unstable. Li et al. starting from human adult dermal fibroblasts (HADFs), succeeded in producing tripotent protein-induced CPC (piCPCs) which gave rise to mixed populations of contractile CMs and ECs [88]. They used GMTH proteins, instead of mRNA, and added ascorbic acid, activin A, BMP4, and FGF2. In rats with myocardial infarction, transplanted piCPCs differentiated into CMs and ECs. Cardiac function was improved and scar size was reduced.

Primed Conversion into Induced Cardiac Progenitor Cells

Stable and tripotent induced expandable CPCs (ieCPCs) were obtained from mouse fibroblasts by Zhang et al. by transient OSKM overexpression, followed by cardiac induction through GSK3 inhibition (CHIR99021) and further treatment with BMP4, activin A, CHIR99021, and SU5402 (BACS) [89]. ieCPCs produced were spontaneously beating CMs which responded to pharmacological agents. They had no potential for noncardiogenic fates. The ieCPCs differentiated into CMs, ECs, and SMCs, when transplanted and decreased cardiac remodelling and improved function in mice with myocardial infarction. There was no evidence of teratoma formation. Interestingly, phenotype maintenance by treatment with BACS could be repeated in CPCs derived from tripotent mouse embryonic stem cells.

Xu et al. generated induced cardiospheres (iCSs) from mouse and human fibroblasts in vitro [90]. They attempted retroviral overexpression of Oct4, Sox2, and Klf4 (OSK), and subsequent treatment with BIO and oncostatin-M (OSM). Spheroid aggregates had a mixture of MESP1+, NKX2.5+, or ISL1+ cells which differentiated into CMs, ECs, and SMCs. Human CMs derived from iCS were however not contractile.

Improving Efficiency of Direct Reprogramming

Transcriptomic studies using single cells have revealed the complexity of the reprogramming process and the key role of diverse factors in improving the efficiency of reprogramming [63]. The effects of addition of transcription factors, inhibitors/cytokines, noncoding RNAs and epigenetic modifiers have been investigated in search of strategies to improve the efficiency of reprogramming (Fig. 19.2).

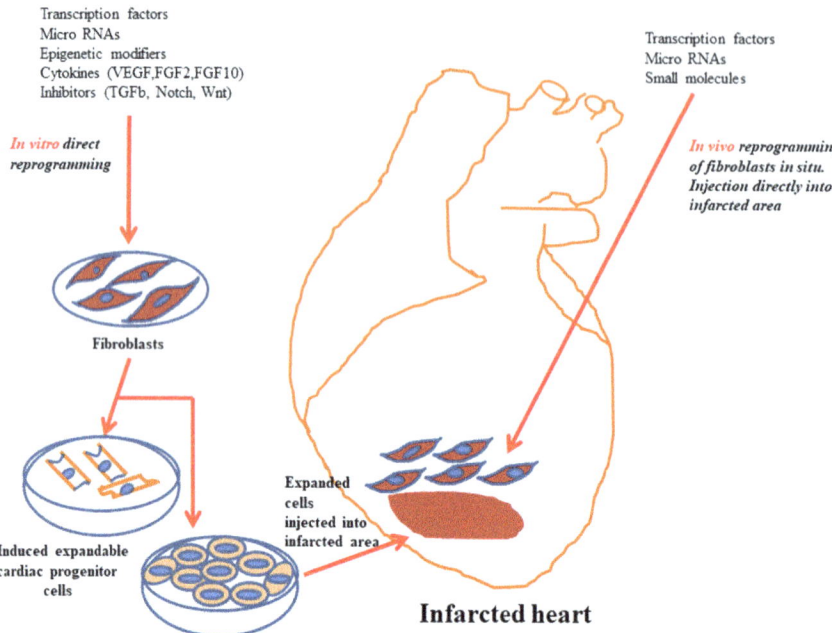

Fig. 19.2 Various approaches used for direct reprogramming of fibroblasts to cardiomyocytes (see text for details)

Transcription Factors

GMT-based conversion of mouse fibroblasts has been improved by the addition of other TFs, such as HAND2, HAND2 and NKX2.5 or MYOCD, SRF (serum response factor), MESP1, and BAF60C [5, 8]. Conversion of human fibroblasts has also been improved by adding MYOCD and ZPFM2 to GMT, ESRRG, and MESP1 [33]. Addition of one more TF, such as GATA6, HAND2, NFATC2 (nuclear factor of activated T cells 2), NKX2.5, RXRG (retinoid X receptor g), SMYD1 (SET and MYND domain containing 1) and TBX20, improved iCM quality, but not the number of cells [91]. In silico Studies have predicted ANKRD1 (ankyrin repeat domain 1), GATA6, HAND2, MEF2A, MYOCD, NKX2.5, PPARGC1A (PPARG coactivator 1a), TBX20, TEAD4 (TEA domain transcription 4) and WT1 (Wilms tumor 1) as candidate TFs which can enhance conversion of fibroblasts to CMs [92–95].

Inhibitors

Transforming growth factor- (TGF-) β pathway through SMADs finally activates TFs that regulate the expressions of genes important for fibroblast activation and proliferation. Given that inhibition of TGF-β enhances differentiation of mouse embryonic stem cells to CMs, TGF-β inhibitors have been added to various reprogramming combinations and found to result in an increase in reprogramming efficiency [96–98]. A fivefold increase in reprogramming efficiency was seen in both mouse embryonic fibroblasts and mouse adult cardiac fibroblasts when GMHT was combined with SB431542, a selective and potent inhibitor of the TGF-β pathway [98]. SB431542 suppresses the activin A receptors ALK5, ALK4, and ALK7. A83-01 is another selective and more potent inhibitor of ALK5, ALK4, and ALK7 than SB431542. When mouse embryonic fibroblasts were reprogrammed with GMT and A83-01, there was enhanced reprogramming efficiency with increased expression of cardiac contractile proteins [3]. Reprogramming efficiency was 8 times more in cardiac fibroblasts when GMT and SB431542 were combined with WNT inhibition, than when GMT was used alone [34].

Signalling Pathways and Growth Factors

Other profibrotic intracellular signalling pathways such as the Rho-associated kinase, JAK/STAT, Notch and Akt pathways have also been studied to find ways to improve reprogramming [35, 36, 54, 99]. Most of the molecules investigated are those known to be important for heart development in mammals and those employed for the differentiation of CMs from pluripotent stem cells.

Fibroblast growth factor-2 (FGF2), FGF10, and vascular endothelial growth factor (VEGF) in combination with GMT or GMHT in mouse embryonic fibroblasts and mouse tail-tip fibroblasts, increase the number of spontaneously beating iCMs [37]. These approaches also accelerate maturation of iCMs in vitro.

Noncoding RNAs

Given their key roles in the posttranscriptional regulation of the expression of cardiac genes during various stages of heart development, miRNAs are an attractive option for reprogramming. miRNA-1, miRNA-133, miRNA-208, and miRNA-499 are known to be cardiac- and muscle specific. miRNA-1 which forms about 40% of total miRNAs in the heart promotes proliferation of CMs and suppress apoptosis. miRNA-133 also promotes proliferation of CMs and silences fibroblast gene signatures during reprogramming [39, 40].

When miRNA-1, miRNA-133, miRNA-208, and miRNA-499 were combined with JAK inhibitor I, there was a tenfold increase in cardiac fibroblast reprogramming [54]. When miRNA-133 was used in combination with GMT, MESP1, and MYOCD or GHT, MYOCD, and miRNA-1, the reprogramming efficiency was improved in both human and mouse fibroblasts [40, 100]. Zhao et al. used a combination of GMHT, miRNA-1, miRNA-133, miRNA208, miRNA-499, Y-27632, and A83-01 in embryonic and adult fibroblasts from mice and obtained about 60% cardiac troponin T+ and 60% α-actinin+ iCMs [35]. HAND2 and MYOCD can be substituted with miRNA-590, known to induce proliferation of adult CMs in direct reprogramming experiments using GMT in human and porcine fibroblasts [41, 44].

A combination of miRNA-1, miRNA-133, miRNA208, and miRNA-499 without other factors has been found to convert mouse cardiac fibroblasts into iCMs in vivo [75]. This effect is thought to be mediated by alterations in the epigenetic machinery of the fibroblasts resulting from changes in the expression of H3K27 methyltransferase and demethylase.

Modified mRNAs

Modified mRNAs (modRNAs) are being investigated for direct cardiac reprogramming [101]. They are noncytopathic, do not integrate into the host genome, have short half-life and thus are safer than viral vectors. ModRNAs mediates dose titratable expression of key TFs over a defined time and in a specific sequence.

Long Noncoding RNAs

It is as yet unclear whether long noncoding RNAs (lncRNAs) can be used for direct reprogramming. Several lncRNAs are involved in cardiomyocyte differentiation, heart development and contractile function [102]. Braveheart (Bvht) and Fendrr activate important cardiac development genes and TFs and thus control the transition from mesoderm to cardiac progenitor cells [103–105]. Hotair, Chaer, and other lncRNAs regulate proteins involved in histone modification at targeted promoters and thus influence the epigenetic landscape during development of the heart [106, 107].

Epigenetic Modifiers

There have been attempts to improve direct cardiac reprogramming using modulators of chromatin environment at cardiomyocyte-specific gene loci [48, 108, 109]. Epigenetic landscape influences reprogramming efficiency as it determines accessibility of TFs to their DNA targets. Epigenetic modifications remodel chromatin structure, modulate the access of TFs to their target genes and can result in the suppression of the expression of fibroblast genes and activation of the cardiomyocyte genes.

During cardiac direct reprogramming, the trimethylated histone H3 of lysine 27 (H3K27me3), a mark of inactivated chromatin, increases at fibroblast promoters and decreases at cardiac promoters. An opposite pattern at important loci is seen at the activated chromatin mark H3K4me3 [22, 62]. The activating H3K4me2 histone mark increases at the regulatory regions of miRNA-1 and miRNA-133 [35]. When Bmi1 activity was knocked down, there was increase in the active histone mark, H3K4me3, and reduction in the repressive H2AK119ub mark, resulting in increased cardiac gene expression at major loci [108].

Epigenetic landscape can also be modulated indirectly via WNT or NOTCH signalling inhibition or via 3D topographic cues [34, 99, 110, 111]. to achieve enhanced reprogramming efficiency.

Small Molecule Cocktails

Small-molecule cocktails have been identified for use in primed conversion by only chemical means [66, 72]. A worry is that as small molecules can enter the blood stream. They may reach organs other than heart and cause unwanted effects.

Compared to TFs and miRNAs, small molecules can be delivered more effectively into cells, are nonimmunogenic, less costly and are generally safer. Reprogramming can be modulated in vitro by altering small molecule dosages and blends. A mixture of ascorbic acid, RepSox (a TGF-β inhibitor), forskolin, valproic acid, and CHIR99021 (a WNT pathway activator) has been noted to reprogram mouse embryonic fibroblasts and mouse tail-tip fibroblasts to iCMs in vitro [66]. Cao et al. used a combination of 9 small molecules (CHIR99021, A83-01, BIX01294, AS8351, SC1, OAC2, Y27632, SU16F, and JNJ10198409) to direct cardiac reprogramming of fibroblasts from human foreskin [72]. Among the small molecules they used, BIX01294 (a methyltransferase inhibitor) and AS8351 (a histone demethylase inhibitor) are epigenetic modifiers. SC1 is an ERK2 and Ras-GAP inhibitor and OAC2 is an Oct4 activator. SU16F is a PDGFRβ inhibitor and JNJ10198409 is a PDGF receptor tyrosine kinase inhibitor.

Transdifferentiation *via* Endothelial Cells

Fibroblasts can also be turned into endothelial cells (ECs) (42–45). They in turn can give rise to beating CMs [112–116]. which can also be converted into induced pacemaker myocytes [117–120].

Altering Fibroblast Transcriptional Signature

Observations made in studies of GMT-based and GMTH-based iCM conversion had indicated potential targets for efficient cell fate reprogramming by silencing the transcriptional signature of the starting cell [35, 62]. The targeted knock down of fibroblast-specific TFs such as TWIST2 (twist family bHLH TF 2), OSR1 (oddskipped related TF 1), PRRX1 (paired related homeobox 1), and LHX9 (LIM homeobox 9) have been tested to see whether that would enhance iCM reprogramming [121].

GMTH-based conversion of mouse fibroblasts has been enhanced by modulating intracellular signalling cascades such as NOTCH signalling [99] and by AKT serine/threonine kinase 1 (AKT1) overexpression [36]. Along with NOTCH inhibition, AKT1 was found to further enhance cardiac reprogramming [99]. The concerns are that long-term AKT overexpression may cause pathologic hypertrophy [122] and that NOTCH signalling inhibition may affect iCM survival by influencing postinfarct neovascularization [123].

In silico predictions have identified PRKD1 (protein kinase D1), PKA, CAMK (calmodulin dependent protein kinases), and PKC as potential enhancers of GMT-based conversion [95]. Janus kinase inhibitor 1 (JI1) has also been found useful in direct cardiac conversion of mouse and human fibroblasts and direct cardiac conversion induced by cardiac TFs, miRNA or a combination of both [30, 40, 43]. Activation of both WNT and NRG1 (neuregulin-1) signalling produces positive effects on iCM induction in human fibroblasts [43].

Repressing Other Cell Fates

iCM induction and maturation compete with the program for adipogenesis [98, 124]. Small molecule inhibition of histone H3K4 methyltransferase MLL1 (mixed-lineage leukemia 1) and its related factor MEN1 (menin) suppresses Ebf1-dependent adipocyte formation but enhance GMT based iCM conversion in mouse fibroblasts.

Conclusion

Despite the current limitations, reprogramming of fibroblasts into cardiomyocyte-like cells and expandable CPCs using transcription factors, small molecules, noncoding RNAs, and other molecules offers great promise for cardiac regeneration. While GMT was initially shown to be sufficient for cardiac reprogramming, further studies have indicated that a multiprong approach may enhance reprogramming efficiency. Continued investigations on key transcription factors, noncoding RNAs, small molecules, reprogramming mechanisms as well as methods for delivery and targeting can advance direct cardiac reprogramming strategies for the treatment of heart failure. For in vivo applications of the reprogramming strategy, relatively safe methods for the delivery of various factors as well as methods for the temporal control and stoichiometric control of TFs need to be explored. Currently available tools for in vivo reprogramming using viral vectors are unable to control dosage and temporal expression of TFs.

References

1. Später D, Hansson EM, Zangi L, Chien KR. How to make a cardiomyocyte? Development. 2014;141:4418–31.
2. Kaur S, Kaur I, Kartha CC. The emerging role of cardiac stem cells in cardiac regeneration. In: Pandey T (editor). Imaging in Stem Cell Transplant and Cell-based Therapy. Humana Press, Cham: 2017. https://doi.org/10.1007/978-3-319-51833-6_73.
3. Engel JL and Ardehali R. Direct cardiac reprogramming: Progress and promise. Stem Cells Int. 2018; Article ID 1435746, 10 pages, https://doi.org/10.1155/2018/1435746.
4. He L, Nguyen NB, Ardehali R, Zhou B. Heart Regeneration by endogenous stem cells and cardiomyocyte proliferation: Controversy, Fallacy, and Progress. Circulation. 2020;142:275–91.
5. Song K, Nam YJ, Luo X, Qi X, Tan W, Huang GN, Acharya A, Smith CL, Tallquist MD, Neilson EG, et al. Heart repair by reprogramming non-myocytes with cardiac transcription factors. Nature. 2012;485:599–604.
6. Qian L, Huang Y, Spencer CI, Foley A, Vedantham V, Liu L, Conway SJ, Fu JD, Srivastava D. In vivo reprogramming of murine cardiac fibroblasts into induced cardiomyocytes. Nature. 2012;485:593–8.
7. Huang C, Tu W, Fu Y, Wang J, Xie X. Chemical-induced cardiac reprogramming in vivo. Cell Res. 2018;28:686–9.

8. Klose K, Gossen M, Stamm C. Turning fibroblasts into cardiomyocytes: technological review of cardiac transdifferentiation strategies. FASEB J. 2019;33:49–70.
9. Gurdon JB. The developmental capacity of nuclei taken from intestinal epithelium cells of feeding tadpoles. J Embryol Exp Morphol. 1962;10:622–40.
10. Campbell KH, McWhir J, Ritchie WA, Wilmut I. Sheep cloned by nuclear transfer from a cultured cell line. Nature. 1996;380:64–6.
11. Miller RA, Ruddle FH. Pluripotent teratocarcinoma thymus somatic cell hybrids. Cell. 1976;9:45–55.
12. Cho HJ, Lee CS, Kwon YW, Paek JS, Lee SH, Hur J, Lee EJ, Roh TY, Chu IS, Leem SH, Kim Y, Kang HJ. Park YB, Kim HS. Induction of pluripotent stem cells from adult somatic cells by protein-based reprogramming without genetic manipulation. Blood. 2010;116:386–95.
13. Takahashi K, Yamanaka S. Induction of pluripotent stem cells from mouse embryonic and adult fibroblast cultures by defined factors. Cell. 2006;126:663–76.
14. Takahashi K, Tanabe K, Ohnuki M, Narita M, Ichisaka T, Tomoda K, Yamanaka S. Induction of pluripotent stem cells from adult human fibroblasts by defined factors. Cell. 2007;131:861–72.
15. Davis RL, Weintraub H, Lassar AB. Expression of a single transfected cDNA converts fibroblasts to myoblasts. Cell. 1987;51:987–1000.
16. Murry CE, Kay MA, Bartosek T, Hauschka SD, Schwartz SM. Muscle differentiation during repair of myocardial necrosis in rats via gene transfer with MyoD. J Clin Invest. 1996;98:2209–17.
17. Eghbali M, Tomek R, Woods C, Bhambi B. Cardiac fibroblasts are predisposed to convert into myocyte phenotype: specific effect of transforming growth factor beta. Proc Natl Acad Sci USA. 1991;88:795–9.
18. Genovese JA, Spadaccio C, Langer J, Habe J, Jackson J, Patel AN. Electrostimulation induces cardiomyocyte predifferentiation of fibroblasts. Biochem Biophys Res Commun. 2008;370:450–5.
19. Zaglia T, Dedja A, Candiotto C, Cozzi E, Schiaffino S, Ausoni S. Cardiac interstitial cells express GATA4 and control dedifferentiation and cell cycle re-entry of adult cardiomyocytes. J Mol Cell Cardiol. 2009;46:653–62.
20. Van Tuyn J, Pijnappels DA, de Vries AA, de Vries I, van der Velde-van DI, Knaan-Shanzer S, van der Laarse A, Schalij MJ, Atsma DE. Fibroblasts from human postmyocardial infarction scars acquire properties of cardiomyocytes after transduction with a recombinant myocardin gene. FASEB J. 2007;21:3369–79.
21. Takeuchi JK, Bruneau BG. Directed transdifferentiation of mouse mesoderm to heart tissue by defined factors. Nature. 2009;459:708–11.
22. Ieda M, Fu JD, Delgado-Olguin P, Vedantham V, Hayashi Y, Bruneau BG, Srivastava D. Direct reprogramming of fibroblasts into functional cardiomyocytes by defined factors. Cell. 2010;142:375–86.
23. Hiroi Y, Kudoh S, Monzen K, et al. Tbx5 associates with Nkx2-5 and synergistically promotes cardiomyocyte differentiation. Nat Genet. 2001;28:276–80.
24. Perrino C, Rockman HA. GATA4 and the two sides of gene expression reprogramming. Circ Res. 2006;98:715–6.
25. Dodou E, Verzi MP, Anderson JP, Xu SM, Black BL. Mef2c is a direct transcriptional target of ISL1 and GATA factors in the anterior heart field during mouse embryonic development. Development. 2004;131:3931–42.
26. Nam YJ, Lubczyk C, Bhakta M, et al. Induction of diverse cardiac cell types by reprogramming fibroblasts with cardiac transcription factors. Development. 2014;141:4267–78.
27. Hirai H, Katoku-Kikyo N, Keirstead SA, Kikyo N. Accelerated direct reprogramming of fibroblasts into cardiomyocyte-like cells with the MyoD transactivation domain. Cardiovasc Res. 2013;100:105–13.
28. Addis RC, Ifkovits JL, Pinto F, et al. Optimization of direct fibroblast reprogramming to cardiomyocytes using calcium activity as a functional measure of success. J Mol Cell Cardiol. 2013;60:97–106.

29. Protze S, Khattak S, Poulet C, Lindemann D, Tanaka EM, Ravens U. A new approach to transcription factor screening for reprogramming of fibroblasts to cardiomyocyte-like cells. J Mol Cell Cardiol. 2012;53:323–32.
30. Christoforou N, Chellappan M, Adler AF, et al. Transcription factors MYOCD, SRF, Mesp1 and SMARCD3 enhance the cardio-inducing effect of GATA4, TBX5, and MEF2C during direct cellular reprogramming. PLoS One. 2013; 8:5 article e63577.
31. Islas JF, Liu Y, Weng KC, et al. Transcription factors ETS2 and MESP1 transdifferentiate human dermal fibroblasts into cardiac progenitors. Proc Natl Acad Sci USA. 2012;109:13016–21.
32. Wada R, Muraoka N, Inagawa K, et al. Induction of human cardiomyocyte-like cells from fibroblasts by defined factors. Proc Natl Acad Sci USA. 2013;110:12667–72.
33. Fu JD, Stone NR, Liu L, et al. Direct reprogramming of human fibroblasts toward a cardiomyocyte-like state. Stem Cell Reports. 2013;1:235–47.
34. Mohamed TM, Stone NR, Berry EC, et al. Chemical enhancement of in vitro and in vivo direct cardiac reprogramming. Circulation. 2017;135:978–95.
35. Zhao Y, Londono P, Cao Y, et al. High-efficiency reprogramming of fibroblasts into cardiomyocytes requires suppression of pro-fibrotic signalling. Nat Commun. 2015;6:8243.
36. Zhou H, Dickson ME, Kim MS, Bassel-Duby R, Olson EN. Akt1/protein kinase B enhances transcriptional reprogramming of fibroblasts to functional cardiomyocytes. Proc Natl Acad Sci USA. 2015;112:11864–9.
37. Yamakawa H, Muraoka N, Miyamoto K, et al. Fibroblast growth factors and vascular endothelial growth factor promote cardiac reprogramming under defined conditions. Stem Cell Reports. 2015;5:1128–42.
38. Rao PK, Toyama Y, Chiang HR, et al. Loss of cardiac microRNA-mediated regulation leads to dilated cardiomyopathy and heart failure. Circ Res. 2009;105:585–94.
39. Liu N, Bezprozvannaya S, Williams AH, et al. MicroRNA133a regulates cardiomyocyte proliferation and suppresses smooth muscle gene expression in the heart. Genes Dev. 2008;22:3242–54.
40. Muraoka N, Yamakawa H, MiyamotoK, et al. MiR-133 promotes cardiac reprogramming by directly repressing Snail and silencing fibroblast signatures. The EMBO Journal. 2014;14:1565–81.
41. Eulalio A, Mano M, Ferro MD, et al. Functional screening identifies miRNAs inducing cardiac regeneration. Nature. 2012;492:376–81.
42. Zhou L, Liu Y, Lu L, Lu X, Dixon RA. Cardiac gene activation analysis in mammalian non-myoblasic cells by Nkx2–5, Tbx5, Gata4 and Myocd. PLoS One. 2012;7:e48028.
43. Christoforou N, Chakraborty S, Kirkton RD, Adler AF, Addis RC, Leong KW. Core transcription factors, microRNAs, and small molecules drive transdifferentiation of human fibroblasts towards the cardiac cell lineage. Sci Rep. 2017;7:40285.
44. Singh VP, Mathison M, Patel V, Sanagasetti D, Gibson BW, Yang J, Rosengart TK. MiR-590 promotes transdifferentiation of porcine and human fibroblasts toward a cardiomyocyte-like fate by directly repressing specificity protein 1. J Am Heart Assoc. 2016;5:e003922.
45. Park G, Yoon BS, Kim YS, Choi SC, Moon JH, Kwon S, Hwang J, Yun W, Kim JH, Park CY, Lim DS, Kim YI, Oh CH, You S. Conversion of mouse fibroblasts into cardiomyocyte-like cells using small molecule treatments. Biomaterials. 2015;54:201–12.
46. Wang H, Cao N, Spencer CI, Nie B, Ma T, Xu T, Zhang Y, Wang X, Srivastava D, Ding S. Small molecules enable cardiac reprogramming of mouse fibroblasts with a single factor, Oct4. Cell Rep. 2014;6:951–60.
47. Kamaraj US, Gough J, Polo JM, Petretto E, Rackham OJ. Computational methods for direct cell conversion. Cell Cycle. 2016;15:3343–54.
48. Hirai H, Kikyo N. Inhibitors of suppressive histone modification promote direct reprogramming of fibroblasts to cardiomyocyte-like cells. Cardiovasc Res. 2014;102:188–90.
49. Chang Y, Guo K, Li Q, Li C, Guo Z, Li H. Multiple directional differentiation difference of neonatal rat fibroblasts from six organs. Cell Physiol Biochem. 2016;39:157–71.

50. Tang CC, Ma GS, Chen JY. Transplantation of 5- azacytidine treated cardiac fibroblasts improves cardiac function of infarct hearts in rats. Chin Med J (Engl.). 2010;123:2586–92.
51. Wang Y, Sun A, Xue J, Jiang Y. Adenovirus-mediated expression of hypoxia-inducible factor 1a double mutant converts neonatal cardiac fibroblasts into (cardio)myocyte phenotype. Cell Biochem Funct. 2012;30:24–32.
52. Maioli M, Rinaldi S, Santaniello S, Castagna A, Pigliaru G, Gualini S, Cavallini C, Fontani V, Ventura C. Radio electric conveyed fields directly reprogram human dermal skin fibroblasts toward cardiac, neuronal, and skeletal muscle-like lineages. Cell Transplant. 2013;22:1227–35.
53. Chen Y, Yang Z, Zhao Z, Shen Z. Direct reprogramming of fibroblasts into cardiomyocytes. Stem Cell Res Ther. 2017;8:118.
54. Jayawardena TM, Egemnazarov B, Finch EA, Zhang L, Payne JA, Pandya K, Zhang Z, Rosenberg P, Mirotsou M, Dzau VJ. MicroRNA-mediated in vitro and in vivo direct reprogramming of cardiac fibroblasts to cardiomyocytes. Circ Res. 2012;110:1465–73.
55. Kim H, Kim D, Ku SH, Kim K, Kim SH, Kwon IC. MicroRNA-mediated non-viral direct conversion of embryonic fibroblasts to cardiomyocytes: comparison of commercial and synthetic non-viral vectors. J Biomater Sci Polym Ed. 2017;28:1070–85.
56. Kim TK, Sul JY, Peternko NB, Lee JH, Lee M, Patel VV, Kim J, Eberwine JH. Transcriptome transfer provides a model for understanding the phenotype of cardiomyocytes. Proc Natl Acad Sci USA. 2011;108:11918–23.
57. Talaei-Khozani T, Heidari F, Esmaeilpour T, Vojdani Z, Mostafavi-Pour Z, Rohani L. Cardiomyocyte marker expression in mouse embryonic fibroblasts by cell-free cardiomyocyte extract and epigenetic manipulation. Iran J Med Sci. 2014;39(Suppl 2):203–12.
58. Umei TC, Yamakawa H, Muraoka N, Sadahiro T, Isomi M, Haginiwa S, Kojima H, Kurotsu S, Tamura F, Osakabe R, Tani H, Nara K, Miyoshi H, Fukuda K, Ieda M. Single construct polycistronic doxycycline-inducible vectors improve direct cardiac reprogramming and can be used to identify the critical timing of transgene expression. Int J Mol Sci. 2017;18:1805.
59. Mathison M, Singh VP, Chiuchiolo MJ, Sanagasetti D, Mao Y, Patel VB, Yang J, Kaminsky SM, Crystal RG, Rosengart TK. In situ reprogramming to transdifferentiate fibroblasts into cardiomyocytes using adenoviral vectors: implications for clinical myocardial regeneration. J Thorac Cardiovasc Surg. 2017;153:329–39.
60. Miyamoto K, Akiyama M, Tamura F, Isomi M, Yamakawa H, Sadahiro T, Muraoka N, Kojima H, Haginiwa S, Kurotsu S, Tani H, Wang L, Qian L, Inoue M, Ide Y, Kurokawa J, Yamamoto T, Seki T, Aeba R, Yamagishi H, Fukuda K, Ieda M. Direct in vivo reprogramming with Sendai virus vectors improves cardiac function after myocardial infarction. Cell Stem Cell. 2018;22:91–103.
61. Lee K, Yu P, Lingampalli N, Kim HJ, Tang R, Murthy N. Peptide-enhanced mRNA transfection in cultured mouse cardiac fibroblasts and direct reprogramming XE "Direct reprogramming" towards cardiomyocyte like cells. Int J Nanomedicine. 2015;10:1841–54.
62. Liu Z, Chen O, Zheng M, Wang L, Zhou Y, Yin C, Liu J, Qian L. Re-patterning of H3K27me3, H3K4me3 and DNA methylation during fibroblast conversion into induced cardiomyocytes. Stem Cell Res (Amst.). 2016;16:507–18.
63. Liu Z, Wang L, Welch JD, Ma H, Zhou Y, Vaseghi HR, Yu S, Wall JB, Alimohamadi S, Zheng M, Yin C, Shen W, Prins JF, Liu J, Qian L. Single-cell transcriptomics reconstructs fate conversion from fibroblast to cardiomyocyte. Nature. 2017;551:100–4.
64. Sauls K, Greco TM, Wang L, Zou M, Villasmil M, Qian L, Cristea IM, Conlon FL. Initiating events in direct cardiomyocyte reprogramming. Cell Rep. 2018;22:1913–22.
65. Efe JA, Hilcove S, Kim J, Zhou H, Ouyang K, Wang G, Chen J, Ding S. Conversion of mouse fibroblasts into cardiomyocytes using a direct reprogramming strategy. Nat Cell Biol. 2011;13:215–22.
66. Fu Y, Huang C, Xu X, Gu H, Ye Y, Jiang C, Qiu Z, Xie X. Direct reprogramming of mouse fibroblasts into cardiomyocytes with chemical cocktails. Cell Res. 2015;25:1013–24.

67. Guo Y, Yu T, Lei L, Duan A, Ma X, Wang H. Conversion of goat fibroblasts into lineage-specific cells using a direct reprogramming strategy. Anim Sci J. 2017;88:745–54.
68. Smith AW, Hoyne JD, Nguyen PK, Mc Creedy DA, Aly H, Efimov IR, Rentschler S, Elbert DL. Direct reprogramming of mouse fibroblasts to cardiomyocyte-like cells using Yamanaka factors on engineered poly (ethylene glycol) (PEG) hydrogels. Biomaterials. 2013;34:6559–71.
69. Kong YP, Carrion B, Singh RK, Putnam AJ. Matrix identity and tractional forces influence indirect cardiac reprogramming. Sci Rep. 2013;3:3474.
70. Talkhabi M, Pahlavan S, Aghdami N, Baharvand H. Ascorbic acid promotes the direct conversion of mousefibroblasts into beating cardiomyocytes. Biochem Biophys Res Commun. 2015;463:699–705.
71. Ghazizadeh Z, Rassouli H, Fonoudi H, Alikhani M, Talkhabi M, Darbandi-Azar A, Chen S, Baharvand H, Aghdami N, Salekdeh GH. Transient activation of reprogramming transcription factors using protein transduction facilitates conversion of human fibroblasts toward cardiomyocyte-like cells. Mol Biotechnol. 2017;59:207–20.
72. Cao N, Huang Y, Zheng J, Spencer CI, Zhang Y, Fu JD, Nie B, Xie M, Zhang M, Wang H, Ma T, Xu T, Shi G, Srivastava D, Ding S. Conversion of human fibroblasts into functional cardiomyocytes by small molecules. Science. 2016;352:1216–20.
73. Bar-Nur O, Verheul C, Sommer AG, Brumbaugh J, Schwarz BA, Lipchina I, Huebner AJ, Mostoslavsky G, Hochedlinger K. Lineage conversion induced by pluripotency factors involves transient passage through an iPSC stage. Nat Biotechnol. 2015;33:761–8.
74. Maza I, Caspi I, Zviran A, Chomsky E, Rais Y, Viukov S, Geula S, Buenrostro JD, Weinberger L, Krupalnik V, Hanna S, Zerbib M, Dutton JR, Greenleaf WJ, Massarwa R, Novershtern N, Hanna JH. Transient acquisition of pluripotency during somatic cell transdifferentiation with iPSC reprogramming factors. Nat Biotechnol. 2015;33:769–74.
75. Jayawardena TM, Finch EA, Zhang L, Zhang H, Hodgkinson CP, Pratt RE, Rosenberg PB, Mirotsou M, Dzau VJ. MicroRNA induced cardiac reprogramming in vivo: evidence for mature cardiac myocytes and improved cardiac function. Circ Res. 2015;116:418–24.
76. Huang C, Tu W, Fu Y, Wang J, Xie X. Chemical induced cardiac reprogramming in vivo. Cell Res. 2018;28:686–9.
77. Mathison M, Gersch RP, Nasser A, Lilo S, Korman M, Fourman M, Hackett N, Shroyer K, Yang J, Ma Y, Crystal RG, Rosengart TK. In vivo cardiac cellular reprogramming efficacy is enhanced by angiogenic preconditioning of the infarcted myocardium with vascular endothelial growth factor. J Am Heart Assoc. 2012;1:e005652.
78. Inagawa K, Miyamoto K, Yamakawa H, Muraoka N, Sadahiro T, Umei T, Wada R, Katsumata Y, Kaneda R, Nakade K, Kurihara C, Obata Y, Miyake K, Fukuda K, Ieda M. Induction of cardiomyocyte-like cells in infarct hearts by gene transfer of Gata4, Mef2c, and Tbx5. Circ Res. 2012;111:1147–56.
79. Wang L, Liu Z, Yin C, Asfour H, Chen O, Li Y, Bursac N, Liu J, Qian L. Stoichiometry of Gata4, Mef2c, and Tbx5 influences the efficiency and quality of induced cardiac myocyte reprogramming. Circ Res. 2015;116:237–44.
80. David R, Brenner C, Stieber J, Schwarz F, Brunner S, Vollmer M, Mentele E, Müller-Höcker J, Kitajima S, Lickert H, Rupp R, Franz WM. MesP1 drives vertebrate cardiovascular differentiation through Dkk-1-mediated blockade of Wnt-signalling. Nat Cell Biol. 2008;10:338–45.
81. Bai F, Ho Lim C, Jia J Santostefano, K, Simmons C, Kasahara H, Wu W, Terada N, Jin S. Directed differentiation of embryonic stem cells into cardiomyocytes by bacterial injection of defined transcription factors. Sci Rep. 2015;5:15014.
82. David R, Stieber J, Fischer E, Brunner S, Brenner C, Pfeiler S, Schwarz F, Franz WM. Forward programming of pluripotent stem cells towards distinct cardiovascular cell types. Cardiovasc Res. 2009;84:263–72.
83. Hartung S, Schwanke K, Haase A, David R, Franz WM, Martin U, Zweigerdt R. Directing cardiomyogenic differentiation of human pluripotent stem cells by plasmid-based transient overexpression of cardiac transcription factors. Stem Cells Dev. 2013;22:1112–25.

84. Belian E, Noseda M, Abreu Paiva MS, Leja T, Sampson R, Schneider MD. Forward programming of cardiac stem cells by homogeneous transduction with MYOCD plus TBX5. PLoS One. 2015;10:e0125384.

85. Morez C, Noseda M, Paiva MA, Belian E, Schneider MD, Stevens MM. Enhanced efficiency of genetic programming toward cardiomyocyte creation through topographical cues. Biomaterials. 2015;70:94–104.

86. Lalit PA, Salick MR, Nelson DO, et al. Lineage reprogramming of fibroblasts into proliferative induced cardiac progenitor cells by defined factors. Cell Stem Cell. 2016;18:354–67.

87. Lalit PA, Rodriguez AM, Downs KM, Kamp TJ. Generation of multipotent induced cardiac progenitor cells from mouse fibroblasts and potency testing in ex vivo mouse embryos. Nat Protoc. 2017;12:1029–54.

88. Li XH, Li Q, Jiang L, Deng C, Liu Z Fu Y, Zhang M, Tan H, Feng Y, Shan Z, Wang J, Yu XY. Generation of functional human cardiac progenitor cells by high-efficiency protein transduction. Stem Cells Transl Med. 2015;4:1415–24.

89. Zhang Y, Cao N, Huang Y, et al. Expandable cardiovascular progenitor cells reprogrammed from fibroblasts. Cell Stem Cell. 2016;18:368–81.

90. Xu JY, Lee YK, Ran X, Liao SY, Yang J, Au KW, Lai WH, Esteban MA, Tse HF. Generation of induced cardiospheres via reprogramming of skin fibroblasts for myocardial regeneration. Stem Cells. 2016;34:2693–706.

91. Bektik E, Dennis A, Prasanna P, Madabhushi A, Fu JD. Single cell qPCR reveals that additional HAND2 and microRNA-1 facilitate the early reprogramming progress of seven factor-induced human myocytes. PLoS One. 2017;12:e0183000.

92. Bian Q, Cahan P. Computational tools for stem cell biology. Trends Biotechnol. 2016;34:993–1009.

93. Cahan P, Li H, Morris SA, Lummertz da Rocha E, Daley GQ, Collins JJ. Cell Net: network biology applied to stem cell engineering. Cell. 2014;158:903–15.

94. Rastegar-Pouyani S, Khazaei N, Wee P, Yaqubi M, Mohammadnia A. Meta-analysis of transcriptome regulation during induction to cardiac myocyte fate from mouse and human fibroblasts. J Cell Physiol. 2017;232:2053–62.

95. Talkhabi M, Razavi SM, Salari A. Global transcriptomic analysis of induced cardiomyocytes predicts novel regulators for direct cardiac reprogramming. J Cell Commun Signal. 2017;11:193–204.

96. Ao A, Hao J, Hopkins CR, Hong CC. DMH1, a novel BMP small molecule inhibitor, increases cardiomyocyte progenitors and promotes cardiac differentiation in mouse embryonic stem cells. PLoS One. 2012;7:article e41627, 2012.

97. Cai W, Guzzo RM, Wei K, Willems E, Davidovics H, Mercola M. A nodal-to-TGFβ cascade exerts biphasic control over cardiopoiesis. Circ Res. 2012;111:876–81.

98. Ifkovits JL, Addis RC, Epstein JA, Gearhart JD. Inhibition of TGFβ signaling increases direct conversion of fibroblasts to induced cardiomyocytes. PLoS One. 2014;9:2, article e89678.

99. Abad M, Hashimoto H, Zhou H, et al. Notch inhibition enhances cardiac reprogramming by increasing MEF2C transcriptional activity. Stem Cell Rep. 2017;8:548–60.

100. Nam YJ, Song K, Luo X, et al. Reprogramming of human fibroblasts toward a cardiac fate. Proc Natl Acad Sci USA. 2013;110:5588–93.

101. Warren L, Manos PD, Ahfeldt T, et al. Highly efficient reprogramming to pluripotency and directed differentiation of human cells with synthetic modified mRNA. Cell Stem Cell. 2010;7:618–30.

102. Gomes CPC, Spencer H, Ford KL, et al. The function and therapeutic potential of long non-coding RNAs in cardiovascular development and disease. Molec Therapy Nucleic Acids. 2017;8:494–507.

103. Klattenhoff CA, Scheuermann JC, Surface LE, et al. Braveheart, a long noncoding RNA required for cardiovascular lineage commitment. Cell. 2013;152:570–83.

104. Xue Z, Hennelly S, Doyle B, et al. A G-rich motif in the lncRNA Braveheart interacts with a zinc-finger transcription factor to specify the cardiovascular lineage. Mol Cell. 2016;64:37–50.

105. Grote P, Wittler L, Hendrix D, et al. The tissue-specific lncRNA Fendrr is an essential regulator of heart and body wall development in the mouse. Dev Cell. 2013;24:206–14.
106. Wang Z, Zhang XJ, Ji YX, et al. The long noncoding RNA Chaer defines an epigenetic checkpoint in cardiac hypertrophy. Nat Med. 2016;22:1131–9.
107. Wang Z, Wang Y. Dawn of the epi-LncRNAs: new path from My heart. Circ Res. 2015;116:235–6.
108. Zhou Y, Wang L, Vaseghi HR, Liu Z, Lu R, Alimohamadi S, Yin C, Fu JD, Wang GG, Liu J, Qian L. Bmi1 is a key epigenetic barrier to direct cardiac reprogramming. Cell Stem Cell. 2016;18:382–95.
109. Dal-Pra S, Hodgkinson CP, Mirotsou M, Kirste I, Dzau VJ. Demethylation of H3K27 is essential for the induction of direct cardiac reprogramming by miR combo. Circ Res. 2017;120:1403–13.
110. Sia J, Yu P, Srivastava D, Li S. Effect of biophysical cues on reprogramming to cardiomyocytes. Biomaterials. 2016;103:1–11.
111. Yoo J, Chang Y, Kim H, Baek S, Choi H, Jeong GJ, Shin J, Kim H, Kim BS, Kim J. Efficient direct lineage reprogramming of fibroblasts into induced cardiomyocytes using nanotopographical cues. J Biomed Nanotechnol. 2017;13:269–79.
112. Lee S, Park C, Han JW, Kim JY, Cho K, Kim EJ, Kim S, Lee SJ, Oh SY, Tanaka Y, Park IH, An HJ, Shin CM, Sharma S, Yoon YS. Direct reprogramming of human dermal fibroblasts into endothelial cells using ER71/ETV2. Circ Res. 2017;120:848–61.
113. Margariti A, Winkler B, Karamariti E, Zampetaki A, Tsai TN, Baban D, Ragoussis J, Huang Y, Han JD, Zeng L, Hu Y, Xu Q. Direct reprogramming of fibroblasts into endothelial cells capable of angiogenesis and reendothelialization in tissue engineered vessels. Proc Natl Acad Sci USA. 2012;109:13793–8.
114. Sayed N, Wong WT, Ospino F, Meng S, Lee J, Jha A, Dexheimer P, Aronow BJ, Cooke JP. Transdifferentiation of human fibroblasts to endothelial cells: role of innate immunity. Circulation. 2015;131:300–9.
115. Wong WT, Cooke JP. Therapeutic transdifferentiation of human fibroblasts into endothelial cells using forced expression of lineage-specific transcription factors. J Tissue Eng. 2016;7:2041731416628329.
116. Condorelli G, Borello U, De Angelis L, Latronico M, Sirabella D, Coletta M, Galli R, Balconi G, Follenzi A, Frati G, Cusella De Angelis MG, Gioglio L, Amuchastegui S, Adorini L, Naldini L, Vescovi A, Dejana E, Cossu G. Cardiomyocytes induce endothelial cells to trans-differentiate into cardiac muscle: Implications for myocardium regeneration. Proc Natl Acad Sci USA. 2001;98:10733–738.
117. Bakker ML, Boink GJ, Boukens BJ, Verkerk AO, van den Boogaard M, den Haan AD, Hoogaars WM, Buermans HP, de Bakker JM, Seppen J, Tan HL, Moorman AF, 't Hoen PA, Christoffels VM. T-box transcription factor TBX3 reprogrammes mature cardiac myocytes into pacemaker-like cells. Cardiovasc Res. 2012;94:439–49.
118. Hu YF, Dawkins JF, Cho HC, Marb´an E, Cingolani E. Biological pacemaker created by minimally invasive somatic reprogramming in pigs with complete heart block. Sci Transl Med. 2014;6:245ra94.
119. Kapoor N, Liang W, Marban E, Cho HC. Direct conversion of quiescent cardiomyocytes to pacemaker cells by expression of Tbx18. Nat Biotechnol. 2013;31:54–62.
120. Rentschler S, Yen AH, Lu J, Petrenko NB, Lu MM, Manderfield LJ, Patel VV, Fishman GI, Epstein JA. Myocardial Notch signaling reprograms cardiomyocytes to a conduction-like phenotype. Circulation. 2012;126:1058–66.
121. Tomaru Y, Hasegawa R, Suzuki T, Sato T, Kubosaki A, Suzuki M, Kawaji H, Forrest A R, Hayashizaki Y, Shin JW, Suzuki H; FANTOM Consortium. A transient disruption of fibroblastic transcriptional regulatory network facilitates transdifferentiation. Nucleic Acids Res. 2014;42:8905–13.
122. O'Neill BT, Abel ED. Akt1 in the cardiovascular system: friend or foe? J Clin Invest. 2005;115:2059–64.

123. Li Y, Hiroi Y, Liao JK. Notch signaling as an important mediator of cardiac repair and regeneration after myocardial infarction. Trends Cardiovasc Med. 2010;20:228–31.
124. Liu L, Lei I, Karatas H, Li Y, Wang L, Gnatovskiy L, Dou Y, Wang S, Qian L, Wang Z. Targeting Mll1 H3K4 methyltransferase activity to guide cardiac lineage specific reprogramming of fibroblasts. Cell Discov. 2016;2:16036.

Part V
Translational Aspects of Cardiomyocyte Biology

Chapter 20
Clinical Translation of Discoveries in Cardiomyocyte Biology

Abstract During the last three decades, there has been significant progress in our understanding of the cellular and molecular basis of normal functions in cardiomyocytes as well as the alterations in the homeostatic mechanisms that lead to pathological states in these cells. Thanks to these advances in our knowledge, there has been great gains in attempts to identify disease biomarkers and therapeutic targets for several heart diseases. Biomarkers are indicators of structural and functional changes in pathological conditions. They are useful to detect disease associated changes in tissues or body fluids. Thus, they are important tools for disease diagnosis, assessing severity of the disease, identifying possible treatment strategies and predicting prognosis. Several biochemical, molecular and genetic biomarkers of heart diseases are in clinical use or under evaluation. Novel therapeutic targets and strategies which enhance clinical efficacy in patients with heart diseases are also currently available. The treatment approaches under trial for ischemic heart disease and heart failure include the use of molecules such as micro RNAs, transcription factors and growth factors, cell therapy, gene therapy, epigenetic therapy, modulation of cardiomyocyte death, metabolic therapies and senolytics. CRISPR-Cas9-based gene editing technology is gaining much attention for its potential therapeutic use in congenital heart defects.

Keywords Heart diseases · Biomarkers · Cell therapy · Gene therapy · Cell death modulators · Senolytics · Metabolic therapy · Gene editing

Introduction

Thanks to the significant advances in our knowledge pertaining to the cellular and molecular mechanisms of normal functions in cardiomyocytes (CMs) as well as the alterations in the homeostatic mechanisms lead to pathological states in them, there has been much progress in attempts to identify disease biomarkers and therapeutic targets for several heart diseases. Several biochemical, molecular and genetic biomarkers of heart diseases are in clinical use or are under evaluation. Novel therapeutic targets and strategies which enhance clinical efficacy in patients with heart diseases are also currently available. Many more are under clinical trial.

© The Author(s), under exclusive license to Springer Nature Switzerland AG 2021   315
C. C. Kartha, *Cardiomyocytes in Health and Disease*,
https://doi.org/10.1007/978-3-030-85536-9_20

Biomarkers:
Ischemic Heart Disease

Cardiac Troponins

All three isoforms of the muscle protein complex troponin are found in CMs. Cardiac troponin T (cTnT) and cardiac troponin I (cTnI) isoforms are specific for the heart muscle [1]. Cardiomyocyte necrosis releases troponin I or T into circulation. They are useful to detect myocardial ischemia. Cardiac troponins were included as indicators of cardiomyocyte injury in acute myocardial infarction (AMI) two decades ago and are currently part of the diagnostic work-up of patients with suspected acute coronary syndromes (ACS) [2, 3]. Troponin is also used for risk stratification in patients with suspected ACS.

There is some data which indicate that measurement of cardiac troponin using high sensitivity assays could be helpful to follow the effect of statin therapy in reducing cardiovascular risk [4]. cTnI and cTnT levels in blood strongly correlate with the risk of development of heart failure (HF) and deaths in patients with ischemic coronary artery disease [5–7].

Heart-Type Fatty Acid-Binding Protein

Heart-type fatty acid-binding protein (hFABP), responsible for the transportation of fatty acids through membranes is most commonly found in the heart. The protein is released into the circulation following acute myocardial ischemia. hFABP levels in serum may provide additional information in the early diagnosis of AMI [8]. When combined with measurements of cTnI and cTnT, hFABP level is considered useful to rule out AMI [9]. Clinical significance is greatest in those who present with chest pain of less than 6 h duration, as hFABP has a rapid clearance rate from blood. Measurement of hFABP may also be useful to distinguish AMI from conditions such as renal or heart failure with chronically elevated levels of cardiac troponin [10]. Serum levels of hFABP are predictive of adverse events in patients with ACS [11, 12], as well as in patients with suspected ACS and normal levels of cardiac troponin [13].

Cardiac Myosin Binding Protein-C

Several studies have suggested that cardiac myosin binding protein-C (c-MyC) is a sensitive marker of cardiomyocyte necrosis. After injury, the plasma levels of c-MyC may rise quicker than the levels of cardiac troponin [14]. The rapid increases in plasma levels of c-MyC after AMI have prompted the use of c-MyC levels to exclude suspected AMI [15, 16].

Growth Differentiation Factor 15

Growth differentiation factor 15 (GDF-15), which is part of the transforming growth factor-β cytokine superfamily is expressed in various tissues including the myocardium. A rise in GDF-15 levels indicates oxidative stress and inflammation [17]. The levels of GDF-15 increase in AMI [18] and acute HF [19], which are associated with cardiomyocyte stress [20]. Upregulation of GDF-15 expression is possibly a defense mechanism in both acute and chronic injury to the heart [20–22]. While GDF-15 is inferior to natriuretic peptides and cardiac troponins for the diagnosis of AMI, GDF-15 is a strong predictor of adverse events in patients with ACS [23] and HF [19, 24].

Secreted Frizzled Related Proteins

Secreted frizzled related proteins (sFRPs) are secreted at the early stages of myocardial infarction (MI) and function as Wnt antagonists [25]. When the Wnt pathway is antagonized by sFRP3, a pro-apoptotic pathway typical of MI and HF is activated [25]. MI, HF and their adverse outcomes are associated with high circulating levels of sFRP3 [26].

Micro RNAs

miRNAs are emerging as biomarkers for cardiovascular diseases and have promising potential [27]. miR-208 has been investigated as a marker of myocardial injury [28, 29]. MiRNA array revealed that miR-208 is produced exclusively by the myocardium, and studies using real-time PCR (polymerase chain reaction) confirm that myocardial injury is significantly associated with raised miR-208 levels in the plasma [28]. In patients with MI, the plasma levels of miR-208b and miR-499 are significantly elevated and are associated with increase in cTnT and creatine phosphokinase levels [30]. Thus, miR-208b and miR-499 are potential candidate biomarkers for AMI. Significantly, plasma levels of miRNA levels are not influenced by age, gender, renal function and body mass index [30].

Soluble Suppressor of Tumorigenicity 2

Suppression of tumorigenecity-2 (ST2) is a member of the Toll-like/interleukin-1 receptor family. ST2 has two isoforms: soluble ST2 (sST2) and a transmembrane receptor form ST2 (ST2L) [31]. Levels of sST2 are of prognostic value in patients with AMI [32, 33] and stable coronary artery disease [34]. They are additive to the prognostic value of the levels of biomarkers such as NT-proBNP in patients AMI [35]. sST2 assay may also be used for guidance in the treatment of patients, post-AMI [36, 37].

Heart Failure

Management of patients with heart failure (HF) has been revolutionized by the use of biomarkers, which aid both diagnosis and prognosis. Natriuretic peptides, soluble suppression of tumorigenecity-2 (sST2) and high sensitivity troponin are the most validated biomarkers for use in the care of patients with HF [38].

Natriuretic Peptides

Natriuretic peptides (NPs), viz., BNP and NTproBNP are the most widely studied and validated biomarkers in HF. They are synthesized by the myocardium in response to pressure or volume overload [39, 40]. Measurement of NPs aids in the diagnosis of HF in patients with acute dyspnoea as well as in risk stratification for both short and long-term prognostication [41–45]. Higher levels of NP at the time of hospital admission and at discharge are associated with both rehospitalization and death [46–56].

Suppression of Tumorigenecity-2 (ST2)

Interleukin -33 (IL-33) /ST2 signalling is known to be involved in T-cell medi-ated immune responses [57, 58]. In the failing heart, upregulation of the IL-33/ST2 signalling pathway is considered to lessen adverse remodelling [59–62]. This bene-ficial effect is reduced by sST2 which may compete with ST2L for IL-33 [63, 64]. sST2 is a mechanically induced cardiomyocyte protein. Myocardial stress as in HF induces sST2 leading to adverse cardiac remodelling [65]. The blood levels of sST2 increase greatly with ventricular distension as seen in both acute and chronic HF. sST2 levels are of prognostic value in patients with HF [66, 67] and additive to the prognostic value of the level of NT-proBNP [68]. With other biomarkers in HF, sST2 levels are useful to improve risk prediction [69]. The usefulness of sST2 is not only as a prognostic marker and complementary to that of NT-proBNP; sST2 levels are better than NTproBNP levels for reclassify risk of death [64]. sST2 measured serially during HF is predictive of cardiovascular outcomes as well [70–74]. sST2 level has been found to strongly predict death at one year in dyspnoeic patients with and without HF [64]. sST2 assay may also be used for guidance in the treatment of patients with HF [75]. In patients with HF, sST2 seems to have a better prognostic value than galectin-3 [76]. The diagnostic value of sST2 in HF is however less than that of NTproBNP [66].

Troponins

In addition to AMI, release of cardiac troponin (cTn) is also seen in HF. Both cTnI and cTnT levels are elevated during episodes of HF and are associated with higher in-hospital deaths as well as adverse outcome on long term follow up [77, 78]. There is increasing evidence that patients with HF and elevated levels of troponin, even without obvious myocardial ischemia, have a higher risk for adverse events. cTnT levels have been found to be an independent prognostic marker with respect to death and readmission [79]. Highly sensitive cTnT (hsTnT) is detected in significantly large number of patients with HF. When compared to cTnT, hsTnT is however not superior in predicting the risk of death up to 406 days [80]. Among those patients with an undetectable cTnT, hsTnT may improve the prediction of death [80, 81]. Patients in HF with increasing hsTnI during treatment have increased risk for death compared with those with stable or decreasing hsTnI [82].

Cystatin C

All cells including CMs produce cystatin C (CysC). Pressure overload as well as cardiac injury promotes secretion of CysC. CysC is useful as a prognostic biomarker in patients with HF [83–85]. CysC was identified as an independent predictor of 1-year mortality among patients with aute diastolic heart failure (ADHF) [83] and is also of prognostic value of in predicting readmission [84].

Heart-Type Fatty Acid-Binding Protein

A strong association between high levels of heart-type fatty acid-binding protein (hFABP) and poor prognosis have been shown in patients with HF [86–88]. hFABP levels are parallel to the levels of natriuretic peptides and may be exploited to assess treatment response in patients with HF [89, 90].

Arrhythmias

Both natriuretic peptides and cardiac troponins are considered useful to predict sudden cardiac death caused by cardiac arrhythmias [91–93]. NT-proBNP may also help in assessing prognosis after cardiac arrest [94]. They however lack specificity for sudden cardiac death. Biomarkers which echo disturbances in calcium homeostasis in CMs may be useful to predict sudden cardiac death [95].

High-sensitivity cTnT (hs-cTnT) has been suggested as a promising independent predictor of incident atrial fibrillation in healthy subjects without apparent heart diseases [96–98]. In a recent meta-analysis, a significant association was found between cTnT levels during ablation of conduction pathways and risk for incidence and recurrence of atrial fibrillation [99]. Reliable markers to assess the risk of ventricular arrhythmias are yet to be identified. ST2 does not strongly predict the risk of ventricular arrhythmias [100]. Potential additional diagnostic and risk stratification biomarkers are those which indicate the severity of inflammation, myocardial stretch, fibrosis and B-cell activation [101].

Several pathogenic autoantibodies with either agonistic or antagonistic effects on specific ion channels or targeted against other components of the cardiomyocyte have also been identified [101]. Autoimmune cardiac channelopathies are associated with ventricular arrhythmias or with brady arrhythmias or conduction disturbances [102]. Anti-Ro/SSA IgGs (particularly anti-Ro/SSA-52 kD) have been shown to be arrhythmogenic in both neonates and adults. The suggested pathogenic mechanisms involve an inhibitory cross-reaction of these antibodies either with cardiac potassium-channels conducting the rapid component of the delayed rectifier K current, with L-type calcium channels or with T-type $Ca2+$ channels resulting in dysfunction of the sinus node and atrioventricular conduction fibres [102].

Autoantibodies against adrenergic receptors, acetylcholine receptor, voltage-gated potassium channel complex, cardiac lipid raft-associated proteins, and angiotensin II receptor (AT2R) have been reported in patients with postural orthostatic tachycardia syndrome (POTS) [103, 104].

High levels of antibodies directed against the myosin heavy chain are more prevalent in patients with paroxysmal atrial fibrillation refractory to therapy [105]. Their pathogenic role is unclear. Increased levels of autoantibodies against muscarinic 2 (M2) receptors are found in patients with atrial fibrillation [106].

Hyperthyroidism is a well-known predisposing condition for atrial fibrillation. In a case–control study, the levels of anti β1 adrenoreceptors and anti-M2 autoantibodies levels were found elevated in patients with atrial fibrillation and hyperthyroid conditions, in comparison with individuals in sinus rhythm. The risk of developing atrial fibrillation was significantly higher when both antibodies were present [107].

Congenital Heart Defects

In patients with congenital heart diseases (CHD), different types of biomarkers aid in the detection of specific defects and also in their management [108, 109]. They also facilitate screening for CHD in foetuses, neonates and asymptomatic patients. Assay of biomarkers in body fluids are used for assessing myocardial damage secondary to haemodynamic changes associated with CHD and also for predicting the prognosis of patients with HF, pulmonary hypertension and cardiomyopathy secondary to CHD [110, 111].

Biochemical Markers

Altered levels of troponin T, troponin I, creatin kinase MB, inflammatory cytokines, caspase 3 and acute phase proteins have been reported in children with CHD [112, 113]. The NT-proBNP test is considered useful in infants and children with various types of CHD [114–123]. Plasma BNP measurements may be useful to differentiate CHD and lung disease in addition to assessing the response to treatment for HF [124]. NT-proBNP, creatine kinase-MB, heart-type fatty acid binding protein, and troponins I and T are suggested as markers to detect postoperative acute kidney injury [125]. The levels of BNP and its N-terminal-pro-fragment are used to detect pulmonary arterial hypertension in patients with CHD. High-sensitive troponin-T has a potential to predict myocardial damage in adult patients with CHD [126]. C-reactive protein, cystatin C and surfactant type B are markers considered useful for post-operative follow up of children with CHD [127–130].

Genetic Markers

Karyotyping, gene sequencing and analyzing gene expression in tissues aid in iden- tification of both syndromic and non-syndromic heart defects [131]. Karyotyping is of utility to detect foetal CHD. CHD is also associated with copy number varia- tions, single gene mutations or single nucleotide polymorphisms [131]. Differential expression network approach has identified HUB genes UBC, APP, HUWE1 and SRC as potential biomarkers for CHD in Down syndrome [132].

Treatment:
Protection from Ischemia/Reperfusion Damage

Thrombolytic therapy or primary percutaneous coronary intervention and angio- plasty is the treatment for MI. The resulting reperfusion is effective in reducing infarct size and improving clinical outcome. Myocardial reperfusion however, can itself aggravate myocardial injury. An approach to reduce myocardial injury induced by ischaemia and reperfusion is cardioprotection using drugs or other interventions which act on endogenous mechanisms of myocardial injury [133].

The mechanisms considered to be involved in reperfusion injury are (i) 'calcium paradox', i.e., the re-entry of calcium after a short duration of calcium-free or ischaemic perfusion (ii) damage to the sarcolemma and the sarcoplasmic reticulum with secondary intracellular and mitochondrial calcium overload [134], (iii) oxidative stress [135, 136], (iv) injurious effects of rapid restoration of pH [137, 138], and (v) inflammation which activates cell-adhesion molecules [139–141]. They all probably have a synergistic effect. Interventions targeted at these mechanisms have been subjected to clinical studies. These studies have however not demonstrated any cardioprotective effects [133].

Cyclosporin A, an immunosuppressive agent has been found to inhibit the opening of mitochondrial permeability transition pore (mPTP), an important mechanism for reperfusion injury and cardiomyocyte death [142–147]. Interestingly, a few groups are investigating the potential of non-pharmacological remote preconditioning and post conditioning protocols for cardiac protection [148–151].

Genetic and Cell-Based Therapies

In animal models, several gene- and cell-based strategies have shown efficacy in the protection and rescue of the ischemic myocardium. Some of them have progressed to small scale clinical trials for evaluation of safety and feasibility. The discovery of cardiotropic vectors suitable for stable, long-term protein expression [152] and success in isolation and expansion of progenitor cells with regenerative potential (see Chap. 18) indicate opportunities to develop therapies for the rescue of the ischemic myocardium. Gene therapy strategies with potential for myocardial protection employ cytoprotective genes such as antioxidant genes, survival genes and proapoptotic genes [153–162].

Heart Failure

β-blockers, loop diuretics, angiotensin-converting enzyme (ACE) inhibitors or ACE receptor blockers and mineralocorticoid receptor blockers are the drugs commonly used for treatment of patients with chronic HF with reduced left ventricular function. They are found to reduce deaths in patients.

Thanks to progress in our understanding of the cellular and molecular mechanisms of cardiac remodelling associated with HF in different settings, several novel treatment targets have been identified. These include cardiomyocyte viability and hypertrophy. Mechanisms of action of new drugs comprise of vasodilatation and counteracting neurohormonal activation, improving contractility of CMs via increasing the sensitivity of myofilament proteins to calcium or myosin activation, modulating the efficiency of calcium signalling and EC-coupling, enhancing cGMP signalling, protecting CMs against oxidative-stress mediated injury, rectifying dysfunction in substrate metabolism and energetics, inhibiting detrimental signalling pathways or enhancing beneficial pathways and inducing cardiomyocyte growth [163–165]. Though certain challenges remain, another novel option on the horizon is miRNA-based therapy using antagomirs [166–169].

Cell Therapies

Resident progenitors in the heart are known to contribute to cardiac homeostasis. Cell therapy using resident progenitors has been found to have beneficial effects via paracrine mechanisms [170–173]. Success in isolation and expansion of progenitor cells with regenerative potential (see chapter 18) has spurred hopes to develop therapies for regeneration of CMs in the failing heart. Yet, several key issues are to be addressed for significant progress to be made with application of cell-based therapy in improving the quality of life of patients [174].

Gene Therapy

Gene therapy with an engineered catalytically inactive G-protein receptor kinase-2 (β-ARKct) enhances the effects of β-blockers in rodents with HF and in failing human CMs [175–177]. Promising results have been reported with strategies which combined the use of transcription factors and miRs. Antago-miR via inactivation of miR-25 improves calcium reuptake and myocardial contractility during HF [178]. SERCA2a gene therapy restores microRNA-1 expression in HF [179, 180]. Gene therapy has been explored for neocardiomyogenesis as well. Proliferation of CMs could be induced by overexpression of the oncogenic miR-17 to miR-92 cluster [181]. Gene transfer of transcription factors such as GATA-4, heart and neural crest derivatives-expressed protein (HAND)-2, T-box transcription factor (TBX)-5, and MEF-2 has been found to reduce fibrosis and cardiac dysfunction via direct reprogramming of fibroblasts into beating cardiomyocyte-like cells [182].

Other Drugs

Epigenetic drugs such as the DNA methylation inhibitor 5-azacytidine and HDAC inhibitors are also being investigated for treating cardiac hypertrophy and fibrosis [183].

Senolytics

Removal of senescent cells or senolysis is considered a promising therapy to reverse aging phenotype in age associated ischemic heart disease and HF [184, 185]. Several senolytics have been identified [186]. They may selectively deplete senescent cells (senolysis) and reverse age-related pathological alterations [184, 187–190]. Senolytics were seen to significantly improve systolic cardiac function and reduce end-systolic left ventricular dimension in two–year old mice [190]. In aged senolytic mouse models, elimination of senescent cells was associated with activation of resident cardiac progenitor cells and increase in proliferative CMs, indicating improvement in regenerative capacity [191]. Selective use of senolytics could be a promising therapy for heart diseases [186, 191–194]. Honokiol (a lignan isolated from the bark, seed cones, and leaves of trees belonging to the genus Magnolia) has also been found to be protective against cardiomyocyte senescence. The protective effect is mediated via the inhibition of TXNIP expression and suppression of the NLRP3 inflammasome [195].

Modulators of Cardiomyocyte Death

Autophagy, apoptosis, and necrosis are important mechanisms in the pathogenesis of heart diseases. Modulation of the autophagic pathway is a promising strategy for treatment of heart diseases. Several agents that target cardiomyocyte death have been investigated for potential therapeutic use [196].

AMPK activator metformin, β-blockers, calcium channel blockers, $\alpha 1$-adrenergic receptor agonist phenylephrine, histone deacetylase (HDAC) inhibitors as well as caloric restriction are all modulators of autophagy [197–201]. Specific agents that can induce and block autophagy are sparse. Further studies are warranted before concluding on the beneficial role of autophagy inducers or inhibitors in patients with ischemic heart disease and HF.

In chronic HF, encouraging results were obtained in pilot clinical trials, using a recombinant human-soluble TNF receptor or a recombinant human-murine chimeric monoclonal antibody, both of which can neutralize TNF-α [202, 203]. Significant clinical benefit was however not found in the ATTACH trial [202, 204]. A single center study in patients with dilated cardiomyopathy found that pentoxifylline, a xanthine derivative that directly modulates the expression of TNF-α mRNA improves left ventricular function [205, 206].

There is some evidence that inhibition of mPTP may protect against necrotic death during ischemia- reperfusion [207]. Cyclosporin administration during percutaneous coronary intervention was observed to decrease infarct size as well as levels of creatine kinase and troponin I [207]. Necrostatin-1 is a selective inhibitor of necroptosis and it targets the kinase RIP1 [208]. Necrostatin-1 was found to remarkably decrease infarct size after ischemia reperfusion injury in mice [208].

Metabolic Therapy for Heart Failure

Significant changes in cardiomyocyte metabolism occur in failing hearts. Energy production is reduced secondary to impairments in substrate utilization as well as in mitochondrial biogenesis and function. Metabolic remodelling is also associated with alterations in metabolic pathways which regulate growth, redox homeostasis and autophagy. Modulating cardiomyocyte metabolism could improve the function of the heart.

The targets of metabolic therapy in patients with HF are fatty acid (FA) utilization, glucose metabolism, anaplerosis and AMPK [209]. Strategies aimed at reducing FA utilization were found beneficial in some studies [210]. There are also contrary findings in animal studies [211–213]. Also, in a study in patients with end stage HF, inhibition of FA oxidation was found to significantly worsen left ventricular function [214].

Animal studies indicate that modulation of glucose utilization could be beneficial [215]. Dichloroacetate (DCA) which indirectly activates glucose oxidation, has been investigated in animals as well as in humans [216–219]. These studies suggest that enhancing glucose utilization could be a useful approach to improve cardiac energetics and attenuate oxidative stress and adverse remodelling in HF.

Anaplerotic flux through pyruvate carboxylation is increased in compensated hypertrophy. Some studies in cardiac hypertrophy have observed improvement of contractile function after inhibition of pyruvate carboxylation or supplementation with an anaplerotic substrate [220, 221]. The mechanistic role of anaplerosis and effects of long-term inhibition of anaplerotic pathways in HF are not yet clear.

AMPK activity is augmented in cardiac hypertrophy and failure. Drugs which further increase AMPK activity are found to lessen adverse ventricular remodelling and improve cardiac function [222–224]. AMPK activation may increase energy production via stimulation of both glucose and FA oxidation [225–227] or improve energy metabolism by increasing PGC-1α activity and induction of the genes of FA utilization [228]. In mice with MI, PGC-1α is considered to mediate increase in mitochondrial biogenesis [222]. Whether PGC-1α can alleviate mitochondrial damage and loss in HF is unclear. Given that AMPK can inhibit ROS signalling and regulate autophagy and inflammation, the impact of AMPK activation on adverse ventricular remodelling in HF deserves scrutiny. Several drugs have been used to activate AMPK. These include the agonist AICAR (5-aminoimidazole-4carboxamide ribonucleotide) [229] and the anti-diabetic drug metformin [222, 230–233].

Receptors for glucagon-like peptide-1 (GLP-1 receptors) are expressed in the heart. Their activation seems to improve contractile function in models of ischemia reperfusion [234], in pacing-induced HF in dogs [235], in spontaneously hypertensive rats [236], and in rats with pressure overload [237]. These effects are thought to be insulin-independent [237]. In HF models, the beneficial effects of GLP-1 are predominantly linked to improved utilization of glucose. The molecular mechanisms that connect GLP-mediated alterations in glucose metabolism and improvement in heart function need to be elucidated.

Arrhythmias

The classic antiarrhythmic drugs variously modify Na+, K+, and Ca2+ channel function and intracellular mechanisms regulated by adrenergic activity [238]. New drugs have become available for the treatment of atrial and ventricular arrhythmias thanks to the identification of novel targets. The new drugs include (i) those which act on the automaticity of the sino-atrial node, (ii) those acting on the voltage sensing components of the sodium channel, preferentially expressed in CMs of the atrium, Purkinje fibres and atrio-ventricular node, (iii) selective and non-selective β1 -adrenergic receptor inhibitors or those that act on parasympathetic targets, (iv) non-selective and selective blockers of voltage-dependent K + channel (v) drugs that open metabolically dependent Kir6.2: I KATP, (vi) blockers of transmitter-dependent K+ channels, (vii) modulators of Ca2+ homeostasis and transport (viii) drugs that target connexin-associated channels or mechanosensitive channels and (ix) drugs acting on upstream modulatory targets [239].

Congenital Heart Defects

Genome Editing

CRISPR-Cas strategies have been studied for application in several diseases [240–243]. Zebra fish, mice, human pluripotent stem cell and patient -derived cell model systems are used for developing the delivery systems and test the efficiency of the strategy [244–250]. The aim of the CRISPR-Cas approach is to repair genetic mutations through either nonhomologous end joining or homologous dependent repair with wild-type alleles. CRISPR-Cas9 recognizes target sequences to make a break in the double stranded DNA. A crRNA and tracrRNA are combined or a fused sgRNA is used to edit specific sequences in the genome by insertion, deletion or substitution [251–253].

Genomic editing could be beneficial in congenital heart defects which are caused by genomic variations [254]. Several syndromic and non-syndromic CHD phenotypes have been considered for intervention using CRISPRCas9. They include Wolff-Parkinson-White syndrome [255], Duchenne muscular dystrophy [256, 257] and Holt-Oram syndrome [258].

CRISPR-Cas9 studies have aided deciphering of causative gene candidates and the molecular genetic mechanisms for heart defects associated with heterotaxy syndrome [259], CAKUTHED syndrome [260], Barth syndrome [261], Noonan syndrome [262] and DiGeorge syndrome [263–266].

Mutations in GATA4, TBX5, and MyH6/7 have been identified as causative for non-syndromic human CHDs [109, 110]. CRISPR-Cas9 has been employed in model systems to correct or confirm the mutations or to demonstrate phenotypically-defective CMs [267–270].

Conclusions

There has been significant advance in attempts to identify disease biomarkers and therapeutic targets for several heart diseases. These have been possible thanks to the progress made in elucidating the homeostatic mechanisms in cardiomyocytes as well as the pathophysiological changes associated with various heart diseases. More biomarkers for disease diagnosis and prognosis as well as novel therapeutic targets and strategies can be expected as new biological pathways are discovered. The challenge would be to establish the reliability and accuracy of the biomarkers and safety and efficacy of new treatment approaches. The demand can be met only through improvements in the quality of preclinical research and convergent efforts in laboratories and clinics.

References

1. Mohammed AA, Januzzi JL. Clinical applications of highly sensitive troponin assays. Cardiol Rev. 2010;18:12–9.
2. Antman E, Bassand J-P, Klein W, et al. Myocardial infarction redefined—a consensus document of The Joint European Society of Cardiology/American College of Cardiology committee for the redefinition of myocardial infarction. J Am Coll Cardiol. 2000;36:959–69.
3. Roffi M, Patrono C, Collet JP, et al. 2015 ESC Guidelines for the management of acute coronary syndromes in patients presenting without persistent ST-segment elevation: Task Force for the Management of Acute Coronary Syndromes in Patients Presenting without Persistent ST-Segment Elevation of the European Society of Cardiology (ESC). Eur Heart J. 2016;37:267–315.
4. Ford I, Shah AS, Zhang R, et al. High-sensitivity cardiac troponin, statin therapy, and risk of coronary heart disease. J Am Coll Cardiol. 2016;68:2719–28.
5. Omland T, de Lemos JA, Sabatine MS, et al. A sensitive cardiac troponin T assay in stable coronary artery disease. N Engl J Med. 2009;361:2538–47.
6. Vavik V, Pedersen EKR, Svingen GFT, et al. Usefulness of higher levels of cardiac troponin T in patients with stable angina pectoris to predict risk of acute myocardial infarction. Am J Cardiol. 2018;122:1142–7.
7. McQueen MJ, Kavsak PA, Xu L, et al. Predicting myocardial infarction and other serious cardiac outcomes using high-sensitivity cardiac troponin T in a high risk stable population. Clin Biochem. 2013;46:5–9.
8. McCann CJ, Glover BM, Menown IB, et al. Novel biomarkers in early diagnosis of acute myocardial infarction compared with cardiac troponin T. Eur Heart J. 2008;29:2843–50.
9. Young JM, Pickering JW, George PM, et al. Heart Fatty Acid Binding Protein and cardiac troponin: development of an optimal rule-out strategy for acute myocardial infarction. BMC Emerg Med. 2016;16:34.
10. Kavsak PA, Roy C, Malinowski P, et al. Macrocomplexes and discordant high-sensitivity cardiac troponin concentrations. Ann Clin Biochem. 2018;55:500–4.
11. Ishii J, Ozaki Y, Lu J, et al. Prognostic value of serum concentration of heart-type fatty acid-binding protein relative to cardiac troponin T on admission in the early hours of acute coronary syndrome. Clin Chem. 2005;51:1397–404.
12. O'Donoghue M, de Lemos JA, Morrow DA, et al. Prognostic utility of heart-type fatty acid binding protein in patients with acute coronary syndromes. Circulation. 2006;114:550–7.

13. Viswanathan K, Kilcullen N, Morrell C, et al. Heart type fatty acid-binding protein predicts long-term mortality and re-infarction in consecutive patients with suspected acute coronary syndrome who are troponin-negative. J Am Coll Cardiol. 2010;55:2590–8.
14. Baker JO, Tyther R, Liebetrau C, et al. Cardiac myosin-binding protein C: a potential early biomarker of myocardial injury. Basic Res Cardiol. 2015;110:23.
15. Kaier TE, Anand A, Shah AS, et al. Temporal relationship between cardiac myosin-binding protein C and cardiac troponin I in type 1 myocardial infarction. Clin Chem. 2016;62:1153–5.
16. Kaier TE, Stengaard C, Marjot J, et al. Cardiac myosin-binding protein C as alternative to cardiac troponin T for the diagnosis of acute myocardial infarction in the very early phase. J Am Coll Cardiol. 2017;69:221.
17. Bottner M, Laaff M, Schechinger B, et al. Characterization of the rat, mouse, and human genes of growth/differentiation factor-15/macrophage inhibiting cytokine-1 (GDF-15/MIC-1). Gene. 1999;237:105–11.
18. Wollert KC, Kempf T, Peter T, et al. Prognostic value of growth-differentiation factor-15 in patients with non-ST-elevation acute coronary syndrome. Circulation. 2007;115:962–71.
19. Kempf T, von Haehling S, Peter T, et al. Prognostic utility of growth differentiation factor-15 in patients with chronic heart failure. J Am Coll Cardiol. 2007;50:1054–60.
20. Kempf T, Eden M, Strelau J, et al. The transforming growth factor-beta superfamily member growth-differentiation factor-15 protects the heart from ischemia/reperfusion injury. Circ Res. 2006;98:351–60.
21. Xu J, Kimball TR, Lorenz JN, et al. GDF15/MIC-1 functions as a protective and antihypertrophic factor released from the myocardium in association with SMAD protein activation. Circ Res. 2006;98:342–50.
22. Kempf T, Zarbock A, Widera C, et al. GDF-15 is an inhibitor of leukocyte integrin activation required for survival after myocardial infarction in mice. Nat Med. 2011;17:581–8.
23. Hagstrom E, James SK, Bertilsson M, et al. Growth differentiation factor-15 level predicts major bleeding and cardiovascular events in patients with acute coronary syndromes: results from the PLATO study. Eur Heart J. 2016;37:1325–33.
24. Chan MM, Santhanakrishnan R, Chong JP, et al. Growth differentiation factor 15 in heart failure with preserved vs. reduced ejection fraction. Eur J Heart Fail. 2016; 18:81–8.
25. Huang A, Huang Y. Role of sfrps in cardiovascular disease. Ther Adv Chronic Dis. 2020; 11.
26. Askevold ET, Aukrust P, Nymo SH, Lunde IG, Kaasboll OJ, Aakhus S, Florholmen G, Ohm IK, Strand M, Attramadal H, et al. The cardiokine secreted frizzled-related protein 3, a modulator of wnt signalling, in clinical and experimental heart failure. J Intern Med. 2014;275:621–30.
27. Cheow ES, Cheng WC, Lee CN, de Kleijn D, Sorokin V, Sze SK. Plasma-derived extracellular vesicles contain predictive biomarkers and potential therapeutic targets for myocardial ischemic injury. Mol Cell Proteom. 2016;15:2628–40.
28. Ji X, Takahash R, Hiura Y, Hirokawa G, Fukushima Y, Iwai N. Plasma mir-208 as a biomarker of myocardial injury. Clin Chem. 2009;55:1944–9.
29. Wang F, Yuan Y, Yang P, Li X. Extracellular vesicles-mediated transfer of mir-208a/b exaggerates hypoxia/reoxygenation injury in cardiomyocytes by reducing qki expression. Mol Cell Biochem. 2017;431:187–95.
30. Corsten MF, Dennert R, Jochems S, Kuznetsova T, Devaux Y, Hofstra L, Wagner DR, Staessen JA, Heymans S, Schroen B. Circulating microrna-208b and microrna-499 reflect myocardial damage in cardiovascular disease. Circ Cardiovasc Genet. 2010;3:499–506.
31. Sanada S, Hakuno D, Higgins LJ, Schreiter ER, McKenzie ANJ, Lee RT. IL-33 and ST2 comprise a critical biomechanically induced and cardioprotective signaling system. J Clin Invest. 2007;117:1538–49.
32. Kohli P, Bonaca MP, Kakkar R, et al. Role of ST2 in non-ST-elevation acute coronary syndrome in the MERLIN-TIMI 36 trial. Clin Chem. 2012;58:257–66.
33. Jenkins WS, Roger VL, Jaffe AS, et al. Prognostic value of soluble ST2 after myocardial infarction: a community perspective. Am J Med. 2017; 130:1112 e9–1112. (e15).
34. Dieplinger B, Egger M, Haltmayer M, et al. Increased soluble ST2 predicts long-term mortality in patients with stable coronary artery disease: results from the Ludwigshafen risk and cardiovascular health study. Clin Chem. 2014;60:530–40.

35. Sabatine MS, Morrow DA, Higgins LJ, et al. Complementary roles for biomarkers of biomechanical strain ST2 and N-terminal prohormone B-type natriuretic peptide in patients with ST-elevation myocardial infarction. Circulation. 2008;117:1936–44.

36. Weir RA, Miller AM, Murphy GE, et al. Serum soluble ST2: a potential novel mediator in left ventricular and infarct remodeling after acute myocardial infarction. J Am Coll Cardiol. 2010;55:243–50.

37. Members WC, Yancy CW, Jessup M, Bozkurt B, et al. ACCF/AHA guideline for the management of heart failure: a report of the American College of Cardiology Foundation/ American Heart Association Task Force on practice guidelines. Circulation. 2013;2013(128):e240–327.

38. Darden D, Nishimura M, Sharim J, Maisel A. An update on the use and discovery of prognostic biomarkers in acute decompensated heart failure. Expert Rev Mol Diagn. 2019;19:1019–29.

39. Daniels LB, Maisel AS. Natriuretic peptides. J Am Coll Cardiol. 2007;50:2357–68.

40. Suttner SW, Boldt J. Natriuretic peptide system: physiology and clinical utility. Curr Opin Crit Care. 2004;10:336–41.

41. Maisel AS, Krishnaswamy P, Nowak RM, et al. Rapid measurement of B-type natriuretic peptide in the emergency diagnosis of heart failure. N Engl J Med. 2002;347:161–7.

42. Moe GW, Howlett J, Januzzi JL, Zowall H. Canadian multi-center improved management of patients with congestive heart failure (IMPROVE-CHF) study investigators. N-terminal pro-B-type natriuretic peptide testing improves the management of patients with suspected acute heart failure: primary results of the Canadian prospective randomized multi-center IMPROVE-CHF study. Circulation. 2007; 115:3103–10.

43. Mueller C, Scholer A, Laule-Kilian K, et al. Use of B-type natriuretic peptide in the evaluation and management of acute dyspnea. N Engl J Med. 2004;350:647–54.

44. Yancy CW, Jessup M, Bozkurt B, et al. ACC/AHA/HFSA focused update of the 2013 ACCF/AHA guideline for the management of heart failure: a report of the American College of Cardiology/ American heart association task force on clinical practice guidelines and the Heart Failure Society of America. J Card Fail. 2017;23:628–51.

45. Januzzi JL, Camargo CA, Anwaruddin S, et al. The N-terminal Pro-BNP investigation of dyspnea in the emergency department (PRIDE) study. Am J Cardiol. 2005;95:948–54.

46. Fonarow GC, Peacock WF, Phillips CO, et al. Investigators ASACa. Admission B-type natriuretic peptide levels and in-hospital mortality in acute decompensated heart failure. J Am Coll Cardiol. 2007; 49:1943–50.

47. Maisel A, Hollander JE, Guss D, et al. Primary results of the rapid emergency department heart failure outpatient trial (REDHOT). A multi-center study of B-type natriuretic peptide levels, emergency department decision making, and outcomes in patients presenting with shortness of breath. J Am Coll Cardiol. 2004; 44:1328–33.

48. Harrison A, Morrison LK, Krishnaswamy P, et al. B-type natriuretic peptide predicts future cardiac events in patients presenting to the emergency department with dyspnea. Ann Emerg Med. 2002;39:131–8.

49. Doust JA, Pietrzak E, Dobson A, et al. How well does B-type natriuretic peptide predict death and cardiac events in patients with heart failure: systematic review. Br Med J. 2005;330:625.

50. Januzzi JL, van Kimmenade R, Lainchbury J, et al. NT-proBNP testing for diagnosis and short-term prognosis in acute destabilized heart failure: an international pooled analysis of 1256 patients: the International Collaborative of NT-proBNP Study. Eur Heart J. 2006;27:330–7.

51. Januzzi JL Jr, Sakhuja R, O'Donoghue M, et al. Utility of amino-terminal pro-brain natriuretic peptide testing for prediction of 1-year mortality in patients with dyspnea treated in the emergency department. Arch Intern Med. 2006;166:315–20.

52. Knebel F, Schimke I, Pliet K, et al. NT-ProBNP in acute heart failure: correlation with invasively measured hemodynamic parameters during recompensation. J Card Fail. 2005;5(Suppl):S38–41.

53. Logeart D, Thabut G, Jourdain P, et al. Predischarge B-type natriuretic peptide assay for identifying patients at high risk of re-admission after decompensated heart failure. J Am Coll Cardiol. 2004;43:635–41.

54. Kociol RD, Horton JR, Fonarow GC, et al. Admission, discharge, or change in B-type natriuretic peptide and long-term outcomes: data from Organized program to initiate lifesaving treatment in hospitalized patients with heart failure (OPTIMIZE-HF) linked to medicare claims. Circ Heart Fail. 2011;4:628–36.

55. Omar HR, Guglin M. Discharge BNP is a stronger predictor of 6-month mortality in acute heart failure compared with baseline BNP and admission-to-discharge percentage BNP reduction. Int J Cardiol. 2016;221:1116–22.

56. Sudharshan S, Novak E, Hock K, et al. Use of biomarkers to predict readmission for congestive heart failure. Am J Cardiol. 2017;119:445–51.

57. De la Fuente M, MacDonald TT, Hermoso MA. The IL-33/ST2 axis: role in health and disease. Cytokine Growth Factor Rev. 2015;26:615–23.

58. Kakkar R, Lee RT. The IL-33/ST2 pathway: therapeutic target and novel biomarker. Nat Rev Drug Discov. 2008;7:827–40.

59. Tseng CCS, Huibers MMH, van Kuik J, et al. The Interleukin-33/ST2 pathway is expressed in the failing Human Heart and Associated with pro-fibrotic remodeling of the myocardium. J Cardiovasc Trans Res. 2017;11:15–21.

60. Weinberg EO, Shimpo M, De Keulenaer GW, et al. Expression and regulation of ST2, an interleukin-1 receptor family member, in cardiomyocytes and myocardial infarction. Circulation. 2002;106:2961–6.

61. Sanada S, Hakuno D, Higgins LJ, et al. IL-33 and ST2 comprise a critical biomechanically induced and cardioprotective signaling system. J Clin Invest. 2007;117:1538–49.

62. Maisel AS, Di Somma S. Do we need another heart failure biomarker: focus on soluble suppression of tumorigenicity 2 (sST2). Eur Heart J. 2017;38:2325–32.

63. Seki K, Sanada S, Kudinova AY, et al. Interleukin-33 prevents apoptosis and improves survival after experimental myocardial infarction through ST2 signaling. Circulation. 2009;2:684–91.

64. Januzzi JL, Peacock WF, Maisel AS, et al. Measurement of the interleukin family member ST2 in patients with acute dyspnea. Results from the PRIDE (Pro-Brain Natriuretic Peptide Investigation of Dyspnea in the Emergency Department) Study. J Am Coll Cardiol. 2007;50:607–13.

65. Shah RV, Chen-Tournoux AA, Picard MH, et al. Serum levels of the interleukin-1 receptor family member ST2, cardiac structure and function, and long-term mortality in patients with acute dyspnea. Circulation. 2009;2:311–9.

66. Januzzi JL, Peacock WF, Maisel AS, et al. Measurement of the interleukin family member ST2 in patients with acute dyspnea: results from the PRIDE (Pro-Brain Natriuretic Peptide Investigation of Dyspnea in the Emergency Department) study. J Am Coll Cardiol. 2007;50:607–13.

67. Pascual-Figal DA, Ordonez-Llanos J, Tornel PL, et al. Soluble ST2 for predicting sudden cardiac death in patients with chronic heart failure and left ventricular systolic dysfunction. J Am Coll Cardiol. 2009;54:2174–9.

68. Ky B, French B, Levy WC, et al. Multiple biomarkers for risk prediction in chronic heart failure. Circ Heart Fail. 2012;5:183–90.

69. Lassus J, Gayat E, Mueller C, et al. GREAT-Network. Incremental value of biomarkers to clinical variables for mortality prediction in acutely decompensated heart failure: The Multinational Observational Cohort on Acute Heart Failure (MOCA) study. Int J Cardiol. 2013;168:2186–94.

70. Aimo A, Vergaro G, Ripoli A, et al. Meta-analysis of soluble suppression of Tumorigenicity-2 and prognosis in acute heart failure. JACC Heart Fail. 2017;5:287–96.

71. Boisot S, Beede J, Isakson S, et al. Serial sampling of ST2 predicts 90-day mortality following destabilized heart failure. J Card Fail. 2008;14:732–8.

72. Manzano-Fernández S, Januzzi JL, Pastor-Pérez FJ, et al. Serial monitoring of soluble interleukin family member ST2 in patients with acutely decompensated heart failure. Cardiology (Switzerland). 2012;122:158–66.

73. Breidthardt T, Balmelli C, Twerenbold R, et al. Heart failure therapy-induced early ST2 changes may offer long-term therapy guidance. J Card Fail. 2013;19:821–8.

74. van Vark LC, Lesman-Leegte I, Baart SJ, et al. Prognostic value of serial ST2 measurements in patients with acute heart failure. J Am Coll Cardiol. 2017;70:2378–88.
75. Gaggin HK, Motiwala S, Bhardwaj A, et al. Soluble concentrations of the interleukin receptor family member ST2 and beta-blocker therapy in chronic heart failure. Circ Heart Fail. 2013;6:1206–13.
76. Bayes-Genis A, de Antonio M, Vila J, et al. Head-to head comparison of 2 myocardial fibrosis biomarkers for long-term heart failure risk stratification: ST2 versus galectin-3. J Am Coll Cardiol. 2014;63:158–66.
77. Peacock WFI, De Marco T, Fonarow GC, et al. Cardiac troponin and outcome in acute heart failure. N Engl J Med. 2008;358:2117–26.
78. Ilva T, Lassus J, Siirila-Waris K, et al. Clinical significance of cardiac troponins I and T in acute heart failure. Eur J Heart Fail. 2008;10:772–9.
79. Perna ER, Macín SM, Cimbaro, et al. Minor myocardial damage detected by troponin T is a powerful predictor of long-term prognosis in patients with acute decompensated heart failure. Int J Cardiol. 2005; 99:253–61.
80. Pascual-Figal DA, Casas T, Ordonez-Llanos J, et al. Highly sensitive troponin T for risk stratification of acutely destabilized heart failure. Am Heart J. 2012;163:1002–10.
81. Parissis JT, Papadakis J, Kadoglou NP, et al. Prognostic value of high sensitivity troponin T in patients with acutely decompensated heart failure and non-detectable conventional troponin T levels. Int J Cardiol. 2013;168:3609–12.
82. Xue Y, Clopton P, Peacock WF, et al. Serial changes in high-sensitive troponin I predict outcome in patients with decompensated heart failure. Eur J Heart Fail. 2011;13:37–42.
83. Lassus J, Harjola V, Sund R, et al. Prognostic value of cystatin C in acute heart failure in relation to other markers of renal function and NT-proBNP. Eur Heart J. 2007;28:1841–7.
84. Tang WHW, Dupont M, Hernandez AF, et al. Comparative assessment of short-term adverse events in acute heart failure with Cystatin C and other estimates of renal function. JACC Hear Fail. 2015;3:40–9.
85. Inazumi H, Koyama S, Tanada Y, et al. Prognostic significance of changes in cystatin C during treatment of acute cardiac decompensation. J Cardiol. 2016;67:98–103.
86. Arimoto T, Takeishi Y, Shiga R, et al. Prognostic value of elevated circulating heart-type fatty acid binding protein in patients with congestive heart failure. J Card Fail. 2005;11:56–60.
87. Niizeki T, Takeishi Y, Arimoto T, et al. Persistently increased serum concentration of heart-type fatty acid-binding protein predicts adverse clinical outcomes in patients with chronic heart failure. Circ J. 2008;72:109–14.
88. Hoffmann U, Espeter F, Weiss C, et al. Ischemic biomarker heart-type fatty acid binding protein (hFABP) in acute heart failure – diagnostic and prognostic insights compared to NT-proBNP and troponin I. BMC Cardiovasc Disord. 2015;15:50.
89. Niizeki T, Takeishi Y, Arimoto T, et al. Combination of heart-type fatty acid binding protein and brain natriuretic peptide can reliably risk stratify patients hospitalized for chronic heart failure. Circ J. 2005;69:922–7.
90. Goto T, Takase H, Toriyama T, et al. Circulating concentrations of cardiac proteins indicate the severity of congestive heart failure. Heart. 2003;89:1303–7.
91. Patton KK, Ellinor PT, Heckbert SR, et al. N-terminal pro-B-type natriuretic peptide is a major predictor of the development of atrial fibrillation: The Cardiovascular Health Study. Circulation. 2009;120:1768–74.
92. Korngold EC, Januzzi JL, Gantzer ML, et al. Aminoterminal pro-B-type natriuretic peptide and high-sensitivity C-reactive protein as predictors of sudden cardiac death among women. Circulation. 2009;119:2868–76.
93. Hussein AA, Gottdiener JS, Bartz TM, et al. Cardiomyocyte injury assessed by a highly sensitive troponin assay and sudden cardiac death in the community: the Cardiovascular Health Study. J Am Coll Cardiol. 2013;62:2112–20.
94. Myhre PL, Tiainen M, Pettila V, et al. NT-proBNP in patients with out-of-hospital cardiac arrest: results from the FINNRESUSCI Study. Resuscitation. 2016;104:12–8.

95. Baumeister P, Quinn TA. Altered calcium handling and ventricular arrhythmias in acute ischemia. Clin Med Insights Cardiol. 2016;10:61–9.
96. Anegawa T, Kai H, Adachi H, Hirai Y, Enomoto M, Fukami A, Otsuka M, et al. High-sensitive troponin T is associated with atrial fibrillation in a general population. Int J Cardiol. 2012;156:98–100.
97. Filion KB, Agarwal SK, Ballantyne CM, Eberg M, Hoogeveen RC, Huxley RR, Loehr LR, et al. High-sensitivity cardiac troponin T and the risk of incident atrial fibrillation: the Atherosclerosis Risk in Communities (ARIC) study. Am Heart J. 2015;169:31–8.
98. Hussein AA, Bartz TM, Gottdiener JS, Sotoodehnia N, Heckbert SR, Lloyd Jones D, Kizer JR, et al. Serial measures of cardiac troponin T levels by a highly sensitive assay and incident atrial fibrillation in a prospective cohort of ambulatory older adults. Heart Rhythm. 2015;12:879–85.
99. Bai Y, Guo SD, Liu Y, Ma CS, Lip GYH. Relationship of troponin to incident atrial fibrillation occurrence, recurrence after radiofrequency ablation and prognosis: a systematic review, meta-analysis and meta-regression. Biomarkers. 2018;23:512–7.
100. Skali H, Gerwien R, Meyer TE, et al. Soluble ST2 and risk of arrhythmias, heart failure, or death in patients with mildly symptomatic heart failure: results from MADIT-CRT. J Cardiovasc Trans Res. 2016;9:421–8.
101. Hammerer-Lerchera A, Namdarb M, Vuilleumier N. Emerging biomarkers for cardiac arrhythmias. Clin Biochem. 2020;75:1–6.
102. Lazzerini PE, Capecchi PL, Laghi-Pasini F, Boutjdir M. Autoimmune channelopathies as a novel mechanism in cardiac arrhythmias. Nat Rev Cardiol. 2017;14:521–35.
103. Yu X, Li H, Murphy TA, Nuss Z, Liles J, Liles C, Aston CE, et al. Angiotensin II type 1 receptor autoantibodies in postural tachycardia syndrome. J Am Heart Assoc. 2018; 7:e008351.
104. Watari M, Nakane S, Mukaino A, Nakajima M, Mori Y, Maeda Y, Masuda T, et al. Autoimmune postural orthostatic tachycardia syndrome. Ann Clin Transl Neurol. 2108; 5:486–92.
105. Maixent JM, Paganelli F, Scaglione J, Levy S. Antibodies against myosin in sera of patients with idiopathic paroxysmal atrial fibrillation. J Cardiovasc Electrophysiol. 1998;9:612–7.
106. Baba AT, Yoshikawa Y, Fukuda T, Sugiyama T, Shimada M, Akaishi M, Tsuchimoto K, et al. Autoantibodies against M2-muscarinic acetylcholine receptors: new upstream targets in atrial fibrillation in patients with dilated cardiomyopathy. Eur Heart J. 2004;25:1108–15.
107. Stavrakis S, Yu X, Patterson E, Huang S, Hamlett SR, Chalmers L, Pappy R, et al. Activating autoantibodies to the beta-1 adrenergic and m2 muscarinic receptors facilitate atrial fibrillation in patients with Graves' hyperthyroidism. J Am Coll Cardiol. 2009;54:1309–16.
108. Nawaytou H, Bernstein HS. Biomarkers in pediatric heart disease. Biomark Med. 2014; 8:943–63.
109. Dobson R, Walker HA, Walker NL. Biomarkers in congenital heart disease. Biomark Med. 2014;8:965–75.
110. Sugimoto M, Kuwata S, Kurishima C, et al. Cardiac biomarkers in children with congenital heart disease. World J Pediatr. 2015;11:309–15.
111. Sun L, Sun S, Li Y, et al. Potential biomarkers predicting risk of pulmonary hypertension in congenital heart disease: the role of homocysteine and hydrogen sulfide. Chin Med J (Engl). 2014;127:893–9.
112. Nassef YE, Hanan F Aly, Manal A Hamed. Inflammatory cytokines, apoptotic, tissue injury and remodelling biomarkers in children with congenital heart disease. Indian J Clin Biochem. 2014; 29:145–49.
113. Zhang X, Wang K, Yang Q, et al. Acute phase proteins altered in the plasma of patients with congenital ventricular septal defect. Proteomics Clin Appl. 2015;9:1087–96.
114. Cantinotti M, Giordano R, Scalese M, et al. Prognostic role of BNP in children undergoing surgery for congenital heart disease: analysis of prediction models incorporating standard risk factors. Clin Chem Lab Med. 2015;53:1839–46.
115. Cantinotti M, Storti S, Ripoli A, et al. Diagnostic accuracy of B-type natriuretic hormone for congenital heart disease in the first month of life. Clin Chem Lab Med. 2010;48:1333–8.
116. Cantinotti M, Walters HL, Crocetti M, Marotta M, Murzi B, Clerico A. BNP in children with congenital cardiac disease: is there now sufficient evidence for its routine use? Cardiol Young. 2015;25:424–37.

117. Davlouros PA, Karatza AA, Xanthopoulou I, et al. Diagnostic role of plasma BNP levels in neonates with signs of congenital heart disease. Int J Cardiol. 2011;147:42–6.
118. El-Khuffash A, Molloy EJ. Are B-type natriuretic peptide (BNP) and N-terminal-pro-BNP useful in neonates? Arch Dis Child Fetal Neonatal Ed. 2007;92:F320–4.
119. Kulkarni M, Gokulakrishnan G, Price J, Fernandes CJ, Leeflang M, Pammi M. Diagnosing significant PDA using natriuretic peptides in preterm neonates: a systematic review. Pediatrics. 2015;135:e510–25.
120. Lowenthal A, Camacho BV, Lowenthal S, et al. Usefulness of B-type natriuretic peptide and N-terminal pro-Btype natriuretic peptide as biomarkers for heart failure in young children with single ventricle congenital heart disease. Am J Cardiol. 2012;109:866–72.
121. Moses EJ, Mokhtar SAI, Hamzah A, Abdullah BS, Yusoff NM. Usefulness of N-terminal-pro-B-type natriuretic peptide as a screening tool for identifying pediatric patients with congenital heart disease. Lab Med. 2011;42:75–80.
122. Massimiliano C, Simona S, Bruno M, Aldo C, Das BB. Clinical relevance of different B-type natriuretic peptide decisional cut off values for the diagnosis of congenital heart disease in the first weeks of life. Pediatr Cardiol. 2011;32:537.
123. Clausen H, Norén E, Valtonen S, Koivu A, Sairanen M, Liuba P. Evaluation of circulating cardiovascular biomarker levels for early detection of congenital heart disease in newborns in Sweden. JAMA Network Open. 2020; 3: e2027561.
124. Sahingozlu T, Karadas U, Eliacik K, et al. Brain natriuretic peptide: the reason of respiratory distress is heart disease or lung disease? Am J Emerg Med. 2015;33:697–700.
125. Bucholz EM, Whitlock RP, Zappitelli M, et al. Cardiac biomarkers and acute kidney injury after cardiac surgery. Pediatrics. 2015;135:945–56.
126. Eindhoven JA, Roos Hesselink JW, Van den Bosch AE, et al. High– sensitive troponin–T in adult congenital heart disease. Int J Cardiol. 2015;184:405–11.
127. Xu Z, Zhang M, Zhu L, et al. Elevated plasma B–type natriuretic peptide and C–reactive protein levels in children with restrictive right ventricular physiology following tetralogy of Fallot repair. Congenit Heart Dis. 2014;9:521–8.
128. Jaworski R, Haponiuk I, Irga Jaworska N, et al. Kinetics of C–reactive protein in children with congenital heart diseases in the early period after cardiosurgical treatment with extracorporeal circulation. Adv Med Sci. 2014;59:19–22.
129. Zappitelli M, Greenberg JH, Coca SG, et al. Association of definition of acute kidney injury by cystatin C rise with biomarkers and clinical outcomes in children undergoing cardiac surgery. JAMA Pediatr. 2015;169:583–91.
130. IsiK O, Disli OM, Bas T, et al. High postoperative serum levels of surfactant type B as novel prognostic markers for congenital heart surgery. Rev Bras Cir Cardiovasc. 2014;29:186–91.
131. Lebo MS, Baxter SM. New molecular genetic tests in the diagnosis of heart disease. Clin Lab Med. 2014;34:137–56.
132. Yu S, Yi H, Wang Z, et al. Screening key genes associated with congenital heart defects in Down syndrome based on differential expression network. Int J Clin Exp Pathol. 2015;8:8385–93.
133. Ferrari R, Biscaglia S, Malagù M, Bertini M, Campo G. Can we improve myocardial protection during ischaemic injury? Cardiology. 2016; 135:14–26.
134. Ferrari R, Ceconi C, Curello S, Alfieri O, Visioli O. Myocardial damage during ischaemia and reperfusion. Eur Heart J. 1993;14:25–30.
135. Downey JM. Free radicals and their involvement during long-term myocardial ischemia and reperfusion. Annu Rev Physiol. 1990;52:487–504.
136. Curello S, Ceconi C, de Giuli F, Panzali AF, Milanesi B, Calarco M, Pardini A, Marzollo P, Alfieri O, Ferrari R, Visioli O. Oxidative stress during reperfusion of human hearts: potential sources of oxygen free radicals. Cardiovasc Res. 1995;29:118–25.
137. Bond JM, Herman B, Lemasters JJ. Protection by acidotic pH against anoxia/reoxygenation injury to rat neonatal cardiac myocytes. Biochem Biophys Res Commun. 1991;179:798–803.
138. Zeymer U, Suryapranata H, Monassier JP, et al. The Na+/H+ exchange inhibitor eniporide as an adjunct to early reperfusion therapy for acute myocardial infarction: results of the

Evaluation of the Safety and Cardioprotective Effects of Eniporide in Acute Myocardial Infarction (ESCAMI) trial. J Am Coll Cardiol. 2001;38:1644–50.

139. Litt MR, Jeremy RW, Weisman HF, Winkelstein JA, Becker LC. Neutrophil depletion limited to reperfusion reduces myocardial infarct size after 90 min of ischemia: evidence for neutrophil-mediated reperfusion injury. Circulation. 1989;80:1816–27.

140. Hayward R, Campbell B, Shin YK, Scalia R, Lefer AM. Recombinant soluble P-selectin glycoprotein ligand-1 protects against myocardial ischemic reperfusion injury in cats. Cardiovasc Res. 1999;41:65–76.

141. Ma XL, Tsao PS, Lefer AM. Antibody to CD18 exerts endothelial and cardiac protective effects in myocardial ischemia and reperfusion. J Clin Invest. 1991;88:1237–43.

142. Crompton M, Ellinger H, Costi A. Inhibition by cyclosporin A of a Ca2+-dependent pore in heart mitochondria activated by inorganic phosphate and oxidative stress. Biochem J. 1988;255:357.

143. Halestrap AP, Davidson AM. Inhibition of Ca2+-induced large-amplitude swelling of liver and heart mitochondria by cyclosporin is probably caused by the inhibitor binding to mitochondrial-matrix peptidyl-prolyl cis-trans isomerase and preventing it interacting with the adenine nucleotide translocase. Biochem J. 1990;268:153.

144. Grifths EJ, Halestrap AP. Further evidence that cyclosporin A protects mitochondria from calcium overload by inhibiting a matrix peptidylprolyl cis-trans isomerase. Implications for the immunosuppressive and toxic efects of cyclosporin. Biochem J. 1991;274(Pt 2):611.

145. Javadov S, Jang S, Parodi-Rullan R, Khuchua Z, Kuznetsov AV. Mitochondrial permeability transition in cardiac ischemia-reperfusion: whether cyclophilin D is a viable target for cardioprotection? Cell Mol Life Sci. 2017;74:2795.

146. Nighoghossian N, Ovize M, Mewton N, Ong E, Cho TH. Cyclosporine A, a potential therapy of ischemic reperfusion injury. A common history for heart and brain. Cerebrovasc Dis. 2016; 42:309.

147. Rahman FA, Abdullah SS, Manan W, Tan LT, Neoh CF, Ming LC, et al. Efficacy and safety of cyclosporine in acute myocardial infarction: a systematic review and meta-analysis. Front Pharmacol. 2018;9:238.

148. Yellon DM, Opie LH. Postconditioning for protection of the infarcting heart. Lancet. 2006;367:456–8.

149. Hahn JY, Song YB, Kim EK, et al. Ischemic postconditioning during primary percutaneous coronary intervention: the effects of postconditioning on myocardial reperfusion in patients with ST-segment elevation myocardial infarction (POST) randomized trial. Circulation. 2013;128:1889–96.

150. Heusch G, Botker HE, Przyklenk K, Redington A, Yellon D. Remote ischemic conditioning. J Am Coll Cardiol. 2015;65:177–95.

151. Sloth AD, Schmidt MR, Munk K, et al. Improved long-term clinical outcomes in patients with ST-elevation myocardial infarction undergoing remote ischaemic conditioning as an adjunct to primary percutaneous coronary intervention. Eur Heart J. 2013;35:168–75.

152. Svensson EC, Marshall DJ, Woodard K, et al. Efficient and stable transduction of cardiomyocytes after intramyocardial injection or intracoronary perfusion with recombinant adeno-associated virus vectors. Circulation. 1999;99:201–5.

153. Melo LG, Agrawal R, Zhang L, et al. Gene therapy strategy for long term myocardial protection using adeno-associated virus-mediated delivery of heme oxygenase gene. Circulation. 2002;105:602–7.

154. Morishita R, Sugimoto T, Aoki M, et al. In vivo transfection of cis element "decoy" against nuclear factor B binding sites prevents myocardial infarction. Nat Med. 1997;3:894–9.

155. Herttuala SY, Alitalo K. Gene transfer as a tool to induce therapeutic vascular growth. Nat Med. 2003;9:694–700.

156. Melo LG, Pachori AS, Zhang L, et al. Preemptive gene therapy by AAV-mediated delivery of heme oxygenase-1 results in long-term normalization of left ventricular function and chamber dimensions. Circulation. 2003;108(suppl IV): IV-144.

157. Agrawal RS, Muangman S, Melo LG, et al. Recombinant adeno associated virus mediated antioxidant enzyme delivery as preventive gene therapy against ischemia-reperfusion injury of the rat myocardium. Mol Ther. 2001;3:A837.

158. Woo YZ, Zhang JC, Vijayasarathy C, et al. Recombinant adenovirus mediated cardiac gene transfer of superoxide dismutase and catalase attenuates postischemic contractile dysfunction. Circulation. 1998; 98 (suppl): II-255–II-260.

159. Zhu HL, Stewart AS, Taylor MD. Blocking free radical production via adenoviral gene transfer decreases cardiac ischemia-reperfusion injury. Mol Ther. 2000;2:470–5.

160. Suzuki K, Sawa Y, Kaneda Y. In vivo gene transfer of heat shock protein 70 enhances myocardial tolerance to ischemia-reperfusion injury in rat. J Clin Invest. 1997;99:1645–50.

161. Chatterjee S, Stewart AS, Bish LT, et al. Viral gene transfer of the antiapoptotic factor Bcl-2 protects against chronic ischemic heart failure. Circulation. 2002;106(suppl):I212–7.

162. Matsui T, Li L, Del Monte F, et al. Adenoviral gene transfer of activated phosphatidylinositol 3-kinase and Akt inhibits apoptosis of hypoxic cardiomyocytes in vitro. Circulation. 1999;100:2373–9.

163. Tarone G, Balligand JL, Bauersachs J, et al. Targeting myocardial remodelling to develop novel therapies for heart failure: a position paper from the Working Group on Myocardial Function of the European Society of Cardiology. Eur J Heart Fail. 2014;16:494–508.

164. von Lueder TG, Krum H. New medical therapies for heart failure. Nat Rev Cardiol. 2015;12:730–40.

165. Nabeebaccus A, Zheng S, Shah AM. Heart failure—potential new targets for therapy. Br Med Bull. 2016;119:99–110.

166. Vegter EL, van der Meer P, de Windt LJ, et al. MicroRNAs in heart failure: from biomarker to target for therapy. Eur J Heart Fail. 2016;18:457–68.

167. Karakikes I, Chaanine AH, Kang S, et al. Therapeutic cardiac-targeted delivery of miR-1 reverses pressure overload-induced cardiac hypertrophy and attenuates pathological remodeling. J Am Heart Assoc. 2013; 2: e000078.

168. Icli B, Dorbala P, Feinberg MW. An emerging role for the miR-26 family in cardiovascular disease. Trends Cardiovasc Med. 2014;24:241–8.

169. Bernardo BC, Gao XM, Winbanks CE, et al. Therapeutic inhibition of the miR-34 family attenuates pathological cardiac remodeling and improves heart function. Proc Natl Acad Sci USA. 2012;109:17615–20.

170. Chimenti I, Forte E, Angelini F, Giacomello A, Messina E. From ontogenesis to regeneration: learning how to instruct adult cardiac progenitor cells. Prog Mol Biol Transl Sci. 2012;111:109–37.

171. Gaetani R, Barile L, Forte E, et al. New perspectives to repair a broken heart. Cardiovasc Hematol Agents Med Chem. 2009;7:91–107.

172. Peruzzi M, De Falco E, Abbate A, et al. State of the art on the evidence base in cardiac regenerative therapy: overview of 41 systematic reviews. BioMed Res Int. 2015; 7, Article ID 613782.

173. Chimenti I, Smith RR, Li TS, et al. Relative roles of direct regeneration versus paracrine effects of human cardiosphere-derived cells transplanted into infarcted mice. Circ Res. 2010;106:971–80.

174. Chamuleau SAJ, van der Naald M, Climent AM, Kraaijeveld AO, Wever KE, Duncker DJ, Fernández-Avilés F, Bolli R. On behalf of the Transnational Alliance for Regenerative Therapies in Cardiovascular Syndromes (TACTICS) Group. Transl Res Cardiovasc Repair Call Paradig Shift Circ Res. 2018; 122:310–18.

175. Petrofski JA, Koch WJ. The beta-adrenergic receptor kinase in heart failure. J Mol Cell Cardiol. 2003;35:1167–74.

176. Rockman HA, Koch WJ, Lefkowitz RJ. Seven transmembrane-spanning receptors and heart function. Nature. 2002;415:206–12.

177. Williams Ml, Hata JA, Schroder J, et al. Targeted beta-adrenergic receptor kinase (betaARK1) inhibition by gene transfer in failing human hearts. Circulation. 2004; 109: 1590–159.

178. Wahlquist C, Jeong D, Rojas-Muñoz A, et al. Inhibition of miR-25 improves cardiac contractility in the failing heart. Nature. 2014; 531–35.
179. Kumarswamy R, Lyon AR, Volkmann I, et al. SERCA2a gene therapy restores microRNA-1 expression in heart failure via an Akt/FoxO3A-dependent pathway. Eur Heart J. 2012;33:1067–75.
180. Zsebo K, Yaroshinsky A, Rudy JJ, et al. Long-term effects of AAV1/SERCA2a gene transfer in patients with severe heart failure: analysis of recurrent cardiovascular events and mortality. Circ Res. 2014;114:101–8.
181. Chen J, Huang ZP, Seok HY, et al. Mir-17-92 cluster is required for and sufficient to induce cardiomyocyte proliferation in postnatal and adult hearts. Circ Res. 2013;112:1557–66.
182. Song K, Nam YJ, LuoX, et al. Heart repair by reprogramming non-myocytes with cardiac transcription factors. Nature. 2012; 485:599–604.
183. Watson CJ, Horgan S, Neary R, et al. Epigenetic therapy for the treatment of hypertension-induced cardiac hypertrophy and fibrosis. J Cardiovasc Pharmacol Ther. 2016;21:127–37.
184. Tchkonia T, Zhu Y, van Deursen J, Campisi J, Kirkland JL. Cellular senescence and the senescent secretory phenotype: therapeutic opportunities. J Clin Invest. 2013;123:966–72.
185. Shimizu I, Minamino T. Cellular senescence in cardiac diseases. J Cardiol. 2019;74:313–9.
186. Katsuumi G, Shimizu I, Yoshida Y, Minamino T. Vascular senescence in cardiovascular and metabolic diseases. Front Cardiovasc Med. 2018;5:18. https://doi.org/10.3389/fcvm.2018.00018.
187. Baar MP, Brandt RM, Putavet DA, Klein JD, Derks KW, Bourgeois BR, et al. Targeted apoptosis of senescent cells restores tissue homeostasis in response to chemotoxicity and aging. Cell. 2017;169:132–47.
188. Chang JH, Wang YY, Shao LJ, Laberge RM, Demaria M, Campisi J, et al. Clearance of senescent cells by ABT263 rejuvenates aged hematopoietic stem cells in mice. Nat Med. 2016; 22:78
189. Xu M, Palmer AK, Ding H, Weivoda MM, Pirtskhalava T, White TA, et al. Targeting senescent cells enhances adipogenesis and metabolic function in old age. Elife. 2015;4:e12997. https://doi.org/10.7554/eLife.12997.
190. Zhu Y, Tchkonia T, Pirtskhalava T, Gower AC, Ding HS, Giorgadze N, et al. The Achilles' heel of senescent cells: from transcriptome to senolytic drugs. Aging Cell. 2015;14:644–58.
191. Lewis-McDougall FC, Ruchaya PJ, Domenjo-Vila E, Shin Teoh T, Prata L, Cottle BJ, et al. Aged-senescent cells contribute to impaired heart regeneration. Aging Cell. 2019; 18(3):e12931.
192. Walaszczyk A, Dookun E, Redgrave R, Tual-Chalot S, Victorelli S, Spyridopoulos I, et al. Pharmacological clearance of senescent cells improves survival and recovery in aged mice following acute myocardial infarction. Aging Cell. 2019; 18(3): e12945.
193. Kirkland JL, Tchkonia T. Cellular senescence: a translational perspective. EBioMedicine. 2017;21:21–8.
194. Kirkland JL, Tchkonia T, Zhu Y, Niedernhofer LJ, Robbins PD. The clinical potential of senolytic drugs. J Am Geriatr Soc. 2017;65:2297–301.
195. Huang PP, Fu J, Liu L, Wu K, Liu H, Qi B, Liu Y, Qi B. Honokiol antagonizes doxorubicin-induced cardiomyocyte senescence by inhibiting TXNIP-mediated NLRP3 inflammasome activation. Int J Mol Med. 2020;45:186–94.
196. Chiong M, Wang ZV, Pedrozo Z, Cao DJ, Troncoso R, Ibacache M, Criollo A, Nemchenko A, Hill JA, Lavandero S. Cardiomyocyte death: mechanisms and translational implications. Cell Death Dis. 2011;2:e244. https://doi.org/10.1038/cddis.2011.130.
197. Yin M, van der Horst IC, van Melle JP, Qian C, van Gilst WH, Sillje HH, et al. Metformin improves cardiac function in a nondiabetic rat model of post-MI heart failure. Am J Physiol. 2011;301:H459–68.
198. Bahro M, Pfeifer U. Short-term stimulation by propranolol and verapamil of cardiac cellular autophagy. J Mol Cell Cardiol. 1987;19:1169–78.
199. Pfeifer U, Fohr J, Wilhelm W, Dammrich J. Short-term inhibition of cardiac cellular autophagy by isoproterenol. J Mol Cell Cardiol. 1987;19:1179–84.

200. Cao DJ, Wang ZV, Battiprolu PK, Jiang N, Morales CR, Kong Y, et al. Histone deacetylase (HDAC) inhibitors attenuate cardiac hypertrophy by suppressing autophagy. Proc Natl Acad Sci USA. 2011;108:4123–8.

201. Weiss EP, Fontana L. Caloric restriction–powerful protection for the aging heart and vasculature. Am J Physiol. 2011;301:H1205–19.

202. Balakumar P, Singh M. Anti-tumour necrosis factor-a therapy in heart failure: future directions. Basic Clin Pharmacol Toxicol. 2006;99:391–7.

203. Bozkurt B, Torre-Amione G, Warren MS, Whitmore J, Soran OZ, Feldman AM, et al. Results of targeted anti-tumor necrosis factor therapy with etanercept (ENBREL) in patients with advanced heart failure. Circulation. 2001;103:1044–7.

204. Chung ES, Packer M, Lo KH, Fasanmade AA, Willerson JT. Randomized, double-blind, placebo-controlled, pilot trial of infliximab, a chimeric monoclonal antibody to tumor necrosis factor-alpha, in patients with moderate-to-severe heart failure: results of the antiTNF Therapy Against Congestive Heart Failure (ATTACH) trial. Circulation. 2003;107:3133–40.

205. Bergman MR, Holycross BJ. Pharmacological modulation of myocardial tumor necrosis factor a production by phosphodiesterase inhibitors. J Pharmacol Exp Ther. 1996;279:247–54.

206. Sliwa K, Woodiwiss A, Candy G, Badenhorst D, Libhaber C, Norton G, et al. Effects of pentoxifylline on cytokine profiles and left ventricular performance in patients with decompensated congestive heart failure secondary to idiopathic dilated cardiomyopathy. Am J Cardiol. 2002;90:1118–22.

207. Piot C, Croisille P, Staat P, Thibault H, Rioufol G, Mewton N, et al. Effect of cyclosporine on reperfusion injury in acute myocardial infarction. N Engl J Med. 2008;359:473–81.

208. Lim SY, Davidson SM, Mocanu MM, Yellon DM, Smith CC. The cardioprotective effect of necrostatin requires the cyclophilin-D component of the mitochondrial permeability transition pore. Cardiovasc Drugs Ther. 2007;21:467–9.

209. Doenst T, Nguyen TD, Abel ED. Cardiac metabolism in heart failure-implications beyond ATP production. Circ Res. 2013;113:709–24.

210. Lionetti V, Stanley WC, Recchia FA. Modulating fatty acid oxidation in heart failure. Cardiovasc Res. 2011;90:202–9.

211. Stanley WC, Dabkowski ER, Ribeiro RF Jr. O'Connell KA. Dietary fat and heart failure: Moving from lipotoxicity to lipoprotection. Circ Res. 2012; 110:764–76.

212. Raher MJ, Thibault HB, Buys ES, Kuruppu D, Shimizu N, Brownell AL, Blake SL, Rieusset J, Kaneki M, Derumeaux G, Picard MH, Bloch KD, Scherrer-Crosbie M. A short duration of highfat diet induces insulin resistance and predisposes to adverse left ventricular remodeling after pressure overload. Am J Physiol Heart Circ Physiol. 2008;295:H2495-2502.

213. Okere IC, Chess DJ, McElfresh TA, Johnson J, Rennison J, Ernsberger P, Hoit BD, Chandler MP, Stanley WC. High-fat diet prevents cardiac hypertrophy and improves contractile function in the hypertensive dahl salt-sensitive rat. Clin Exp Pharmacol Physiol. 2005;32:825–31.

214. Tuunanen H, Engblom E, Naum A, Nagren K, Hesse B, Airaksinen KE, Nuutila P, Iozzo P, Ukkonen H, Opie LH, Knuuti J. Free fatty acid depletion acutely decreases cardiac work and efficiency in cardiomyopathic heart failure. Circulation. 2006;114:2130–7.

215. Liao R, Jain M, Cui L, D'Agostino J, Aiello F, Luptak I, Ngoy S, Mortensen RM, Tian R. Cardiac-specific overexpression of glut1 prevents the development of heart failure attributable to pressure overload in mice. Circulation. 2002;106:2125–31.

216. Kato T, Niizuma S, Inuzuka Y, Kawashima T, Okuda J, Tamaki Y, Iwanaga Y, Narazaki M, Matsuda T, Soga T, Kita T, Kimura T, Shioi T. Analysis of metabolic remodeling in compensated left ventricular hypertrophy and heart failure. Circ Heart Fail. 2010;3:420–30.

217. Atherton HJ, Dodd MS, Heather LC, Schroeder MA, Griffin JL, Radda GK, Clarke K, Tyler DJ. Role of pyruvate dehydrogenase inhibition in the development of hypertrophy in the hyperthyroid rat heart: a combined magnetic resonance imaging and hyperpolarized magnetic resonance spectroscopy study. Circulation. 2011;123:2552–61.

218. Bersin RM, Wolfe C, Kwasman M, Lau D, Klinski C, Tanaka K, Khorrami P, Henderson GN, de Marco T, Chatterjee K. Improved hemodynamic function and mechanical efficiency in congestive heart failure with sodium dichloroacetate. J Am Coll Cardiol. 1994;23:1617–24.

219. Lewis JF, DaCosta M, Wargowich T, Stacpoole P. Effects of dichloroacetate in patients with congestive heart failure. Clin Cardiol. 1998;21:888–92.
220. Pound KM, Sorokina N, Ballal K, Berkich DA, Fasano M, Lanoue KF, Taegtmeyer H, O'Donnell JM, Lewandowski ED. Substrate-enzyme competition attenuates upregulated anaplerotic flux through malic enzyme in hypertrophied rat heart and restores triacylglyceride content: Attenuating upregulated anaplerosis in hypertrophy. Circ Res. 2009;104:805–12.
221. Nguyen TD, Shingu Y, Amorim PA, Schwarzer M, Doenst T. Triheptanoin diet alleviates hypertrophy and diastolic dysfunction and preserves cardiac glucose oxidation in rats with pressure overload. Clin Res Cardiol. 2013;102(Suppl 1):V577.
222. Gundewar S, Calvert JW, Jha S, Toedt-Pingel I, Ji SY, Nunez D, Ramachandran A, AnayaCis-neros M, Tian R, Lefer DJ. Activation of amp-activated protein kinase by metformin improves left ventricular function and survival in heart failure. Circ Res. 2009;104:403–11.
223. Beauloye C, Bertrand L, Horman S, Hue L. Ampk activation, a preventive therapeutic target in the transition from cardiac injury to heart failure. Cardiovasc Res. 2011;90:224–33.
224. McGaffin KR, Witham WG, Yester KA, Romano LC, O'Doherty RM, McTiernan CF, O'Donnell CP. Cardiac-specific leptin receptor deletion exacerbates ischaemic heart failure in mice. Cardiovasc Res. 2011;89:60–71.
225. Russell RR 3rd, Bergeron R, Shulman GI, Young LH. Translocation of myocardial glut-4 and increased glucose uptake through activation of ampk by aicar. Am J Physiol. 1999;277:H643-649.
226. Luiken JJ, Coort SL, Willems J, Coumans WA, Bonen A, van der Vusse GJ, Glatz JF. Contraction-induced fatty acid translocase/cd36 translocation in rat cardiac myocytes is mediated through amp-activated protein kinase signaling. Diabetes. 2003;52:1627–34.
227. Kudo N, Barr AJ, Barr RL, Desai S, Lopaschuk GD. High rates of fatty acid oxidation during reperfusion of ischemic hearts are associated with a decrease in malonyl-coa levels due to an increase in 5'-amp-activated protein kinase inhibition of acetyl-coa carboxylase. J Biol Chem. 1995;270:17513–20.
228. Jager S, Handschin C, St-Pierre J, Spiegelman BM. Amp-activated protein kinase (ampk) action in skeletal muscle via direct phosphorylation of pgc-1alpha. Proc Natl Acad Sci USA. 2007;104:12017–22.
229. Zaha VG, Young LH. Amp-activated protein kinase regulation and biological actions in the heart. Circ Res. 2012;111:800–14.
230. Cittadini A, Napoli R, Monti MG, Rea D, Longobardi S, Netti PA, Walser M, Sama M, Aimaretti G, Isgaard J, Sacca L. Metformin prevents the development of chronic heart failure in the shhf rat model. Diabetes. 2012;61:944–53.
231. Aguilar D, Chan W, Bozkurt B, Ramasubbu K, Deswal A. Metformin use and mortality in ambulatory patients with diabetes and heart failure. Circ Heart Fail. 2011;4:53–8.
232. El-Mir MY, Nogueira V, Fontaine E, Averet N, Rigoulet M, Leverve X. Dimethyl biguanide inhibits cell respiration via an indirect effect targeted on the respiratory chain complex i. J Biol Chem. 2000;275:223–8.
233. Zhou G, Myers R, Li Y, Chen Y, Shen X, Fenyk-Melody J, Wu M, Ventre J, Doebber T, Fujii N, Musi N, Hirshman MF, Goodyear LJ, Moller DE. Role of amp-activated protein kinase in mechanism of metformin action. J Clin Invest. 2001;108:1167–74.
234. Ravassa S, Zudaire A, Diez J. Glp-1 and cardioprotection: From bench to bedside. Cardiovasc Res. 2012;94:316–23.
235. Nikolaidis LA, Elahi D, Hentosz T, Doverspike A, Huerbin R, Zourelias L, Stolarski C, Shen YT, Shannon RP. Recombinant glucagon-like peptide-1 increases myocardial glucose uptake and improves left ventricular performance in conscious dogs with pacing-induced dilated cardiomyopathy. Circulation. 2004;110:955–61.
236. Poornima I, Brown SB, Bhashyam S, Parikh P, Bolukoglu H, Shannon RP. Chronic glucagon-like peptide-1 infusion sustains left ventricular systolic function and prolongs survival in the spontaneously hypertensive, heart failure-prone rat. Circ Heart Fail. 2008;1:153–60.
237. Nguyen TD, Shingu Y, Amorim PA, Schwarzer M, Doenst T. Chronic glp-1 treatment preserves diastolic function and improves survival in rats with pressure overload. Eur Heart J. 2011;32(783):P4391.

238. Vaughan Williams EM. Classification of antidysrhythmic drugs. Pharmacol Ther B. 1975;1:115–38.
239. Lei M, Wu L, Terrar DA, Huang Christopher L.-H. Circulation. 2018; 138:1879–96.
240. van Kampen SJ, van Rooij E. CRISPR craze to transform cardiac biology. Trends Mol Med. 2019;25:791–802.
241. Motta BM, Pramstaller PP, Hicks AA, Rossini A. The impact of CRISPR/Cas9 technology on cardiac research: from disease modelling to therapeutic approaches. Stem Cells Int. 2017;2017:8960236. https://doi.org/10.1155/2017/8960236.
242. Vermersch E, Jouve C, Hulot JS. CRISPR/Cas9 gene-editing strategies in cardiovascular cells. Cardiovasc Res. 2020;116:894–907.
243. Nguyen Q, Lim KRQ, Yokota T. Genome editing for the understanding and treatment of inherited cardiomyopathies. Int J Mol Sci. 2020;21:733.
244. Carroll KJ, Makarewich CA, McAnally J, Anderson DM, Zentilin L, Liu N, et al. A mouse model for adult cardiac-specific gene deletion with CRISPR/Cas9. Proc Natl Acad Sci USA. 2016;113:338–43.
245. Guo Y, VanDusen NJ, Zhang L, Gu W, Sethi I, Guatimosim S, et al. Analysis of cardiac myocyte maturation using CASAAV, a platform for rapid dissection of cardiac myocyte gene function in vivo. Circ Res. 2017;120:1874–88.
246. Guo Y, Jardin BD, Zhou P, Sethi I, Akerberg BN, Toepfer CN, et al. Hierarchical and stagespecific regulation of murine cardiomyocyte maturation by serum response factor. Nat Commun. 2018;9:3837.
247. Ogasawara T, Shiba Y. iPS cells as a source of cardiac regeneration. Nihon Rinsho. 2016;74(Suppl 6):287–92.
248. Rikhtegar R, Pezeshkian M, Dolati S, Safaie N, Afrasiabi Rad A, Mahdipour M, et al. Stem cells as therapy for heart disease: iPSCs, ESCs, CSCs, and skeletal myoblasts. Biomed Pharmacother. 2019;109:304–13.
249. Martins AM, Vunjak-Novakovic G, Reis RL. The current status of iPS cells in cardiac research and their potential for tissue engineering and regenerative medicine. Stem Cell Rev Rep. 2014;10:177–90.
250. Porteus M. Genome editing: a new approach to human therapeutics. Annu Rev Pharmacol Toxicol. 2016;56:163–90.
251. Gaj T, Gersbach CA, Barbas CF 3rd. ZFN, TALEN, and CRISPR/Cas-based methods for genome engineering. Trends Biotechnol. 2013;31:397–405.
252. Li B, Niu Y, Ji W, Dong Y. Strategies for the CRISPR-based therapeutics. Trends Pharmacol Sci. 2020;41:55–65.
253. Adli M. The CRISPR tool kit for genome editing and beyond. Nat Commun. 2018;9:1911.
254. Seok H, Deng R, Cowan DB, Wang D. Application of CRISPR-Cas9 gene editing for congenital heart disease. Clin Exp Pediatr. 2021; Mar 2. https://doi.org/10.3345/cep.2020. 02096.
255. Xie C, Zhang YP, Song L, Luo J, Qi W, Hu J, et al. Genome editing with CRISPR/Cas9 in postnatal mice corrects PRKAG2 cardiac syndrome. Cell Res. 2016;26:1099–111.
256. Long C, McAnally JR, Shelton JM, Mireault AA, Bassel-Duby R, Olson EN. Prevention of muscular dystrophy in mice by CRISPR/Cas9-mediated editing of germline DNA. Science. 2014;345:1184–8.
257. El Refaey M, Xu L, Gao Y, Canan BD, Adesanya TMA, Warner SC, et al. In vivo genome editing restores dystrophin expression and cardiac function in dystrophic mice. Circ Res. 2017;121:923–9.
258. Boyle Anderson EAT, Ho RK. A transcriptomics analysis of the Tbx5 paralogues in zebrafish. PLoS One. 2018; 13:e0208766.
259. Liu C, Cao R, Xu Y, Li T, Li F, Chen S, et al. Rare copy number variants analysis identifies novel candidate genes in heterotaxy syndrome patients with congenital heart defects. Genome Med. 2018;10:40.
260. Alankarage D, Szot JO, Pachter N, Slavotinek A, Selleri L, Shieh JT, et al. Functional characterization of a novel PBX1 de novo missense variant identified in a patient with syndromic congenital heart disease. Hum Mol Genet. 2020;29:1068–82.

261. Wang S, Li Y, Xu Y, Ma Q, Lin Z, Schlame M, et al. AAV gene therapy prevents and reverses heart failure in a murine knockout model of Barth syndrome. Circ Res. 2020;126:1024–39.
262. Hanses U, Kleinsorge M, Roos L, Yigit G, Li Y, Barbarics B, et al. Intronic CRISPR repair in a preclinical model of Noonan syndrome-associated cardiomyopathy. Circulation. 2020;142:1059–76.
263. Cirino A, Aurigemma I, Franzese M, Lania G, Righelli D, Ferrentino R, et al. Chromatin and transcriptional response to loss of TBX1 in early differentiation of mouse cells. Front Cell Dev Biol. 2020; 8:571501.
264. Watanabe S, Sakurai T, Nakamura S, Miyoshi K, Sato M. The combinational use of CRISPR/Cas9 and targeted toxin technology enables efficient isolation of bi-allelic knockout nonhuman mammalian clones. Int J Mol Sci. 2018;19:1075.
265. Molinard-Chenu A, Dayer A. The candidate schizophrenia risk gene DGCR2 regulates early steps of corticogenesis. Biol Psychiatry. 2018;83:692–706.
266. Mugikura SI, Katoh A, Watanabe S, Kimura M, Kajiwara K. Abnormal gait, reduced loco-motor activity and impaired motor coordination in Dgcr2-deficient mice. Biochem Biophys Rep. 2016;5:120–6.
267. Garg V, Kathiriya IS, Barnes R, Schluterman MK, King IN, Butler CA, et al. GATA4 muta-tions cause human congenital heart defects and reveal an interaction with TBX5. Nature. 2003;424:443–7.
268. Zeng Y, Li J, Li G, Huang S, Yu W, Zhang Y, et al. Correction of the Marfan syndrome pathogenic FBN1 mutation by base editing in human cells and heterozygous embryos. Mol Ther. 2018;26:2631–7.
269. Gifford CA, Ranade SS, Samarakoon R, Salunga HT, de Soysa TY, Huang Y, et al. Oligogenic inheritance of a human heart disease involving a genetic modifier. Science. 2019;364:865–70.
270. Tomita-Mitchell A, Stamm KD, Mahnke DK, Kim MS, Hidestrand PM, Liang HL, et al. Impact of MYH6 variants in hypoplastic left heart syndrome. Physiol Genomics. 2016;48:912–21.

Index

Lightning Source UK Ltd.
Milton Keynes UK
UKHW022358210922
409193UK00002B/11